Advances in Data-Based Approaches for Hydrologic Modeling and Forecasting

Bellie Sivakumar
The University of New South Wales, Sydney, Australia
and
University of California, Davis, USA

Ronny Berndtsson
Lund University, Sweden

NEW JERSEY · LONDON · SINGAPORE · BEIJING · SHANGHAI · HONG KONG · TAIPEI · CHENNAI

Published by

World Scientific Publishing Co. Pte. Ltd.
5 Toh Tuck Link, Singapore 596224
USA office: 27 Warren Street, Suite 401-402, Hackensack, NJ 07601
UK office: 57 Shelton Street, Covent Garden, London WC2H 9HE

British Library Cataloguing-in-Publication Data
A catalogue record for this book is available from the British Library.

ADVANCES IN DATA-BASED APPROACHES FOR HYDROLOGIC
MODELING AND FORECASTING
Copyright © 2010 by World Scientific Publishing Co. Pte. Ltd.
All rights reserved. This book, or parts thereof, may not be reproduced in any form or by any means, electronic or mechanical, including photocopying, recording or any information storage and retrieval system now known or to be invented, without written permission from the Publisher.

For photocopying of material in this volume, please pay a copying fee through the Copyright Clearance Center, Inc., 222 Rosewood Drive, Danvers, MA 01923, USA. In this case permission to photocopy is not required from the publisher.

ISBN-13 978-981-4307-97-0
ISBN-10 981-4307-97-1

Printed by FuIsland Offset Printing (S) Pte Ltd. Singapore

PREFACE

The last few decades have witnessed an enormous growth in hydrologic modeling and forecasting. Many factors have contributed to this growth: sheer necessity, pure curiosity, and others. Population explosion and its associated effects (e.g. increase in water demands, degradation in water quality, increase in human and economic impacts of floods and droughts) have certainly necessitated better understanding, modeling, and forecasting of hydrologic systems and processes and increased funding for hydrologic teaching, research, and practice. Technological developments (e.g. invention of powerful computers, remote sensors, geographic information systems, worldwide web and networking facilities) and methodological advances (e.g. novel concepts, data analysis tools, pattern recognition techniques) have largely facilitated extensive data collection, better data sharing, formulation of sophisticated mathematical methods, and development of complex hydrologic models, which have led to new directions in hydrology. The widespread availability of technology, hydrologic data, and analysis tools have also aroused a certain level of curiosity in studying hydrologic systems, both by trained engineers/scientists and by others. All these have brought about a whole different dimension to hydrologic teaching, research, and practice. There is no doubt that we today have a far greater ability to mimic real hydrologic systems and processes and possess a much better understanding of almost all of their salient properties as well as finer details (e.g. determinism, stochasticity, linearity, nonlinearity, complexity, scale, thresholds, sensitivity to initial conditions) when compared to not so long ago.

An examination of hydrologic literature reveals that there has been, in recent years, an exponential increase in the number of scientific approaches and their applications for hydrologic modeling and forecasting. Among these, the so-called 'data-based' or 'data-driven' approaches have become particularly popular, not only for their utility in developing, by simply taking advantage of the available hydrologic data, sophisticated models to represent more realistically the salient features of hydrologic systems and processes but also for their grandiose nature (and sometimes mainly because of it). The existing data-based approaches are very many, and there have already been a good number of books on almost all of these approaches, not to mention the numerous journal articles and other scientific publications. These books indeed play important roles in advancing our knowledge of data-based approaches in hydrology. However, they also have one glaring drawback, since any of these books focuses on one specific approach or another (e.g. stochastic methods, artificial neural networks), they are largely useful only for the 'specialist' in a particular approach or for those who have already decided (for whatever reason) to pursue a particular approach. Therefore, despite their excellence, none of the existing books, it is fair to say, is adequate enough to learn about the overall progress and the state-of-the-art of data-based approaches in hydrologic modeling and forecasting.

It may be appreciated that a major challenge a student/researcher (even an experienced one, let alone a novice) often encounters is on the selection, in the first place, of an appropriate concept/approach for addressing the hydrologic problem at hand. With the existing situation just mentioned above, this challenge becomes much more difficult and almost impossible, since it will be a highly cumbersome, and often fruitless, exercise to attempt to study all of the existing books on specific approaches. What is needed to overcome this problem, therefore, is a book that covers details of as many major data-based approaches as possible, if not all of them, and thus could save the reader a significant amount of time and effort that would otherwise be spent simply searching.

Motivated by this, an attempt is made in this book to present a comprehensive account of the advances in data-based approaches for modeling and forecasting hydrologic systems and processes. Among the many existing data-based approaches, eight major and most popular ones are selected for inclusion, with a chapter for each. These eight approaches are (roughly in chronological order in which they started to become an attraction in hydrology): stochastic methods, parameter estimation techniques, scaling and fractal methods, remote sensing techniques, artificial neural networks, evolutionary programming techniques, wavelets, and nonlinear dynamics and chaos methods. These approaches are chosen so as to address a wide range of hydrologic system characteristics (e.g. determinism, stochasticity, linearity, nonlinearity, complexity, scale, thresholds, sensitivity to initial conditions), processes (e.g. rainfall, river flow, rainfall-runoff, sediment transport, groundwater contamination), and the associated problems (e.g. system identification and prediction, synthetic data generation, data disaggregation, data measurement, model parameter estimation and uncertainty, regionalization).

For each of the eight approaches, the presentation includes a comprehensive review of the fundamental concepts, their applications in hydrology thus far, and a discussion on potential future directions. Therefore, the book provides a much more efficient and effective way for the reader to get familiarized with the role of data-based approaches in hydrology. The organization of each chapter, with coverage of first the basic concepts and early days of the approach and then the subsequent advances and future directions, makes this book suitable for both new and experienced academicians, researchers, and practitioners. The book is also expected to facilitate dialog towards finding common grounds and achieving generalization in hydrologic theory, research, and practice. The chapters are contributed by leading researchers in the applications of the respective data-based approaches in hydrology, which certainly adds to the importance of the book. It is our sincere hope that readers would find this book unique, just as we, the editors, expected it to be.

Our sincere thanks, first of all, to World Scientific Publishing Company, for their invitation to us to publish a timely and much-needed book in the field of hydrology and water resources, although the choice of the topic on data-based approaches was our own. In particular, Kimberly Chua at World Scientific deserves our special thanks, for her encouragement, patience, and constant correspondence throughout the course of the preparation of this book. Our sincere thanks also to all the contributors for their interest and willingness to contribute to this book. Their hard work and communications with us throughout the chapter preparation, revision (more than once), and proof-reading stages, amid their many other commitments and busy schedules, have been crucial to complete this book in a timely manner. The chapters of this book have been subjected to a peer-review process, and our sincere appreciation and thanks to the following reviewers for their timely and constructive reviews, which have paved the way for a much better product than it was in the initial stage: H. Aksoy, Y. Hong, A. W. Jayawardena, L. Marshall, N. Muttil, J. Olsson, P. Saco, A. Sharma, K. P. Sudheer, B. Troutman, and T. Wagener.

Bellie Sivakumar
The University of New South Wales, Sydney, Australia
and
University of California, Davis, USA

Ronny Berndtsson
Lund University, Lund, Sweden

CONTENTS

Preface	v
List of Contributors	xvii

1. Setting the Stage — 1
  1.1. Background and Motivation — 1
  1.2. Organization — 4
  References — 10

2. Stochastic Methods for Modeling Precipitation and Streamflow — 17
  2.1. Introduction — 18
  2.2. Stochastic Simulation of Precipitation — 19
    2.2.1. Continuous Precipitation Models — 19
    2.2.2. Models of Cumulative Precipitation over Non-overlapping Time Intervals — 21
      2.2.2.1. Markov Chain Models — 21
      2.2.2.2. Alternating Renewal Models — 23
      2.2.2.3. Models for Precipitation Amount — 24
    2.2.3. Nonparametric Models for Simulating Precipitation — 25
      2.2.3.1. Kernel Density Estimators — 25
      2.2.3.2. K-NN Bootstrap Models — 30
    2.2.4. Precipitation Disaggregation Models — 31
  2.3. Stochastic Simulation of Streamflow — 32
    2.3.1. Continuous Time to Hourly Simulation — 32
    2.3.2. Weekly, Monthly, and Seasonal Streamflow Simulation at a Single Site — 33

2.3.3. Annual Streamflow Simulation at a Single Site	35
2.3.4. Multisite Streamflow Simulation	36
2.3.5. Temporal and Spatial Disaggregation Models	37
2.3.6. Nonparametric Streamflow Simulation Models	39
2.3.6.1. Single Site	39
2.3.6.2. Extension to Multisite	41
2.4. Extensions of K-NN Resampling Approach	41
2.5. Summary and Future Directions	43
References	47

3. Model Calibration in Watershed Hydrology   53

3.1. Introduction	54
3.2. Approaches to Parameter Estimation for Watershed Models	56
3.2.1. Model Calibration	58
3.2.1.1. Overview of the Manual Calibration Approach	61
3.2.1.2. Overview of Automated Calibration Approaches	62
3.3. Single-criterion Automatic Calibration Methods	63
3.3.1. A Historical Perspective	64
3.3.2. The Shuffled Complex Evolution – University of Arizona (SCE–UA) Algorithm	67
3.3.3. Limitations of Single-criterion Methods	70
3.4. Multi-criteria Automatic Calibration Methods	71
3.4.1. Simultaneous Multi-criteria Calibration Approach	72
3.4.2. Stepwise Multi-criteria Calibration Approach	76
3.4.3. Multi-criteria Constraining Approach	79
3.5. Automated Calibration of Spatially Distributed Watershed Models	81
3.6. Treatment of Parameter Uncertainty	87
3.6.1. Random Walk Metropolis (RWM) Algorithm	88
3.6.2. DiffeRential Evolution Adaptive Metropolis (DREAM)	90

	3.6.3. Sequential Data Assimilation	93
3.7.	Parallel Computing	95
3.8.	Summary and Conclusions	95
References		99

4. Scaling and Fractals in Hydrology — 107

- 4.1. Introduction — 108
- 4.2. Fractal and Scale-invariant Sets — 109
  - 4.2.1. Fractal Sets and their Fractal Dimensions — 110
  - 4.2.2. Scale-invariant Sets and their Generation — 112
- 4.3. Scale Invariance for Random Processes — 115
  - 4.3.1. Self-similar Processes — 116
  - 4.3.2. Scalar Multifractal Processes — 118
  - 4.3.3. Multifractal Fields and Vector-valued Processes — 122
- 4.4. Generation of Scale-invariant Random Processes — 124
  - 4.4.1. Relationship Between Scale-invariant and Stationary Processes — 124
  - 4.4.2. Scale-invariant Processes as Renormalization Limits — 126
  - 4.4.3. Scale-invariant Processes as Weighted Sums and Products — 128
    - 4.4.3.1. Self-similar Processes — 128
    - 4.4.3.2. Multifractal Processes — 131
  - 4.4.4. Scale-invariant Processes from Fractional Integration of α-stable Measures — 138
    - 4.4.4.1. H-sssi Processes — 139
    - 4.4.4.2. Multifractal Processes — 141
  - 4.4.5. Processes with Limited Scale Invariance — 143
- 4.5. Properties of Scale-invariant Processes — 148
  - 4.5.1. H-sssi Processes — 148
  - 4.5.2. Moment Scaling of Multifractal Processes — 151
  - 4.5.3. Existence, Moments, and Distributions of Stationary Multifractal Measures — 153
  - 4.5.4. Extremes of Stationary Multifractal Measures and Origin of "Multifractality" — 159

	4.5.4.1. A Large Deviation Result	159
	4.5.4.2. Implications on the Bare Cascades and Extension to the Dressed Cascades	160
	4.5.4.3. Cascade Extremes	161
	4.5.4.4. Origin of the Term "Multifractality"	168
4.6.	Forecasting and Downscaling of Stationary Multifractal Measures	169
	4.6.1. Forecasting	169
	4.6.2. Downscaling	175
4.7.	Inference of Scaling from Data	181
	4.7.1. Estimation of $K(q)$, $q > 0$, for Stationary Multifractal Measures	182
	4.7.2. Estimation of $K(q)$, $q < 0$, for Stationary Multifractal Measures in the Presence of Noise or Quantization	190
	4.7.3. Estimation of $K(q)$, $q > 0$, for Multi-affine Processes by the Gradient Amplitude Method	195
	4.7.4. Estimation of Universal Parameters by the Double Trace Moment Method	197
4.8.	Selected Applications in Hydrology	199
	4.8.1. Rainfall	199
	4.8.2. Fluvial Topography and River Networks	208
	4.8.3. Floods	220
	4.8.4. Flow Through Porous Media	228
4.9.	Concluding Remarks	233
	References	234

5. Remote Sensing for Precipitation and Hydrologic Applications — 245

5.1.	Introduction	246
5.2.	Precipitation Nowcasting from Space-based Platforms	250
5.3.	Data Uses in Precipitation Forecasting	254
5.4.	Data Uses in Hydrology	256
	5.4.1. Soil Moisture	256

	5.4.2. Flood Forecasting and Water Management	259
5.5.	Conclusions and Outlook	262
References		263

6. Nearly Two Decades of Neural Network Hydrologic Modeling — 267
- 6.1. The Long and Winding Road — 268
- 6.2. From Little Acorns Grow Mighty Oaks — 275
- 6.3. Establishing the Field (1990–2000) — 280
- 6.4. Evaluating the Field (2000–2001) — 282
- 6.5. Enlarging the Field (2001–present) — 298
  - 6.5.1. Traditional Regression-type Applications — 300
  - 6.5.2. The Merging of Technologies — 302
    - 6.5.2.1. Neuro-fuzzy Products — 303
    - 6.5.2.2. Neuroevolution Products — 305
  - 6.5.3. Modular and Ensemble Modeling — 308
  - 6.5.4. Neuro-hybrid Modeling — 315
    - 6.5.4.1. Error Correction/Forecast Updating — 317
    - 6.5.4.2. Embedded Products — 319
    - 6.5.4.3. Neuroemulation Products — 322
- 6.6. Taking Positive Steps to Deliver the Goods — 325
  - 6.6.1. Building a Collective Intelligence — 327
  - 6.6.2. Deciphering of Internal Components — 330
  - 6.6.3. Estimating Confidence and Uncertainty — 332
- 6.7. Final Thoughts: Searching for a 'Killer App' — 334
- References — 337

7. Evolutionary Computing in Hydrology — 347
- 7.1. Introduction — 348
- 7.2. Evolutionary Computing — 349
  - 7.2.1. Evolution Principle — 349
  - 7.2.2. Evolutionary Computing Techniques — 350
- 7.3. Evolutionary Computing in Hydrology: An Overview — 351
- 7.4. Genetic Programming and its Scope in Hydrology — 352
  - 7.4.1. Modeling the Observations: GP as a Modeling Tool — 355

	7.4.2.	Modeling the Model Error: GP as a Data Assimilation Tool	358
	7.4.3.	Modeling the Model: GP as a Model Emulator Tool	361
		7.4.3.1. Model Order Reduction using GP Models as Spatial Interpolators	362
		7.4.3.2. GP as a Representative Model for Model Emulation	363
7.5.	Important Issues Pertaining to Genetic Computing	365
7.6.	Conclusions	366
	References	367

8. Wavelet Analyses in Hydrology — 371
- 8.1. From Fourier to Wavelets — 371
- 8.2. The Continuous Wavelet Transform — 373
- 8.3. Discrete-time Wavelet Transform and Multi-resolution Analysis — 377
  - 8.3.1. Discrete Wavelet Transform — 377
  - 8.3.2. Orthogonal Wavelet Bases and Orthogonal Wavelet Expansions — 383
- 8.4. Signal Energy Repartition in the Wavelet Frame — 385
- 8.5. Statistical Entropy in the Wavelet Frame — 391
- 8.6. Wavelet Analysis of the Time-scale Relationship Between Two Signals — 394
- 8.7. Wavelet Cross Spectrum and Coherence — 398
- 8.8. Applications of Wavelet Transforms in Hydrology and Earth Sciences — 400
- 8.9. Challenges and Future Directions — 405
- References — 406

9. Nonlinear Dynamics and Chaos in Hydrology — 411
- 9.1. Introduction — 412
- 9.2. Chaos Identification and Prediction — 415
  - 9.2.1. Autocorrelation Function and Power Spectrum — 415
  - 9.2.2. Phase Space Reconstruction — 416

	9.2.3.	Correlation Dimension Method	420
	9.2.4.	Nonlinear Prediction Method and Inverse Approach	421
	9.2.5.	Some Other Methods	423
	9.2.6.	Remarks on Chaos Identification and Prediction Methods	425
9.3.	Chaos Theory in Hydrology: A Review	425
	9.3.1.	Rainfall	426
	9.3.2.	River Flow	429
	9.3.3.	Rainfall-Runoff	436
	9.3.4.	Lake Volume	437
	9.3.5.	Sediment Transport	439
	9.3.6.	Groundwater	440
	9.3.7.	Other Hydrologic Problems	442
		9.3.7.1. Scale and Scale Invariance	442
		9.3.7.2. Parameter Estimation	443
		9.3.7.3. Catchment Classification	444
		9.3.7.4. Sundry	445
9.4.	Chaos Theory in Hydrology: Current Status and Challenges	446
9.5.	Closing Remarks: Chaos Theory as a Bridge Between Deterministic and Stochastic Theories	452
References	456

10. Summary and Future

10.1.	Summary of Findings	463
10.2.	Data vs. Physics	468
10.3.	Integration of Different Approaches	470
10.4.	Communication	472
10.5.	Conclusion	474
References	475

Index   479

LIST OF CONTRIBUTORS

Robert J. Abrahart
School of Geography
University of Nottingham
Nottingham NG7 2RD
UK
E-mail: bob.abrahart@nottingham.ac.uk

Emmanouil N. Anagnostou
Department of Civil and Environmental Engineering
University of Connecticut
Storrs, CT 06269
USA
E-mail: manos@engr.uconn.edu
    and
Institute of Inland Waters
Hellenic Center for Marine Research
Anavissos Attica
GREECE 19013
E-mail: manos@ath.hcmr.gr

Vladan Babovic
Singapore-Delft Water Alliance
Department of Civil Engineering
National University of Singapore
SINGAPORE 117576
E-mail: cvebv@nus.edu.sg

Ronny Berndtsson
Department of Water Resources Engineering
Lund University
Box 118, S-22100 Lund
SWEDEN
E-mail: ronny.berndtsson@tvrl.lth.se

Christian W. Dawson
Department of Computer Science
Loughborough University
Loughborough LE11 3TU
UK
E-mail: c.w.dawson1@lboro.ac.uk

Hoshin Gupta
Department of Hydrology and Water Resources
University of Arizona
Tucson, AZ 85721
USA
E-mail: hoshin.gupta@hwr.arizona.edu

David Labat
Laboratoire des Mécanismes et Transferts en Géologie
(CNRS-Univ. P. Sabatier-IIRD)
14 Avenue Edouard Belin, 31400 Toulouse
FRANCE
E-mail: labat@lmtg.obs-mip.fr

Upmanu Lall
Department of Earth and Environmental Engineering and
    International Research Institute for Climate Prediction
Columbia University
New York, NY 10027
USA
E-mail: ula2@columbia.edu

Andreas Langousis
Department of Civil and Environmental Engineering
Massachusetts Institute of Technology
Cambridge, MA 02139
USA
E-mail: andlag@alum.mit.edu

Balaji Rajagopalan
Department of Civil, Environmental and Architectural Engineering
University of Colorado
Boulder, CO 80309
USA
E-mail: rajagopalan.balaji@colorado.edu

Raghuraj Rao
Singapore-Delft Water Alliance
Department of Civil Engineering
National University of Singapore
SINGAPORE 117576
E-mail: cverr@nus.edu.sg

Jose D. Salas
Department of Civil Engineering
Colorado State University
Fort Collins, CO 80523-1372
USA
E-mail: jsalas@engr.colostate.edu

Linda M. See
School of Geography
University of Leeds
Leeds LS2 NJT
UK
E-mail: l.m.see@leeds.ac.uk

Asaad Y. Shamseldin
Department of Civil and Environmental Engineering
University of Auckland
Private Bag 92019, Auckland
NEW ZEALAND
E-mail: a.shamseldin@auckland.ac.nz

Bellie Sivakumar
School of Civil and Environmental Engineering
The University of New South Wales
Sydney, NSW 2052
AUSTRALIA
E-mail: s.bellie@unsw.edu.au
    and
Department of Land, Air and Water Resources
University of California
Davis, CA 95616
USA
E-mail: sbellie@ucdavis.edu

Soroosh Sorooshian
Center for Hydrometeorology and Remote Sensing (CHRS)
The Henry Samueli School of Engineering
University of California
Irvine, CA 92697
USA
E-mail: soroosh@uci.edu

Daniele Veneziano
Department of Civil and Environmental Engineering
Massachusetts Institute of Technology
Cambridge, MA 02139
USA
E-mail: venezian@mit.edu

Jasper A. Vrugt
Department of Civil and Environmental Engineering
University of California
Irvine, CA 92697
USA
E-mail: jasper@uci.edu
    and
Center for Nonlinear Studies
Los Alamos National Laboratory
Los Alamos, NM 87545
USA
E-mail: vrugt@lanl.gov

Robert L. Wilby
Department of Geography
Loughborough University
Loughborough LE11 3TU
UK
E-mail: r.l.wilby@lboro.ac.uk

Koray K. Yilmaz
Earth System Science Interdisciplinary Center
University of Maryland
College Park, MD 20742
USA
    and
NASA Goddard Space Flight Center
Laboratory for Atmospheres
Greenbelt, MD 20771
USA
E-mail: yilmaz@agnes.gsfc.nasa.gov

CHAPTER 1

SETTING THE STAGE

Bellie Sivakumar

School of Civil and Environmental Engineering,
The University of New South Wales, Sydney, NSW 2052, Australia
E-mail: s.bellie@unsw.edu.au
and
Department of Land, Air and Water Resources,
University of California, Davis, CA 95616, USA
E-mail: sbellie@ucdavis.edu

Ronny Berndtsson

Department of Water Resources Engineering, Lund University,
Box 118, S-221 00, Lund, Sweden
E-mail: ronny.berndtsson@tvrl.lth.se

1.1. Background and Motivation

Hydrology (as the 'study of water') is as old as the human civilization, and yet (as a 'science') is also very young. Some people may opine that 'hydrology' (as we are accustomed to refer to these days) is still at a crossroads as a scientific field; some others may argue that it is a full-fledged one; and still others may point out that even discussions and debates on whether it belongs to 'science' or 'engineering' are continuing. Amid all this confusion, one thing is clear: it is growing, and growing fast. If nothing else, the rapidly increasing number of text books, scientific journals, research articles, and conferences related to hydrology presents testimony to this.

The growth in hydrology may be seen from two broad angles. One, in terms of the growing number of sub-fields: surface hydrology, subsurface hydrology (or groundwater hydrology or hydrogeology), hydrometeorology, hydroclimatology, paleohydrology, snow hydrology, urban hydrology, physical hydrology, chemical hydrology, isotope hydrology, ecohydrology (or hydroecology), vadose zone hydrology, and hydroinformatics, among others. The other, in terms of the growing number of scientific theories and mathematical techniques for measurements, modeling, and forecasting: for example, stochastic methods, scaling and fractal theories, remote sensing, geographic information systems, artificial neural networks, evolutionary computing techniques, wavelets, nonlinear dynamics and chaos, fuzzy logic and fuzzy set theory, support vector machines, and principal component analysis.

Either type of this growth is certainly good for hydrology. On one hand, the creation of sub-fields of hydrology facilitates 'compartmentalization' of the large hydrologic space (e.g. hydrologic cycle) into many smaller sub-spaces (e.g. hydrologic components) to better identify specific mechanisms of interest to be understood and issues that need to be resolved. On the other hand, the import of various scientific theories and mathematical techniques helps to better measure and understand such mechanisms and resolve the associated issues.

If, and when, these two types of growth go side by side, progress in hydrology can be remarkable. However, even this is only a partial job done, since the focus likely is still on a specific sub-field and a theory/technique (i.e. 'specialization'). Consequently, the central issues in the larger hydrologic space are either completely unaddressed or at least temporarily pushed to the background. Another increasingly recognizable danger in this kind of 'specialization' is the potential lack of communication among the different sub-groups and, consequently, the lack of knowledge about the broader hydrologic issues.

These dangers make it clear that, in order to achieve true progress in hydrology, there needs to be a good balance between the way we tend to pursue 'specialization' and the way 'generalization' must be pursued in hydrology. This balancing act, however, is a great challenge, especially since it also often requires knowledge of *all* the sub-fields for their salient characteristics and of *all* the theories/techniques for their advantages and limitations. In other words, for example, in order to choose the most appropriate technique to solve a given hydrologic problem, knowledge of *all* the available techniques is essential in the first place (Although this makes the purpose of 'specialization' meaningless on an individual basis, when considered collectively it does not).

As for the mathematical techniques in hydrology, the current hydrologic literature is not only too vast but also increasingly specialized. For example, there are many books purely on stochastic methods or artificial neural networks. Although, this situation is a reflection of how far we have advanced in our scientific theories and understanding, it also creates a major problem: that is, there is not even a single book that can help one to gain knowledge of at least a few, let alone *all*, of the major techniques in hydrology. An obviously related problem is that the 'search process' for finding details about the different techniques is oftentimes strenuous. What is urgently needed, therefore, is a unique book that presents details of all the major techniques in hydrology, or at least a few of them. This is the motivation for this book.

It is relevant to note that, one of the main reasons for the proliferation of mathematical techniques in hydrology (and in other fields) is our advancements in measurement and computer technologies. The availability of more and better quality data (both in space and in time) and the increase in computational power and speed have certainly facilitated (and, sometimes, even 'forced' or 'coerced') the development of numerous 'data-based' approaches. Although data is central to almost all scientific endeavors (pattern recognition, modeling, forecasting, verification), our interpretation of 'data-based' approaches herein refers to 'data-driven' approaches (and, sometimes, also termed as 'data-mining' approaches; in a broader sense, 'time series' approaches). With this interpretation, the primary question we try to address in this book is: What is the best possible way for both experienced researchers and students of hydrology to learn the different data-based theories and techniques?

With numerous data-based approaches in existence and applied in hydrology, and more and more emerging day by day, it is near-impossible to present details of *all* the approaches in a single book. Therefore, the best possible way to achieve the above goal is to select a few 'major' data-based approaches and present comprehensive accounts of those approaches. Obviously, such a selection is subjective (upon us).

Nevertheless, our selection is based on the 'popularity' of an approach, the extent of its applications in hydrology thus far, and the potential for its future.

As per this, we select eight major data-based approaches for presentation in this book. These approaches are: stochastic methods, parameter estimation techniques, scaling and fractal methods, remote sensing techniques, artificial neural networks, evolutionary computing techniques, wavelets, and nonlinear dynamics and chaos methods. These approaches are chosen, as they have found key applications in the study of hydrologic system characteristics (e.g. determinism, stochasticity, linearity, nonlinearity, complexity, scale, self-organized criticality, thresholds, sensitivity to initial conditions), processes (e.g. rainfall, river flow, rainfall-runoff, evaporation and evapotranspiration, sediment transport, water quality, groundwater contamination), and the associated problems (e.g. system identification and prediction, synthetic data generation, data disaggregation, data measurement, model parameter estimation and uncertainty, regionalization).

1.2. Organization

The above eight approaches are presented in the following eight chapters, with a chapter for each approach. The approaches are also arranged (in the above order), so as to reflect roughly the chronological order in which they started to become an attraction in hydrology. For each approach, the presentation includes a comprehensive review of the fundamental concepts, their applications in hydrology thus far, and a discussion on potential future directions. Therefore, the book provides a much more efficient and effective way for the reader to get familiarized with the role of these major data-based approaches in hydrology. The final chapter summarizes the highlights of these eight chapters and also offers some future directions for further advancement of hydrology. Before we move on to details of the individual approaches, it would be helpful to present herein a brief account of these approaches.

In Chapter 2 (by Rajagopalan, Salas, and Lall), a detailed account of stochastic methods is presented, with particular emphasis placed on precipitation and streamflow modeling. The term 'stochastic' was derived from the Greek word 'Στόχος,' which means 'random.' Stochastic methods in hydrology emerged during 1960s–early 1970s,[1-6] and have been dominating hydrologic studies ever since.[7-18] They started off with parametric techniques and then advanced to nonparametric techniques; methods that combine parametric and nonparametric techniques have also been proposed. There exist numerous stochastic models in hydrology, and software programs as well. Popular among these models include ARMA (autoregressive moving average) models, alternating renewal models, Markov chain models, kernel density estimators, and K-nearest neighbor bootstrap models. Since their inception, stochastic methods have been applied to numerous hydrologic processes and problems, including for rainfall and streamflow data generation and disaggregation and for groundwater flow and contaminant transport studies. A particularly notable application of stochastic methods these days is for the downscaling of coarse-scale outputs from Global Climate Models or General Circulation Models (GCMs) to regional- and catchment-scale hydrologic variables.[19-24] Since global climate change is anticipated to increase the number and magnitude of extreme hydrologic events (e.g. floods, droughts), assessment of climate change impacts on hydrology and water resources would critically depend on our ability to accurately downscale the coarse-scale GCM outputs to catchment-scale hydrologic data. This suggests an even far greater role of stochastic methods in hydrologic studies in the future.

Parameter estimation is an important problem in our modeling endeavor, regardless of the type of model (e.g. physical-based, conceptual-based, black-box). Research on parameter estimation in hydrologic models and on the associated uncertainties gained importance in the 1980s,[25-29] and has advanced tremendously during the last two decades,[30-44] largely contributed by the development of more and more complex hydrologic models. Among the popular parameter estimation and uncertainty methods are the shuffled complex evolution (SCE) algorithm, genetic algorithm (GA), Monte Carlo Markov chain (MCMC) method, Bayesian recursive estimation, and the generalized likelihood uncertainty estimation (GLUE). Although the necessity of more complex models for a more adequate and accurate representation of our hydrologic systems has been clearly recognized in certain situations, our increasing tendency to develop more and more complex models than perhaps that may actually be needed has also come to light and been questioned in terms of data and other limitations.[45-54] Nevertheless, for both right and wrong reasons, we will be witnessing, at least in the near future, even more and more complex models and, hence, the model parameter estimation and uncertainty issues will continue to be important topics of research in hydrology. Chapter 3 (by Yilmaz, Vrugt, Gupta, and Sorooshian) presents details of the model parameter estimation and uncertainty issues, with special focus on model calibration in watershed hydrology.

Chapter 4 (by Veneziano and Langousis) presents a comprehensive account of scaling and fractal concepts in hydrology. Generally speaking, 'scaling' refers to a situation where the system/process properties are independent of the scale of observation, and the term 'fractals' is used to refer to 'scaling with non-integer' dimensions. The scaling and fractal ideas in hydrology go back to as early as the 1930s–1940s,[55,56] but their importance was brought forth to better light in the 1960s.[57-59] Further significant advancements have been made only since the 1980s.[60-80] During this period, these concepts have found their applications in studying rainfall and river flow distributions, rainfall disaggregation, river networks, width functions, and groundwater flow and contaminant transport, among others. In addition to these, of particular current applications of the scaling and fractal ideas in hydrology include closely-related problems of catchment regionalization, catchment classification, and predictions in ungaged basins (PUBs).[81-83] With increasing calls in recent times for a generalized modeling framework in hydrology[84-87] together with the need to make predictions in ungaged basins to deal with extreme hydrologic events, especially floods of far greater magnitudes potentially resulting from climate change, the scaling and fractal concepts will likely play a far more crucial role in future hydrologic and water resources studies.

Apart from computer power and speed, one of the main reasons that led to the development of sophisticated data-based approaches and their applications in hydrology is the availability of more and better quality hydrologic data made through remote sensing technology. Remote sensing (e.g. using geostationary satellites) facilitates measurements of precipitation and other hydrometeorologic variables (e.g. land surface temperature, near-surface soil moisture, landscape roughness and vegetation cover, snow cover and water equivalent) at high resolutions both in space and in time. Although the origin of the remote sensing technology may be tracked down to more than 150 years ago with the invention of the camera (that too only from the perspective of an 'equipment'), the revolution in 'remote sensing' technology (as we refer to these days) first began in the 1960s with the deployment of space satellites, advanced further in the 1970s with the deployment of the Landsat satellites, and has been continuing since then. Remotely sensed data have been studied and used for many different purposes in hydrology, including forecasting of precipitation and floods and estimation of soil moisture.[88-106] With space missions and satellite technology growing at a very fast pace, remotely-sensed hydrologic and related data will be used to study numerous hydrologic problems at different spatial and temporal scales. One of the areas where remote sensing can play a vital role in the near future is the transboundary waters, where ground measurements are oftentimes extremely difficult, both for physical and for other reasons. Chapter 5 (by Anagnostou) presents a detailed account of remote sensing in hydrology, particularly focusing on precipitation nowcasting/forecasting and hydrologic applications.

Artificial neural networks (ANNs), developed based on our knowledge of the human nervous system, are probably the most widely used tool in hydrologic studies these days, perhaps next only to stochastic methods. The ability of ANNs to represent the nonlinear relationship between the input variable(s) and the output variable(s) only through a learning procedure with a transfer (activation) function, without the necessary knowledge of the underlying system, is arguably their strongest suit, which makes them often more attractive than the other methods. Applications of ANNs in hydrology started only as recently as the early 1990s,[107-111] but have skyrocketed since then.[112-126] Thus far, artificial neural networks have been applied for numerous purposes in hydrology, including forecasting (rainfall, river flow, river stage, groundwater table depth, suspended sediment, evapotranspiration), data infilling or missing data estimation, and downscaling of GCM outputs. There have and continue to be strong criticisms on the use of ANNs in hydrology on the basis of them being weak extrapolators, their lack of physical explanation, their inability to provide information on input/parameter selection, and the possibility of over-parameterization.[113,127,128] Many studies have indeed addressed these (and many other) issues and also brought to light their utility and appropriateness for hydrologic modeling and forecasting.[121,123,129-131] Nevertheless, there is also a clear recognition on the need to do much more in this respect. If and when this can be successfully done, the important role of ANNs in hydrologic modeling and forecasting will be beyond questionable. A comprehensive account of the state-of-the-art of ANNs in hydrology is presented in Chapter 6 (by Abrahart, See, Dawson, Shamseldin, and Wilby).

Chapter 7 (by Babovic and Rao) presents details of applications of evolutionary computation methods in hydrology, in particular on the role of genetic programming in hydrologic studies. Evolutionary computation generally refers to computation based on principles of natural evolution. There are many different methods under the umbrella of 'evolutionary computation,' depending upon the purpose of their use and the way the evolution principles are interpreted/used. These methods include genetic algorithms, genetic programming, ant colony algorithm, and particle swarm method, among others. Evolutionary computation studies in hydrology emerged in the early 1990s, and there has been a significant growth since then.[132-146] Evolutionary computation methods have already found their applications in several areas and problems of hydrology, including precipitation, rainfall-runoff, water quality, evapotranspiration, soil moisture, groundwater, and sediment transport. However, these methods are still in the state of infancy and, thus, our knowledge of these methods remains very limited. As a result, we have thus far not been able to explore them enough for hydrologic modeling and forecasting, but this situation will change soon as we vigorously and rigorously continue our research in this direction. At the current time, use of these methods for calibration of rainfall-runoff models is gaining particular importance. Since the increasing availability of hydrologic data and, thus, the growing complexity of hydrologic models facilitate/necessitate data processing, parameter estimation, and many other tasks, evolutionary computation methods will play a crucial role in dictating how we will go about hydrologic modeling in the future.

Wavelets are mathematical functions that cut up data into different frequency components and then study each component with a resolution matched to its scale. They have advantages over traditional Fourier methods in analyzing physical situations where the signal contains discontinuities and sharp spikes, which are commonplace in hydrology. Applications of wavelets in hydrology started in the 1990s and have been continuing at a very fast pace.[147-161] Among the areas and problems of hydrology where wavelets have been employed are precipitation fields and variability, river flow forecasting, streamflow simulation, rainfall-runoff relations, drought forecasting, suspended sediment discharge, and water quality. The outcomes of these studies are certainly encouraging, as they indicate the utility of wavelets for studying hydrologic signals as well as their superiority over some traditional signal processing methods (e.g. Fourier transforms). Nevertheless, there remain serious concerns on the lack of physical interpretation of the results from wavelet analysis. Development of wavelet-based models that take into account the intrinsic multiscale nature of physical relationships of hydrologic processes would help address these concerns. Another possible means may be by coupling wavelets with other methods that can represent, at least to a certain degree, the physical relationships. The literature on wavelets in hydrology provides good indications as to the positive direction in which we are moving. Chapter 8 (by Labat) presents an excellent account of wavelets and their applications in hydrology (and in earth sciences).

Hydrologic phenomena are inherently nonlinear and interdependent, and also possess hidden determinism and order. Their 'complex' and 'random-looking' behaviors need not always be the outcome of 'random' systems but can also arise from simple deterministic systems with sensitive dependence on initial conditions, called 'chaos.' Although the discovery of 'chaos theory' in the 1960s brought about a noticeable change in our perception of 'complex' systems,[162] not much happened in 'chaos' research in hydrology until the 1980s due to the absence of powerful computers and nonlinear mathematical tools. Early studies on applications of nonlinear dynamics and chaos concepts in hydrology were conducted during the late 1980s–early 1990s.[163-168] Since then, there has been an enormous growth in chaos studies in hydrology.[73,82,169-186] The application areas and problems include rainfall, river flow, rainfall-runoff, sediment transport, groundwater contaminant transport, modeling, prediction, noise reduction, scaling, disaggregation, missing data estimation, reconstruction of system equations, parameter estimation, and catchment classification. These studies and their outcomes certainly provide different perspectives and new avenues to study hydrologic systems and processes; simplification in hydrologic modeling is just one of them. In fact, arguments as to the potential of chaos theory to serve as a bridge between our traditional and dominant deterministic and stochastic theories have also been put forward.[186] A comprehensive account of nonlinear dynamics and chaos studies in hydrology is presented in Chapter 9 (by Sivakumar and Berndtsson).

There is no question that the above eight data-based approaches have significantly advanced hydrologic theory, research, and practice. A similar conclusion may also be arrived in regards to the other data-based approaches that are not part of this book, including fuzzy logic and fuzzy set theory, support vector machines, principal component analysis, and singular spectrum analysis. Despite these advances, however, there are also growing concerns on at least two important points: (i) our tendency to 'specialize' in these individual scientific theories and mathematical techniques, rather than to find ways to integrate them to better address the larger hydrologic issues; and (ii) the lack of 'physical' explanation of these concepts and the parameters involved in the methods to real catchments and their salient properties.[52] Although there have certainly been some efforts in advancing research in this direction,[2,7,9,87,121,123,130,187-189] they are few and far between. To achieve true progress in hydrology, this situation needs to change, and change quickly. These issues and the associated challenges are highlighted in the final chapter (by Sivakumar and Berndtsson).

References

1. H. A. Thomas and M. B. Fiering, in *Design of Water Resource Systems*, Eds. Mass *et al.* (Harvard University Press, Cambridge, Massachusetts, 1962), p. 459.

2. V. M. Yevjevich, *Hydrology Paper 1* (Colorado State University, Fort Collins, Colorado, 1963), p. 1.
3. M. B. Fiering, *Streamflow Synthesis* (Harvard University Press, Cambridge, Massachusetts, 1967).
4. A. A. Harms and T. H. Campbell, *Water Resour. Res.*, 653 (1967).
5. V. M. Yevjevich, *Stochastic Processes in Hydrology* (Water Resources Publications, Littleton, Colorado, 1972).
6. D. Valencia and J. C. Schaake, *Water Resour. Res.*, 580 (1973).
7. V. Klemeŝ, *Adv. Hydrosci.*, 285 (1978).
8. M. L. Kavvas and J. W. Delleur, *Water Resour. Res.*, 1151 (1981).
9. J. D. Salas and R. A. Smith. *Water Resour. Res.*, 428 (1981).
10. R. Srikanthan and T. A. McMahon, *Trans. ASAE*, 754 (1983).
11. R. L. Bras and I. Rodriguez-Iturbe, *Random Functions and Hydrology* (Dover Publications, New York, 1985).
12. L. W. Gelhar, *Stochastic Subsurface Hydrology* (Prentice-Hall, New Jersey, 1983).
13. J. D. Salas, J. W. Delleur, V. M. Yevjevich and W. L. Lane, *Applied Modeling of Hydrologic Time Series* (Water Resources Publications, Littleton, Colorado, 1995).
14. U. Lall and A. Sharma, *Water Resour. Res.*, 679 (1996).
15. A. Sharma, D. G. Tarboton and U. Lall, *Water Resour. Res.*, 291 (1997).
16. O. E. Barndorff-Nielsen, V. K. Gupta, V. Perez-Abreu and E. C. Waymire, Eds., *Stochastic Methods in Hydrology: Rain, Landforms and Floods* (World Scientific, Singapore, 1998).
17. R. S. Govindaraju, Ed., *Stochastic Methods in Subsurface Contaminant Hydrology* (American Society of Civil Engineers, New York, 2002).
18. M. L. Kavvas, *ASCE J. Hydrol. Eng.*, 44 (2003).
19. R. L. Wilby, T. M. L. Wigley, D. Conway, P. D. Jones, B. C. Hewitson, J. Main and D. S. Wilks, *Water Resour. Res.*, 2995 (1998).
20. R. Huth, *Clim. Res.*, 91 (1999).
21. R. L. Wilby, L. E. Hay, W. J. J. Gutowski, R. W. Arritt, E. S. Takle, Z. Pan, G. H. Leavesley and M. P. Clark, *Geophys. Res. Lett.*, 1199 (2000).
22. C. Prudhomme, N. Reynard and S. Crooks, *Hydrol. Process.*, 1137 (2002).
23. S. P. Charles, B. C. Bates, I. N. Smith and J. P. Hughes, *Hydrol. Process.*, 1373 (2004).
24. R. Mehrotra and A. Sharma, *J. Geophys. Res.-Atmos.*, 111(D15101), doi:10.1029/2005JD006637 (2006).
25. S. Sorooshian and J. A. Dracup, *Water Resour. Res.*, 430 (1980).
26. G. M. Hornberger and R. C. Spear, *J. Env. Manag.*, 7 (1981).
27. S. Sorooshian and V. K. Gupta, *Water Resour. Res.*, 251 (1983).
28. V. K. Gupta and S. Sorooshian, *Water Resour. Res.*, 473 (1985).
29. M. B. Beck, *Water Resour. Res.*, 1393 (1987).
30. K. J. Beven and A. M. Binley, *Hydrol. Process.*, 279 (1992).
31. Q. Duan, S. Sorooshian and V. K. Gupta, *Water Resour. Res.*, 1015 (1992).

32. K. J. Beven, *Adv. Water Resour.*, 41 (1993).
33. H. V. Gupta, S. Sorooshian and P. O. Yapo, *Water Resour. Res.*, 751 (1998).
34. G. Kuczera and E. Parent, *J. Hydrol.*, 69 (1998).
35. M. G. Anderson and P. D. Bates, Eds., *Model Validation: Perspectives in Hydrological Science* (Wiley, Chichester, 2001).
36. B. C. Bates and E. P. Campbell, *Water Resour. Res.*, 937 (2001).
37. Q. Duan, H. V. Gupta, S. Sorooshian, A. N. Rousseau and R. Turcotte, Eds., *Calibration of Watershed Models* (American Geophysical Union, Washington, 2002).
38. J. A. Vrugt, W. Bouten, H. V. Gupta and S. Sorooshian, *Water Resour. Res.*, W1312, doi:10.1029/2001WR001118 (2002).
39. A. Montanari, *Water Resour. Res.*, W08406, doi:10.1029/2004WR003826 (2005).
40. K. J. Beven, *Hydrol. Process.*, 3141 (2006).
41. D. Kavetski, G. Kuczera and S. W. Franks, *Water Resour. Res.*, W03407, doi:10.1029/2005WR004368 (2006).
42. D. Kavetski, G. Kuczera and S. W. Franks, *Water Resour. Res.*, W03408, doi:10.1029/2005WR004376 (2006).
43. P. Mantovan and E. Todini, *J. Hydrol.*, 368 (2006).
44. J. A. Vrugt, C. J. F. ter Braak, H. V. Gupta and B. A. Robinson, *Stoch. Environ. Res. Risk Assess.*, 1011 (2009).
45. L. F. Konikow and J. D. Bredehoeft, *Adv. Water Resour.*, 75 (1992).
46. A. J. Jakeman and G. M. Hornberger, *Water Resour. Res.*, 2637 (1993).
47. P. C. Young, S. D. Parkinson and M. Lees, *J. Appl. Stat.*, 165 (1996).
48. C. Perrin, C. Michel and V. Andréassian, *J. Hydrol.*, 275 (2001).
49. K. J. Beven, in *Environmental Foresight and Models: A Manifesto*, Ed. M. B. Beck (Elsevier, The Netherlands, 2002), p. 227.
50. J. C. Refsgaard and H. J. Henriksen, *Adv. Water Resour.*, 71 (2004).
51. J. W. Kirchner, *Water Resour. Res.*, W03S04, doi:10·1029/2005WR004362 (2006).
52. B. Sivakumar, *Stoch. Environ. Res. Risk Assess.*, 737 (2008).
53. B. Sivakumar, *Hydrol. Process.*, 893 (2008).
54. B. Sivakumar, *Hydrol. Process.*, 4333 (2008).
55. R. E. Horton, *Trans. Am. Geophys. Union*, 350 (1932).
56. R. E. Horton, *Bull. Geol. Soc. Am.*, 275 (1945).
57. R. L. Schreve, *J. Geol.*, 17 (1966).
58. B. B. Mandelbrot and J. R. Wallis, *Water Resour. Res.,* 909 (1968).
59. B. B. Mandelbrot and J. R. Wallis, *Water Resour. Res.*, 321 (1969).
60. V. K. Gupta and E. C. Waymire, *J. Hydrol.*, 95 (1983).
61. V. Klemeŝ, *J. Hydrol.*, 1 (1983).
62. B. B. Mandelbrot, *The Fractal Geometry of Nature* (W. H. Freedman and Company, New York, 1983).
63. S. Lovejoy and B. B. Mandelbrot, *Tellus*, 209 (1985).
64. J. C. I. Dooge, *Water Resour. Res.*, 46S (1986).

65. D. Schertzer and S. Lovejoy, *J. Geophys. Res.*, 9693 (1987).
66. S. W. Wheatcraft and S. W. Tyler, *Water Resour. Res.*, 566 (1988).
67. R. Ababou and L. W. Gelhar, in *Dynamics of Fluids in Hierarchical Porous Media* (Academic Press, San Diego, California, 1990), p. 393.
68. V. K. Gupta and E. C. Waymire, *J. Geophys. Res.* (D3), 1999 (1990).
69. S. P. Neuman, *Water Resour. Res.*, 1749 (1990).
70. R. Rigon, A. Rinaldo, I. Rodriguez-Iturbe, R. L. Bras and E. Ijjasz-Vasquez, *Water Resour. Res.*, 1635 (1993).
71. G. Blöschl and M. Sivapalan, *Hydrol. Process.*, 251 (1995).
72. J. D. Kalma and M. Sivapalan, Eds., *Scale Issues in Hydrological Modelling* (Wiley, London, 1996).
73. C. E. Puente and N. Obregon, *Water Resour. Res.*, 2825 (1996).
74. D. G. Tarboton, *J. Hydrol.*, 105 (1996).
75. I. Rodriguez-Iturbe and A. Rinaldo, *Fractal River Basins: Chance and Self-organization* (Cambridge University Press, Cambridge, 1997).
76. D. Veneziano and V. Iacobellis, *J. Geophys. Res.*, **B6**, 12797 (1999).
77. C. E. Puente, O. Robayo, M. C. Díaz and B. Sivakumar, *Stoch. Environ. Res. Risk Assess.*, 357 (2001).
78. C. E. Puente, O. Robayo and B. Sivakumar, *Stoch. Environ. Res. Risk Assess.*, 372 (2001).
79. V. K. Gupta, *Chaos Solitons Fractals*, 357 (2004).
80. B. Dodov and E. Foufoula-Georgiou, *Water Resour. Res.*, W05005, doi:10.1029/2004WR003408 (2005).
81. M. Sivapalan, K. Takeuchi, S. W. Franks, V. K. Gupta, H. Karambiri, V. Lakshmi, X. Liang, J. J. McDonnell, E. M. Mendiondo, P. E. O'Connell, T. Oki, J. W. Pomeroy, D. Schertzer, S. Uhlenbrook and E. Zehe, *Hydrol. Sci. J.*, 857 (2003).
82. B. Sivakumar, A. W. Jayawardena and W. K. Li, *Hydrol. Process.*, 2713 (2007).
83. T. Wagener, M. Sivapalan, P. A. Troch and R. A. Woods, *Geography Compass*, 901 (2007).
84. R. B. Grayson and G. Blöschl, Eds., *Spatial Patterns in Catchment Hydrology: Observations and Modelling* (Cambridge University Press, Cambridge, 2000).
85. M. Sivapalan, G. Blöschl, L. Zhang and R. Vertessy, *Hydrol. Process.*, 2101 (2003).
86. J. J. McDonnell and R. A. Woods, *J. Hydrol.*, 2 (2004).
87. M. Sivapalan, in *Encyclopedia of Hydrological Sciences* (Wiley, London, 2005), p. 193.
88. E. C. Barret and D. W. Martin, *The Use of Satellite Data in Rainfall Monitoring* (Academic Press, New York, 1981).
89. J. A. Smith and W. F. Krajewski, *Water Resour. Res.*, 1893 (1987).
90. R. F. Adler and A. J. Negri, *J. Appl. Meteorol.*, 30 (1988).
91. J. C. Ritchie, *Hydrol. Sci. J.*, 625 (1996).
92. E. N. Anagnostou and W. F. Krajewski, *Water Resour. Res.*, 1419 (1997).

93. K.-L. Hsu, X. Gao, S. Sorooshian and H. V. Gupta, *J. Appl. Meteorol.*, 1176 (1997).
94. P. A. Townsend and S. J. Walsh, *Geomorphology*, 295 (1998).
95. E. N. Anagnostou and W. F. Krajewski, *J. Atmos. Ocean. Tech.*, 189 (1999).
96. E. N. Anagnostou and W. F. Krajewski, *J. Atmos. Ocean. Tech.*, 198 (1999).
97. M. Borga, E. N. Anagnostou and E. Frank, *J. Geophys. Res.-Atmos.*, **105**(D2), 2269 (2000).
98. W. F. Krajewski and J. A. Smith, *Adv. Water Resour.*, 1387 (2002).
99. T. J. Schmugge, W. P. Kustas, J. C. Ritchie, T. J. Jackson and A. Rango, *Adv. Water Resour.*, 1367 (2002).
100. S. Sorooshian, X. Gao, K. Hsu, R. A. Maddox, Y. Hong, H. V. Gupta and B. Imam, *J. Climate*, 983 (2002).
101. M. Gebremichael and W. F. Krajewski, *J. Appl. Meteorol.*, 1180 (2004).
102. F. Hossain, E. N. Anagnostou, M. Borga and T. Dinku. *Hydrol. Process.*, 3277 (2004).
103. F. Hossain and E. N. Anagnostou. *Adv. Water Resour.*, 1336 (2005).
104. W. F. Krajewski, M. C. Anderson, W. E. Eichinger, D. Entekhabi, B. K. Hornbuckle, P. R. Houser, G. G. Katul, W. Kustas, J. M. Norman, C. Peters-Lidard and E. F. Wood, *Water Resour. Res.*, W07301, doi:10.1029/2005WR004435 (2006).
105. H. Moradkhani, K. Hsu, Y. Hong and S. Sorooshian, *Geophys. Res. Lett.*, L12107 (2006).
106. E. A. Smith, G. Asrar, Y. Furuhama, A. Ginati, C. Kummerow, V. Levizzani, A. Mugnai, K. Nakamura, R. Adler, V. Casse, M. Cleave, M. Debois, J. Durning, J. Entin, P. Houser, T. Iguchi, R. Kakar, J. Kaye, M. Kojima, D. Lettenmaier, M. Luther, A. Mehta, P. Morel, T. Nakazawa, S. Neeck, K. Okamoto, R. Oki, G. Raju, M. Shepherd, E. Stocker, J. Testud and E. F. Wood, in *Measuring Precipitation from Space – EURAINSAT and the Future* (Springer, New York, 2007), p. 611.
107. M. N. French, W. F. Krajewski and R. R. Cuykendall, *J. Hydrol.*, 1 (1992).
108. K.-L. Hsu, H. V. Gupta and S. Sorooshian, *Water Resour. Res.*, 2517 (1995).
109. H. R. Maier and G. C. Dandy, *Water Resour Res.*, 1013 (1996).
110. A. W. Minns and M. J. Hall, *Hydrol. Sci. J.*, 399 (1996).
111. A. Y. Shamseldin, *J. Hydrol.*, 272 (1997).
112. American Society of Civil Engineers, *ASCE J. Hydrol. Eng.*, 115 (2000).
113. American Society of Civil Engineers, *ASCE J. Hydrol. Eng.*, 124 (2000).
114. H. R. Maier and G. C. Dandy, *Environ. Model. Softw.*, 101 (2000).
115. L. M. See and S. Openshaw, *Hydrol. Sci. J.*, 523 (2000).
116. R. J. Abrahart and S. M. White, *Phys. Chem. Earth B*, 19 (2001).
117. P. Coulibaly, F. Anctil, R. Aravena and B. Bobée, *Water Resour. Res.*, 885 (2001).
118. C. W. Lawson and R. L. Wilby, *Prog. Phys. Geog.*, 80 (2001).
119. M. Khalil, U. S. Panu and W. C. Lennox, *J. Hydrol.*, 153 (2001).
120. R. S. Govindaraju and A. R. Rao, Eds., *Artificial Neural Networks in Hydrology* (Kluwer Academic Publishers, Dordrecht, The Netherlands, 2002).

121. R. L. Wilby, R. J. Abrahart and C. W. Dawson, *Hydrol. Sci. J.*, 163 (2003).
122. R. J. Abrahart, P. E. Kneale and L. M. See, Eds., *Neural Networks for Hydrological Modelling* (A. A. Balkema Publishers, Rotterdam, 2004).
123. A. Jain, K. P. Sudheer and S. Srinivasulu, *Hydrol. Process.*, 571 (2004).
124. P. Coulibaly, Y. B. Dibike and F. Anctil, *J. Hydromet.*, 483 (2005).
125. C. W. Dawson, R. J. Abrahart, A. Y. Shamseldin and R. L. Wilby, *J. Hydrol.*, 391 (2006).
126. Ö. Kisi, *Hydrol. Process.*, 1925 (2007).
127. E. Gaume and R. Gosset, *Hydrol. Earth Syst. Sci.*, 693 (2003).
128. D. Koutsoyiannis, *Hydrol. Sci. J.*, 832 (2007).
129. R. J. Abrahart, L. M. See and P. E. Kneale, *J. Hydroinform.*, 103 (1999).
130. K. P. Sudheer and A. Jain, *Hydrol. Process.*, 833 (2004).
131. J. Chen and B. J. Adams, *J. Hydrol.*, 232 (2006).
132. Q. J. Wang, *Water Resour. Res.*, 2467 (1991).
133. D. C. McKinney and M. D. Lin, *Water Resour. Res.*, 1897 (1994).
134. B. J. Ritzel, J. W. Eheart and S. Ranjithan, *Water Resour. Res.*, 1589 (1994).
135. V. Babovic, *Emergence, Evolution, Intelligence: Hydroinformatics - A Study of Distributed and Decentralised Computing using Intelligent Agents* (A. A. Balkema Publishers, Rotterdam, Holland, 1996).
136. M. Franchini, *Hydrol. Sci. J.*, 21 (1996).
137. D. A. Savic, G. A. Walters and G. W. Davidson, *Water Resour. Management*, 219 (1999).
138. V. Babovic, *Computer-Aided Civil and Infrastructure Engg.*, 383 (2000).
139. Z. Sen and A. Oztopal, *Hydrol. Sci. J.*, 255 (2001).
140. S. Y. Liong, T. R. Gautam, S. T. Khu, V. Babovic and N. Muttil, *J. Am. Water Resour. Assoc.*, 705 (2002).
141. V. Babovic and M. Keijzer, *Nord. Hydrol.*, (2003).
142. A. Makkeasorn, N.-B. Chang, M. Beaman, C. Wyatt and C. Slater, *Water Resour. Res.*, W09401, doi:10.1029/2005WR004033 (2006).
143. K. Parasuraman, A. Elshorbagy and S. Carey, *Hydrol. Sci. J.*, 563 (2007).
144. A. Aytek and Ö. Kişi, *J. Hydrol.*, 288 (2008).
145. G. Tayfur and T. Moramarco, *J. Hydrol.*, 77 (2008).
146. X. Zhang, R. Srinivasan and D. Bosch, *J. Hydrol.*, 307 (2009).
147. P. Kumar and E. Foufoula-Georgiou, *Water Resour. Res.*, 2515 (1993).
148. P. Kumar, *J. Geophys. Res.*, 393 (1996).
149. V. Venugopal and E. Foufoula-Georgiou, *J. Hydrol.*, 3 (1996).
150. P. Kumar and E. Foufoula-Georgiou, *Rev. Geophys.*, 385 (1997).
151. D. Labat, R. Ababou and A. Mangin, *J. Hydrol.*, 149 (2000).
152. M. Bayazit, B. Önöz and H. Aksoy, *Hydrol. Sci. J.*, 623 (2001).
153. T. W. Kim and J. B. Valdes, *ASCE J. Hydrol. Eng.*, 319 (2003).
154. H. Aksoy, T. Akar and N. E. Unal, *Nord. Hydrol.*, 165 (2004).
155. J. Sujono, S. Shikasho and K. Hiramatsu, *Hydrol. Process.*, 403 (2004).

156. P. Coulibaly and D. H. Burn, *Water Resour. Res.*, W03105, doi: 10.1029/2003WR002667 (2006).
157. S. Kang and H. Lin, *J. Hydrol.*, 1 (2007).
158. S. N. Lane, *Hydrol. Process.*, 586 (2007).
159. B. Schaefli, D. Maruan and M. Holschneider, *Adv. Water Resour.*, 2511 (2007).
160. J. F. Adamkowski, *Hydrol. Process.*, 4877 (2008).
161. D. Labat, *Adv. Water Resour.*, 109 (2008).
162. E. N. Lorenz, *J. Atmos. Sci.*, 130 (1963).
163. A. Hense, *Beitr. Phys. Atmos.*, 34 (1987).
164. I. Rodriguez-Iturbe, F. B. De Power, M. B. Sharifi and K. P. Georgakakos, *Water Resour. Res.*, 1667 (1989).
165. M. B. Sharifi, K. P. Georgakakos and I. Rodriguez-Iturbe, *J. Atmos. Sci.*, 888 (1990).
166. B. P. Wilcox, M. S. Seyfried and T. H. Matison, *Water Resour. Res.*, 1005 (1991).
167. R. Berndtsson, K. Jinno, A. Kawamura, J. Olsson and S. Xu, *Trends Hydrol.*, 291 (1994).
168. A. W. Jayawardena and F. Lai, *J. Hydrol.*, 23 (1994).
169. H. D. I. Abarbanel and U. Lall, *Climate Dyn.*, 287 (1996).
170. A. Porporato and L. Ridolfi, *Water Resour. Res.*, 1353 (1997).
171. Q. Liu, S. Islam, I. Rodriguez-Iturbe and Y. Le, *Adv. Water Resour.*, 463 (1998).
172. Q. Wang and T. Y. Gan, *Water Resour. Res.*, 2329 (1998).
173. N. Lambrakis, A. S. Andreou, P. Polydoropoulos, E. Georgopoulos and T. Bountis, *Water Resour. Res.*, 875 (2000).
174. B. Sivakumar, *J. Hydrol.*, 1 (2000).
175. B. Sivakumar, R. Berndttson, J. Olsson and K. Jinno, *Hydrol. Sci. J.*, 131 (2001).
176. B. Sivakumar, S. Sorooshian, H. V. Gupta and X. Gao, *Water Resour. Res.*, 61 (2001).
177. A. Elshorbagy, S. P. Simonovic and U. S. Panu, *J. Hydrol.*, 123 (2002).
178. B. Faybishenko, *Adv. Water Resour.*, 793 (2002).
179. B. Sivakumar and A. W. Jayawardena, *Hydrol. Sci. J.*, 405 (2002).
180. Y. Zhou, Z. Ma and L. Wang, *J. Hydrol.*, 100 (2002).
181. S. K. Regonda, B. Sivakumar and A. Jain, *Hydrol. Sci. J.*, 373 (2004).
182. B. Sivakumar, *Chaos Solitons Fractals*, 441 (2004).
183. J. D. Salas, H. S. Kim, R. Eykholt, P. Burlando and T. R. Green, *Nonlinear Proc. Geophys.*, 557 (2005).
184. B. Sivakumar, T. Harter and H. Zhang, *Nonlinear Proc. Geophys.*, 211 (2005).
185. F. Hossain and B. Sivakumar, *Stoch. Environ. Res. Risk Assess.*, 66 (2006).
186. B. Sivakumar, *Stoch. Env. Res. Risk Assess.*, 1027 (2009).
187. M. B. Parlange, G. G. Katul, R. H. Cuenca, M. L. Kavvas, D. R. Nielsen and M. Mata, *Water Resour. Res.* 2437 (1992).
188. B. Sivakumar, *Hydrol. Process.*, 2349 (2004).
189. J. Hill, F. Hossain and B. Sivakumar, *Stoch. Environ. Res. Risk Assess.*, 47 (2008).

CHAPTER 2

STOCHASTIC METHODS FOR MODELING PRECIPITATION AND STREAMFLOW

Balaji Rajagopalan

Civil, Environmental and Architectural Engineering,
University of Colorado, Boulder, CO 80309, USA
and
Cooperative Institute for Research in Environmental Sciences,
University of Colorado, Boulder, CO 80309, USA
E-mail: rajagopalan.balaji@colorado.edu

Jose D. Salas

Department of Civil and Environmental Engineering,
Colorado State University, Fort Collins, CO 80523, USA
E-mail: jsalas@engr.colostate.edu

Upman Lall

Department of Earth and Environmental Engineering,
Columbia University, New York, NY 10027, USA
E-mail: ula2@columbia.edu

Stochastic simulation and forecasting of hydroclimatic processes, such as precipitation and streamflow, are vital tools for risk-based management of water resources systems. Stochastic hydrology has a long and rich history in this area. The traditional approaches have been based on mathematical models with assumed or derived structure representing the underlying mechanisms and processes involved. The model generally includes several variables and a parameter set. Such "parametric models" have been quite useful in practice for analyzing and synthesizing hydrologic time series at various timescales. A lot of experience has been gained using such "traditional techniques" with large complex systems, such as the Colorado River, the Great Lakes, the Ottawa River, and the Nile River. Further, over the last two decades, the field of stochastic hydrology has been enriched by the emergence of nonparametric data-driven methods. Nonparametric methods are gaining wide prominence and are being applied to a variety of hydrologic and climatologic applications. In fact, in many cases, the proper combination of both parametric and nonparametric techniques has been quite useful and beneficial. Also, the availability of software and computational power has paved the way for more efficient applicability of both techniques. In this chapter, we attempt to provide an overview of both parametric and nonparametric techniques for modeling hydrologic time series, particularly precipitation and streamflow.

2.1. Introduction

Risk-based planning and management of water resources systems generally require knowledge of the variability of hydroclimatic processes, such as precipitation, temperature, and streamflow. Stochastic simulation of these processes provides input scenarios that may be used to drive process models, such as crop models, hydrologic models, and water resources management models, which provide distributions of various decision variables of interest and aid in devising effective planning and management strategies. Due to only sparse temporal and spatial data available, which is often encountered in practice, it is difficult to obtain a robust understanding of the underlying variability from limited data. Consequently, the risks may not be accurately reflected, thus leading to sub-optimal planning and management decisions. To this end, it is important to be able to apply proper techniques that can reflect the underlying physical and stochastic mechanisms of the variables involved.

Synthetically generated sequences of daily hydroclimatic variables, especially precipitation and streamflow, are often used for efficient short-term and long-term operation and management of water resources systems. Clearly, stochastic models that generate the sequences should be able to faithfully capture the distributional and dependence properties of the historical data. Furthermore, they should be able to generate sequences conditionally (e.g. conditional on seasonal climate forecast), so as to provide realistic scenarios for use in seasonal to interannual planning.

In this chapter, we attempt to provide an overview of selected stochastic techniques as simulation and forecasting tools. The chapter is organized as follows. Traditional parametric methods for precipitation modeling and simulation are presented first, followed by nonparametric approaches. Use of these methods in the context of weather generators is then described. We follow this by descriptions of parametric and nonparametric tools for stochastic streamflow simulation. We conclude the presentation with a brief summary of extensions for forecasting and for other applications.

2.2. Stochastic Simulation of Precipitation

2.2.1. *Continuous Precipitation Models*

The theory of *point processes* has been one of the earliest tools for modeling precipitation as a continuous process.[1] In this, the number of storms $N(t)$ in a time interval $(0, t)$ arriving at a location is assumed to be Poisson-distributed with parameter λt (λ = storm arrival rate). If n storms arrived in the interval $(0, t)$ at times $t_1,...,t_n$, then the number of storms in any time interval T is also Poisson-distributed with parameter λT. It is further assumed that the rainfall amount R associated with a storm arrival is *white noise* (e.g. R may be gamma-distributed) and that $N(t)$ and R are independent. Thus, rainfall amounts $r_1,...,r_n$ correspond to storms occurring at times $t_1,...,t_n$. Such a rainfall generating process has been called *Poisson white noise* (PWN).

Under this formulation, the cumulative rainfall in the interval $(0, t)$, $Z(t) = \sum_{j=1}^{N(t)} R_j$ is a *compound Poisson process*. In addition, the cumulative rainfall over successive non-overlapping time intervals T is given by $Y_i = Z(iT) - Z(iT - T)$, $i = 1, 2, ...$. The basic statistical properties of Y_i, assuming that $Z(t)$ is generated by a PWN model, has been widely studied.[2,3] For example, its autocorrelation function $\rho_k(Y)$ is equal to zero for all lags greater than zero, which contradicts actual observations (e.g. $\hat{\rho}_1(Y) = 0.446$ for hourly precipitation at Denver Airport station for the month of June based on the 1948–83 record). Nevertheless, the PWN model has been useful for predicting annual precipitation[4] and extreme precipitation events.[5] Modifications to the PWN model consider rainfall as an occurrence with a random duration D and intensity I, called the *Poisson rectangular pulse* (PRP) model.[6] Commonly, D and I are assumed to be independent and exponentially distributed. In this formulation, n storms may occur at times $t_1,...,t_n$ with associated intensities and durations $(i_1, d_1),...,(i_n, d_n)$, and the storms may overlap so that the aggregated process Y_i becomes autocorrelated. Although the PRP model is better conceptualized than the PWN, it is still limited when applied to actual rainfall data.[6]

Neyman and Scott[7] originally suggested the concept of *clusters* in modeling the spatial distribution of galaxies. This concept of space clustering has been applied to model continuous time rainfall.[8-12] The *cluster process* can be described as a two-level mechanism for generating rainfall: first, the storm arrival is assumed to be Poisson-distributed with a given parameter, and then each storm is associated with a number of precipitation bursts, which are distributed as Poisson or Geometric. In general, m_j precipitation bursts are associated with the storm that arrived at time t_j. In addition, the time of occurrence of bursts τ relative to the storm origin t_j may be assumed to be exponentially distributed. Then, if the precipitation burst is described by an instantaneous random precipitation depth R, the resulting precipitation process is known as *Neyman-Scott white noise* (NSWN), while if the precipitation burst is a rectangular pulse the precipitation process is known as *Neyman-Scott rectangular pulse* (NSRP).

Parameter estimation of Neyman-Scott models has been extensively studied[8,9,13-17] using the method of moments and other approaches. A major estimation problem is that parameters estimated based on data for one level of aggregation, say hourly, may be significantly different from those estimated from data for another level of aggregation, say daily.[2,6,9,16] Weighted moments estimates of various timescales in a least squares fashion is an alternative.[5,14] Constraints may be set on the parameters based on the physical understanding of the process that can improve parameter estimation,[18] as shown in a space-time cluster model,[19,20] but the difficulty in estimating the parameters even when using physical considerations persists.

Besides the Poisson and Neyman-Scott cluster processes, other types of temporal precipitation models have also been suggested, such as those based on Cox processes,[21] renewal processes,[22,23] and Barlett-Lewis processes.[11,24] Likewise, alternative space-time multi-dimensional precipitation models have also been developed.[25] All these precipitation models based on point and cluster processes that have been developed thus far are severely limited for modeling convective rainfall, where daily periodicity is observed.[16,26] Nonlinear dynamics and chaotic behavior of rainfall process[27] further highlights the limitations of the point and cluster process-based models. Discussions on rainfall analysis, modeling, and predictability have been made in various reviews[28-32] and special issues of journals in the Applied Meteorology discipline.[33,34]

2.2.2. Models of Cumulative Precipitation over Non-overlapping Time Intervals

While point process and cluster models provide a general framework for modeling rainfall as a continuous process, models may also be formulated directly for aggregate rainfall at the desired timescale (e.g. hourly, daily, weekly, monthly). Examples of such models are presented next.

2.2.2.1. Markov Chain Models

While *Markov chain* models have been widely suggested in the literature for simulating precipitation (mainly at daily timescale),[22,26,35-39] they have also been used for many other hydrologic processes, such as streamflow, soil moisture, temperature, solar radiation, and water storage in reservoirs. Markov chain models are used for modeling the precipitation occurrence (i.e. wet or dry), and a probability density function (PDF) is used for generation of the rainfall amount on a wet day, as described below.

Let $X(t)$ be a discrete-valued process that starts at time 0 and develops through time. Then, $P[X(t) = x_t | X(0) = x_0, X(1) = x_1, ..., X(t-1) = x_{t-1}]$

is the probability that the process $X(t) = x_t$, given its entire history. If this probability simplifies to $P[X(t) = x_t | X(t-1) = x_{t-1}]$, the process is a *first-order Markov chain* or a *simple Markov chain*. Because $X(t)$ is discrete-valued, we use here the notation $X(t) = j$, $j = 0, 1,..., r$ instead of $X(t) = x_t$, where j represents a *state* and $r + 1$ is the number of states; for example, in modeling daily rainfall, one may consider two states, $j = 0$ for a dry day (no rain) and $j = 1$ for a wet day. A simple Markov chain is defined by its *transition probability matrix* $P(t)$, a square matrix with elements $p_{ij}(t) = P[X(t) = j | X(t-1) = i]$ for all i, j pairs. Furthermore, $q_j(t) = P[X(t) = j]$, $j = 0, 1,..., r$, is the marginal probability distribution of the chain being at any state j at time t and $q_j(0)$ is the distribution of the initial states. Moreover, the Markov chain is a *homogeneous* or *stationary chain* if $P(t)$ does not depend on time, and, in this case, the notations P and p_{ij} are used. The probabilities that are useful for simulation and forecasting of precipitation events are: the n-step transition probability $p_{ij}^{(n)}$, the marginal distribution $q_j(t)$ given the distribution $q_j(0)$, and the steady-state probability vector q^*. These probabilities can be determined from well-known relations available in the literature.[38,39]

Estimation of probabilities for a simple Markov chain amounts to estimating the transition probability matrix, which is usually obtained by the method of moments and the maximum likelihood method.[40] Other methods, such as the Akaike Information Criteria (AIC), have also been proposed to test the adequacy of the Markov chain and to help in the selection of the order of the Markov chain.[40-42] Simple Markov chains may be adequate for representing many processes, although sometimes more complex models may be necessary. For instance, in modeling daily rainfall processes throughout the year, the parameters of the Markov chain must vary with time to capture the seasonality. Thus, for a two-state Markov chain, the transition probabilities p_{ij} vary along the year and the estimates can be fitted with Fourier series to smooth-out sample variations.[37,43] Also, higher-order Markov chains that vary seasonally may be necessary; for instance, analyzing daily precipitation records across the continental United States, it was concluded that generally second- and third-order models were preferred for the winter months while first-order model for the summer months.[40] Generic higher-order formulation that allows incorporation of dependence on aggregate continuous variables has also been proposed[41] as well as Markov chain models for daily precipitation conditioned on the total monthly precipitation.[44]

Multistate Markov chain models that also consider the dependence between transition probabilities and rainfall amounts may be necessary to capture better the extreme rainfall.[45-48] Also, models with periodic Markov chains for hourly rainfall that account for the effect of daily periodicity have been suggested.[26] Further, a hierarchical Markov chain model to describe the daily precipitation process, given the heterogeneous generating mechanisms, has been proposed.[49] Generally, Markov chain models do not reproduce long-term persistence and event-clustering.[29,33] Despite some well-known limitations, Markov chain models are attractive because of their simple structure, ease of application and interpretability, and well-developed literature.

2.2.2.2. *Alternating Renewal Models*

The term 'renewal' stems from the implied independence between the wet and dry period lengths, while the term 'alternating' refers to the fact that wet and dry states alternate; no transition to the same state is possible. An advantage of this representation is that it allows direct consideration of a composite precipitation event. A geometric or a negative binomial distribution may be used as a model for spell length, where a daily time step is of interest.[37] A probability distribution for precipitation amount also needs to be developed. The primary difficulties with the wet/dry spell approach for daily rainfall modeling are: (i) the need for disaggregating the wet spell precipitation into daily or event precipitation (this is not an issue if independence in daily precipitation amounts is assumed, since that is typically assumed in Markov chain models); (ii) the justification of the independence between the wet and dry spell lengths at short timescales; and (iii) the effective reduction in the sample size by considering spells, rather than days.

2.2.2.3. *Models for Precipitation Amount*

Markov chain and renewal models, described above, simulate the precipitation occurrence (wet or dry) and the wet and dry spells. The rainfall amount on wet days must be determined, which involves fitting a PDF for each month or season from the observed data and using them to simulate the amounts. Typically, a parsimonious member of the exponential family (Gamma, Lognormal, Weibull) that best fits the data is used and the goodness of fit of the PDF is determined using traditional tests, such as the Kolmogrov-Smirnov test and the Chi-square test.[50]

This approach may be extended for modeling precipitation at longer timescales, such as monthly, seasonal, and annual. In such cases, modeling precipitation at a given site amounts to finding the probability distribution for each time interval (e.g. monthly). Generally, different distributions are needed for each month. Precipitation data in semi-arid and arid regions may include zero values for some months; hence, precipitation is a mixed random variable. Consider that $X_{v,\tau}$ is the precipitation for year v and season τ, and define $P_\tau(0) = P(X_{v,\tau} = 0)$, $\tau = 1,..., 12$. Then, $F_{X\tau|X\tau>0}(x) = P(X \leq x | X > 0)$ is the conditional distribution of monthly precipitation. Thus, prediction of monthly precipitation requires estimating $P_\tau(0)$ and $F_{X\tau|X\tau>0}(x)$. Several distributions, such as the Gamma, Lognormal, and Log-Pearson, have been used for fitting the empirical distribution of monthly precipitation. Modeling annual precipitation is similar to modeling seasonal precipitation; that is, determining either the marginal distribution $F_X(x)$ or the conditional distribution $F_{X|X>0}(x)$ depending on the particular case at hand.

Modeling precipitation at several sites is not trivial. Inter-site cross-correlations and the marginal distribution (at each site) must be considered in a multivariate framework. The data is transformed into a Normal distribution using power transforms and then a lag-zero multivariate model is applied for modeling the transformed precipitation (an approach similar to modeling streamflow, described below). Likewise, modeling annual precipitation at several sites is generally based on transforming the data into Normal and using a multivariate Normal model. In all cases where Normal transformation is utilized, after generating data in the Normal domain they must be inverted back into the original precipitation domain. Likewise, where modeling the occurrence of precipitation (zeros and non-zeros), an appropriate multivariate Markov chain model must be applied.

2.2.3. *Nonparametric Models for Simulating Precipitation*

Fitting a best PDF model to the precipitation data is often difficult because real data may exhibit a variety of features, such as bimodality, unusual skew, and long tail, that cannot be easily captured by a limited set of traditional probability density functions. Furthermore, the parameters of the traditional PDF models can be unduly influenced by outliers, leading to high variance in the selection of the best parametric model, which, in turn, impacts the ability to properly estimate the behavior of the underlying probability distribution in the tails and in the body of the distribution. Consequently, simulations from such models may not faithfully represent the observed data. Nonparametric methods offer an attractive alternative in this regard. Nonparametric models for simulating precipitation differ from the traditional methods, described above, in the manner in which the precipitation amount is modeled. The tail behavior of the data does not unduly influence the probability distribution in the main body of the data, and serial dependence is preserved in a more general sense. As a result, the representation of individual extreme events may not be any better than that achieved through the traditional parametric models. However, the properties of sequences, including the statistics of a run of extreme events, may be better represented. Given that much of the vulnerability of water systems to climate derives from exposure to persistent extremes, nonparametric methods may provide an effective tool. Kernel density estimators, described next, are one of the methods used in this regard.

2.2.3.1. *Kernel Density Estimators*

Nonparametric estimation of probability and regression functions now has a nearly 20-year history in stochastic hydrology, and in computationally intensive statistics.[51] A function approximation method is considered nonparametric if: (i) it is capable of approximating a large number of target functions; (ii) it is "local" in that estimates of the target function at a point use only observations located within some small neighborhood of the point; and (iii) no prior assumptions are made as to the overall functional form of the target function. A histogram is a familiar example of such a method. Note that such methods do have parameters (e.g. the bin width of the histogram) that influence the estimate at a point. However, they are different from "parametric" methods, where the entire function is indexed by a finite set of parameters (e.g. mean, standard deviation) and a prescribed functional form. Kernel density estimation is a nonparametric method of estimating a PDF from data that is related to the histogram. Expository monographs, which develop these ideas in detail and provide an interesting and robust presentation, are available in the literature.[52-54]

Given a set of observations $x_1, x_2,\ldots, x_n$ (in general, x may be a scalar or a vector), the kernel density estimate (KDE) is defined as

$$\hat{f}(x) = \frac{1}{Nh} \sum_{i=1}^{N} K((x-x_j)/h) \qquad (2.1)$$

where $K(\ )$ is a weight or kernel function and h is a bandwidth. This can be explained using a histogram. Consider the definition of probability as a relative frequency of event occurrence. Now, an estimate of the probability density at a point x may be obtained if we consider a box or window of width $2h$ centered at x and count the number of observations x_i that fall in such a box. The estimate $\hat{f}(x)$ is then the number of observations x_i that lie within $[x_i - h, x_i + h)]/(2hN)$. In this example, a histogram, we have used a rectangular kernel $K(t) = 1/2$ for $|t| < 1$ and 0 otherwise; and $t = (x - x_i)/h$ for the estimate in the locale of x. As the sample size n grows, one could shrink the bandwidth h such that asymptotically the underlying PDF is well approximated. Note that for a finite sample this is much like describing a histogram, except that the "bins" are centered at each observation or at each point of estimate, as desired.

From the point of view of simulation, one can treat each observation x_i as being equally likely to occur in the window $x_i \pm h$ and resample it uniformly in that interval (for this example). Clearly, one is not restricted to rectangular kernels. The "parameters" of this method are the kernel function or "local density" and the bandwidth h. A valid PDF estimate is obtained for any $K(\ )$ that is itself a valid PDF. Symmetry of $K(\ )$ is assumed for unbounded data to ensure pointwise unbiasedness of the estimate. Finite variance of $K(\ )$ is assumed to ensure that $\hat{f}(x)$ also has finite variance. This still leads to a wide choice of functions for $K(\ )$. It turns out that, in terms of the mean square error (MSE) of $\hat{f}(x)$, the choice of $K(\ )$ is not crucial. Different kernels can be made equivalent under rescaling by choosing appropriate bandwidths. A Gaussian kernel with a large bandwidth can give MSE of $\hat{f}(x)$ comparable to that using a rectangular kernel with a smaller bandwidth. Thus, given a kernel function, the focus shifts to appropriate specification or estimation of the bandwidth. It is important to note that specifying a kernel function does not have the same implications as that of choosing a parametric model for the whole density because the focus remains on a *good pointwise or local approximation* of the density rather than on fitting the whole curve directly. Different choices of $K(\ )$ still yield a local approximation of the underlying curve point by point. One can understand this by thinking of a weighted Taylor series approximation to $f(x)$ at a point x. The interplay between h and $K(\ )$ can be thought of in terms of the interval of approximation and a weight sequence used to localize the approximation. The length of the interval (or bandwidth in this case) is more important in terms of approximation error. However, the tail behavior of $K(\ )$ is important in a simulation context, since it relates to the likely degree of extrapolation of the process. Some typically used kernels are standard Normal, Quadratic (or Epanechnikov), and Bisquare.[53]

The consensus in the statistics literature[52,53] is that the choice of kernel is secondary in estimating $f(x)$, and research has focused on choosing an appropriate bandwidth optimally (in a likelihood or MSE sense) from the data. Reference bandwidth that minimizes the MSE, assuming a Normal kernel and an underlying Normal PDF of the data, provides an optimal *reference* bandwidth and consequently a smoothed PDF.[53] As mentioned above, this optimal reference bandwidth does not result in a Normal PDF of the data, because the estimation of $f(x)$ is performed locally using the KDE — this *local estimation* is an important aspect of this approach. This provides a quick and easy estimate to the bandwidth, and is widely used in software implementations.[54,55] Data-driven bandwidth selection methods, which use recursive method to minimize the average mean integrated square error (MISE)[54] of $\hat{f}(x)$ or other objective criteria (e.g. cross validation), have also been suggested.[52-54] The bandwidth may vary by location (i.e. value of x), being larger where the data are sparser. Typically, this is achieved by perturbing the global bandwidth h obtained from one of the foregoing methods.[52,53]

The KDE approach extends to discrete variables and also to multiple variables. In the case of discrete variables, the probability mass function estimator is

$$\hat{f}(L) = \sum_{j=1}^{L\max} K_d((L-j)/h)\overline{p}_j \qquad (2.2)$$

where L is the discrete value of interest for the probability mass function, $\overline{p}_j$ is the sample relative frequency (n_j/n), and $Lmax$ is the maximum discrete value in the observed data. The kernel function used in the case of continuous variables can be also used here, but they need to be normalized such that their weights are concentrated at discrete points. Discrete quadratic kernels, which have been developed and extensively tested, obviate the need for such scaling,[56] besides being consistent with a discrete random variable.

One of the annoying aspects of KDE is the increased bias within the bandwidth of the boundary (e.g. for precipitation and streamflow, the boundary is at 0) of the sample space. The bias is a consequence of the increasingly asymmetric distribution of the random variable as one approaches the boundary; hence, modifications to kernel density estimate are needed within this region. The problem is aggravated if a kernel with infinite support (i.e. Normal) is used. Several methods are available to deal with this problem, but the boundary kernels are more effective at alleviating this.[57] In this, a set of equivalent boundary kernels for a given kernel, used in Eq. (2.1), is developed when evaluating within one bandwidth from the boundary. These boundary kernels pose their own set of problems: (i) boundary kernel estimates can lead to negative values for the PDF, which is unrealistic; (ii) boundary kernels are mathematically difficult to obtain for all kernels; and (iii) the bandwidth obtained for the regular kernel may not be valid for the boundary kernel. For most applications, this is not an issue and so neglected, but the problem is acute for data with heavy concentration near the boundary, such as the case of precipitation where most of the data are close to 0. A logarithmic transform of the precipitation data prior to density estimation is often considered, and the KDE in Eq. (2.1) is now modified as

$$\hat{f}(x) = \frac{1}{hx} \sum_{i=1}^{N} K\{[\log(x) - \log(x_j)]/h\} \qquad (2.3)$$

where h is the bandwidth of the logarithm of the data. This approach works extremely well and detailed investigation of this and the boundary kernels bear this out.[58] Comparisons of kernels, bandwidth selection schemes, and boundary treatments in the context of rainfall simulation have already been performed.[59]

Simulation from the KDE simply proceeds by randomly selecting an observation x_j and generating a random value from the selected kernel (note that the kernels themselves are valid PDFs, as mentioned before). The simulated value is $x_j + e_j \times h$, and this approach is also known as *smoothed* bootstrap.[53] Slight modifications to this are proposed to reproduce the variance.[53] The simulated values will have the same PDF as that estimated by the KDE.

The rainfall simulation involves generating rainfall occurrences (wet or dry) from an appropriate Markov chain model (described in Sec. 2.2.2.1) and generating rainfall amount on wet days from the KDE, as discussed above. Nonparametric alternate renewal models have been proposed,[60] wherein wet and dry spells are alternatively simulated from the discrete kernel probability mass estimates, and the KDE is used to simulate the rainfall amounts on wet days.

Extension of the KDE to multivariate and conditional PDF estimation is straightforward in that the scalars are replaced by vectors in the equations above.[52,53] The conditional PDFs have been constructed with a similar smooth bootstrap approach to simulate daily rainfall.[59,61] This enables the ability to capture long-term persistence (i.e. persistence at seasonal and interannual timescales).

2.2.3.2. K-NN Bootstrap Models

The kernel density estimation, described above, suffers from boundary problems that get worse in the multivariable case and, thus, the density estimates and simulations tend to be biased. The problem is even more acute for variables with heavy concentration of data close to the boundary, such as the case with precipitation. The K-nearest neighbor (K-NN) approach offers a flexible and robust alternative. It is simple, intuitive, and robust, and can be applied to a variety of time series modeling problems. A brief description is presented here, and specific applications to time series modeling in the context of streamflow simulation are discussed in the following sections. A basic K-NN density estimator is given as[53]

$$\hat{f}_{NN}(\mathbf{x}) = \frac{k/N}{V_k(\mathbf{x})} = \frac{k/N}{C_d r_k^d(\mathbf{x})} \qquad (2.4)$$

where k is the number of nearest neighbors to $\mathbf{x}$, d is the dimension of the space, C_d is the volume of a unit sphere in the d dimension, and r_k is the Euclidean distance to the kth-nearest neighbor. The density estimator in Eq. (2.1) can also be modified as[53]

$$\hat{f}_{NN}(x) = \frac{1}{r_k^d(x)N} \sum_{i=1}^{N} K((x-x_i)/r_k(x)) \qquad (2.5)$$

The bandwidth is distance to the kth-nearest neighbor; that is, the number of nearest neighbors, k, is the 'smoothing' parameter. The kernel has the role of a weight function (x_i closer to the point of estimate x are weighted more), and can be chosen to be any valid probability density function. Under optimal MSE arguments, k should be chosen proportional to $n^{4/(d+4)}$ for any probability density that is twice differentiable. The sensitivity to the choice of k is somewhat lower as a kernel that is monotonically decreasing with $r_k(x)$ is used. This automatically achieves the effect of bandwidth changing with x, thus providing the ability to adapt to the data variability. Equations (2.4) and (2.5) can be used to readily construct multivariate and conditional PDFs, and the simulation uses the same *smooth bootstrapped* approach described earlier, or one of the nearest neighbors can be resampled using the kernel function. This is a modification to the straight bootstrap.[62,63] Theoretical basis for using the *K*-NN density estimators for time series forecasting has been well developed.[64-68] This approach has been used for simulating rainfall and other weather variables in the context of stochastic weather generators.[69] Details on the application of this for time series modeling and simulation are discussed later in this chapter.

2.2.4. *Precipitation Disaggregation Models*

Often, precipitation generated at certain timescales (e.g. daily, multi-days) needs to be disaggregated to smaller timescales (e.g. hourly or daily within a wet spell). This disaggregation is generally done empirically.[70] For instance, using proportions, one can disaggregate 24-hr (daily) precipitation into 6-hr precipitation. Also, disaggregation of daily rainfall has been suggested by modeling the number of rain showers, and the magnitudes, duration, and arrival time within a day.[71,72] In addition, disaggregation schemes of short-term rainfall based on a specified model structure for continuous rainfall has been developed[2] as well as using neural networks,[73] but they can be computationally intensive. Although the foregoing models are innovative, they are complex and require many transformations of the original data to obtain reasonable results. Other shortcomings include the lack of flexibility in the number of intervals considered and the incompatibility of parameter estimates at different aggregation levels. Alternatively, nonparametric methods for disaggregation have been developed,[60] in which a wet spell length is first simulated and the total rainfall magnitude is obtained from a nonparametric density estimator. Subsequently, a vector of proportions to distribute the rainfall to the individual days of the wet spell is simulated and applied to the spell rainfall. In addition, disaggregation methods for spatial and temporal streamflow simulation, described in the following section, also provide alternatives that can be modified and applied for rainfall simulation.

2.3. Stochastic Simulation of Streamflow

2.3.1. *Continuous Time to Hourly Simulation*

Streamflow simulation on a continuous timeframe requires the formulation of a model structure that is capable of reproducing the streamflow fluctuations on a wide dynamic range. The application of stochastic approaches to continuous time and short timescale streamflows has been limited, because of the complex nonlinear relations that characterize the precipitation-streamflow processes at those temporal scales. The early attempts to model hourly and daily streamflows were based on autoregressive (AR) models after standardization and transformation. However, such models, essentially based on process persistence, do not properly account for the rising limb and recession characteristics that are typical of hourly and daily flow hydrographs. Also, shot noise or Markov processes and transfer function models have been proposed for daily flow simulation, with some limited success in reproducing the rising limb and recessions.[74]

Nevertheless, interesting work has been done with some success using conceptual-stochastic models. Conceptual representation of a watershed, considering the effects of direct runoff and surface and groundwater storages, has been applied.[75-78] Direct runoff is modeled by a periodic AR order one (PAR(1)) model (see below) with an indicator function to produce intermittence, and the other components are modeled using linear reservoirs. The conceptual stochastic model produced reasonable results in the generation of daily flows for the Powell River, Tennessee. Other conceptual-stochastic models for short time runoff have been proposed. For example, assuming an independent Poisson process for rainfall, a three-level conceptual runoff model represents surface runoff to estimate the daily response of a watershed, and the base flow modeled by three linear reservoirs that represent the contribution of deep aquifers with over-year response, aquifers with annual renewal, and subsurface runoff, the foregoing scheme leads to a multiple shot noise streamflow process. The model is effective in reproducing streamflow variability. In addition, intermittent daily streamflow process has been modeled by combining conceptual approach with product models[79,80] and gamma AR models,[81] and by using a three-state Markov chain describing the onset of streamflow and an exponential decay of streamflow recession.[82]

2.3.2. Weekly, Monthly, and Seasonal Streamflow Simulation at a Single Site

Stationary stochastic models may be applied for modeling weekly, monthly, and seasonal streamflows after *seasonal standardization*. This approach may be useful where the temporal correlations do not vary throughout the year. In general, however, models with periodic correlation structure, such as periodic autoregressive (PAR)[83] and periodic ARMA (PARMA) are more applicable.[84,85] An example is the PARMA(1,1) model[85]

$$y_{v,\tau} = \mu_\tau + \phi_{1,\tau}(y_{v,\tau-1} - \mu_{\tau-1}) + \varepsilon_{v,\tau} - \theta_{1,\tau}\varepsilon_{v,\tau-1} \tag{2.6}$$

where $\tau = 1,...,\omega$ (ω is the number of seasons), μ_τ, $\phi_{1,\tau}$, $\theta_{1,\tau}$, and $\sigma_\tau(\varepsilon)$ are the parameters. When the θ's are zeros, this becomes the PARMA(1,0) or PAR(1) model. Low-order models, such as PARMA(1,0) and PARMA(1,1), have been widely used for simulating monthly and weekly flows.[85-91] The method of moments is typically used for estimation of the model parameters. The model in Eq. (2.6) is based on the traditional linear regression in that the past values are linearly related to the current values and the residuals (ε) are Normally distributed with mean 0 and standard deviation $\sigma_\tau(\varepsilon)$. Consequently, the data must be Normally distributed; otherwise, they have to be transformed to Normality before fitting the model. In some cases, this may pose additional problems (see below).

Periodic ARMA models can be derived from physical/conceptual principles. Considering all hydrologic processes and parameters in the watershed varying along the year, it has been shown that seasonal streamflow falls within the family of PARMA models.[76] Alternatively, a constant parameter ARMA(2,2) model with periodic independent residuals was suggested.[77,78] A desirable property of stochastic models of seasonal streamflows is the preservation of seasonal and annual statistics.[92] However, such dual preservation of statistics has been difficult to get with simple models, such as the PAR(1) and PAR(2). For this reason, in the 1970s, hydrologists turned to the so-called *disaggregation* models (see below). Periodic ARMA models, having more flexible correlation structure than PAR models, offer the possibility of preserving seasonal and annual statistics. Some hydrologists have argued that PARMA models have too many parameters. However, it may also be possible to reduce the number of parameters by keeping some of them constant. An alternative for reproducing both seasonal and annual statistics is the family of multiplicative models.

Multiplicative models were first suggested by Box and Jenkins.[93] These models have the characteristic of linking the variable $y_{v,\tau}$ with $y_{v,\tau-1}$ and $y_{v-1,\tau}$. Multiplicative models, after differencing the logarithms of the original series, have been applied for simulating and forecasting monthly streamflows.[94] However, they were not able to reproduce the seasonality in the covariance structure. This problem occurred because the referred multiplicative model did not include periodic parameters. A model (with periodic parameters) that can overcome these limitations is the multiplicative PARMA model.[92] For instance, the multiplicative PARMA(1,1) x (1,1)$_\omega$ model is written as

$$z_{v,\tau} = \Phi_{1,\tau} z_{v-1,\tau} + \phi_{1,\tau} z_{v,\tau-1} - \Phi_{1,\tau} \phi_{1,\tau} z_{v-1,\tau-1} + \varepsilon_{v,\tau} \\ - \Theta_{1,\tau} \varepsilon_{v-1,\tau} - \theta_{1,\tau} \varepsilon_{v,\tau-1} + \Theta_{1,\tau} \theta_{1,\tau} \varepsilon_{v-1,\tau-1} \quad (2.7)$$

in which $z_{v,\tau} = y_{v,\tau} - \mu_\tau$ and $\Phi_{1,\tau}$, $\Theta_{1,\tau}$, $\phi_{1,\tau}$, $\theta_{1,\tau}$, and $\sigma_\tau(\varepsilon)$ are the model parameters. This model has been applied successfully for simulating the Nile River flows.[92]

A particular limitation of the foregoing PARMA and multiplicative PARMA models for modeling streamflow time series is the requirement that the underlying series be transformed into Normal. An alternative that circumvents this problem is the PGAR(1) model for modeling seasonal flows with periodic correlation structure and periodic gamma marginal distribution.[95] Consider that $y_{v,\tau}$ is a periodic correlated variable with gamma marginal distribution with location λ_τ, scale α_τ, and shape β_τ parameters varying with τ, and $\tau = 1,...,\omega$ (ω is the number of seasons). Then, the variable $z_{v,\tau} = y_{v,\tau} - \lambda_\tau$ is a two-parameter gamma that can be represented by the model $z_{\nu,\tau} = \phi_\tau z_{\nu,\tau-1} + (z_{\nu,\tau-1})^{\delta_\tau} w_{\nu,\tau}$, where ϕ_τ is the periodic autoregressive coefficient, δ_τ is the periodic autoregressive exponent, and $w_{\nu,\tau}$ is the noise process. This model has a periodic correlation structure equivalent to that of the PAR(1) process. It has been applied to weekly streamflows for several rivers in the United States.[95] Results obtained indicate that such PGAR model compares favorably with respect to the Normal-based models (e.g. the PAR model after logarithmic transformation) in reproducing the basic statistics usually considered for streamflow simulation.

Periodic ARMA and PGAR models are less useful for modeling flows in ephemeral streams because the flows are intermittent, a characteristic that is not represented by these models. Instead, periodic product models, such as $y_{\nu,\tau} = x_{\nu,\tau} z_{\nu,\tau}$,[80] are more realistic, where $x_{\nu,\tau}$ is a periodic correlated Bernoulli (1,0) process, $z_{\nu,\tau}$ may be either an uncorrelated or correlated periodic process with a given marginal distribution, and x and z are mutually uncorrelated. Properties and applications of these models for simulating intermittent monthly flows of some ephemeral streams have already been reported in the literature.[80,96]

2.3.3. *Annual Streamflow Simulation at a Single Site*

Generally, stationary stochastic models have been proposed and utilized for simulating annual streamflow processes.[83,85,91] Various alternative models have been suggested and applied, depending on the particular case at hand, such as the ARMA models,[85,91,97] gamma AR models,[83] fractional Gaussian noise,[98] broken line,[99] and shifting mean.[100] Among these, the ARMA models have found wider applicability. For example, the ARMA(1,1) model may be expressed as in Eq. (2.6), except that all the parameters are constant values. Also, ARMA models arise from conceptual considerations of the precipitation-runoff process of a watershed.[101,102] Significant amount of work and experience has been gained applying ARMA models for simulation and forecasting of hydroclimatic processes in general, and streamflows in particular.[84,85,91,97,103,104]

2.3.4. *Multisite Streamflow Simulation*

Some of the foregoing models can be extended to multiple sites. Typically, at the annual timeframe, simple models, such as multivariate AR(1) and ARMA(1,1), are usually adequate. We illustrate here multisite models using a periodic model. The model equations now contain vectors, as opposed to scalars, and the model parameters are matrices, as opposed to scalars.[84,85,103] For example, the multivariate PARMA(1,1) model is

$$Z_{\nu,\tau} = \Phi_\tau Z_{\nu,\tau-1} + \underline{\varepsilon}_{\nu,\tau} - \Theta_\tau \underline{\varepsilon}_{\nu,\tau-1} \tag{2.8}$$

in which $Z_{\nu,\tau} = Y_{\nu,\tau} - \underline{\mu}_\tau$; $\underline{\mu}_\tau$ is a column parameter vector with elements $\mu_\tau^{(1)},...,\mu_\tau^{(n)}$ the seasonal mean vector, Φ_τ and Θ_τ are $n \times n$ periodic parameter matrices, the noise term $\underline{\varepsilon}_{\nu,\tau}$ is a column vector Normally distributed with: $E(\underline{\varepsilon}_{\nu,\tau}) = \underline{0}$, $E(\underline{\varepsilon}_{\nu,\tau}\underline{\varepsilon}_{\nu,\tau}^T) = \Gamma_\tau$, and $E(\underline{\varepsilon}_{\nu,\tau}\underline{\varepsilon}_{\nu,\tau-k}^T) = 0$ for $k \neq 0$, and n is the number of sites. In addition, it is assumed that $\underline{\varepsilon}_{\nu,\tau}$ is uncorrelated with $Z_{\nu,\tau-1}$. Parameter estimation of this model can be made by the method of moments, although the solution is not straightforward. Dropping the moving average term, i.e. $\Theta_\tau = 0$ for all τ's, yields a simpler multivariate PARMA(1,0) or PAR(1) model. This simpler model has been widely used for generating seasonal hydrologic processes. Further simplifications of the foregoing models can be made in order to facilitate parameter estimation. Assuming that Φ_τ and Θ_τ of Eq. (2.8) are diagonal matrices, the multivariate PARMA(1,1) model can be decoupled into univariate models for each site. To maintain the cross-correlation among sites, $\underline{\varepsilon}_{\nu,\tau}$ is modeled as $\underline{\varepsilon}_{\nu,\tau} = B_\tau \underline{\xi}_{\nu,\tau}$, where $E(\underline{\xi}_{\nu,\tau}\underline{\xi}_{\nu,\tau}^T) = 1$ and $E(\underline{\xi}_{\nu,\tau}\underline{\xi}_{\nu,\tau-k}^T) = 0$ for $k \neq 0$. This modeling scheme is a *contemporaneous* PARMA(1,1) or CPARMA(1,1) model. Useful references on this type of models are available in the literature.[84,85,89,103] Furthermore, contemporaneous multiplicative models are also available, and have been applied for simulating the Nile River system.[105]

2.3.5. *Temporal and Spatial Disaggregation Models*

Disaggregation models, i.e. downscaling models in time and/or space, have been an important part of stochastic hydrology, not only because of the scientific interest in understanding and describing the temporal and spatial variability of hydrologic processes but also because of practical engineering applications. For example, many hydrologic design and operational problems require hourly precipitation data. Because hourly precipitation data are not as commonly available as daily data, a typical problem has been to downscale or disaggregate daily data into hourly data. Similarly, for simplifying the modeling of large-scale systems involving a large number of precipitation and streamflow stations, temporal and spatial disaggregation procedures are needed. A brief overview of several empirical and mathematical models and procedures for temporal and spatial disaggregation of precipitation and streamflow is presented below. The multiscale statistical disaggregation model by Valencia and Schaake[106] is the starting point for any disaggregation effort given by

$$Y = A X + B \varepsilon \qquad (2.9)$$

where Y is a vector of disaggregated values (e.g. monthly streamflows), X is a vector or scalar of aggregate flows (e.g. annual streamflows), A and B are parameter matrices, and ε is a vector of independent standard Normal deviates. Parameter estimation, based on the method of moments, leads to the preservation of the first- and second-order moments at all levels of aggregation.

The shortcoming of low-order PAR models when applied for simulation of seasonal flows in reproducing the annual flow statistics led to the development of disaggregation models. In this model, the simulation of seasonal flows is accomplished in two or more steps. First, the annual flows (or aggregate flows) are modeled so as to reproduce the desired annual statistics (e.g. based on the ARMA(1,1) model). Then, synthetic annual flows are generated, which, in turn, are disaggregated into the seasonal flows by means of the model presented in Eq. (2.9). While the variance-covariance properties of the seasonal flow data are preserved and the generated seasonal flows also add up to the annual flows, it does not preserve the covariances of the first season of a year and any preceding season. To circumvent this shortcoming, this model has been modified as $Y = A\ X + B\ \underline{\varepsilon} + C\ Z$, where C is an additional parameter matrix and Z is a vector of seasonal values from the previous year (usually only the last season of the previous year) for each site.[107] Further refinements and corrections, assuming an annual model that reproduces S_{XX} and S_{XZ}, have been suggested[108,109] as well as a scheme that does not depend on the annual model's structure yet reproduces the moments S_{YY}, S_{YX}, and S_{XX}.[110] The parameter estimation and appropriate adjustments, so that the seasonal values add exactly to the annual values at each site, can be found in the literature.[85,109,111]

The foregoing disaggregation models have too many parameters, a problem that may be significant, especially when the number of sites is large and the available historical sample size is small. The estimation problem can be simplified if the disaggregation is done in steps (stages or cascades), so that the size of the matrices and, consequently, the number of parameters involved decrease. For instance, annual flows can be disaggregated into monthly flows directly in one step (this is the usual approach), or they can be disaggregated in two or more steps (e.g. into quarterly flows in a first step, then each quarterly flow is further disaggregated into monthly flows in a second step). However, even in the latter approach, considerable size of the matrices will result when the number of seasons and the number of sites are large. A *stepwise disaggregation scheme* has been proposed[112] in such a way that, at each step, the disaggregation is always made into two parts or two seasons. This scheme leads to a maximum parameter matrix size of 2×2 for single site disaggregation and $2n \times 2n$ for multisite. Condensed disaggregation models that reproduce seasonal statistics and the covariance of seasonal flows with annual flows assuming lognormal seasonal and annual flows have also been suggested.[113,114]

Disaggregation models are useful for modeling complex systems. For example, an 'index' time series can be created by adding all the individual time series and the disaggregation approach is applied to model and simulate the time series collectively, thereby capturing the statistical properties of the individual time series and also their dependence.

2.3.6. *Nonparametric Streamflow Simulation Models*

Nonparametric methods, as discussed earlier, offer an attractive alternative. In the nonparametric framework, the time series modeling is considered as simulation from 'conditional PDF,' e.g. $f(X|Y)$ or $f(X_t|X_{t-1}, X_{t-2}, ..., X_{t-p})$. In the parametric linear modeling framework, since the data is assumed to be Normal (as discussed above), these conditional PDFs are also Normal and the relationship captured is also linear (as the correlation coefficient — which is a measure of linear association — is a variable that is part of the joint, marginal, and conditional Normal PDF). In the nonparametric framework, the "local" estimation provides the capability to capture local nonlinear and non-Normal features that might be present in the data.

2.3.6.1. *Single Site*

A kernel density-based seasonal periodic model for streamflow simulation at a single site has been proposed.[115] In this, a KDE for conditional PDF of $f(X_t|X_{t-1})$ is constructed using a Normal kernel in two dimensions, where t and $t-1$ are successive months (or successive seasons). Consequently, the *smooth bootstrap* is modified in that the conditional PDF provides a weight to each data point that is used to bootstrap from. A good description of this approach[103] also demonstrates the capability to capture bimodality and nonlinearity in comparison to traditional linear models.

The KDE can be used to perform data transformation in that, first the PDF of the data is estimated using KDE and, consequently, the cumulative distribution function (CDF). The CDF values of the data points are mapped on to a Normal distribution $F(x_i) = c_i$ computed by integrating the KDE, and this is inverted using a Normal distribution, $F^{-1}_N(c_i) = y_i$ to obtain the transformed value in the Normal space. The time series model is fitted to the transformed data and the simulations are back-transformed to the original space. Since the transformation is done from the PDF estimated by the KDE, the final simulations are likely to capture the non-Normal and other features that might be present in the data.[116]

As noted before, a lag-1 seasonal model (parametric or nonparametric) does not capture the annual and interannual statistics well. The interannual properties are important for simulating long wet and dry periods that are critical for drought management and planning. The lag-1 model was modified to include the sum of the streamflow from the past 18 months as an additional conditional variable,[117] so that the model is a conditional PDF simulation of $f(X_t|X_{t-1},Z_t)$, where Z_t is the sum of flows of the 18 months prior to time t. This modification was shown to be effective at capturing the interannual statistics in the simulation of rainfall[61,117] and streamflow.[118] Also, another modification has been proposed by using a pilot variable to lead the generation of the seasonal flows.[119] This offers the possibility of using either a parametric or a nonparametric model to generate the pilot variable (flows) to assure that long-term variability and low frequency are reproduced.

The KDE-based methods suffer from boundary bias (described before), which gets exaggerated in higher dimensions. While reference bandwidths are typically used, as they are easy, it is not trivial to estimate optimal bandwidths in higher dimensions. To address this, K-NN time series bootstrap was developed.[120] In this, the simulation is performed as follows: (i) a conditioning 'feature vector' D_t is developed — if it is a lag-1 model, then it is a scalar value; (ii) K nearest neighbors of D_t are identified from the data using Euclidean distance metric; (iii) each neighbor is assigned a weight, with largest weight to the nearest and least to the farthest, using a weight function

$$W(k) = \frac{1}{k \sum_{i=1}^{K} \frac{1}{i}}, \ k = 1, 2, \ldots, K;$$ (iv) one of the neighbors is resampled

using this weight metric, say j; and (v) the successor value x_{j+1} is the simulated value for the current time. The procedure is repeated. This approach is simple and robust that it can easily simulate from conditional PDF of any dimension. It has been shown that the simulations are insensitive to the choice of the weight function, as long as the selected weight function weighs the nearest neighbor the most relative to the farthest.[107] The only parameter is K, the number of nearest neighbors, and, based on asymptotic heuristic arguments,[120,121] it is shown that the choice $K = \sqrt{N}$ is quite robust. This approach was applied to seasonal streamflow simulation[120] and found to perform better than the kernel density-based model at much less computation cost and in a simple and parsimonious manner.[115] One shortcoming of this approach is that the variance may be underestimated, especially where the historical data are correlated. However, this can be fixed by using a gamma perturbation.[119]

2.3.6.2. *Extension to Multisite*

Kernel-based methods have been extended for disaggregation of annual streamflows to monthly flows. The conditional PDF $f(X|Z)$, where Z is the aggregate (annual) flow and X the vector of dimension d, the disaggregated monthly streamflows, forms the basis of the simulation.[122] Due to the summability criteria (i.e. the values of X should sum to Z), the conditional PDF resides in a $d - 1$ space. The steps involved are as follows: (i) the monthly data is transformed to a space such that they are orthonormal to the conditioning plane, $Y = X R$, where the last column of Y is the conditioning plane, $Y_d = Z / \sqrt{d} = Z'$ and the other columns are orthonormal to this, thus $Y = [U, Z']$, R is a rotational matrix obtained from Gram Schmidt orthonormalization procedure; (ii) the conditional PDF $f(U|Z')$ is estimated using KDE on the same lines as in the single site case;[115] and (iii) a vector U is generated by smooth bootstrap and it is back-transformed to obtain the disaggregated vector of monthly flows X.[115]

As mentioned before, kernel methods have serious limitations when applied to higher dimensions; therefore, the use of KDE was replaced by K-NN time series resampling,[123] which provides the ability to easily apply this to space and time disaggregation together and also for a number of spatial locations simultaneously. This was demonstrated to be simple, robust, and efficient, like the single site simulation.[119,120]

2.4. Extensions of *K*-NN Resampling Approach

The *K*-NN resampling approach for multivariate streamflow simulation, described above, has been applied to a variety of other problems, most notably for multisite daily weather simulation. A lag-1 multivariate resampling model was implemented to simulate daily rainfall and other weather variables at a single site.[69] This was subsequently extended to simulate daily rainfall and weather variables at multiple locations simultaneously.[124] A semiparametric approach to rainfall and weather generation was proposed, in which a Markov chain model was fitted for each month separately to simulate the precipitation occurrence, and then K-NN lag-1 resampling approach was used to generate the vector of weather variables conditionally on the transitional state of the precipitation occurrence; for example, if the simulated current day's occurrence was wet and the following day dry, then the neighbors are obtained from the historical data that have the same transition.[125] The semiparametric approach has also been demonstrated for simulating daily weather conditional on seasonal forecast, which are very useful for driving process models, such as crop and water resources for planning and management.

The K-NN resampling method is an extremely versatile method. Once K nearest neighbors of a feature vector are identified, a number of possibilities are available. If the feature vector is a set of predictors of streamflow, then the identified neighbors can be used to provide an ensemble streamflow forecast or a weighted average mean forecast. This was applied skillfully for the multisite seasonal ensemble streamflow forecast in a watershed in Brazil.[126] This approach, with its simplicity and robustness, is gaining prominence, and is being used for a variety of forecasting applications. Using a feature vector consisting of spatial information (e.g. latitude, longitude), the resampling approach can be used to generate ensembles of variables at locations with no observations; i.e. as a means of spatial filling. This approach is being successfully introduced in the simulation of water quality variables of desired locations with limited or no data, based on data from other locations.[127] A similar approach has been used for reconstructing ensembles of streamflows based on tree ring information.[128]

The K nearest neighbors identified can also be used to fit 'local polynomials.'[129,130] This allows the ability to extrapolate values beyond the range of the observations and also provides smoothness in recovering the underlying functional relationship. There have been a number of successful applications of this approach for hydroclimatic simulation[131] and forecast.[132-138] Applications of the resampling approach have also been extended to groundwater and other hydrologic problems with good success.[51] The *K*-NN-based methods are being adapted for downscaling of climate projections (seasonal, decadal, and climate change) to regional- and point-scale hydrologic scenarios for use in various process models, such as crop and water resources, for efficient planning and management of resources.

2.5. Summary and Future Directions

Stochastic models have been utilized in actual practice over several decades for planning and management studies of water resources systems. For example, stochastic models may be applied for simulating possible hydroclimatic scenarios that may occur in the future, which, in turn, are used for sizing hydraulic structures, such as dams, for re-evaluating the capacity of existing reservoirs under uncertain streamflow scenarios, or for investigating the possible occurrence and frequency of extreme events, such as droughts. Also, stochastic models may be useful for forecasting hydroclimatic events that may occur in short or long timeframes.

A variety of stationary and nonstationary models for single and multiple sites has been developed in the past 50 years, and much experience and successful applications have been gained. Parametric and nonparametric modeling schemes, as well as combined approaches, have been developed. The traditional modeling tools have been those that are defined in the form of mathematical models with a parameter set (parametric models) that must be estimated from historical data. As in any other modeling framework, they are not free from shortcomings. For example, most of the parametric time series models are *linear* in their form, in that they can only capture linear relationships between the variables. Also, since the residuals are assumed to be Normally distributed and identical and independently distributed (i.i.d.), the underlying variables in the model are also Normally distributed. Thus, historical data must be transformed to Normality before the model is fitted, and the model fitted in the transformed space does not guarantee the reproduction of statistical properties in the original space. In addition, simulations can only reproduce Normal distribution, and features, such as bimodality, cannot be captured; and in the disaggregation models, the transformation can destroy the summability of the disaggregations to the aggregate values.

However, some, if not most, of these shortcomings can be circumvented. For example, there are a variety of mathematical functions that can be quite useful for transforming data to Normal. Thus, building the models in the transformed Normal domain also has advantages in that well-known modeling and testing procedures can be applied. Also, if models built in the Normal domain do not reproduce historical statistics in the real (original) domain, appropriate estimation procedures are also available for many of the models to circumvent this concern. Furthermore, there are some models available where the marginal distribution of the underlying variable is not necessarily Normal. It has been argued that, in some cases, one or more months of the historical data shows some evidence of bimodality and that the parametric models may not faithfully reproduce such bimodality. However, the effect of reproducing such particularity may not be important.

Nonparametric methods, while being data-driven, have higher variance of estimation for the underlying probability distribution, and near the edges of the data are biased. Especially, the resampling methods tend to be biased when the sample size is small, which also limits the variety of sequences that can be generated. Obviously, the resampling methods cannot generate values outside the range of the historical data. While this is not an issue in some cases, in many applications a variety of values outside the observations are highly desirable. The sample size also impacts the ability of the nonparametric methods in general to simulate the tails of the distribution well. This can be alleviated by fitting local functions, but the sample size issue is persistent. A semiparametric approach that combines the advantages of the parametric approach with those of the nonparametric framework is an attractive alternative. Such combined approaches have been developed for weather generator,[125] for streamflow forecasting,[126] and for conditional flood frequency estimation.[139]

Generalized linear modeling (GLM)-based approaches seem to offer attractive alternatives to traditional parametric methods. In this approach, variables with different characteristics (i.e. continuous, binary, discrete, skewed) and distributions can be easily combined, providing the flexibility and the ability to capture a range of features. This approach has recently been applied to develop a stochastic daily weather generator[140] and, coupled with an extreme value distribution,[141] it can also generate extreme value precipitation characteristics, which the traditional and nonparametric generators, described earlier, have difficulty with. The GLM can also be implemented in a nonparametric (local polynomial) framework, whereby the functional estimation is performed 'locally' — thus, combining the advantages of both the approaches. Recent application of this modification to water quality forecasting shows promising results.[142] Bayesian extension of the GLM approach is readily possible providing robust estimates of uncertainty.[143]

Overall, the field of stochastic hydrology has been well-developed, offering a wide variety of parametric and nonparametric models and techniques. Often, modeling of complex systems may gain from the appropriate combination of both techniques, especially because simulation and forecasting of complex systems involving several sites, more often than not, involve the application of several models. Also, the field has been developed to a point where many of the stochastic models that we have discussed in this chapter can be found in well-known statistical packages, and particularly in specialized softwares, such as SPIGOT[114] and SAMS.[144]

However, despite the significant advances made in the field, as pointed out above, there are many challenges ahead. First of all, we would like to alert that, in many cases, the models built particularly for stochastic simulation require that the data be naturalized to remove the effect of upstream derivations, lake regulations, etc; often, errors in such data base could be significant. Also, applying many of the techniques above require some reasonable sample size (data length). The easy availability of softwares often leads to applying stochastic techniques with very short data lengths, disregarding the effect of the uncertainty in estimating the parameters involved. As a matter of fact, techniques are available to account for such uncertainties.[145]

Perhaps an area where most challenges appear is how to account for changes in land use and climate change. This is especially important, since both parametric and nonparametric models/techniques generally utilize historical data that have been gathered during many years in the past. This is an issue that does not have general answers, since it depends very much on the system at hand. Several hydrologic systems exhibit low-frequency variations arising from fluctuations of large-scale climate forcings, such as El Niño Southern Oscillation (ENSO), Pacific Decadal Oscillation (PDO), and Atlantic Multidecadal Oscillation (AMO). Traditional stochastic methods need to be modified to capture the low-frequency variability. Steps in this direction are the shifting mean models[100] and hidden Markov models.[146-148] However, this generally requires long records of the variables involved to identify whether such changing episodes of the mean occurs, as mentioned above. At the same time, in places where there may be some important effects of climatic signals, as highlighted above, short records may suggest either upward or downward trends, misleading the real cause of such significant changes and the way how such episodes may be accounted in developing possible, say streamflow, scenarios that may occur in the future. Wavelet- and Bayesian-estimation methods are being developed to simulate hydrologic scenarios that reproduce realistic low-frequency variability.[143,149,150] These methods have the ability to capture nonstationarity and also provide realistic estimates of uncertainty. Another case where judgment and experience are needed is how to take into account the effect of the two major El Niños of the 1982–83 and 1997–98 (15 years apart) if one uses stochastic simulation for predicting streamflows, say for the next 50 years, in places such as the north of Peru, where many of the streams showed annual streamflows during those years that were several orders of magnitude greater than the average flows based on the other records. Incorporating such mega-Niños in the stochastic simulation of streamflow might require information from other sources of data (e.g. paleo-proxy data, such as tree rings and ice cores), in which information from the proxy and observational data have to be combined. Nonhomogeneous Markov chain-based methods are being developed, wherein the proxy data is used to generate the state of the system and the observational data for the magnitude (e.g. streamflow).[151] These methods are proving to be effective at generating realistic low-frequency variability that can be of immense use in long-term water resources planning. These methods can also be used for stochastic simulation of regional hydroclimatology based on future projections of climate. We hope that this review article provides the readers with a comprehensive overview of the variety of stochastic simulation techniques for hydrologic applications and their potential modifications for a suite of applications.

References

1. L. A. Le Cam, in *Proceedings of Fourth Berkeley Symposium on Mathematical Statistics and Probability*, Ed. J. Newman (University of California Press, Berkeley, 1961), p. 165.
2. L. G. Cadavid, J. D. Salas and D. C. Boes, *Water Resour. Papers* (Colorado State University, Fort Collins, Colorado, 1992), p. 106.
3. A. Carsteanu and E. Foufoula-Georgiou, *J. Geophys. Res.*, **D21**, 26363 (1996).
4. P. Eagleson, *Water Resour. Res.*, 713 (1978).
5. P. Burlando and R. Rosso, in *Stochastic Hydrology and its Use in Water Resources Systems Simulation and Optimization, Proc. NATO-ASI, Peñiscola, Spain,* Eds. J. Marco-Segura, R. Harboe and J. D. Salas (Kluwer, 1993), p. 137.
6. I. Rodriguez-Iturbe, V. K. Gupta and E. C. Waymire, *Water Resour. Res.*, 1611 (1984).
7. J. Neyman and E. L. Scott, *J. Roy. Stat. Soc.*, **Ser. B**, 1 (1958).
8. M. L. Kavvas and J. W. Delleur, *Water Resour. Res.*, 1151 (1981).
9. E. Foufoula-Georgiou and P. Guttorp, *Water Resour. Res.*, 1316 (1986).
10. J. A. Ramirez and R. L. Bras, *Water Resour. Res.*, 317 (1985).
11. I. Rodriguez-Iturbe, D. R. Cox and V. Isham, *Proc. R. Soc. London*, **Ser. A**, 269 (1987).
12. I. Rodriguez-Iturbe, F. B. De Power and J. B. Valdes, *J. Geophys. Res.*, **D8**, 9645 (1987).
13. P. Burlando and R. Rosso, *J. Geophys. Res.*, **D5**, 9391 (1991).
14. D. Entekhabi, I. Rodriguez-Iturbe and P. S. Eagleson, *Water Resour. Res.*, 295 (1989).
15. S. Islam, D. Entekhabi and R. L. Bras, *J. Geophys. Res.*, 2093 (1990).
16. J. T. B. Obeysekera, G. Tabios and J. D. Salas, *Water Resour. Res.*, 1837 (1987).
17. P. S. P. Cowpertwait, *Proc. R. Soc. London*, **Ser. A**, 163 (1995).
18. P. S. P. Cowpertwait and P. E. O'Connell, *Hydrol. Earth Syst. Sci.*, 71 (1997).
19. R. W. Koepsell and J. B. Valdes, *ASCE J. Hydr. Eng.*, 832 (1991).
20. E. C. Waymire, V. K. Gupta and I. Rodriguez-Iturbe, *Water Resour. Res.*, 1453 (1984).
21. J. A. Smith and A. F. Karr, *Water Resour. Res.*, 95 (1983).

22. T. A. Buishand, *J. Hydrol.*, 295 (1977).
23. E. Foufoula-Georgiou and D. P. Lettenmaier, *Water Resour. Res.*, 875 (1987).
24. Y. Gyasi-Agyei and G. R. Willgoose, *Water Resour. Res.*, 1699 (1997).
25. J. A. Smith and W. F. Krajewski, *Water Resour. Res.*, 1893 (1987).
26. R. W. Katz and M. B. Parlange, *Water Resour. Res.*, 1331 (1995).
27. I. Rodriguez-Iturbe, F. B. De Power, M. B. Sharifi and K. P. Georgakakos, *Water Resour. Res.*, 1667 (1989).
28. E. Foufoula-Georgiou and W. F. Krajewski, *Rev. Geophys.*, 1125 (1995).
29. E. Foufoula-Georgiou and K. P. Georgakakos, *Recent Advances in the Modelling of Hydrologic Systems* (NATO ASI Series, 1988).
30. K. P. Georgakakos and M. L. Kavvas, *Rev. Geophys.*, 163 (1987).
31. E. C. Waymire and V. K. Gupta, *Water Resour. Res.*, 1261 (1981).
32. E. C. Waymire and V. K. Gupta, *Water Resour. Res.*, 1273 (1981).
33. Special issue, *J. Appl. Meteorol.*, **32** (1993).
34. Special issue, *J. Geophys. Res.*, **104 D24** (1999).
35. T. J. Chang, M. L. Kavvas and J. W. Delleur, *Water Resour. Res.*, 565 (1984).
36. C. W. Richardson and D. A. Wright, *USDA-ARS, ARS-8* (USDA, 1984), p. 1.
37. J. Roldan and D. A. Woolhiser, *Water Resour. Res.*, 1451 (1982).
38. D. S. Wilks, *Statistical Methods in the Atmospheric Sciences* (Academic Press, San Diego, USA, 1995).
39. P. Guttorp, *Stochastic Modeling of Scientific Data* (Chapman Hall, London, 1995).
40. E. H. Chin, *Water Resour. Res.*, 949 (1977).
41. R. Mehrotra and A. Sharma, *J. Hydrol.*, 102 (2007).
42. R. W. Katz, *Technometrics*, 243 (1981).
43. A. M. Feyerherm and L. D. Bark, *J. Appl. Meteorol.*, 320 (1965).
44. D. S. Wilks, *Water Resour. Res.*, 1429 (1989).
45. J. W. Hopkins and P. Robillard, *J. Appl. Meteorol.*, 600 (1964).
46. C. T. Haan, D. M. Allen and J. O. Street, *Water Resour. Res.*, 443 (1976).
47. A. G. Guzman and C. W. Torrez, *J. Clim. Appl. Meteorol.*, 1009 (1985).
48. R. Srikanthan and T. A. McMahon, *Trans. ASAE*, 754 (1983).
49. L. L. Wilson and D. P. Lettenmaier, *J. Geophys. Res.*, 2791 (1993).
50. C. T. Haan, *Statistical Methods in Hydrology* (Blackwell Publishers, 2002).
51. U. Lall, *Rev. Geophys.*, 1093 (1995).
52. D. W. Scott, *Multivariate Density Estimation: Theory, Practice and Visualization* (John Wiley, New York, 1992).
53. B. W. Silverman, *Density Estimation for Statistics and Data Analysis* (Chapman and Hall, New York, 1986).
54. W. Härdle, *Smoothing Techniques with Implementation in S* (Springer-Verlag, New York, 1991).
55. A. W. Bowman and A. Azzalini, *Applied Smoothing Techniques for Data Analysis* (Clarendon, 1997).
56. S. J. Sheather and M. C. Jones, *J. R. Stat. Soc.*, **Ser. B**, 683 (1991).

57. B. Rajagopalan and U. Lall, *J. Nonparam. Stats.*, 409 (1995).
58. H. G. Muller, *Biometrika*, 521 (1992).
59. B. Rajagopalan, U. Lall and D. G. Tarboton, *Stoch. Hydrol. Hydrau.*, 523 (1997).
60. U. Lall, B. Rajagopalan and D. G. Tarboton, *Water Resour. Res.*, 2803 (1996).
61. T. I. Harrold, A. Sharma, and S. J. Sheather, *Water Resour. Res.*, 1300 (2003).
62. B. Efron, *Ann. Stat.*, 1 (1979).
63. B. Efron and R. Tibishirani, *An Introduction to the Bootstrap* (Chapman and Hall, New York, 1993).
64. S. Yakowitz, *Water Resour. Res.*, 1271 (1973).
65. S. Yakowitz, *Ann. Stat.*, 671 (1979).
66. S. Yakowitz, *Stoch. Proc. Appl.*, 311 (1993).
67. S. Yakowitz and M. Karlsson, in *Stochastic Hydrology*, Eds. J. B. Macneil and G. J. Humphries (Reidel, Norwell, MA, 1987), p. 149.
68. S. Yakowitz, *J. Am. Stat. Assoc.*, 215 (1985).
69. B. Rajagopalan and U. Lall, *Water Resour. Res.*, 3089 (1999).
70. L. E. Ormsbee, *ASCE J. Hydrau. Eng.*, 507 (1989).
71. J. Hershenhorn and D. A. Woolhiser, *J. Hydrol.*, 299 (1987).
72. D. A. Woolhiser and H. B. Osborn, *Water Resour. Res.*, 511 (1985).
73. S. J. Burian, S. R. Durrans, S. Tomic, R. L. Pimentel and C. N. Wai, *ASCE J. Hydrol. Eng.*, 299 (2000).
74. B. Treiber and E. J. Plate, *Hydrol. Sci. Bull.*, 175 (1977).
75. J. Kelman, *J. Hydrol.*, 235 (1980).
76. J. D. Salas and J. T. B. Obeysekera, *ASCE J. Hydraul. Eng.*, 1186 (1992).
77. P. Claps, F. Rossi and C. Vitale, *Water Resour. Res.*, 2545 (1993).
78. F. Murrone, F. Rossi and P. Claps, *Stoch. Hydrol. Hydrau.*, 483 (1997).
79. N. D. Evora and J. R. Rousselle, *ASCE J. Hydraul. Eng.*, 33 (2000).
80. J. D. Salas and M. Chebaane, *Proc. Intern. Symp. HY&IR Div. ASCE* (San Diego, CA, 1990), p. 749.
81. B. Fernandez and J. D. Salas, *ASCE J. Hydraul. Eng.*, 1403 (1990).
82. H. Aksoy and M. Bayazit, *Hydrol. Process.*, 1725 (2000).
83. M. B. Fiering and B. B. Jackson, *Synthetic Streamflows*, Water Resources Monograph 1 (American Geophysical Union, Washington, DC, 1971).
84. J. D. Salas, J. W. Delleur, V. M. Yevjevich and W. L. Lane, *Applied Modeling of Hydrologic Time Series* (Water Resources Publications, Littleton, Colorado, 1980).
85. J. D. Salas, in *Handbook of Hydrology*, Ed. D. R. Maidment (McGraw Hill, New York, 1993), p. 1424.
86. P. Bartolini and J. D. Salas, *Water Resour. Res.*, 2573 (1993).
87. J. W. Delleur and M. L. Kavvas, *J. Appl. Meteorol.*, 1528 (1978).
88. R. M. Hirsch, *Water Resour. Res.*, 1603 (1979).
89. P. F. Rasmussen, J. D. Salas, L. Fagherazzi, J. C. Rassam and B. Bobee, *Water Resour. Res.*, 3151 (1996).

90. V. M. Yevjevich, *Stochastic Processes in Hydrology* (Water Resources Publications, Littleton, Colorado, 1972).
91. D. P. Loucks, J. R. Stedinger and D. A. Haith, *Water Resources Planning and Analysis* (Prentice-Hall, Englewood Cliffs, NJ, 1981).
92. J. D. Salas and M. W. Abdelmohsen, *Proc. U.S.-PRC Bilateral Symp. on Droughts and Arid-Region Hydrology*, (Tucson, Arizona, 1991), *USGS Open-File Report* No. 91-244 (1991).
93. G. E. P. Box and G. M. Jenkins, *Time Series: Analysis, Forecasting and Control* (Holden-Day Press, San Francisco, 1976).
94. A. I. McKerchar and J. W. Delleur, *Water Resour. Res.*, 246 (1974).
95. B. Fernandez and J. D. Salas, *Water Resour. Res.*, 1385 (1986).
96. M. Chebaane, J. D. Salas and D. C. Boes, *Water Resour. Res.*, 1513 (1995).
97. R. L. Bras and I. Rodriguez-Iturbe, *Random Functions and Hydrology* (Dover Publications, New York, 1985).
98. B. B. Mandelbrot and J. W. Van Ness, *SIAM Rev.*, 422 (1968).
99. J. M. Mejia, I. Rodriguez-Iturbe and D. R. Dawdy, *Water Resour. Res.*, 931 (1972).
100. O. G. B. Sveinsson, J. D. Salas, D. C. Boes and R. A. Pielke, *J. Hydromet.*, 489 (2003).
101. M. B. Fiering, *Streamflow Synthesis* (Harvard University Press, Cambridge, MA, 1967).
102. J. D. Salas and R. A. Smith, *Water Resour. Res.*, 428 (1981).
103. K. W. Hipel and A. I. McLeod, *Time Series Modeling of Water Resources and Environmental Systems* (Elsevier, Amsterdam, 1994).
104. P. E. O'Connell, in *Mathematical Models in Hydrology, Symposium, Warsaw* (IAHS Pub. No. 100, 1971), p. 169.
105. J. D. Salas, N. Saada and C. H. Chung, *Computing Hydrology Laboratory, Tech. Report No. 25* (Engineering Research Center, Colorado State University, Fort Collins, CO, 1995).
106. R. D. Valencia and J. C. Schaake, *Water Resour. Res.*, 580 (1973).
107. J. M. Mejia and J. R. Rousselle, *Water Resour. Res.*, 185 (1976).
108. W. L. Lane, in *Statistical Analysis of Rainfall and Runoff*, Ed. V. P. Singh (Water Resources Publications, Littleton, Colorado, 1982), p. 700.
109. W. L. Lane, *Applied Stochastic Techniques (Last Computer Package), User Manual* (Bureau of Reclamation, Denver, Colorado, 1979).
110. J. R. Stedinger and R. M. Vogel, *Water Resour. Res.*, 47 (1984).
111. J. C. Grygier and J. R. Stedinger, *Water Resour. Res.*, 1574 (1988).
112. E. Santos and J. D. Salas, *ASCE J. Hydraul. Eng.*, 765 (1992).
113. J. R. Stedinger, D. Pei and T. A. Cohn, *Water Resour. Res.*, 665 (1985).
114. J. C. Grygier and J. R. Stedinger, SPIGOT, A Synthetic Streamflow Generation Software Package, Technical Description, Version 2.6 (School of Civil and Environmental Engineering, Cornell Unversity, Ithaca, New York, 1991).
115. A. Sharma, D. G. Tarboton and U. Lall, *Water Resour. Res.*, 291 (1997).

116. D. G. Tarboton, *J. Hydrol.*, 31 (1994).
117. T. I. Harrold, A. Sharma and S. J. Sheather, *Water Resour. Res.*, 1343 (2003).
118. A. Sharma and R. O'Neill, *Water Resour. Res.*, 5.1 (2002).
119. J. D. Salas and T. Lee, *ASCE J. Hydrol. Eng.*, in press (2010).
120. U. Lall and A. Sharma, *Water Resour. Res.*, 679 (1996).
121. K. Fukunaga, *Introduction to Statistical Pattern Recognition* (Academic Press, San Diego, 1990).
122. D. G. Tarboton, A. Sharma and U. Lall, *Water Resour. Res.*, 107 (1998).
123. J. B. Prairie, B. Rajagopalan, U. Lall and T. Fulp, *Water Resour. Res.*, W03432 (2007).
124. D. Yates, S. Gangopadhyay, B. Rajagopalan and K. Strzepek, *Water Resour. Res.*, 1199 (2003).
125. S. Apipattanavis, G. Podestá, B. Rajagopalan and R. W. Katz, *Water Resour. Res.*, W11401 (2007).
126. F. De Souza and U. Lall, *Water Resour. Res.*, WR01373 (2002).
127. E. Towler, B. Rajagopalan, S. Summers and C. Seidel, *Env. Sci. Tech.*, 1407 (2009).
128. S. Gangopadhyay, B. Harding, B. Rajagopalan and T. Fulp, *Water Resour. Res.*, (2009).
129. C. Loader, *Local Regression and Likelihood* (Springer, New York, 1999).
130. J. Prairie, B. Rajagopalan, T. Fulp and E. Zagona, *ASCE J. Hydrol. Eng.*, 371 (2006).
131. B. Rajagopalan, K. Grantz, S. Regonda, M. Clark and E. Zagona, in *Advances in Water Science Methodologies*, Ed. U. Aswathanarayana (Taylor and Francis, Netherlands, 2005), p. 230.
132. S. Regonda, B. Rajagopalan, U. Lall, M. Clark and Y. Moon, *Nonlin. Proc. Geophys.*, 397 (2005).
133. S. Regonda, B. Rajagopalan, M. Clark and E. Zagona, *Water Resour. Res.*, W09494 (2006).
134. S. Regonda, B. Rajagopalan and M. Clark, *Water Resour. Res.*, W09501 (2006).
135. N. Singhrattna, B. Rajagopalan, M. Clark and K. Krishna Kumar, *Int. J. Climatol.*, 649 (2005).
136. K. Grantz, B. Rajagopalan, M. Clark and E. Zagona, *Water Resour. Res.*, W10410 (2005).
137. P. Block and B. Rajagopalan, *J. Hydrometet.*, 327 (2007).
138. S. Stapleton, S. Gangopadhyay and B. Rajagopalan, *J. Hydrol.*, 131 (2007).
139. A. Sankarasubramanian and U. Lall, *Water Resour. Res.*, 1134 (2003).
140. E. M. Furrer and R. W. Katz, *Clim. Res.*, 129 (2007).
141. E. M. Furrer and R. W. Katz, *Water Resour. Res.*, W12439 (2008).
142. E. Towler, B. Rajagopalan, R. S. Summers and D. Yates, *Water Resour. Res.*, in review (2010).
143. C. H. R. Lima and U. Lall, *Water Resour. Res.*, W07422 (2009).

144. O. G. B. Sveinsson, T. S. Lee, J. D. Salas, W. L. Lane and D. K. Frevert, *Stochastic Analysis, Modeling and Simulation* (SAMS-2009), *Users Manual*, (Department of Civil and Environmental Engineering, Colorado State University, Fort Collins, Colorado, 2009).
145. D. J. Lee, J. D. Salas and D. C. Boes, in *Uncertainty Analysis for Synthetic Streamflow Generation* (ASCE/EWRI World Environmental and Water Resources Congress, Tampa, Florida, 2007).
146. V. Fortin, L. Perreault and J. D. Salas, *J. Hydrol.*, 135 (2004).
147. M. Thyer and G. Kuczera, *Water Resour. Res.*, 3301 (2003).
148. B. Akintug and P. F. Rasmussen, *Water Resour. Res.*, W09424 (2005).
149. H.-H. Kwon, U. Lall and A. F. Khalil, *Water Resour. Res.*, W05407 (2007).
150. H.-H. Kwon, U. Lall, Y.-I. Moon, A. F. Khalil and H. Ahn, *Water Resour. Res.*, W11404 (2006).
151. J. Prairie, K. Nowak, B. Rajagopalan, U. Lall and T. Fulp, *Water Resour. Res.*, W06423 (2008).

CHAPTER 3

MODEL CALIBRATION IN WATERSHED HYDROLOGY

Koray K. Yilmaz

Earth System Science Interdisciplinary Center, University of Maryland, College Park, MD 20742, USA
and
NASA Goddard Space Flight Center, Laboratory for Atmospheres, Greenbelt, MD 20771, USA
E-mail: yilmaz@agnes.gsfc.nasa.gov

Jasper A. Vrugt

Department of Civil and Environmental Engineering, University of California, Irvine, CA 92697, USA
E-mail: jasper@uci.edu
and
Center for Nonlinear Studies, Los Alamos National Laboratory, Los Alamos, NM 87545, USA
E-mail: vrugt@lanl.gov

Hoshin V. Gupta

Department of Hydrology and Water Resources, The University of Arizona, Tucson, AZ 85721, USA
E-mail: hoshin.gupta@hwr.arizona.edu

Soroosh Sorooshian

Center for Hydrometeorology and Remote Sensing (CHRS), The Henry Samueli School of Engineering, University of California, Irvine, CA 92697, USA
E-mail: soroosh@uci.edu

Hydrologic models use relatively simple mathematical equations to conceptualize and aggregate the complex, spatially distributed, and highly interrelated water, energy, and vegetation processes in a watershed. A consequence of process aggregation is that the model parameters often do not represent directly measurable entities and must, therefore, be estimated using measurements of the system inputs and outputs. During this procedure, known as model calibration, the parameters are adjusted so that the behavior of the model approximates, as closely and consistently as possible, the observed response of the hydrologic system over some historical period of time. This Chapter reviews the current state-of-the-art of model calibration in watershed hydrology with special emphasis on our own contributions in the last few decades. We discuss the historical background that has led to current perspectives and review different approaches for manual and automatic single- and multi-objective parameter estimation. In particular, we highlight the recent developments in the calibration of distributed hydrologic models using parameter dimensionality reduction sampling, parameter regularization, and parallel computing. Finally, this chapter concludes with a short summary of methods for assessment of parameter uncertainty, including recent advances in Markov chain Monte Carlo sampling and sequential data assimilation methods based on the Ensemble Kalman Filter.

3.1. Introduction

Hydrologic models serve as important tools for improving our knowledge of watershed functioning (understanding), for providing critical information in support of sustainable management of water resources (decision making), and for prevention of water-related natural hazards, such as flooding (forecasting/prediction). Hydrologic models consist of a general structure, which mathematically represents the coupling of dominant hydrologic processes perceived to control hydrologic behavior of "many (similar) watersheds." Traditionally, this general model structure is then used for simulation and/or prediction of the hydrologic behavior of a "particular watershed," simply by estimating the unknown coefficients, known as "parameters," of the mathematical expressions embodied within.

No matter how sophisticated and spatially explicit, all hydrologic models aggregate (at some level of detail) the complex, spatially distributed vegetation and subsurface properties into much simpler homogeneous storages with transfer functions that describe the flow of water within and between these different compartments. These conceptual storages correspond to physically identifiable control volumes in real space, even though the boundaries of these control volumes are generally not known. A consequence of this aggregation process is that most of the parameters in these models cannot be inferred through direct observation in the field, but can only be meaningfully derived by calibration against an input-output record of the watershed response. In this process, the parameters are adjusted in such a way that the model approximates, as closely and consistently as possible, the response of the watershed over some historical period of time. The parameters estimated in this manner are, therefore, effective conceptual representations of spatially and temporally heterogeneous properties of the watershed. Therefore, successful application of any hydrologic model depends critically on the chosen values of the parameters.

In this chapter, we review the current state-of-the-art of model calibration in watershed hydrology. We discuss manual and automatic parameter estimation techniques for calibration of lumped and spatially distributed hydrologic models. Specific methods include the widely used SCE–UA (Shuffled Complex Evolution – University of Arizona) and MOCOM–UA (Multi Objective COMplex evolution – University of Arizona) approaches for single- and multi-objective model calibration, step-wise (MACS: Multistep Automatic Calibration Scheme) and sequential parameter estimation methods (DYNIA: DYNamic Identifiability Analysis; and PIMLI: Parameter Identification Method based on the Localization of Information), and emerging simultaneous multi-method evolutionary search methods (AMALGAM: A Multi-ALgorithm Genetically Adaptive Multiobjective). We highlight recent developments in the calibration of distributed hydrologic models containing spatially distributed parameter fields, using parameter dimensionality reduction sampling, parameter regularization, and parallel computing. The chapter concludes with a short summary on methodologies for parameter uncertainty assessment, including Markov chain Monte Carlo sampling and sequential data assimilation using the Ensemble Kalman Filter (EnKF); here we discuss the RWM (Random Walk Metropolis), SCEM–UA (Shuffled Complex Evolution Metropolis – University of Arizona), DREAM (DiffeRential Evolution Adaptive Metropolis), and SODA (Simultaneous Optimization and Data

Assimilation) sampling algorithms. Note that, although our discussion is limited to watershed models, the ideas and methods presented herein are applicable to a wide range of modeling and parameter estimation problems.

3.2. Approaches to Parameter Estimation for Watershed Models

There are two major approaches to parameter estimation: "*a priori* estimation" and "*a posteriori* estimation" via model calibration. In *a priori* estimation, values of the model parameters are specified without recourse to the observed dynamic hydrologic response (e.g. streamflow) of the watershed under study. Calibration, on the other hand, involves the selection of a parameter set that generates model responses that reproduce, as closely as possible, the historically observed hydrologic response of a particular watershed. Therefore, calibration can only be performed when long-term historical measurements of input-state-output behavior of the watershed (including streamflow, precipitation, and potential evapotranspiration) are available.

Parameter estimation strategies are intimately tied to the degree of hydrologic process representation embedded within the model. Hydrologic models can broadly be classified accordingly as *conceptual* or *physically-based*. Most hydrologic models in use today are of the *conceptual* type, that conceptualize and aggregate the complex, spatially distributed, and highly interrelated water, energy, and vegetation processes in a watershed into relatively simple mathematical equations without exact treatment of the detailed underlying physics and basin-scale heterogeneity. Typical examples of conceptual type models are the SAC–SMA (SACramento Soil Moisture Accounting) model,[1,2] and the HBV (Hydrologiska Byråns Vattenbalansmodell) model.[3] Due to hydrologic process aggregation, the parameters in these models cannot generally be measured directly in the field at the desired scale of interest. Instead, when using conceptual-type models, only the ranges of feasible parameter values can generally be specified *a priori* (perhaps with the combined knowledge of model structure and of dominant watershed processes). Calibration is then employed to select parameter estimates (from within the *a priori* defined ranges) that capture, as closely and consistently as possible, the historical record of the measured (target) hydrologic response of the watershed the model is intended to represent.

Spatially distributed *physically-based* hydrologic models contain a series of partial differential equations describing physical principles related to conservation of mass and momentum (and energy). Typical examples of physically-based models are MIKE–SHE[4] (Systeme Hydrologique European) and KINEROS[5] (KINematic Runoff and EROSion). Their spatially distributed physically-based structure provides two potential strengths: (i) the ability to account for the spatial variability of runoff producing mechanisms; and (ii) the ability to infer model parameter values directly from spatio-temporal data by establishing physical or conceptual relationships between observable watershed characteristics (e.g. geology, topography, soils, land cover) and the parameters for the hydrologic processes represented in the model.[6-9] The latter is defined as the "local" *a priori* parameter estimation approach and is particularly valuable for implementing hydrologic models in poorly-gaged and ungaged watersheds where local response data is sparse or non-existent. In general, parameters estimated via the local *a priori* approach will still require some degree of fine-tuning via a calibration approach to obtain effective values that account for the influencing factors, such as heterogeneity, emergent processes, and differences in scale between model (larger scale) and the embedded hydrologic theory (developed at point/small scale). This refinement process ensures proper consistency between the model input-state-output behavior and the available response data.[10-12]

Another parameter estimation strategy, developed mainly for the implementation of conceptual type of models in ungaged basins, is called the "regional" *a priori* approach.[13-17] The "regionalized" approach involves the development of regional regression relationships between the model parameter values estimated for a large number of gaged basins (via calibration) and observable watershed characteristics (i.e. land cover and soil properties) at those locations. The idea is that these relationships can be used to infer parameter estimates for "hydrologically similar" ungaged basins, given knowledge of their observable watershed characteristics. A major assumption of the regional *a priori* approach is that the calibrated model parameters are uniquely and clearly related to observable watershed properties. This assumption can be difficult to justify when many combinations of parameters are found to produce similar model responses due to parameter interaction,[18] measurement uncertainty,[19] and model structural uncertainty,[20] and can, therefore, result in ambiguous and biased relationships between the parameters and the watershed characteristics.[17] One way to improve the efficiency of the regionalized approach is to impose conditions (via watershed characteristics) on the calibrated parameters.[21] In an alternative approach, Yadav *et al.*[22] proposed regionalization of streamflow indices. In this approach, relationships between streamflow indices and physical watershed characteristics are established at the gaged locations. The regionalized flow indices, providing dynamic aspects of the ungaged watersheds, are then used to constrain hydrologic model predictions (and parameters). One advantage of this approach is that, the regionalized indices are independent of model structure and, therefore, can be used to constrain any watershed model.

The above paragraphs provide a broad overview of approaches commonly used by the hydrologic community to specify values of the parameters in hydrologic models; namely "*a priori*" approach and "calibration." The specific focus of this chapter is on calibration of hydrologic models, and we provide, in the following sections, a more detailed overview of various calibration strategies that have been developed within the water resources context and have found widespread use in the hydrologic community. We discuss these methods within the context of their historical development, including current and future perspectives.

3.2.1. *Model Calibration*

For a model to be useful in prediction, the values of the parameters need to accurately reflect the invariant properties of the components of the underlying system they represent. Unfortunately, in watershed hydrology, many of the parameters cannot be generally measured directly, but can only be meaningfully derived through calibration against historical record of dynamic response (traditionally streamflow) data. Calibration is an iterative process in which the model parameters are adjusted so that the dynamic response of the model represents, as closely as possible, the observed target response (e.g. outlet streamflow) of the watershed. Figure 3.1 provides a schematic representation of the resulting model calibration problem. In this figure, ⊕ represents observations of the forcing (rainfall) and streamflow response that are subject to measurement errors and uncertainty, and, therefore, may be different from the true values. Similarly, Φ represents the hydrologic model with functional response □ to indicate that the model is at best only an approximation of the underlying watershed. The label "output" on the y-axis of the plot on the right hand side can represent any dynamic time series of response data; here, this is considered to be the streamflow response.

Using *a priori* values of the parameters, the predictions of the model shown in this figure (indicated with solid-gray line) are behaviorally consistent with the observations (dotted line), but demonstrate a significant bias towards lower streamflow values. The common approach is to ascribe this mismatch between model and data to parameter uncertainty. The goal of model calibration then becomes one of finding those values of the parameters that provide the "best" possible fit to the observed behavior using either manual or computerized iterative search methods. A model calibrated by such means can generally be used for the simulation or prediction of hydrologic events outside of the historical record used for model calibration, provided that it can be reasonably assumed that the physical characteristics of the watershed and the hydrologic/climatic conditions remain similar.

Mathematically, the model calibration problem depicted in Fig. 3.1 can be formulated as follows. Let $\tilde{S} = \Phi(\psi, X, \theta)$ denote the streamflow predictions $\tilde{S} = \{\tilde{s}_1,...,\tilde{s}_n\}$ of the model Φ with observed forcing X (rainfall and potential evapotranspiration), state vector ψ, and model parameters θ. Let $S = \{s_1,...,s_n\}$ represent a vector with n observed streamflow values. The difference between the model-predicted and measured streamflow can be represented by the residual vector E as:

$$E(\theta) = \{T(\tilde{S}) - T(S)\} = \{T(\tilde{s}_1) - T(s_1),...,T(\tilde{s}_n) - T(s_n)\} = \{e_1(\theta),...,e_n(\theta)\} \quad (3.1)$$

where $T(.)$ allows for various monotonic (such as logarithmic) transformations of the model outputs.

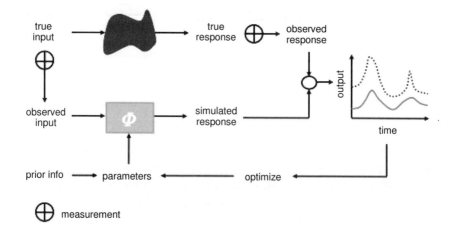

Fig. 3.1. A schematic representation of the general model calibration problem. The model parameters are iteratively adjusted so that the predictions of the model, Φ, (represented with the solid line) approximate as closely and consistently as possible the observed response (represented with dotted line).

Traditionally, we seek values for the model parameters that result in a minimal discrepancy between the model predictions and observations. This can be done by minimizing an aggregate summary statistic of the residuals:

$$F(\theta) = F\{e_1(\theta), e_2(\theta), ..., e_n(\theta)\} \quad (3.2)$$

where the function $F(\theta)$ allows for various user-selected linear and nonlinear transformations of the residuals and is interchangeably called a "criterion," "measure" or "objective function" in the watershed modeling literature. Note that, in typical time series analysis, the influence of the initial condition, ψ, on the model output diminishes with increasing distance from the start of the simulation. In those situations, it is common to use a spin-up period to reduce sensitivity to state-value initialization.

Boyle et al.[23] classified the process of model calibration into three levels of increasing sophistication. In level zero, approximate ranges for the parameter estimates are specified by physical reasoning that incorporates available data about static watershed characteristics (e.g. geology, soil, land cover, slope), via lookup tables or by borrowing values from similar watersheds. At this level, only crude *a priori* ranges of the parameters are estimated without conditioning on the input-output streamflow response data. In level one, the parameter ranges are refined by identifying and analyzing the characteristics of specific segments of the response data that are thought to be controlled by a single or a group of parameter(s). In this level, the effects of parameter interactions are ignored. Finally, in level two, a detailed analysis of parameter interactions and performance trade-offs is performed using a carefully chosen representative period of historical data (i.e. calibration data) to further refine the parameter ranges or select a representative parameter set. This level involves a complex, iterative process of parameter adjustments to bring the simulated response as closely and consistently as possible to the observed watershed response.

Calibration can be further divided into two types, depending on whether this iterative process is being guided manually by an expert hydrologist or automatically by a computer following pre-defined algorithmic rules. These approaches are called, respectively, "*manual calibration*" and "*automated calibration*."

3.2.1.1. *Overview of the Manual Calibration Approach*

Manual calibration is a guided trial-and-error procedure performed by an expert hydrologist, involving complex knowledge-based analyses to match the perceived hydrologic processes in the watershed with their conceptual equivalents represented in the model structure. This interactive procedure can involve a variety of graphical interfaces and a multitude of performance measures to transform the historical data into information that will aid the hydrologist in decision making. Although progressive steps during manual calibration of a model are generally established via pre-defined guidelines, the actual sequence of procedures will vary based on the experience and training of the modeler, the properties of the data, and characteristics of the watershed system being modeled.[23] Successful manual calibration, therefore, requires a good knowledge of the physical and response characteristics of the watershed, as well as a good understanding of the structure and functioning of the various model components and parameters. The major aim is to find values for the parameters that are consistent with the hydrologic processes they were designed to represent.[24] The consistency between model simulations of hydrologic behaviors for which observations are available is examined at various timescales and time periods to try and isolate individual effects of each parameter. Hydrologic behaviors include, for example, annual averages to understand the long-term water balance dynamics, seasonal and monthly averages to identify the trends and low-high flow periods, extended recession periods to understand watershed baseflow characteristics, and event-based measures to analyze the shape and timing of floods.[2,25] Manual calibration is expected to produce process-based (i.e. conceptually realistic) and reliable predictions. The National Weather Service (NWS) of the United States, for example, primarily uses a manual approach (assisted with automatic techniques) to estimate the parameters of their lumped hydrologic models used in operational streamflow forecasting. The manual calibration, however, is a time- and labor-intensive process involving many subjective decisions. Due to this subjectivity, different modelers will most likely produce different model parameter values for the same watershed. Another difficulty in manual calibration is the ever-increasing complexity of watershed models. As more parameters are added, it becomes more difficult to estimate them accurately; parameter interaction and compensating effects on the modeled variables make it difficult to estimate individual parameters. Note that Bayesian methods (see Sec. 3.6) provide a general framework for explicit use of expert knowledge/belief in the form of priors. An excellent and comprehensive discussion of manual calibration is given in three recent reviews.[23-25]

3.2.1.2. *Overview of Automated Calibration Approaches*

Automated parameter identification (calibration) methods rely on an *a priori* model structure, optimization algorithm, and one or more mathematical measures of model performance (often called objective function, criterion or measure) to estimate the model parameters using a historical record of observed response data. The advantages of the automated calibration approach are not difficult to enumerate. Such methods use objective, rather than visual and subjective, measures for performance evaluation, and exploit the power and speed of computers to efficiently and systematically search the feasible model parameter space. Within this context, the goal has been to develop an objective strategy for parameter estimation that provides consistent and reproducible results independent of the user.[23] A potential disadvantage is that automatic calibration algorithms can, if not properly designed, return values of the model parameters that are deemed to be hydrologically unrealistic. It is, therefore, the modeler's responsibility to examine the robustness of the optimized parameters. A traditional approach, accepted as a minimum requirement for "evaluation" of parameter robustness, is to compare the modeled and observed response for an independent time period that is not used for calibration. This can be done using a split-sample approach and should cover a long enough time period so as to contain various ranges of hydrologic conditions (e.g. wet and dry periods), and, if possible, using multiple response data.[26,27]

In the following sections, we discuss single- and multi-objective model calibration strategies. Single-objective methods utilize a single measure of closeness between the simulated and observed variables, whereas multi-objective approaches attempt to emulate the strengths of the manual calibration approach by simultaneously using multiple different measures of performance.

3.3. Single-criterion Automated Calibration Methods

Traditional automated approaches to model calibration utilize a single mathematical criterion (sometimes defined as objective function or measure) to search for a unique "best" parameter set. An objective function (Eq. (3.2)) can be defined as an aggregate statistical summary quantifying the "closeness" or "goodness-of-fit" between the simulated and observed hydrologic variable(s). The objective function largely influences the outcome of the automated calibration procedure and, hence, should be carefully selected based on the goal of modeling. Traditionally, automated calibration problems seek to minimize the discrepancy between the model predictions and observations by minimizing the following additive simple least square (SLS) objective function (or its variations) with respect to θ:

$$F_{SLS}(\theta) = \sum_{i=1}^{n} e_i(\theta)^2 \qquad (3.3)$$

Often, the square root form, such as the root mean squared error (RMSE) criterion, is used because it has the same units as that of the variable being estimated (e.g. streamflow discharge). An RMSE objective function puts strong emphasis on simulating the high-flow events; to increase the emphasis on the low flows, it can be used after performing a log-transformation of the output variables.

The popularity of the SLS function stems from its statistical properties; the SLS is an unbiased estimator under strong assumptions related to the distribution of the residuals, i.e. the residuals are pair-wise independent, have constant variance (homogeneous), and normally distributed with a mean value of zero.[28-29] However, the validity of these assumptions is questionable within a hydrologic context, as most hydrograph simulations published to date in the literature show significant nonstationarity of error residuals.[30,31]

3.3.1. *A Historical Perspective*

A powerful optimization algorithm is a major requirement to ensure finding parameter sets that best fit the data. Optimization algorithms iteratively explore the response surface (a surface mapped out by the objective function values in the parameter space) to find the "best" or "optimum" parameter set. Automatic optimization algorithms developed in the past to solve the nonlinear SLS optimization problem, stated in Eq. (3.3), may be classified as local search methodologies if they are designed to seek a systematic improvement of the objective function using an iterative search starting from a single arbitrary initial point in the parameter space, or as stochastic global search methods if multiple concurrent searches from different starting points are conducted within the parameter space. Local optimization approaches can be classified into two categories, based on the type of search procedure employed, namely derivative-based (gradient) and derivative-free (direct) methods.

Gradient-based methods make use of the estimates of the local downhill direction based on the first and/or second derivative of the response surface with respect to each individual model parameter.[32] The simplest gradient-based algorithm is that of the steepest descent, which searches along the first-derivative direction for iterative improvement of the objective function. Newton-type methods, such as the Gauss-Newton family of algorithms, are examples of second-derivative algorithms. Gupta and Sorooshian[33] and Hendrickson *et al.*[34] demonstrate how analytical or numerical derivatives can be computed for conceptual watershed models.

Derivative-based algorithms will evolve towards the global minimum in the parameter space in situations where the objective function exhibits a topographical convex shape in the entire parameter domain. Unfortunately, numerous contributions to the hydrologic literature have demonstrated that the response surface seldom satisfies these restrictive conditions and exhibits multiple optima in the parameter space. Local gradient-based search algorithms are not designed to handle these peculiarities and, therefore, often prematurely terminate their search at a final solution that is dependent on the starting point in the parameter space. Another related problem is that many of the hydraulic parameters typically demonstrate significant interactions, because of an inability of the observed experimental data to properly constrain all of the calibration parameters.

Direct search methods sample the value of the objective function in a systematic manner without computing derivatives of the response surface with respect to each parameter. Popular examples of direct search methods include the Simplex Method,[35] the Pattern Search Method,[36] and the Rotating Directions Method of Rosenbrock.[37] Many studies have focused on comparative performance analysis of local search methods for calibration of watershed models.[38-40] Their general conclusion was that local search methods were not powerful enough to reliably find the best (global optimum) values of the watershed model parameters. The main limitation of local search methods is that, like gradient-based algorithms, their outcome is highly dependent on their initial starting point. For example, local search algorithms are prone to getting trapped in local basins of attraction (local minima) and, as argued above, may become confused in finding the preferred direction of improvement in the presence of threshold structures and other undesirable irregularities of the response surface.

Among others, Moore and Clarke[41] and Sorooshian and Gupta[42] pointed out that the causes of the above difficulties were mainly difficulties concerned with the underlying model structure and that local search methods were not powerful to do the job. Gupta and Sorooshian,[43] for example, focused on model structural inadequacies in the SAC–SMA (Sacramento Soil Moisture Accounting) model and showed that parameter identifiability can be improved by a careful re-parameterization of the percolation function. Other seminal contributions[30,42] concluded that a properly chosen objective function, which can better recognize the stochastic nature of the errors in the calibration data (such as those derived using Maximum Likelihood Theory), can result in smoother response surfaces for which the global optimum is easier to identify. They[30,42] pointed out that streamflow measurements contain errors that are temporally autocorrelated and heteroscedastic (having non-constant, magnitude-dependent variance) and also introduced a Heteroscedastic Maximum Likelihood Estimator (HMLE) that accounts for nonstationary error variance arising from various sources, including the rating curve used to convert the stage measurements (in units of height) into runoff rate (in units of discharge, e.g. cubic feet per day). In a parallel work, Kuczera[44] proposed a methodology based on Bayesian statistics to properly account for the measurement error properties of the data while predicting the confidence bounds for the parameter estimates. Recently, Kavetski et al.[45,46] and Vrugt et al.[47,48] have extended that approach to account for error in rainfall depths as well. Another commonly used methodology applies a parameterized power transformation to the streamflow data to stabilize the heteroscedastic measurement error:[49]

$$\hat{y} = \frac{(y+1)^\lambda - 1}{\lambda} \quad (3.4)$$

where y and $\hat{y}$ represent flows in the original and transformed spaces, respectively, and λ is the transformation parameter (a commonly used value is $\lambda \sim 0.0$ to 0.3).

Other researchers have investigated the requirements for calibration data and pointed out that the quantity and quality of calibration data play a critical role in controlling the success of the calibration procedure. These studies concluded that the informativeness of the data is far more important than the length and amount used for calibration.[50-54]

The convergence problems encountered with local search algorithms have inspired researchers to develop and test global search algorithms for calibration of watershed models. While local optimization methods rely on a single initial point within the feasible parameter space to start the search, global optimization methods utilize multiple concurrent searches from different starting points to reduce the chance of getting stuck in a single basin of attraction. Examples of global optimization algorithms applied to watershed model calibration include the Random Search (RS) method,[55] Adaptive Random Search (ARS),[56] ARS coupled with direct local search methods,[57] Controlled Random Search,[58,59] Simulated Annealing,[60,61] the multi-start simplex,[62] and genetic algorithm.[62-64] For a detailed overview of global search algorithms, please see Duan.[65] With the advent of computational power, Duan et al.[62] conducted a seminal study; focusing first on a detailed analysis of the properties of the response surface, they identified five major characteristics that complicate the optimization problem in watershed models:

• it contains more than one main region of attraction.
• it has many local optima within each region of attraction.
• it is rough with discontinuous derivatives.
• it is flat near the optimum with poor and varying degrees of parameter sensitivities.
• its shape is non-convex and includes long and curved ridges.

In an effort to design an optimization strategy capable of dealing with these difficulties in single-objective calibration problems, Duan et al.[62] introduced a novel procedure called the Shuffled Complex Evolution – University of Arizona (SCE–UA).

3.3.2. *The Shuffled Complex Evolution–University of Arizona (SCE–UA) Algorithm*

The Shuffled Complex Evolution Algorithm developed at the University of Arizona (SCE–UA)[62,66,67] is a global search strategy that synthesizes the features of the simplex procedure, controlled random search,[58] and competitive evolution[68] with the newly introduced concept of complex shuffling. The SCE–UA algorithm has since been employed in a number of studies and proved to be consistent, effective, and efficient in locating the global optimum to the parameter estimation problems for watershed models.[67,69-72] In brief, SCE–UA algorithmic steps can be listed as follows:[62,66,73]

(i) Generate initial population: sample s points randomly in the feasible (*a priori*) parameter space (using uniform distribution, unless prior information exists) and compute the objective function value at each point.

(ii) Ranking: sort the s points in order of increasing objective function value so that the first point represents the smallest criterion value and the last point represents the largest criterion value (assuming that the goal is to minimize the criterion value).

(iii) Partitioning into complexes: partition the s points into p complexes, each containing m points. The complexes are partitioned such that the first complex contains every $p(k-1)+1$ ranked point, the second complex contains every $p(k-1)+2$ ranked point, and so on, where $k = 1, 2, \ldots, m$.

(iv) Complex evolution: evolve each complex according to the competitive complex evolution algorithm, which is based on the Simplex downhill search scheme.[35] The evolution procedure generates new points called "offspring" that, on average, lie within the improvement region.

(v) Complex shuffling: combine the points in the evolved complexes into a single sample population; sort the sample population in order of increasing criterion value; shuffle (i.e. re-partition) the sample population into p complexes according to the procedure specified in Step (iii).

(vi) Check for convergence: if any of the pre-specified termination criteria are satisfied, stop; otherwise, continue. Termination criteria can be specified as maximum number of iterations (or maximum number of shuffling) or parameter convergence.

Experience with the method has indicated that the effectiveness and efficiency of the SCE–UA algorithm is influenced by the choice of a few algorithmic parameters. Duan et al.[66,73] performed sensitivity studies and suggested practical guidelines for selecting these algorithmic parameters according to the degree of difficulty of the optimization problem. The primary parameter to be selected is the number of complexes, p. The above studies showed that the dimension of the calibration problem (i.e. number of parameters to be optimized), n, is the primary factor determining the proper choice of p; practically, if no other information is available, p is set to the greater value between 2 or n (see also Kuczera[72]). The size of a complex, m, is generally chosen to be equal to $2n + 1$. Accordingly, the sample (population) size, s, becomes the product $p \cdot m$. The number of offsprings, β, that can be generated by each independently evolving complex between two consecutive shuffles is the same as the complex size ($2n + 1$), the size of each sub-complex selected for generation of an offspring (via Simplex scheme) is $n + 1$ and defines a first order approximation to the objective function space. The number of consecutive offsprings generated by each sub-complex, α, is equal to 1. In selecting these algorithmic parameters, a balance between algorithm effectiveness and efficiency should be sought. For instance, selecting a large number of complexes increases the probability of converging to the global optimum, however at the expense of a larger number of simulations (and hence longer computational time). The SCE–UA code is available free of charge from the following web address: www.sahra.arizona.edu/software.

While significant progress has been made in the use of global optimization algorithms for parameter estimation, the current generation of optimization algorithms typically implements a single operator (i.e. Simplex search in the case of SCE–UA) for population evolution. Reliance on a single model of natural selection and adaptation presumes that a single method can efficiently evolve a population of potential solutions through the parameter space and work well for a large range of problems. However, existing theory and numerical benchmark experiments have demonstrated that it is impossible to develop a single universal algorithm for population evolution that is always efficient for a diverse set of optimization problems.[74] This is because, the nature of the response surface often varies considerably between different optimization problems, and often dynamically changes en route to the global optimal solution. It, therefore, seems productive to develop a search strategy that adaptively updates the way it generates offsprings based on the local peculiarities of the response surface.

In light of these considerations, Vrugt and Robinson[75] and Vrugt et al.[76] have recently introduced a new concept of self-adaptive multi-method evolutionary search. This approach, termed as A Multi-Algorithm Genetically Adaptive Method (AMALGAM), runs a diverse set of optimization algorithms simultaneously for population evolution and adaptively favors individual algorithms that exhibit the highest reproductive success during the search. By adaptively changing preference to individual algorithms during the course of the optimization, AMALGAM has the ability to quickly adapt to the specific peculiarities and difficulties of the optimization problem at hand. A brief algorithmic description of AMALGAM for solution of multi-objective optimization problems is given in Sec. 3.4. Synthetic benchmark studies covering a diverse set of problem features, including multimodality, ruggedness, ill-conditioning, non-separability, interdependence (rotation) and high-dimensionality, have demonstrated that AMALGAM significantly improves the efficiency of evolutionary search.[75,76] An additional advantage of self-adaptive search is that the need for algorithmic parameter tuning is reduced, thus increasing applicability to solving search and optimization problems in many different fields of study. An extensive algorithmic description of AMALGAM, including comparison against other state-of-the-art optimization methods, can be found in Vrugt and Robinson[75] and Vrugt et al.[76] The AMALGAM code is available from the second author of this chapter upon request.

3.3.3. *Limitations of Single-criterion Methods*

Despite these algorithmic advances, automated model evaluation strategies that rely on a single regression-based aggregate measure of performance (e.g. RMSE) are, in general, weak and make it unnecessarily difficult to isolate the effects of different parameters on the model output.[10,77,78] Hence, two different parameter combinations might give completely different streamflow responses, but result in very similar values of the objective function. This is undesirable. A major reason for this is the loss (or masking) of valuable information inherent in the process of projecting from the high dimension of the dataset ($\mathcal{R}^{Data}$) down to the single dimension of the residual-based summary statistic ($\mathcal{R}^1$), leading to an ill-posed parameter estimation problem ($\mathcal{R}^{Parameter} < \mathcal{R}^{Data}$).[77] To avoid (or at least minimize) this problem, an optimization strategy must necessarily make use of multiple, carefully selected, measures of model performance, thereby more closely matching the number of unknowns (the parameters) with the number of pieces of information (the measures), resulting in a better-posed identification problem. There is, therefore, an urgent need to develop mathematical theory that more convincingly proves this line of thought and provides ways forward to improve parameter inference. This is especially pressing within the context of spatially distributed models that contain a manifold of parameters for which little compelling *a priori* information is available about appropriate values.

3.4. Multi-criteria Automated Calibration Methods

Multi-criteria analysis can be used to assimilate information from multiple non-commensurable (i.e. not measurable by the same standard) sources.[10] The goal is to increase the extraction of information content from the data (decrease the gap between measure and parameter dimensions), by properly expressing the different important aspects of model performance. For instance, a number of criteria can be formulated, each of which is sensitized to a specific watershed output flux (e.g. water, energy, chemical constituents) for which measurements are available.[27,79] In principle, each criterion can also be designed to isolate a different characteristic behavior of some component of the physical system.[80] Note that the process of interactive manual-expert evaluation and calibration of a model, following a process-guided set of rules, actually follows a powerful (albeit somewhat subjective) multi-criteria approach, wherein a variety of graphical and numerical tools are used to highlight different aspects of model response.[23,25,27] A major advantage of the automated multi-criteria approach is that various aspects of the manual calibration strategy can be absorbed into the calibration process, thus strengthening the physical basis of identified parameters. Many automated (or semi-automated) multi-criteria calibration strategies have been proposed for calibration of watershed models. These strategies can be broadly classified into simultaneous, step-wise, and constraining approaches. These three approaches are discussed next.

3.4.1. *Simultaneous Multi-criteria Calibration Approach*

The simultaneous multi-criteria approach finds a set of solutions (so-called "Pareto optimal" region) that simultaneously optimize (i.e. in one optimization run) and trade-off the performance of several user-selected criteria that measure different aspects of model performance.[10,11,81,82] In general, the multi-criteria model constraining problem can be expressed in the following form:[10]

$$\min_{w.r.t.\theta} F_{MC}(\theta) = \{F_1(\theta), F_2(\theta), ... F_n(\theta)\} \quad \theta \in \Theta \tag{3.5}$$

where $F_1(\theta), F_2(\theta), ..., F_n(\theta)$ represent the different performance criteria summarizing information related to various components of the physical system. To solve this multi-criteria problem, model parameter sets (θ) are systematically sampled from their *a priori* region (Θ) in search of solutions that simultaneously minimize all of these criteria. As is well known, it is generally not possible to satisfy all of the criteria simultaneously. The solution to this minimization problem (Eq. (3.5)) is generally not unique but takes the form of a Pareto surface that characterizes the trade-offs in its ability to satisfy all of the competing criteria.[10,81]

As a commonplace illustration, consider the migration of birds from Scandinavia to Africa and backwards that simultaneously considers flight time and energy use (Fig. 3.2).[83] For this situation, there is no single optimal solution. Rather, there is a family of trade-off solutions along a curve called the "Pareto-optimal front" (surface in three and higher dimensions) in which improvement in one objective (say, reduction in flight time) comes only at the expense of increased energy use per day (second objective). In other words, moving from one solution to another along the Pareto surface will result in the improvement of at least one criterion while deteriorating at least one other. Analysis of the size and properties of the Pareto region can provide insights into possible model improvements as well as degrees of confidence in different aspects of the model predictions.[10,84]

Many computational approaches to deriving efficient estimates of the Pareto solution set have been proposed.[81,82,85,86] A pioneering algorithm that provides an approximation to the Pareto optimal region in a single optimization run is called the MOCOM–UA (Multi-Objective COMplex evolution algorithm) developed at the University of Arizona.[81] The MOCOM–UA algorithm is a general-purpose multi-objective global optimization strategy that is based on an extension of the SCE–UA population evolution method.[62] It combines the strengths of the controlled random search method,[58] Pareto ranking,[87] and a multi-objective downhill simplex search method.

In brief, the MOCOM–UA method starts with an initial sampling of a population of s points distributed randomly throughout the n-dimensional feasible parameter space, Θ. In the absence of prior information about the location of the Pareto region, a uniform sampling distribution is used.

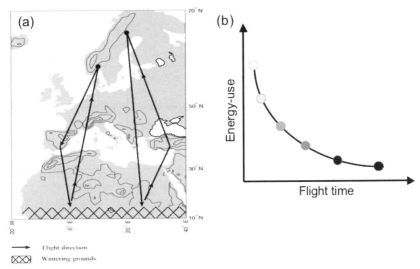

Fig. 3.2. An illustration of the Pareto optimality. a) Map showing migration paths of birds from Scandinavia to Africa and backwards. b) Birds trying to simultaneously optimize flight time and energy use. There exists a family of trade-off solutions along a curve called the "Pareto-optimal front". (after Vrugt et al.[83])

For each sampled point, the multi-criteria vector, $F_{MC}(\theta)$, is calculated and the population is ranked and sorted using the Pareto-ranking procedure presented by Goldberg.[87] Simplexes of $n + 1$ points are then selected from the population according to a robust rank-based selection method suggested by Whitley.[88] A multi-objective extension of the downhill simplex method is used to evolve each simplex in a multi-criteria improvement direction and generate new (better) points. Iterative application of the ranking and evolution procedures causes the entire population to converge towards the Pareto optimum. The procedure terminates automatically when the population converges to the Pareto region with all points mutually non-dominated. The final population provides a fairly uniform approximation of the Pareto solution space, $P(\Theta)$. Details of the MOCOM–UA algorithm can be found in Yapo et al.[81] The MOCOM–UA code is available from the following web address: www.sahra.arizona.edu/software.

The MOCOM–UA has been successfully applied to a number of multi-criteria calibration studies of hydrologic[10,23] and land-surface models.[89,90] However, some studies reported that the MOCOM–UA algorithm has a tendency to cluster the solutions in the center of the Pareto region[91] and may require a very large number of model runs for convergence.[90] The MOSCEM–UA (Multi-Objective Shuffled Complex Evolution Metropolis – University of Arizona) method[82] is especially designed to overcome some of these convergence problems, but its stochastic nature causes the Pareto solution set to contain irregularities with bumpy fronts/surfaces as a consequence.

The AMALGAM evolutionary search algorithm, recently developed by Vrugt and Robinson,[75] has shown to be the method of choice for solving multi-objective optimization problems. This method utilizes self-adaptive multi-method evolutionary search and is more efficient and robust than the existing search approaches, including the commonly used Strength Pareto Evolutionary Algorithm (SPEA2)[92] and Non-dominated Sorting Genetic Algorithm (NSGA–II).[93] The AMALGAM algorithm is initiated using a random initial population P_0 of size N, generated using Latin hypercube sampling. Then, each parent is assigned a rank using the fast non-dominated sorting (FNS) algorithm.[93] A population of offspring

Q_0, of size N, is subsequently created by using the multi-method search concept that lies at the heart of the AMALGAM method. Instead of implementing a single operator for reproduction, we simultaneously use k individual algorithms to generate the offspring, $Q_0 = \{Q_0^1,...,Q_0^k\}$. Each of these algorithms creates a pre-specified number of offspring points, $N = \{N_t^1,...,N_t^k\}$, from P_0 using different adaptive procedures. After creation of the offspring, a combined population $R_0 = P_0 \cup Q_o$ of size $2N$ is created and R_0 ranked using FNS. By comparing the current offspring with the previous generation, elitism is ensured since all previous non-dominated members will always be included in R.[93-95] Finally, members for the next population P_1 are chosen from subsequent non-dominated fronts of R_0 based on their rank and crowding distance.[93] The new population P_1 is then used to create offspring using the method described below, and the aforementioned algorithmic steps are repeated until convergence is achieved.

To ensure that the "best" algorithms are weighted so that they contribute the most offsprings to the new population, we update $\{N_t^1,...,N_t^k\}$ according to:

$$N_t^i = N \frac{P_t^i}{N_{t-1}^i} \left[\sum_{i=1}^{k} \frac{P_t^i}{N_{t-1}^i} \right]^{-1} \quad i=1,...,k \quad (3.6)$$

The term (P_t^i / N_{t-1}^i) is the ratio of the number of offspring points an algorithm contributes to the new population, P_t^i, and the corresponding number the algorithm created in the previous generation (N_{t-1}^i). The rest of the expression scales the reproductive success of an individual algorithm to the combined success of all the algorithms.

Benchmark results using a set of well-known multi-objective test problems show that AMALGAM is three to ten times more efficient than the existing multi-objective optimization algorithms. Initial applications to hydrologic parameter estimation problems have reported similar efficiency improvements.[96,97]

Notwithstanding this progress made, approximation of the entire Pareto surface can be computationally too expensive. This is especially true within the context of distributed hydrologic modeling, for which approximate Pareto optimal solutions can be obtained by lumping the various individual performance criteria within a single aggregate function through scalarization:

$$F_{agg}(\theta) = \sum_{i=1}^{k} \omega_i \tau_i F_i(\theta) \quad (3.7)$$

where k denotes the number of criteria, and ω's and τ's are weights and scaling transformations, respectively, applied to each criterion. Here, the weights define the relative contributions of each criterion to the aggregate function, F_{agg}, such that $\omega_1 + \omega_2 + ... + \omega_k = 1$, and can be subjectively assigned to place greater or lesser emphasis on some aspect of the model. Further, they can also be formalized to reflect our knowledge regarding the relative degree of uncertainty in the various measurement data sources.[98] It can be shown theoretically that if the Pareto region is convex, its shape can be well approximated by systematically varying the weights ω_i assigned to each of the criteria $F_i(\theta)$ over all possible values. Note that the above formulation also accounts for the fact that, in general, the various criteria $F_i(\theta)$ may not be directly commensurable and may, in fact, vary over very different orders of magnitude; the multiplier τ_i is used to compensate for this difference by transforming the criteria onto a commensurable scale.[11,85,99]

3.4.2. Stepwise Multi-criteria Calibration Approach

The stepwise multi-criteria approach aims to reduce the dimension of the optimization problem by breaking it into several sub-problems that are handled in a step-by-step manner; each step considers different aspects of hydrologic response by transforming the flows,[100,101] by focusing on different time periods,[23,102-104] and/or different timescales.[9,105,106] For example, the Multi-step Automatic Calibration Scheme[100,101] (MACS) uses the SCE–UA global search algorithm and a step-by-step procedure to emulate some aspects of the progression of steps followed by an expert hydrologist during manual calibration. The step-by-step procedure is as follows: (i) the lower zone parameters are calibrated to match a logarithmic transformation of the flows, thereby placing a strong emphasis on reproducing the properties of the low-flow portions of the hydrograph; (ii) the lower zone parameters are subsequently fixed and the remaining parameters are optimized with the RMSE objective function to provide a stronger emphasis on simulating high-flow events; and (iii) finally, a refinement of the lower zone parameters is performed using the log-transformed flows while keeping the upper zone parameters fixed at values derived during Step (ii). The method has been tested for a wide variety of hydroclimatic regimes in the United States and has been shown to produce model simulations that are comparable to traditional manual calibration techniques.[100,101]

Other studies that utilize step-wise procedures in the calibration of hydrologic model parameters also exist. Brazil[102] estimated parameters of the SAC–SMA model using a combination of an interactive analysis of observed streamflow time series, an automated global search algorithm, and a local search algorithm for fine-tuning. Similarly, Bingeman *et al.*[27] utilized a hybrid approach combining manual and automatic approaches in a step-wise strategy to calibrate and evaluate the spatially distributed WATFLOOD watershed model. Boyle *et al.*[23] utilized a simultaneous multi-criteria approach to generate a set of Pareto optimal solutions, showing performance trade-off in fitting different segments of the hydrograph, and then selected a parameter set from within the Pareto region that best satisfies two long-time-scale statistical measures of fit (mean and variance). In a more sophisticated approach, Pokhrel *et al.*[107] constructed the Pareto optimal solutions using traditional objective functions (i.e. RMSE and log-RMSE) and then employed "signature measures" of model performance[80] to select a parameter set from within the solutions in close proximity to Pareto region. In an effort to improve the performance of the United States Geological Survey (USGS) Precipitation Runoff Modeling System, Leavesley *et al.*[9] performed a step-wise approach which began by calibrating the parameters affecting water balance, followed by the parameters related to hydrograph timing and soils and vegetation, respectively. Harlin[108] and Zhang and Lindstrom[103] introduced process-oriented step-wise calibration strategies for the HBV hydrologic model. Their procedures partitioned the flow time series into several periods, where specific hydrologic processes dominate, and linked these periods to specific parameter(s) during calibration. Turcotte *et al.*[105] developed a step-wise calibration strategy for the HYDROTEL distributed model. In this procedure, they first calibrated the model parameters sensitive to objectives related to long timescales and then calibrated those parameters sensitive to short timescales. Their calibration steps included parameters related to large-scale water balances, evapotranspiration, infiltration capacity, and routing, respectively. Shamir et al.[106] introduced a step-wise parameter estimation approach based on a set of streamflow descriptors that emphasize the dynamics of the streamflow record at different timescales. Fenicia et al.[84] showed how simultaneous and step-wise multi-criteria optimization strategies can help in understanding model deficiencies and hence guide model development. Wagener et al.[109] (DYNIA) and Choi and Beven[104] developed process-based approaches based on Generalized Sensitivity Analysis[110,111] (GSA) to incorporate the time-varying nature of the hydrologic responses into model/parameter identification. Specifically, the DYNIA approach adopts the GSA procedure within a sequential Monte Carlo framework to locate periods of high identifiability for individual parameters and to diagnose possible failures of model structures. Vrugt et al.[112] presented a similar idea to assess the information content of individual observations with respect to the various model parameters. Their method, called PIMLI, merges the strength of Bayesian Inference with Random Walk Metropolis sampling to resample the parameter space each time the most informative measurement is added to the objective function. Results showed that only a very small number of streamflow measurements was actually needed for reliable calibration of a parsimonious five-parameter rainfall-runoff model. Specifically, about 95% of the discharge observations turned out to contain redundant information. Similar conclusions were drawn for case studies involving subsurface flow and transport models. The development of PIMLI and DYNIA has been inspired by Sequential Monte Carlo (SMC) schemes that provide a generalized treatment of time-varying parameters and states, and are increasingly being used for posterior tracking within the statistical and computational science literature.

3.4.3. Multi-criteria Constraining Approach

The basic concept of the constraining approach differs from automated calibration strategies, described above, in the sense that it considers "consistency" of model structure/parameters as the ultimate goal, while the latter aims at optimality. In other words, the constraining approach seeks parameter estimates that are consistent with some minimal thresholds of performance on several criteria. Rooted in the concept of the Generalized Sensitivity Analysis (GSA),[110,111] models/parameters are separated into behavioral/non-behavioral groups by comparing their performance with a subjectively selected threshold behavior. The models/parameters that present better performance than the selected threshold are accepted as "behavioral" and considered as equally acceptable representation of the system. The remaining models/parameters are rejected as non-behavioral. The Monte Carlo simulation-based GSA approach is at the core of many model identification and uncertainty estimation techniques (e.g. GLUE methodology of Beven and Binley[113] and DYNIA methodology of Wagener et al.[109]).

The main advantages of the constraining approach are its ease of implementation and use and its flexibility in incorporating additional performance criteria into the parameter estimation problem. Similar to the step-wise multi-criteria calibration approach (Sec. 3.4.2), several performance measures can be formulated and behavioral models/parameter sets can be selected considering a single or a group of performance measure(s).[80] The selected set (behavioral set) can be further analyzed using the concept of Pareto optimality to identify a single "best parameter set."

We illustrate the constraining approach with an overview of the step-wise semi-automated methodology, recently proposed in the study by Yilmaz et al.,[80] where parameters of the spatially distributed SAC-SMA model were constrained based on a set of hydrologically meaningful "signature measures." These signature measures target and extract hydrologically relevant (contextual) *information* from the outlet streamflow observations in ways that correspond to the following behavioral functions of a watershed system: (i) maintain overall water balance; (ii) vertically redistribute excess rainfall between fast and slow runoff components; and (iii) redistribute the runoff in time (influencing hydrograph timing and shape). The selected signature measures were defined in terms of percent bias between the following hydrologically meaningful indices (calculated using simulated and observed outlet streamflow): the percent bias in runoff ratio (%BiasRR), various properties of the flow-duration curve (high-flow volume, %BiasFHV; slope of the mid-segment, %BiasFMS; low-flow volume, %BiasFLV; and median flow, %BiasFMM), and a simple index of watershed lag-time (%BiasTLag). One advantage of using signature-based measures in a constraining approach is that they can take on both positive and negative values, thereby indicating the direction of improvement. The procedure starts with establishing relationships between signature measures and parameters of the SAC–SMA model (via a procedure rooted in random sampling). These relationships are then used to constrain the ranges of parameters towards regions of signature measure improvement. In the second step, additional random samples are generated from the constrained parameter ranges and behavioral parameter sets are selected by establishing thresholds on the signature measures. There are various ways in which these thresholds can be defined. In Yilmaz et al.,[80] the performance of the SAC–SMA model with *a priori* parameter values was used as benchmark. The procedure for selection of the behavioral parameter sets is depicted in Fig. 3.3; the names of the signature measures are listed on the *x*-axis, and the *y*-axis shows their corresponding values. Each line along the *x*-axis represents a parameter set. The gray-dashed line represents the performance of the model using *a priori* parameters and the shaded region envelops the signature measure improvement region (i.e. behavioral region), defined as ±1 times the *a priori* model performance. The solid line-triangle parameter combination falls entirely within the gray region and, therefore, represents a behavioral parameter set.

3.5. Automated Calibration of Spatially Distributed Watershed Models

"Distributed" watershed models[4,114-117] have the potential ability to simulate the spatial distribution of hydrologic processes in the watershed of interest, as well as to provide estimates of discharge volume along the entire length of the channel network so that, for instance, the dynamic evolution of flood inundation regions can be estimated.

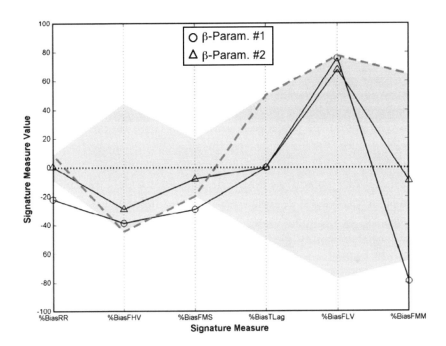

Fig. 3.3. Illustration of the parameter constraining approach proposed by Yilmaz et al.[80] Thresholds are selected based on the signature measure values for the model using *a priori* parameter sets (gray-dashed line). Behavioral region is defined as ±1 times the *a priori* model performance (shaded region). Solid-line-triangle and solid-line-circle represent a behavioral and a non-behavioral parameter set, respectively. (after Yilmaz et al.[80])

Although their potential benefits are clear, the spatial complexity of these processes is perceived to be an obstacle to the proper identification of distributed model components and parameters, translating into significant predictive uncertainty in the model results.[118] Therefore, controversy still persists regarding how such models should be implemented.[119-121] For example, Phase One of the recent Distributed Model Intercomparison Project (DMIP-1)[122] concludes that "reliable and objective procedures for parameter estimation, data assimilation, and error correction" still need to be developed.

An important strength of distributed watershed models is that their parameters (and structure) can, in principle, be inferred from data about watershed physical characteristics via a local *a priori* parameter estimation approach (see Sec. 3.2 for details). However, in general, parameters estimated using a local *a priori* approach continue to require some degree of calibration (adjustment of the model parameter fields) to ensure that the input-state-output behavior of such models remains consistent with the available data.[10-12,123]

In calibration of watershed models having spatially distributed parameter fields, the number of parameter values to be estimated can be quite large, resulting in an ill-posed optimization problem. Note that the watershed outlet streamflow time series, the most widely used hydrologic variable in calibration of watershed models, represents the *integrated response* of the catchment and, therefore, the effects of sub-catchment-(local-)scale variability in hydraulic properties is averaged out to some extent. Therefore, it will typically only contain information about the watershed-scale properties of the parameter fields (the mean values and the broad-scale spatial patterns), rather than the smaller-scale patterns of variation. As a consequence, the number of unknowns (parameters) will be larger than the one that can be identified based on the information content of the streamflow time series and, hence, the estimated parameters will most likely be unrealistic.

Part of the solution to the stabilization of ill-posed problems of this kind is to recognize that the spatially distributed elements of the model parameter fields are not, in fact, independent entities that can take on arbitrary values in the parameter space. Instead, their values are somehow related to the spatial distributions of various hydrologically relevant watershed characteristics, including, for example, geology, soil type, vegetation, and topography.[124] Recognition of these dependencies can facilitate implementation of regularization relationships that help to constrain the dimensionality of the parameter estimation problem.[125,126]

Regularization is a mathematical technique that allows stabilization of the otherwise ill-posed (over-parameterized) parameter estimation problems through introduction of additional information about the parameters.[127-130] Two kinds of regularization techniques are in common use and these help to: (i) improve conditioning of the optimization problem through use of additional information; and/or (ii) reduce the dimensionality of the parameter search space.[125,126,131] An example of the former technique is called Tikhonov Regularization (TR),[127] which employs a penalty function approach[132] wherein the original parameter dimension is retained, while the shape of the global objective function to be optimized, $F_G(\theta, \mu)$, is modified by a regularization objective function, $F_{par}(\theta)$:

$$F_G(\theta,\mu) = F(\theta) + \mu \cdot F_{par}(\theta) \qquad (3.8)$$

where $F_{par}(\theta)$ represents a penalty applied on the solutions that deviate from satisfying the regularization constraints. The regularization (or penalty) parameter, μ, is the weight assigned to the regularization constraints; greater values assigned to μ will result in improved parameter reasonableness and stability, although possibly at the cost of a poorer fit between model outputs and measurements.[133] Sometimes, an appropriate value for μ can be assigned through knowledge of the underlying uncertainty associated with the various components of $F_{par}(\theta)$. Such information, however, is often unavailable and thus alternative methods have been developed.[98,130,133] An example of a regularization technique with the potential to significantly reduce the dimensionality of the estimable parameter space is the Truncated Singular Value Decomposition (TSVD). This technique only focuses on dominant directions and excludes search directions that are associated with negligible eigenvalues with little or no function sensitivity.[134-136] Either separately or together, Tikhonov regularization and TVSD have been used to address high-dimensional inverse modeling problems in hydrology.[126,137-140] Vrugt et al.[141] have recently completed the development of a global optimization algorithm with self-adaptive subspace learning that uses TVSD to continuously update the dominating eigenvectors estimated from an evolving population of solutions. Initial results demonstrate significant efficiency improvements in solving high-dimensional parameter estimation problems (more than 200 parameters). A beta-version of this algorithm has been developed in MATLAB® and can be obtained from the second author upon request.

A simple and commonly used regularization technique to reduce the dimensionality of the spatially distributed parameters of watershed models seeks to characterize and preserve the pattern of relative spatial variation provided by the (local) *a priori* parameter fields (or maps). In its simplest form, one scalar multiplier per *a priori* parameter field is used to vary only the mean level of the parameter field, while constraining the values to remain within their pre-defined physical ranges.[142,143] More sophisticated approaches utilize both additive and multiplicative terms,[9,144] nonlinear transformations via a single parameter,[80] or more complex approaches[125] that are combinations of the above. These techniques simplify the calibration problem to that of finding the parameters of a transformation function (e.g. multipliers) that modifies the parameter fields in such a way that the model performance is improved, while preserving, to some extent, the spatial patterns of the parameters. As an example, Fig. 3.4(a) shows a grid overlay of the distributed version of the SAC–SMA model for Blue River Basin, near Blue, Oklahoma, USA. The grid shown is the *a priori* field for the upper zone free water maximum (UZFWM) parameter (size of the upper layer free water tank) estimated via Koren *et al.*[7] approach. The histogram in Fig. 3.4(b) displays the distribution of *a priori* values of parameter UZFWM within its feasible range (*x*-axis limits). Figure 3.4(c) shows the UZFWM distributions upon transformation via scalar multiplier. It can be seen that multiplication has a large impact on the variance of the distribution of UZFWM in the form of expansion (multiplier > 1) and compression (multiplier < 1). This approach creates problems when any of the individual grid values in each parameter field exceeds its specified (physically reasonable) bounds; either the parameter distribution must be truncated so that any values exceeding the range are fixed at the boundaries or the mean level must be prevented from varying over the entire range. An alternative, and more flexible, approach that removes these restrictions has been proposed by Yilmaz *et al.*,[80] which utilizes a nonlinear transformation with one-parameter (β-parameter) to vary the entire parameter field. Notice, in Fig. 3.4(d), that as the β-parameter is varied towards either of its limiting values, (0,2], the variance of the parameter distribution is compressed so as to keep the entire distribution within the feasible range, while preserving the monotonic relative ordering of parameter values in the field.

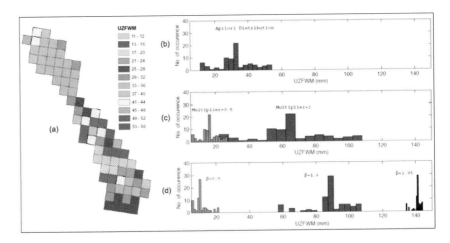

Fig. 3.4. The effect of using transformation functions on the distribution of UZFWM parameter values. (a) Map showing the *a priori* field of the UZFWM parameter (size of the upper layer free water tank) within the distributed version of the SAC–SMA model setup for the Blue River Basin, near Blue, Oklahoma, USA. Histograms showing the distribution of parameter values within UZFWM parameter field: (b) *a priori* specified values, (c) parameter values transformed using scalar multipliers, (d) parameter values after using β-transformation function proposed by Yilmaz *et al.*[80]

Two recent approaches, which have not been discussed so far but show great promise to solving the hydrologic model calibration problem, are the Dynamically Dimensioned Search (DDS) Algorithm[145] and the Constrained Optimization using Response Surfaces CORS) method.[146,147] These two algorithms work well with a limited budget of function evaluations, and are admirably suited for computationally demanding distributed watershed models, which require between minutes and hours to run on a single processor.

The Dynamically Dimensioned Search (DDS) Algorithm[145] is a stochastic, single-objective search algorithm designed to rapidly find "good quality" global solutions. Initially, DDS explores the parameter space globally, but when the search progresses the method settles down with emphasis on local parameter refinements. The transition from global to local search is achieved by dynamically and probabilistically reducing the number of search dimensions (i.e. the number of model parameter values being perturbed) in the neighborhood of the current best estimate. The DDS algorithm has many elements in common with evolutionary strategies, and has only one algorithmic parameter. This is the neighborhood perturbation size, which is calculated from a multi-normal distribution with standard deviation equal to some fraction of the initial range of each decision variable. The user-specified inputs are the initial solution and the maximum number of model evaluations. The DDS algorithm is very simple to implement, and has been used in calibration of a rainfall-runoff model[145] and a land surface scheme.[148] The DDS algorithm can also be used to assess parameter uncertainty by using multiple trails with different starting points and analyze their joint sampling paths.[149] Such an approach violates first-order Markovian properties and can, therefore, only provide a rough estimate of parameter uncertainty (see next section). Recent work has compared the performance of DDS with SCE–UA for a range of studies.[145,150,151] Behrangi et al.[150] have introduced an alternative SCE–UA algorithm that is at least as efficient as DDS when a limited budget of function evaluations is considered.

Constrained Optimization using Response Surfaces (CORS)[146,147] is a response surface approximation method. A response surface model is a multivariate approximation of the black-box objective function and used as surrogate for optimization in situations where function evaluations are computationally expensive. In the CORS method, the next point (parameter combination) is chosen to be the one that minimizes the current response surface model subject to various constraints, including that the new point must be of some minimum distance from previously evaluated points. This distance is sequentially reduced from a high value (global search) at the start of the search to a low value (local search) at termination. This method has been shown to converge nicely to the global optimum for a set of continuous functions.

3.6. Treatment of Parameter Uncertainty

One major weakness of the automated calibration methods, discussed earlier in this chapter, is their underlying treatment of the uncertainty as being primarily attributable to the model parameters, without explicit treatment of input, output, and model structural uncertainties. It is well known, however, that the uncertainties in the watershed modeling procedure stem not only from uncertainties in the parameter estimates but also from measurement errors associated with the system inputs and outputs, and from model structural errors arising from the aggregation of spatially distributed real-world processes into a relatively simple mathematical model. Not properly accounting for these errors results in residuals that exhibit considerable variation in bias (nonstationarity), variance (heteroscedasticity), and correlation structures under different hydrologic conditions and, hence, undermines our efforts to derive meaningful parameter estimates that properly mimic the target hydrologic processes.

One response to the treatment of uncertainty in the parameter estimation problem is to abandon the search for a single "best" parameter combination and adopt a Bayesian viewpoint, which allows the identification of a distribution specifying the uncertainty regarding the values of the model parameters. Bayesian statistics have recently found increasing use in the field of hydrology for statistical inference of parameters, state variables, and model output prediction.[152-157] For a recent review, see Liu and Gupta.[157] The Bayesian paradigm provides a simple way to combine multiple probability distributions using Bayes theorem. In a hydrologic context, this method is suited to systematically address and quantify the various error sources within a single cohesive, integrated, and hierarchical method.

To successfully implement the Bayesian paradigm, sampling methods that can efficiently summarize the posterior probability density function (PDF) are needed. This distribution combines the data likelihood with a prior distribution using Bayes theorem, and contains all the desired information to make statistically sound inferences about the uncertainty of the individual components in the model. Unfortunately, for most practical hydrologic problems, this posterior distribution cannot be obtained by analytical means or by analytical approximation. For this reason, researchers commonly resort to iterative approximation methods, such as Markov chain Monte Carlo (MCMC) sampling, to generate a sample from the posterior PDF.

3.6.1. *Random Walk Metropolis (RWM) Algorithm*

The basis of the MCMC method is a Markov chain that generates a random walk through the search space with stable frequencies stemming from a fixed probability distribution. To visit configurations with a stable frequency, an MCMC algorithm generates trial moves from the current ("old") position of the Markov chain θ_{t-1} to a new state ϑ. The earliest and most general MCMC approach is the Random Walk Metropolis (RWM) algorithm. Assuming that a random walk has already sampled points $\{\theta_0,\ldots,\theta_{t-1}\}$, this algorithm proceeds via the following three steps. First, a candidate point ϑ is sampled from a proposal distribution q that is symmetric, $q(\theta_{t-1}, \vartheta) = q(\vartheta, \theta_{t-1})$ and may depend on the present location, θ_{t-1}. Next, the candidate point is either accepted or rejected using the Metropolis acceptance probability:

$$\alpha(\theta_{t-1}, \vartheta) = \begin{cases} \min\left(\dfrac{\pi(\vartheta)}{\pi(\theta_{t-1})}, 1\right) & \text{if } \pi(\theta_{t-1}) > 0 \\ 1 & \text{if } \pi(\theta_{t-1}) = 0 \end{cases} \qquad (3.9)$$

where $\pi(\bullet)$ denotes the density of the target distribution. Finally, if the proposal is accepted, the chain moves to ϑ; otherwise, the chain remains at its current location, θ_{t-1}.

The original RWM scheme was constructed to maintain detailed balance with respect to the target distribution $\pi(\bullet)$ at each step in the chain:

$$p(\theta_{t-1})p(\theta_{t-1} \to \vartheta) = p(\vartheta)p(\vartheta \to \theta_{t-1}) \qquad (3.10)$$

where $p(\theta_{t-1})(p(\vartheta))$ denotes the probability of finding the system in state $\theta_{t-1}(\vartheta)$, and $p(\theta_{t-1} \to \vartheta)p(\vartheta \to \theta_{t-1})$ denotes the conditional probability of performing a trial move from θ_{t-1} to ϑ (ϑ to θ_{t-1}). The result is a Markov chain which, under certain regularity conditions, has a unique stationary distribution with PDF $\pi(\bullet)$. In practice, this means that if one looks at the values of θ generated by the RWM that are sufficiently far from the starting value, the successively generated parameter combinations will be distributed with stable frequencies stemming from the underlying posterior PDF of θ, $\pi(\bullet)$. Hastings[158] extended Eq. (3.10) to include non-symmetrical proposal distributions, i.e. $q(\theta_{t-1},\vartheta) \neq q(\vartheta,\theta_{t-1})$, in which a proposal jump to ϑ and the reverse jump do not have equal probability. This extension is called the Metropolis Hastings (MH) algorithm, and has become the basic building block of many existing MCMC sampling schemes.

Existing theory and experiments prove convergence of well-constructed MCMC schemes to the appropriate limiting distribution under a variety of conditions. In practice, this convergence is often observed to be impractically slow. This deficiency is frequently caused by an inappropriate selection of the proposal distribution used to generate trial moves in the Markov chain. To improve the search efficiency of MCMC methods, it seems natural to tune the orientation and scale of the proposal distribution during the evolution of the sampler to the posterior target distribution, using the information from past states. This information is stored in the sample paths of the Markov chain.

An adaptive MCMC algorithm that has become popular in the field of hydrology is the Shuffled Complex Evolution Metropolis (SCEM–UA) global optimization algorithm.[155] This method is a modified version of the original SCE–UA global optimization algorithm,[62] and runs multiple chains in parallel to provide a robust exploration of the search space. These chains communicate with each other through an external population of points, which are used to continuously update the size and shape of the proposal distribution in each chain. The MCMC evolution is repeated until the $\hat{R}$ statistic of Gelman and Rubin[159] indicates convergence to a stationary posterior distribution. This statistic compares the between- and within-variance of the different parallel chains.

Numerous studies have demonstrated the usefulness of the SCEM–UA algorithm for estimating (nonlinear) parameter uncertainty. However, the method does not maintain detailed balance at every single step in the chain, casting doubt on whether the algorithm will appropriately sample the underlying PDF. Although various benchmark studies have reported very good sampling efficiencies and convergence properties of the SCEM–UA algorithm, violating detailed balance is a reason for at least some researchers and practitioners not to use this method for posterior inference. An adaptive MCMC algorithm that is efficient in hydrologic applications, and maintains detailed balance and ergodicity, therefore, remains desirable.

3.6.2. *DiffeRential Evolution Adaptive Metropolis (DREAM)*

Vrugt et al.[160] recently introduced the DiffeRential Evolution Adaptive Metropolis (DREAM) algorithm. This algorithm uses differential evolution as genetic algorithm for population evolution, with a Metropolis selection rule to decide whether candidate points should replace their respective parents or not. The DREAM is a follow up to the DE–MC (Differential Evolution Markov Chain) method of ter Braak,[161] but contains several extensions to increase search efficiency and acceptance rate for complex and multimodal response surfaces with numerous local optimal solutions. Such surfaces are frequently encountered in hydrologic modeling. The method is presented below.

(i) Draw an initial population Θ of size N using the specified prior distribution. Typically, N is at least equal to the dimension of θ (denoted by the symbol d). Recommendations for N are given later.

(ii) Compute the density $\pi(\theta^i)$ of each point of Θ, $i = 1,...,N$.

FOR $i \leftarrow 1,..., N$ DO (CHAIN EVOLUTION)

(iii) Generate a candidate point, ϑ^i in chain i,

$$\vartheta^i = \theta^i + (1_d + \mathbf{e})\gamma(\delta, d_{\mathit{eff}})\left[\sum_{j=1}^{\delta}\theta^{r_1(j)} - \sum_{n=1}^{\delta}\theta^{r_2(n)}\right] + \varepsilon \quad (3.11)$$

where δ signifies the number of pairs used to generate the proposal, and $r_1(j), r_2(n) \in \{1,...,N\}$; $r_1(j) \neq r_2(n) \neq i$ for $j = 1,...,\delta$ and $n = 1,...,\delta$. The values of **e** and ε are drawn from $U_d(-b,b)$ and $N_d(0,b^*)$ with $|b| < 1$, and b^* small compared to the width of the target distribution, respectively, and the value of jump-size γ depends on δ and d', the number of dimensions that will be updated jointly (see next step and Eq. (3.12)). By comparison with RWM, a good choice for $\gamma = 2.38/\sqrt{2\delta d_{eff}}$,[161,162] with $d_{eff} = d$, but potentially decreased in the next step. This choice is expected to yield an acceptance probability of 0.44 for $d = 1$, 0.28 for $d = 5$, and 0.23 for large d.

(iv) Replace each element, $j = 1,...,d$ of the proposal ϑ^i_j with θ^i_j using a binomial scheme with crossover probability CR,

$$\vartheta^i_j = \begin{cases} \theta^i_j & \text{if } U \leq 1-CR, \ d_{eff} = d_{eff} - 1 \\ \vartheta^i_j & \text{otherwise} \end{cases} \quad j = 1,...,d \quad (3.12)$$

where $U \in [0,1]$ is a draw from a uniform distribution.

(v) Compute $\pi(\theta^i)$ and accept the candidate point with Metropolis acceptance probability, $\alpha(\theta^i, \vartheta^i)$

$$\alpha(\theta^i, \vartheta^i) = \begin{cases} \min\left(\dfrac{\pi(\vartheta^i)}{\pi(\theta^i)}, 1\right) & \text{if } \pi(\theta^i) > 0 \\ 1 & \text{if } \pi(\theta^i) = 0 \end{cases} \quad (3.13)$$

(vi) If the candidate point is accepted, move the chain, $\theta^i = \vartheta^i$; otherwise remain at the old location, θ^i.

END FOR (CHAIN EVOLUTION)

(vii) Remove potential outlier chains using the Inter-Quartile-Range (IQR) statistic.

(viii) Compute the Gelman-Rubin $\hat{R}_j$ convergence diagnostic[159] for each dimension $j = 1,..., d$ using the last 50% of the samples in each chain.

(ix) If $\hat{R}_j < 1.2$ for all j, stop; otherwise go to CHAIN EVOLUTION.

At every step, the points in Θ contain the most relevant information about the search, and this population of points is used to globally share information about the progress of the search of the individual chains. This information exchange enhances the survivability of individual chains, and facilitates adaptive updating of the scale and orientation of the proposal distribution.

The DREAM algorithm adaptively updates the scale and orientation of the proposal distribution during the evolution of the individual chains to a limiting distribution. If the state of a single chain is given by a single d-dimensional vector θ, then at each generation t, the N chains in DREAM define a population Θ, which corresponds to an $N \times d$ matrix, with each chain as a row. Jumps in each chain are derived from the remaining chains by taking a fixed multiple of the d-dimensional difference vector of the locations of randomly chosen pairs of chains. The Metropolis ratio is used to decide whether to accept candidate points or not. This series of operations results in a MCMC sampler that conducts a robust and efficient search of the parameter space. Because the joint PDF of the N chains factorizes to $\pi(\theta_1) \, x \, ... \, x \, \pi(\theta_N)$, the states $\theta_1 \, ... \, \theta_N$ of the individual chains are independent at any generation after DREAM has become independent of its initial value. After this so-called burn-in period, the convergence of a DREAM run can thus be monitored with the $\hat{R}$ statistic of Gelman and Rubin.[159]

Case studies presented in Vrugt et al.[160] have demonstrated that DREAM is generally superior to existing MCMC schemes, and can efficiently handle multimodality, high-dimensionality, and nonlinearity. In that same paper, recommendations have also been given for some of the values of the algorithmic parameters. The only parameter that remains to be specified by the user before the sampler can be used for statistical inference is the population size, N; a general recommendation is to use $N > d$, although the sub-space sampling strategy allows taking $N \ll d$.

3.6.3. *Sequential Data Assimilation*

In the past few years, ensemble-forecasting techniques based on Sequential Data Assimilation (SDA) methods have become increasingly popular due to their potential ability to explicitly handle the various sources of uncertainty in environmental modeling. Techniques based on the Ensemble Kalman Filter[163] (EnKF) have been suggested as having the power and flexibility required for data assimilation using nonlinear models. In particular, Vrugt *et al.*[164] recently presented the Simultaneous Optimization and Data Assimilation (SODA) method, which uses the EnKF to recursively update model states while estimating time-invariant values for the model parameters using the SCEM–UA[155] optimization algorithm. A novel feature of SODA is its explicit treatment of errors due to parameter uncertainty, uncertainty in the initialization and propagation of state variables, model structural error, and output measurement errors. The development below closely follows that of Vrugt *et al.*[164]

To help facilitate the description of the classical Kalman Filter (KF), we start by writing the model dynamics as a stochastic equation. In keeping with Fig. 3.1, consider a model Φ in which the discrete time evolution of the state vector ψ is described with:

$$\psi_{i+1} = \Phi(\psi_i, X_i, \theta) + q_i \tag{3.14}$$

where X represents the observed forcing (e.g. boundary conditions), θ is the vector of parameter values, i denotes time, and q_i is a dynamic noise term representing errors in the conceptual model formulation. This stochastic forcing term flattens the probability density function of the states during the integration. We assume that the model predictions are related to its internal state according to:

$$\tilde{s}_i = H(\psi_i) + \varepsilon_i \qquad \varepsilon_i \sim N(0, \sigma_i^0) \tag{3.15}$$

where $H(\cdot)$ is the measurement operator, which maps the state space into measurement or model output space, and also has a random additive error ε_i, called the measurement error. Note that σ_i^0 denotes the error deviation of the measurements, and ψ^* represents the true model states. At each measurement time, when an output observation becomes available, the output forecast error z_i is computed:

$$z_i = s_i - H(\psi_i^f) \tag{3.16}$$

and the forecasted states, ψ_i^f, are updated using the standard KF analysis equation:

$$\psi_i^u = \psi_i^f + K_i(s_i - H(\psi_i^f)) \tag{3.17}$$

where ψ_i^u is the updated or analyzed state, and K_i denotes the Kalman gain. The size of the gain directly depends on the size of the measurement and model error.

The analyzed state then recursively feeds the next state propagation step into the model:

$$\psi_{i+1}^f = \Phi(\psi_i^u, X_i, \theta) \tag{3.18}$$

The virtue of the KF method is that it offers a very general framework for segregating and quantifying the effects of input, output, and model structural error in watershed modeling. Specifically, uncertainty in the model formulation and observational data are specified through the stochastic forcing terms q and ε, whereas errors in the input data are quantified by stochastically perturbing the elements of X.

The SODA method is an extension of traditional techniques in that it uses the ensemble Kalman Filter (EnKF) to solve for Eqs. (3.14)–(3.18). The EnKF uses a Monte Carlo (MC) method to generate an ensemble of model trajectories from which the time evolution of the probability density of the model states and related error covariances are estimated.[165] The EnKF avoids many of the problems associated with the traditional extended Kalman filter (EKF) method. For example, there is no closure problem as is introduced in the EKF by neglecting contributions from higher-order statistical moments in the error covariance evolution. Moreover, the conceptual simplicity, relative ease of implementation, and computational efficiency of the EnKF make the method an attractive option for data assimilation in the meteorologic, ocean, and hydrologic sciences.

In summary, the EnKF propagates an ensemble of state vector trajectories in parallel, such that each trajectory represents one realization of generated model replicates. When an output measurement is available, each forecasted ensemble state vector ψ_i^f is updated by means of a linear updating rule in a manner analogous to the Kalman Filter. A detailed description of the EnKF method, including the algorithmic details, is found in Evensen,[163] and so will not be repeated here.

3.7. Parallel Computing

The traditional implementation and application of the local and global optimization methods, discussed herein, involves sequential execution of the algorithm using the computational power of a single Central Processing Unit (CPU). Such an implementation works acceptably well for relatively simple optimization problems, and those optimization problems with models that do not require much computational time to execute. However, for high-dimensional optimization problems involving complex spatially distributed models, such as the ones frequently used in the field of environmental sciences, this sequential implementation needs to be revisited. Much of the computational time required for calibrating parameters in watershed models is spent running the model code and generating the desired output. Thus, there should be large computational efficiency gains from parallelizing the algorithm so that independent model simulations are run on different nodes in a distributed computer system. General-purpose parallel versions of SCE–UA, MOCOM, MOSCEM–UA, AMALGAM, and DREAM have been developed in Octave and can be obtained from the second author upon request.

3.8. Summary and Conclusions

This chapter reviewed the current state-of-the-art of model calibration in watershed hydrology. The current status in watershed modeling is that, regardless of their type (i.e. conceptual/physically-based) and structural complexity (e.g. spatially lumped/distributed), watershed models still need to be calibrated to improve the reliability of their simulation/prediction performance. In calibration, model parameters are systematically adjusted in such a way that the behavior of the model approximates, as closely and consistently as possible, the historical observed response of the watershed system under study. In parallel with advances in computing resources and measurement techniques, hydrologic models are becoming increasingly complex, not only representing the spatial heterogeneity of a watershed but also simulating various internal states and fluxes. This increase in complexity is typically accompanied by a decrease in identifiability (or justification) of the components and parameters in these models due to problems with direct observation and noisy data. This leads to highly uncertain model predictions. There is, therefore, a pressing need for better approaches to model calibration.

We discussed manual and automated parameter estimation techniques for calibration of lumped and spatially distributed watershed models. Historically, manual calibration has been the common practice for model calibration. Its main strength is the use of expert knowledge to match the perceived hydrologic processes in the watershed with their equivalents in the model, and thus they help to prescribe consistent values for the parameters. However, manual calibration is time- and labor-intensive, involving many subjective decisions. The era of digital computing and advances in hardware power and speed has led to the development of automated procedures that are more objective to implement and use, and more efficient in systematically searching for the "best" values of the model parameters. We have discussed the historical development of single-criterion methods and provided details of the widely used SCE–UA global optimization algorithm. The major limitation of single-criterion automated algorithms is that they rely on a single regression-based aggregate measure of model performance, which often leads to an ill-posed parameter estimation problem. This is due to a loss of information when projecting from the high-dimensional data space down to the single dimension of the residual-based summary statistic. For highly complex models with many interacting parameters, this approach is often poor at discriminating between the effects of individual parameters on the simulated model output. This, therefore, unnecessarily introduces non-uniqueness and equifinality in parameter estimation.

More sophisticated automated (or semi-automated) methods for parameter adjustment have been developed within the framework of multi-criteria theory, thereby more closely matching the number of unknowns (the parameters) with the number of pieces of information (the measures). A major advantage of the multi-criteria approach is that various aspects of the manual calibration strategy can be absorbed into the calibration process, thus strengthening the physical basis of the identified parameters. Although multi-criteria methods are widely used in the calibration of watershed models and powerful algorithms have been developed (among these, we discussed MOCOM–UA, MACS, DYNIA, PIMLI, AMALGAM, and MOSCEM) in the past decade or so, how to properly incorporate expert knowledge into automated parameter estimation still remains difficult and an active area of research. To better realize the strengths of multi-criteria optimization requires a shift from statistical regression-based performance measures towards more powerful and hydrologically relevant (contextual) measures of information.[77,80] One possible way to develop these signature-based measures is the detection of characteristic or "signature" patterns in the spatio-temporal data and by relating these to their causal mechanisms. Signature measures that quantitatively summarize these signature patterns can be used to estimate plausible model parameters and further point towards possible causes of model failure and guide model improvement strategies. More research is warranted on how to (i) properly construct different and complementary signature measures of model performance that are diagnostic,[77,80] while being representative of the watershed response behaviors that are deemed important to reproduce with the model; and (ii) integrate these signature measures into automated calibration techniques while keeping the resulting multi-criteria optimization problem computationally tractable.

We have also discussed the calibration of spatially distributed watershed models. Calibration of such models is complicated by their large number of parameters. Part of the solution to the stabilization of ill-posed problems of this kind is to recognize that the spatially distributed elements of the model parameter fields are not, in fact, independent entities that can take on arbitrary values in the parameter space. Instead, their values are somehow related to the spatial distributions of various hydrologically relevant watershed characteristics, including, for example, geology, soil type, vegetation, and topography. Recognition of these dependencies can facilitate implementation of regularization relationships that help to constrain the dimensionality of the parameter estimation problem. We would like to emphasize that, although application of regularization techniques in calibration of complex watershed models is in its infancy,[125] considerable progress has already been reported in adjacent fields, such as groundwater and subsurface hydrology. Methods and approaches developed therein might be of great use and inspiration to solve parameter estimation problems in watershed hydrology. Particularly appealing are approaches that attempt: (i) to improve conditioning of the optimization problem through use of additional information (penalty functions[127]); (ii) to reduce the dimensionality of the optimization problem using concepts of sensitivity and principal component analysis,[139] as discussed briefly herein; and (iii) representer-based calibration methods.[166] Further, the availability of parallel computing and powerful optimization algorithms that can efficiently handle a high number of parameters in complex distributed hydrologic models is of particular importance for developing rigorous model calibration strategies for such models.[75]

Finally, this chapter has discussed emerging methods for parameter uncertainty estimation. Uncertainty quantification is currently receiving a surge in attention in hydrology, as researchers try to better understand what is well and what is not well understood about the watersheds that are being studied and as decision makers push to better quantify accuracy and precision of model predictions. Various methodologies have been developed in the past decade to better treat uncertainty. We have highlighted the standard RWM for posterior exploration, and have discussed recent advances in MCMC sampling by combining multi-chain population evolution with the genetic algorithm Differential Evolution and sub-space sampling. We have also summarized emerging state-space filtering methods that rely on the Ensemble Kalman Filter to specifically treat forcing (input), output, parameter, and state error, within a recursive estimation framework.

Among emerging approaches for uncertainty estimation not discussed in this chapter include methods such as Bayesian Estimation of Structure (BESt),[167] Bayesian Model Averaging (BMA),[168-171] and its maximum likelihood (ML) variant, MLBMA.[172-174] Such methods seem particularly appealing because they help to account for model structural uncertainty. The BMA and MLBMA, in particular, are relatively easy to implement, accounting jointly for conceptual model and parameter uncertainty by considering a set of alternative models that may be very different from each other while being based on a common set of data.

Acknowledgments

The first author was supported by NASA's Applied Sciences Program (Stephen Ambrose) and PMM (Ramesh Kakar) of NASA Headquarters. The second author was supported by a J. Robert Oppenheimer Fellowship from the LANL postdoctoral program. Partial support for the first, third, and fourth authors was provided by SAHRA, the Science and Technology Center for Sustainability of semi-Arid Hydrology and Riparian Areas under NSF-STC grant EAR-9876800 and by the US National Weather Service Office of Hydrology under grant NA04NWS462001.

References

1. R. J. C. Burnash, R. L. Ferral and R. A. McGuire, in *A Generalized Streamflow Simulation System-Conceptual Modeling for Digital Computers*. Technical Report of the Joint Federal-State River Forecast Center, (Department of Water Resources, State of California and National Weather Service, 1973), p. 204.
2. R. J. C. Burnash, in *Computer Models of Watershed Hydrology*, Ed. V. P. Singh (Water Resources Publications, Colorado, 1995), p. 311.
3. S. Bergstrom, in *Computer Models of Watershed Hydrology*, Ed. V. P. Singh (Water Resources Publications, Colorado, 1995), p. 443.
4. J. C. Refsgaard and B. Storm, in *Computer Models of Watershed Hydrology*, Ed. V. P. Singh (Water Resources Publications, Colorado, 1995), p. 809.
5. D. A. Woolhiser, R. E. Smith and D. C. Goodrich, USDA, Agricultural Research Service Report No: ARS-77, p. 130 (1990).
6. J. C. Refsgaard, *J. Hydrol.*, 69 (1997).
7. V. Koren, M. Smith, D. Wang and Z. Zhang, in *15th Conference on Hydrology*, AMS Conference Proceeding (AMS, Long Beach, California, 2000), p. 103.
8. V. Koren, M. Smith and Q. Duan, in *Calibration of Watershed Models*, Eds. Q. Duan, S. Sorooshian, H. V. Gupta, H. Rosseau and R. Turcotte (Water Science and Applications, American Geophysical Union, 2003), p. 239.
9. G. H. Leavesley, L. E. Hay, R. J. Viger and S. L. Markstrom, in *Calibration of Watershed Models*, Eds. Q. Duan, S. Sorooshian, H. V. Gupta, H. Rosseau and R. Turcotte (Water Science and Applications, American Geophysical Union, 2003), p. 255.

10. H. V. Gupta, S. Sorooshian and P. O. Yapo, *Water Resour. Res.*, 751 (1998).
11. H. Madsen, *Adv. Water Resour.*, 205 (2003).
12. S. Reed, V. Koren, M. Smith, Z. Zhang, F. Moreda, D.-J. Seo and DMIP participants, *J. Hydrol.*, 27 (2004).
13. F. A. Abdulla and D. P. Lettenmaier, *J. Hydrol.*, 230 (1997).
14. C. E. M. Sefton and S. M. Howarth, *J. Hydrol.*, 1 (1998).
15. W. Fernandez, R. M. Vogel and S. Sankarasubramanian, *Hydrol. Sci. J.*, 689 (2000).
16. T. Wagener, H. S. Wheater and H. V. Gupta, *Rainfall-runoff Modeling in Gauged and Ungauged Catchments* (Imperial College Press, London, UK, 2004), p. 300.
17. T. Wagener and H. S. Wheater, *J. Hydrol.*, 132 (2006).
18. N. McIntyre, H. Lee, H. S. Wheater, A. Young and T. Wagener, *Water Resour. Res.*, W12434, doi:10.1029/2005WR004289 (2005).
19. D. Kavetski, S. W. Franks, G. Kuczera, in *Calibration of Watershed Models*, Eds. Q. Duan, S. Sorooshian, H. V. Gupta, H. Rosseau and R. Turcotte (Water Science and Applications, American Geophysical Union, 2003), p. 49.
20. K. J. Beven, *Water Sci. Tech.*, 167 (2005).
21. J. Götzinger and A. Bàrdossy, *J. Hydrol.*, 374 (2007).
22. M. Yadav, T. Wagener and H. V. Gupta, *Adv. Water Resour.*, 1756 (2007).
23. D. P. Boyle, H. V. Gupta and S. Sorooshian, *Water Resour. Res.*, 3663 (2000).
24. Hydrologic Research Center (Video Series produced for the National Weather Service, Hydrologic Research Center, 1999).
25. M. B. Smith, D. P. Laurine, V. Koren, S. Reed and Z. Zhang, in *Calibration of Watershed Models*, Eds. Q. Duan, S. Sorooshian, H. V. Gupta, H. Rosseau and R. Turcotte (Water Science and Applications, American Geophysical Union, 2003), p. 133.
26. M. Mroczkowski, G. P. Raper and G. Kuczera, *Water Resour. Res.*, 2325 (1997).
27. A. K. Bingeman, N. Kouwen and E. D. Soulis, *J. Hydrol. Eng.*, 451 (2006).
28. S. Sorooshian and H. V. Gupta, in *Computer Models of Watershed Hydrology*, Ed. V. P. Singh (Water Resources Publications, Colorado, 1995), p. 23.
29. H. V. Gupta, K. J. Beven and T. Wagener, in *Encyclopedia of Hydrological Sciences*, Ed. M. G. Anderson (John Wiley & Sons Ltd., Chichester, UK, 2006).
30. S. Sorooshian and J. A. Dracup, *Water Resour. Res.*, 430 (1980).
31. K. J. Beven, *Rainfall-Runoff Modeling – The Primer* (Wiley, Chichester, 2001), p. 360.
32. Y. Bard, *Nonlinear Parameter Estimation* (Academic Press, 1974), p. 341.
33. V. K. Gupta and S. Sorooshian, *Water Resour. Res.*, 473 (1985).
34. J. D. Hendrickson, S. Sorooshian and L. Brazil, *Water Resour. Res.*, 691 (1988).
35. J. A. Nelder and R. Mead, *Comp. J.*, 308 (1965).
36. R. Hooke and T. A. Jeeves, *J. Assoc. Comput. Mach.*, 212 (1961).
37. H. H. Rosenbrock, *Comp. J.*, 175 (1960).
38. R. P. Ibbitt and T. O'Donnell, *J. Hydraul. Eng-ASCE*, 1331 (1971).

39. P. R. Johnston and D. Pilgrim, *Water Resour. Res.*, 477 (1976).
40. G. Pickup, *Hydrol. Sci. B.*, 257 (1977).
41. R. J. Moore and R. T. Clarke, *Water Resour. Res.*, 1367 (1981).
42. S. Sorooshian and V. K. Gupta, *Water Resour. Res.*, 251 (1983).
43. V. K. Gupta and S. Sorooshian, *Water Resour. Res.*, 269 (1983).
44. G. Kuczera, *J. Hydrol.*, 229 (1988).
45. D. Kavetski, G. Kuczera and S. W. Franks, *Water Resour. Res.*, W03407, doi:10.1029/2005WR004368 (2006).
46. D. Kavetski, G. Kuczera and S. W. Franks, *Water Resour. Res.*, W03408, doi:10.1029/2005WR004376 (2006).
47. J. A. Vrugt, C. J. F. ter Braak, H. V. Gupta and B. A. Robinson, *Stoch. Environ. Res. Risk Assess.*, 1011 (2009).
48. J. A. Vrugt, C. J. F. ter Braak, M. P. Clark, J. M. Hyman and B. A. Robinson, *Water Resour. Res.*, W00B09, doi:10.1029/2007WR006720 (2008).
49. G. E. P. Box and D. R. Cox, *J. Roy. Stat. Soc.* **B**, 211 (1964).
50. G. Kuczera, *Water Resour. Res.*, 146 (1982).
51. S. Sorooshian, V. K. Gupta and J. L. Fulton, *Water Resour. Res.*, 251 (1983).
52. V. K. Gupta and S. Sorooshian, *J. Hydrol.*, 57 (1985).
53. P. O. Yapo, H. V. Gupta and S. Sorooshian, *J. Hydrol.*, 23 (1996).
54. J. A. Vrugt, H. V. Gupta, S. Sorooshian, T. Wagener and W. Bouten, *J. Hydrol.*, 288 (2006).
55. D. C. Karnopp, *Automatica*, 111 (1963).
56. S. F. Masri, G. A. Bekey and F. B. Safford, *Appl. Math. Comput.*, 353 (1980).
57. L. E. Brazil and W. F. Krajewski, presented in *Conference on Engineering Hydrology* (Hydraulics Division American Society of Civil Engineers, Williamsburg, VA, August 3-7, 1987).
58. W. L. Price, *J. Optimiz. Theory App.*, 333 (1983).
59. O. Klepper, H. Scholten and J. P. G. van de Kamer, *J. Forecasting*, 191 (1991).
60. S. Kirkpatrick, C. D. Gelatt Jr. and M. P. Vecchi, *Science*, 671 (1983).
61. M. Thyer, G. Kuczera and B. C. Bates, *Water Resour. Res.*, 767 (1999).
62. Q. Duan, V. K. Gupta and S. Sorooshian, *Water Resour. Res.*, 1015 (1992).
63. Q. J. Wang, *Water Resour. Res.*, 2467 (1991).
64. J. A. Vrugt, M. T. van Wijk, J. W. Hopmans and J. Simunek, *Water Resour. Res.*, 2457 (2001).
65. Q. Duan, in *Calibration of Watershed Models*, Eds. Q. Duan, S. Sorooshian, H. V. Gupta, H. Rosseau and R. Turcotte (Water Science and Applications, American Geophysical Union, 2003), p. 89.
66. Q. Duan, V. K. Gupta and S. Sorooshian, *J. Optimiz. Theory App.*, 501 (1993).
67. S. Sorooshian, Q. Duan and V. K. Gupta, *Water Resour. Res.*, 1185 (1993).
68. J. H. Holland, *Adaptation in Natural and Artificial Systems* (University of Michigan Press, Ann Arbor, 1975).
69. T. Y. Gan and G. F. Biftu, *Water Resour. Res.*, 2161 (1996).

70. C. H. Luce and T. W. Cundy, *Water Resour. Res.*, 1057 (1994).
71. H. Tanakamaru, *Trans. JSIDRE*, 103 (1995).
72. G. Kuczera, *Water Resour. Res.*, 177 (1997).
73. Q. Duan, S. Sorooshian and V. K. Gupta, *J. Hydrol.*, 256 (1994).
74. D. Wolpert and W. G. MacReady, *IEEE Trans. Evol. Comp.*, 67 (1999).
75. J. A. Vrugt and B. A. Robinson, *Proc. Natl. Acad. Sci. USA*, 708 (2007).
76. J. A. Vrugt, B. A. Robinson and J. M. Hyman, *IEEE Trans. Evolut. Comput.*, 10.1109/TEVC.2008.924428 (2008).
77. H. V. Gupta, T. Wagener and Y. Liu, *Hydrol. Process.*, 3802 (2008).
78. T. Wagener and H. V. Gupta, *Stoch. Environ. Res. Risk Assess.*, 378 (2005).
79. T. Meixner, H. V. Gupta, L. A. Bastidas and R. C. Bales, in *Calibration of Watershed Models*, Eds. Q. Duan, S. Sorooshian, H. V. Gupta, H. Rosseau and R. Turcotte (Water Science and Applications, American Geophysical Union, 2003), p. 213.
80. K. K. Yilmaz, H. V. Gupta and T. Wagener, *Water Resour. Res.*, W09417, doi:10.1029/2007WR006716 (2008).
81. P. O. Yapo, H. V. Gupta and S. Sorooshian, *J. Hydrol.*, 83 (1998).
82. J. A. Vrugt, H. V. Gupta, L. A. Bastidas, W. Bouten and S. Sorooshian, *Water Resour. Res.*, 1214, doi:10.1029/2002WR001746 (2003).
83. J. A. Vrugt, J. van Belle and W. Bouten, *J. Avian Biology*, 432 (2007).
84. F. Fenicia, H. H. G. Savenije, P. Matgen and L. Pfister, *Water Resour. Res.*, W03434, doi:10.1029/2006WR 005098 (2007).
85. H. Madsen, *J. Hydrol.*, 276 (2000).
86. P. Reed, B. S. Minsker and D. E. Goldberg, *Water Resour. Res.*, 1096 (2002).
87. D. E. Goldberg, *Genetic Algorithms in Search, Optimization and Machine Learning* (Addison-Wesley Publishing Company, Reading, MA, 1989), pp. 412.
88. D. Whitley, in *Proceedings of the Third International Conference on Genetic Algorithms*, Arlington, VA (Morgan Kaufman Publishers, San Mateo, CA, 1989), p. 116.
89. Y. Liu, *Ph.D. Dissertation* (Department of Hydrology and Water Resources, University of Arizona, Tucson, Arizona, USA, 2003), p. 236.
90. H. V. Gupta, L. A. Bastidas, S. Sorooshian, W. J. Shuttleworth and Z. L. Yang, *J. Geophys. Res.* **D16**, 19491 (1999).
91. H. V. Gupta, L. Bastidas, J. A. Vrugt and S. Sorooshian, in *Calibration of Watershed Models*, Eds. Q. Duan, S. Sorooshian, H. V. Gupta, H. Rosseau and R. Turcotte (Water Science and Applications, American Geophysical Union, 2003), p. 125.
92. E. Zitzler, M. Laumanns, and L. Thiele, Department of Electrical Engineering, Swiss Federal Institute of Technology (ETH) Zurich, TIK-Report 103 (2001).
93. K. Deb, A. Pratap, S. Agarwal and T. Meyarivan, *IEEE Trans. Evol. Comp.*, 182 (2002).
94. E. Zitzler, K. Deb, and L. Thiele, *Evol. Comp.*, 173 (2000).

95. E. Zitzler, and L. Thiele, *IEEE Trans. Evol. Comp.*, 257 (1999).
96. J. A. Vrugt, P. H. Stauffer, T. Wöhling, B. A. Robinson and V. V. Vesselinov, *Vadose Zone J.*, 843 (2008).
97. T. Wöhling and J. A. Vrugt, *Water Resour. Res.*, W12432, doi:10.1029/2008WR007154 (2008).
98. J. Mertens, H. Madsen, L. Feyen, D. Jacques and J. Feyen, *J. Hydrol.*, 251 (2004).
99. M. B. Butts, J. T. Payne, M. Kristensen and H. Madsen, *J. Hydrol.*, 242 (2004).
100. T. S. Hogue, S. Sorooshian, H. V. Gupta, A. Holz and D. Braatz, *J. Hydrometeor.*, 524 (2000).
101. T. S. Hogue, H. V. Gupta and S. Sorooshian, *J. Hydrol.*, 202 (2006).
102. L. E. Brazil, *Multilevel Calibration Strategy for Complex Hydrologic Simulation Models* (Ph.D. Dissertation, Colorado State University, Fort Collins, 1988), p. 217.
103. X. Zhang and G. Lindström, *Hydrol. Process.*, 1671 (1997).
104. T. H. Choi and K. Beven, *J. Hydrol.*, 316 (2007).
105. R. Turcotte, A. N. Rousseau, J.-P. Fortin, J.-P. Villeneuve, in *Calibration of Watershed Models*, Eds. Q. Duan, S. Sorooshian, H. V. Gupta, H. Rosseau and R. Turcotte (Water Science and Applications, American Geophysical Union, 2003), p. 153.
106. E. Shamir, B. Imam, H. V. Gupta and S. Sorooshian, *Water Resour. Res.*, doi:10.1029/2004WR003409 (2005).
107. P. Pokhrel, K. K. Yilmaz and H. V. Gupta, *J. Hydrol.*, doi:10.1016/j.jhydrol.2008.12.004 (2009).
108. J. Harlin, *Nordic Hydrol.*, 15 (1991).
109. T. Wagener, N. McIntyre, M. J. Less, H. S. Wheater and H. V. Gupta, *Hydrol. Process.*, 455 (2003).
110. G. M. Hornberger and R. C. Spear, *J. Environ. Manage.*, 7 (1981).
111. R. C. Spear and G. M. Hornberger, *Water Res.*, 43 (1980).
112. J. A. Vrugt, W. Bouten, H. V. Gupta and S. Sorooshian, *Water Resour. Res.*, 1312, doi:10.1029/2001WR001118 (2002).
113. K. J. Beven and A. M. Binley, *Hydrol. Process.*, 279 (1992).
114. V. Y. Ivanov, E. R. Vivoni, R. L. Bras and D. Entekhabi, *J. Hydrol.*, 80 (2004).
115. T. M. Carpenter and K. P. Georgakakos, *J. Hydrol.*, 61 (2004).
116. R. E. Smith, D. C. Goodrich, D. A. Woolhiser and C. L. Unkrich, in *Computer Models of Watershed Hydrology*, Ed. V. P. Singh (Water Resources Publications, Colorado, 1995), p. 697.
117. V. Koren, S. Reed, M. Smith, Z. Zhang and D. J. Seo, *J. Hydrol.*, 297 (2004).
118. K. J. Beven and J. E. Freer, *J. Hydrol.*, 11 (2001).
119. K. J. Beven, *J. Hydrol.*, 157 (1989).
120. K. J. Beven, *Hydrol. Process.*, 189 (2002).
121. R. B. Grayson, I. D. Moore and T. A. McMahon, *Water Resour. Res.*, 2659 (1992).
122. M. B. Smith, D. J. Seo, V. Koren, S. Reed, Z. Zhang, Q. Duan, S. Cong, F. Moreda and R. Anderson, *J. Hydrol.*, 4 (2004).

123. B. E. Vieux and F. G. Moreda, in *Calibration of Watershed Models*, Eds. Q. Duan, S. Sorooshian, H. V. Gupta, H. Rosseau and R. Turcotte (Water Science and Applications, American Geophysical Union, 2003), p. 267.
124. R. Grayson and G. Bloschl, Eds., *Spatial Patterns in Catchment Hydrology: Observations and Modelling* (Cambridge University Press, Cambridge, UK, 2000).
125. P. Pokhrel, H. V. Gupta and T. Wagener, *Water Resour. Res.*, W12419, doi:10.1029/2007WR006615 (2008).
126. J. Doherty and B. E. Skahill, *J. Hydrol.*, 564 (2006).
127. A. Tikhonov and V. Arsenin, *Solutions of Ill-posed Problems* (V. H. Winston, 1977).
128. V. Cristina, F. Barbosa and J. B. Silva, *Geophysics*, 57 (1994).
129. H. W. Engl, M. Hanke and A. Neubauer, *Regularization of Inverse Problems* (Kluwer, Dordrecht, 1996).
130. J. Doherty, *Ground Water*, 170 (2003).
131. J. Linden, B. Vinsonneau and K. J. Burnham, in *Proceedings of the 18th International Conference on Systems Engineering* (IEEE Computer Society, Las Vegas Nevada, 2005), p. 112.
132. R. Fletcher, *IMA J. Appl. Math.*, 319 (1974).
133. K. K. Yilmaz, *Ph.D. Dissertation* (Department of Hydrology and Water Resources, University of Arizona, Tucson, USA, 2007), p. 263.
134. G. Demoment, *IEEE Trans. Acoustics Speech Signal Proc.*, 2024 (1989).
135. C. L. Lawson and J. R. Hanson, *Classics in Applied Mathematics* (Society for Industrial and Applied Mathematics, Philadelphia, PA, 1995), pp. 337.
136. R. Weiss and L. Smith, *Water Resour. Res.*, 647 (1998).
137. T. H. Skaggs and Z. J. Kabala, *Water Resour. Res.*, 71 (1994).
138. E. E. van Loon and P. A. Troch, *Hydrol. Process.*, 531 (2002).
139. M. J. Tonkin and J. Doherty, *Water Resour. Res.*, W10412, doi:10.1029/2005WR003995 (2005).
140. E. Keating, J. Doherty, J. A. Vrugt and Q. Kang, *Water Resour. Res.*, submitted (2010).
141. J. A. Vrugt, J. Doherty, C. J. F. ter Braak and M. P. Clark, *Proc. Natl. Acad. Sci. USA*, to be submitted (2010).
142. C. Bandaragoda, D. G. Tarboton and R. Woods, *J. Hydrol.*, 178 (2004).
143. L. W. White, B. E. Vieux and D. Armand, *Adv. Water Resour.*, 337 (2003).
144. B. E. Vieux, *Distributed Hydrologic Modeling using GIS* (Water Science and Technology Library, Kluwer Academic Publishers, The Netherlands, 2001), pp. 293.
145. B. A. Tolson and C. A. Shoemaker, *Water Resour. Res.*, W01413, doi:10.1029/2005WR004723 (2007).
146. R. G. Regis and C. A. Shoemaker, *J. Global Optim.*, 113 (2007).
147. R. G. Regis and C. A. Shoemaker, *J. Global Optim.*. 153 (2005).
148. P. F. Dornes, J. W. Pomeroy, A. Pietroniro and D. L. Verseghy, *J. Hydrometeor.*, 789 (2008).

149. B. A. Tolson and C. A. Shoemaker, *Water Resour. Res.*, W04411, doi:10.1029/2007WR005869 (2008).
150. A. Behrangi, B. Khakbaz, J. A. Vrugt, Q. Duan and S. Sorooshian, *Water Resour. Res.*, W12603, doi:10.1029/2007WR006429 (2008).
151. B. A. Tolson and C. A. Shoemaker, *Water Resour. Res.*, W12604, doi:10.1029/2008WR006862 (2008).
152. G. Kuczera and E. Parent, *J. Hydrol.*, 69 (1998).
153. B. C. Bates and E. P. Campbell, *Water Resour. Res.*, 937 (2001).
154. K. Engeland and L. Gottschalk, *Hydrol. Earth Syst. Sci.*, 883 (2002).
155. J. A. Vrugt, H. V. Gupta, W. Bouten and S. Sorooshian, *Water Resour. Res.*, 1201, doi:10.1029/2002WR001642 (2003).
156. L. Marshall, D. Nott and A. Sharma, *Water Resour. Res.*, W02501, doi:10.1029/2003WR002378 (2004).
157. Y. Liu and H. V. Gupta, *Water Resour. Res.*, W07401, doi:10.1029/2006WR005756 (2007).
158. W. K. Hastings, *Biometrika*, 97 (1970).
159. A. Gelman and D. B. Rubin, *Stat. Sci.*, 457 (1992).
160. J. A. Vrugt, C. J. F. ter Braak, M. P. Clark, J. M. Hyman and B.A. Robinson, *Water Resour. Res.*, W00B09, doi:10.1029/2007WR006720 (2008).
161. C. J. F. ter Braak, *Stat. Comput.*, 239 (2006).
162. G. O. Roberts and J. S. Rosenthal, *Stat. Sci.*, 351 (2001).
163. G. Evensen, *J. Geophys. Res.*, 10143 (1994).
164. J. A. Vrugt, C. G. H. Diks, H. V. Gupta, W. Bouten and J. M. Verstraten, *Water Resour. Res.*, W01017, doi:10.1029/204WR003059 (2005).
165. G. Evensen, *J. Geophys. Res.*, 17905 (1992).
166. J. R. Valstar, D. B. McLaughlin, C. B. M. T. Stroet and F. C. van Geer, *Water Resour. Res.*, W05116 (2004).
167. N. Bulygina and H. V. Gupta, *Water Resour. Res.*, W00B13, doi:10.1029/2007WR006749 (2009).
168. J. A. Hoeting, D. Madigan, A. E. Raftery and C. T. Volinsky, *Stat. Sci.*, 382 (1999).
169. A. E. Raftery, T. Gneiting, F. Balabdaoui and M. Polakowski, *Mon. Weather Rev.*, 1155 (2005).
170. J. A. Vrugt, M. P. Clark, C. G. H. Diks, Q. Duan and B. A. Robinson, *Geophys. Res. Lett.*, L19817, doi:10.1029/2006GL027126 (2006).
171. J. A. Vrugt and B. A. Robinson, *Water Resour. Res.*, W01411, doi:10.1029/2005WR004838 (2007).
172. S. P. Neuman, *Stoch. Environ. Res. Risk Assess.*, 291 (2003).
173. M. Ye, S. P. Neuman and P. D. Meyer, *Water Resour. Res.*, W05113, doi:10.1029/2003WR002557 (2004).
174. M. Ye, S. P. Neuman, P. D. Meyer and K. Pohlmann, *Water Resour. Res.*, W12429, doi:10.1029/2005WR004260 (2005).

CHAPTER 4

SCALING AND FRACTALS IN HYDROLOGY

Daniele Veneziano

*Department of Civil and Environmental Engineering,
Massachusetts Institute of Technology,
77 Massachusetts Avenue, Cambridge, MA 02139, USA
E-mail: venezian@mit.edu*

Andreas Langousis

*Department of Civil and Environmental Engineering,
Massachusetts Institute of Technology,
77 Massachusetts Avenue, Cambridge, MA 02139, USA
E-mail: andlag@alum.mit.edu*

Virtually, all areas of hydrology have been deeply influenced by the concepts of fractality and scale invariance. The roots of scale invariance in hydrology can be traced to the pioneering works, as early as in the 1940s, of Horton, Shreve, Hack, and Hurst on the topology and metric properties of river networks and on river flow. This early work uncovered symmetries and laws that only later were recognized as manifestations of scale invariance. Le Cam, who in the early 1960s pioneered the development of multiscale pulse models of rainfall, provided renewed impetus to the use of scale-based models. Fractal approaches in hydrology have become more rigorous and widespread since Mandelbrot systematized fractal geometry and multifractal processes were discovered. This chapter reviews the main concepts of fractality and scale invariance, the construction of scale-invariant processes, their properties, and the inference of scale invariance from data. We highlight the recent developments in four areas of hydrology: rainfall, fluvial erosion topography, river floods, and flow through porous media.

4.1. Introduction

This chapter deals with fractal methods and their importance in hydrology. By fractal methods, we mean models, analysis, and inference procedures that emphasize scale invariance, which is the property that an object reproduces itself under some scale-change transformation.

Like other forms of invariance (invariance under translation called stationarity; invariance under translation, rotation, and reflection called isotropy), invariance under a change of scale is a fundamental property that entities like sets, functions, or measures may have. When they do, this form of invariance often sheds light on the genesis of the phenomenon, reduces the complexity of models and their inference, and allows one to devise special methods to upscale/downscale, characterize extremes, make predictions, etc. Our interest in scale invariance stems from this deep understanding and richness of applications.

In the physical world, scale invariance rarely manifests itself as a deterministic property, whereby an object is made of exact scaled replicas of itself. Rather, nature displays variability that is generally best described through stochastic models. For this reason, here we deal exclusively with random objects that are scale-invariant in the sense that their probability laws (their ensemble statistical properties) do not change under certain scale-change transformations. Specific realizations of the object are not expected to display deterministic scale invariance.

Even in the limited context of hydrology, the use of scale invariance has grown very significantly over the years, making it difficult to provide a comprehensive coverage in book-chapter form. Entire books (e.g. Rodriguez-Iturbe and Rinaldo[1]) exist on just specific application areas. This rapid growth, together with the fact that we, the authors, are more intimately familiar with certain areas, necessarily results in a personal topic selection and presentation style. The material itself is often drawn from our previous publications.

This review emphasizes stochastic (random process) approaches to scale invariance. For an alternative approach through determinism and chaos theory, see Barnsley,[2] and for applications of scale-invariant chaotic models in hydrology see, for example, Puente and Obregón[3] and Puente and Sivakumar.[4]

Sections 4.2, 4.3, and 4.4 contain introductory material. Section 4.2 deals with fractal and scale-invariant sets, their fractal dimension, and generation. Section 4.3 presents various scale-invariance conditions for random processes (ordinary functions and generalized functions or measures), drawing a distinction between self-similarity and multifractality. Section 4.4 is devoted to the characterization and generation of scale-invariant processes, including their relationship with stationary processes and their generation as renormalization limits of other processes or as weighted sums of processes at different scales. Deviations from exact scaling are briefly covered at the end of Sec. 4.4. Section 4.5 deals with basic properties of scale-invariant processes. Some properties, such as moment scaling, hold irrespective of whether the process is stationary or not, whereas other properties (e.g. marginal distributions, extremes) are more specific to stationary random measures; this explains the emphasis on stationary measures in Sec. 4.5. Section 4.6 covers two frequently needed operations with stationary multifractal measures: forecasting using observations from the past and downscaling of coarse measurements. Section 4.7 is devoted to the estimation of scaling properties from data. Rather than covering data analysis in a comprehensive way (again, a downing task given the size limitations of the chapter), we choose to discuss four popular inference techniques, suggesting improvements and corrections. The fact that some popular procedures are inefficient or altogether incorrect is a significant problem, as many published results on scaling are consequently inaccurate or suspicious. Section 4.8 gives a brief overview of the use of scale invariance in four areas of hydrology, namely rainfall, river networks and fluvial erosion topography, river flow, and flow through saturated porous media. Concluding remarks are made in Sec. 4.9.

4.2. Fractal and Scale-invariant Sets

The objects studied by classical geometry have integer dimensions. For example, straight lines have dimension 1, planar figures, such as squares and triangles, have dimension 2, and cubes and other polyhedra in three-dimensional space have dimension 3. All these objects are locally smooth. By contrast, fractal geometry deals with sets that are highly irregular and have non-integer (fractal) dimension,[5] in the sense explained below. One example is one-dimensional Brownian motion, which gives the position of a particle that starts at the origin and, during constant increments of time, displaces by independent and identically distributed Gaussian amounts. Another example is the Sierpinski triangle, which is obtained by first dividing an equilateral triangle of side length l into four triangles of side length $l/2$ and removing the triangle at the center, and then repeating the same operation of sub-division and elimination on each of the remaining triangles of side length $l/2$, then the nine remaining triangles of side length $l/4$, and so on.

In order to be fractal, an object must not just be irregular but the irregularities must also, in turn, depend in a regular way on scale (such that the number of tiles needed to cover the object must be a power function of the tile size; see Sec. 4.2.1). This is why fractality often occurs concurrently with scale invariance, which, loosely speaking, is the property that, under transformations that involve a change of scale, any part of an object looks like the whole. For example, Brownian motion and the Sierpinski triangle are both fractal and scale-invariant objects. In spite of being often used interchangeably (including in the title of this chapter), fractality and scale invariance are not equivalent concepts. For example, a straight line on the (x, y) plane that passes through the origin reproduces itself under isotropic scaling and, therefore, is scale-invariant, but has dimension 1 and, therefore, is non-fractal. Our interest is in scale invariance, but since fractality is a frequent property of scale-invariant objects, we start with a brief review of fractal sets and their fractal dimensions.

4.2.1. *Fractal Sets and their Fractal Dimensions*

There are many definitions of fractal dimension. The most general and mathematically satisfactory one is the Hausdorff dimension, D_H. Its definition is rather technical[6] and is given below in a simplified form, which is sufficient in many cases, including scale-invariant objects.

Consider a set S of R^d and let $s_\delta \subset R^d$ be a measurable set, for example a segment in R^1; a square, a rectangle or a disc in R^2; and so on. The set s_δ has diameter (maximum linear size) δ and area A_δ. Suppose that N_δ translated/rotated versions of s_δ are needed to cover S. Then the total area of the covering set is $N_\delta A_\delta$. If S has topological dimension less than d, for example a line in R^2 or R^3, then typically $\lim_{\delta \to 0}(N_\delta A_\delta) = 0$ (this is however not always the case; for example it is not for "space-filling" curves). Now define M_D as the small-diameter limit of the modified area $(N_\delta A_\delta)\delta^{D-d}$, i.e.

$$M_D = \lim_{\delta \to 0}(N_\delta A_\delta)\frac{\delta^D}{\delta^d} = \lim_{\delta \to 0} N_\delta \left(\frac{A_\delta}{\delta^d}\right)\delta^D \qquad (4.1)$$

For example, if s_δ is a d-dimensional cube, then $A_\delta = \delta^d$ and $M_D = \lim_{\delta \to 0}(N_\delta \delta^D)$. There is a value D_H of D such that

$$M_D = \begin{cases} 0, & \text{if } D > D_H \\ \infty, & \text{if } D < D_H \end{cases} \qquad (4.2)$$

Such D_H is called the Hausdorff dimension of S.

Notice that D_H is a small-scale property and may be different in different local neighborhoods of an object. Therefore, one should not estimate D_H from large-scale properties of S, unless some form of homogeneity and scale invariance applies, whereby D_H is known to be the same at all locations of the object and over the range of scales considered.

To illustrate how one can apply the above definition of fractal dimension to scale-invariant sets S, consider the simple case when S is the union of m non-overlapping subsets $S_1,...,S_m$, each obtained by translation and rigid rotation of S/r, with $r > 1$. This implies that each S_i is, in turn, the union of rescaled sets $S_{i_1},...,S_{i_m}$ obtained by translation and rigid rotation of S/r^2, and so on. Calculation of the Hausdorff dimension of such sets is simple. To cover S, one may use sets s_δ of diameter $\delta = 1/r, 1/r^2,...,1/r^q,...$ and area $A_\delta = c\delta^d$, where c is a finite positive constant. Then for $\delta = 1/r^q$, it is $N_\delta = m^q$, $A_\delta/\delta^d = c$ and $\delta^D = r^{-qD}$, giving

$$M_D = c \lim_{q \to \infty}\left(m^q\, r^{-qD}\right) \qquad (4.3)$$

Since M_D in Eq. (4.3) is finite non-zero for

$$D = D_H = \log_r(m) = \frac{\log(m)}{\log(r)} \quad (4.4)$$

one concludes that D_H in Eq. (4.4) is the Hausdorff dimension of S.

When m isotropically contracted "boxes" of diameter δ (with linear resolution $r = 1/\delta$) are used to cover a set S, the limit $D_B = \lim_{r \to \infty} \log(m)/\log(r)$ is called the box dimension of S. As Eq. (4.4) shows, this is an appropriate way to determine D_H for self-similar sets like those of interest in this chapter.

The previous definition of D_H is extended to random sets by replacing the coverage area $N_\delta A_\delta$ with its expected value $E[N_\delta A_\delta]$. If the covering sets s_δ are deterministic, then one needs only to replace the number of tiles N_δ in the previous analysis with the expected number of tiles, $E[N_\delta]$.

4.2.2. *Scale-invariant Sets and their Generation*

As noted already, the intuitive notion of a scale-invariant object is that any part of the object is similar to the whole. However, what does "similar" mean, and under what scaling operation is the object invariant? For example, for objects embedded in spaces of dimension $d > 1$, does one allow translation, rotation, and specular reflection in addition to scaling? Is contraction or dilation the same (isotropic) or different (anisotropic) in different directions? Is the part equivalent to the whole in the sense of deterministic or stochastic identity, the latter referring to sameness of the probability distributions? An additional consideration is whether, in addition to scale invariance, the set possesses other forms of invariance, for example statistical homogeneity: a homogeneous set (say the pattern of fractures on a rock outcrop) looks statistically the same to an observer located anywhere on the outcrop and may be invariant under contraction and magnification around any given point, whereas a non-homogeneous scale-invariant set (such as a botanical tree) is generally invariant under contraction or magnification around a limited set of points. In some cases, interest is in the self-similarity (*ss*) of only certain features of a set. For example, in the case of drainage networks, there has been much interest in the topology of the branching pattern, at the exclusion of metric considerations.

The variety of possible scale-invariance conditions is, therefore, very large, and in any application one should be clear as to the specific condition assumed or detected. In Secs. 4.2.2 and 4.2.3, we focus on functions and measures, and for them we delve in detail on the different possible notions of scale invariance. In some cases (e.g. curves, surfaces, branching structures, sets of points in R^d), a measure $\mu(S)$ can be associated with a set (for example, μ might count the number of points, the total length of lines or the total surface area in S). Then, one may characterize the scale invariance of the set through the scale invariance of μ. For example, this is what is done when the scale invariance of a fluvial drainage network is studied through the drainage density (see Sec. 4.8.2).

Hutchinson[7] was the first to propose a precise notion of self-similarity for deterministic sets. According to Hutchinson's definition, a set S is self-similar if there exists a finite set $\Phi = (F_1, ..., F_n)$ of contraction similarity maps such that $F_i(S) \cap F_j(S)$ be "almost empty" and $S = \Phi(S) = \bigcup_{i=1}^{n} F_i(S)$. Recall that F is a contraction map if it reduces the distance between any two points, i.e. if $d(F(\underline{x}), F(\underline{y})) < d(\underline{x}, \underline{y})$ for all $\underline{x} \neq \underline{y}$. If the maps F_i contract S isotropically by a factor $r > 1$, then, as we have already seen, S is fractal with fractal dimension $D_H = \ln(n)/\ln(r)$. Hutchinson[7] also proved that the set S is uniquely defined by the set of transformations Φ and is the attractor of the sequence of sets obtained by recursively applying Φ as $S_j = \Phi(S_{j-1})$, starting from any closed bounded set S_0.

The work of Hutchinson[7] on self-similarity is limited to deterministic sets. Extensions to random sets have been proposed independently by Falconer,[8] Mauldin and Williams,[9] and Graf,[10] and further extended by Zahle.[11] In essence, for random sets, self-similarity (*ss*) requires invariance of the probability distributions under appropriate scaling transformations. However, a set may satisfy broader or narrower *ss* conditions. A very mild condition is when contraction or dilation of the set is with respect to the origin. To narrow down the class of *ss* sets, one might further impose stationarity, meaning that the *ss* property should hold relative to all points of the embedding space, not just the origin.

However, this condition is so restrictive that any such set must be either empty or equal to the entire space. A third possibility is to require scale invariance under contraction and dilation relative to any point of the random set. This intermediate condition requires that the conditional distribution of the set with the origin at any of its points (the so-called "Palm distribution") satisfies certain scale invariance conditions.[11]

There are several methods to generate self-similar sets. Most use recursive procedures that at each step add further detail to the set. How this is done depends on the type of set and the method used. For example, one may generate self-similar curves Γ in R^d using Hutchinson's approach, starting from a piece-wise linear curve Γ_0 that connects $n + 1$ given points $A_1, A_2, ..., A_{n+1}$. Then, one recursively applies a set of contractive similarity transformations $\Phi = (F_1, ..., F_n)$, where F_i maps the segment $A_1 A_{n+1}$ into $A_i A_{i+1}$, to obtain increasingly detailed representations $\Gamma_j = \bigcup_{i=1}^{n} F_i(\Gamma_{j-1})$ of $\Gamma = \lim_{j \to \infty} \Gamma_j$.[2,12] Another recursive construction is exemplified by the random midpoint displacement method to generate Brownian motion[13] and fractional Brownian motion.[14] Similar replacement procedures can be used to construct self-similar rooted trees.[15]

A different construction is via the so-called chaos game.[2] One specifies certain rules to recursively move a point in R^d from one location to the next and records the location of each generated point by adding it to the set. Under certain conditions, the generated set has an attractor (the invariant set under the given transformation rules) that is fractal and scale-invariant. A classic example is the construction of the Sierpinski triangle as a chaos-game attractor.[16]

The above are examples of mathematical constructions, which may be useful to generate synthetic scale-invariant sets and elucidate possible generic mechanisms. However, in any given application, it is often more enlightening if one can formulate physically-based algorithms. For example, in the case of fractures in rock, joints are generally formed sequentially in time, so that the joints at level j depend on the geometry of the pre-existing joints up to level $j - 1$ in addition to physical parameters (the local strength of the rock, temperature, stress concentrations, and other fracture initiation factors[17]). Similarly, a river network may initiate at the outlet and then bifurcate and propagate upstream depending on the local erodibility of the soil, the local flow, and the local slope of the terrain.[1,18] In some cases, it is possible to establish that the set-generating rule is scale-invariant and, therefore, admits solutions with similar renormalization properties (however, whether those solutions are actually realized often depends on the initial and boundary conditions; see Dubrulle[19] and Sec. 4.8.2 below). In other cases, it may be difficult to justify self-similarity from the mechanism that generates the set. For example, it has long been unclear in which respects, if any, diffusion limited aggregation (DLA) clusters are fractal and self-similar.[20]

The reason why this review gives only fleeting attention to sets is that they are less pervasive in hydrologic applications than functions and measures. We shall return to sets in Sec. 4.8.2, when we discuss river networks.

4.3. Scale Invariance for Random Processes

This section and the one that follows focus more specifically on random processes (functions and measures). Like for sets, an important first step is to understand what forms scale invariance may take, i.e. under what scale-change operations the probability laws of a random process may remain the same. Many terms are used in the literature to describe such scale-invariance conditions: self-similarity, self-affinity, mono-fractality, multifractality, multi-affinity, dynamic scaling, generalized scale invariance, etc. Some of these terms are synonymous, whereas others refer to different conditions. This may be confusing. Here, we distinguish only between self-similarity (ss) and multifractality (mf), and link this distinction to whether the scale-change transformation under which the function is statistically invariant is deterministic (ss) or stochastic (mf). To reflect this notion of multifractality, the first author[21] proposed to change "multifractality" to "stochastic-self-similarity (sss)", but the term multifractality is so engrained that we shall retain it throughout the chapter.

Section 4.3.1 defines self-similarity and Sec. 4.3.2 introduces multifractality for random functions on the line as a generalization of self-similarity. Section 4.3.3 extends multifractality to vector-valued processes and random fields.

4.3.1. *Self-similar Processes*

Self-similar (*ss*) processes are random functions that are invariant under simultaneous scaling of the support (time, space) and the amplitude. The first paper to give a rigorous general treatment of *ss* processes is Lamperti,[22] where fundamental linkages of self-similarity to stationarity and limit properties of random processes are proved. Below we give the definition of self-similarity and mention some of its variants.

A random process $X(t)$, $t > 0$, is said to be self-similar if, for any $r > 0$ and some $a_r > 0$, its statistical properties remain unchanged under (positive affine) transformations of the type $\{rt \to t, a_r X \to X\}$. This means that

$$X(t) \stackrel{d}{=} a_r X(rt) \qquad (4.5)$$

where $\stackrel{d}{=}$ denotes equality in distribution of the right and left hand sides, viewed as random functions of t. If Eq. (4.5) is applied twice sequentially, first with $r = r_1$ and then with $r = r_2$, the result must be the same as applying Eq. (4.5) once with $r = r_1 r_2$. This gives $a_{r_1 r_2} = a_{r_1} a_{r_2}$ for any $r_1, r_2 > 0$. Since, in addition, it must be $a_1 = 1$, the only admissible amplitude scaling functions a_r have the form $a_r = r^{-H}$ for some real H. Hence, a more explicit definition of self-similarity is the property that, for some real H and any $r > 0$,

$$X(t) \stackrel{d}{=} r^{-H} X(rt), \quad t > 0 \qquad (4.6)$$

The exponent H is called the index of self-similarity and $X(t)$ in Eq. (4.6) is said to be H-ss. An H-ss process with $H \neq 1$ is sometimes said to be self-affine rather than self-similar. This is to stress the fact that, for $H \neq 1$, the right-hand side of Eq. (4.6) corresponds to an affine transformation of the graph of $X(t)$. Here, we use the term self-similarity as synonymous of self-affinity.

The classical definition of self-similarity, given in Eq. (4.6), is for processes on the positive real line. One may extend this definition to processes on the entire real line (just let t in Eq. (4.6) range from $-\infty$ to ∞), to processes in spaces of dimension $d > 1$, and to generalized processes (random measures). One may also weaken Eq. (4.6) by applying it to the increments of the process rather than the process itself and modify the *ss* condition on the increments to apply "locally." We briefly mention below a few of these extensions and weaker variants of Eq. (4.6). Further extensions to random measures, vector processes, and random fields in *d*-dimesional space are made in Secs. 4.3.2 and 4.3.3 in the broader context of multifractality. Those concepts apply also to *ss* processes.

(i) *ss of the increments (ssi)*. Under Eq. (4.6), the increments of $X(t)$ satisfy

$$[X(t_1) - X(t_2)] \stackrel{d}{=} r^{-H} [X(rt_1) - X(rt_2)] \qquad (4.7)$$

for any $r > 0$ and $t_1, t_2 > 0$. However, Eq. (4.7) does not imply Eq. (4.6). Therefore, self-similarity of the increments (*ssi*) is a weaker condition than self-similarity of the process (*ss*). For example, *ssi* is preserved under addition of a constant, $X \to X + a$, whereas the *ss* property is not;

(ii) *ss with stationary increments (sssi)*. Except for trivial cases, such as $X(t)$ = constant, which is *H-ss* with $H = 0$, *ss* processes are nonstationary. For example, under Eq. (4.6), it must be that $X(t)$ vanishes at 0 if $H > 0$ and vanishes at infinity if $H < 0$. However, *H-ss* processes may have stationary increments. We refer to this important class of processes as *H-sssi*. For any positive t_1, t_2, τ and r, an *H-sssi* process satisfies

$$[X(t_1 + \tau) - X(t_1)] \stackrel{d}{=} r^{-H} [X(t_2 + r\tau) - X(t_2)] \qquad (4.8)$$

In particular, setting $t_1 = t_2 = t$ gives

$$[X(t + \tau) - X(t)] \stackrel{d}{=} r^{-H} [X(t + r\tau) - X(t)] \qquad (4.9)$$

where the increments are taken from the same arbitrary point t. Therefore, an *H-sssi* process satisfies Eq. (4.6) after translation of the origin to any point of the process. Important construction methods for *H-sssi* processes are reviewed in Sec. 4.4.4 and basic properties are given in Sec. 4.5.1;

(iii) *local ss of the increments (lssi)*. One may relax the condition in Eq. (4.9) by applying it only to small (technically, infinitesimal) increments τ. We refer to this weakened condition as local self-similarity of the increments (*lssi*). It is interesting that a process may be both *H-ss* and H_L-*lssi* with $H \neq H_L$. For example, $X(t) = Ct^H$ with C any random variable is *H-ss* as well as H_L-*lssi* with $H_L = 1$.

4.3.2. *Scalar Multifractal Processes*

Multifractality is the most general form of scale invariance for a random process. This property has been variously defined, most often with reference to scaling of the moments or to the fractal dimension of the support of singularities and local Holder exponents of different orders (see Sec. 4.5.4). Here, we prefer a different, and arguably more fundamental, definition based on scale invariance. As we shall see in Secs. 4.5.2 and 4.5.4, the moment-scaling property follows from this scale-invariance condition and the fractal dimension of singularities of a given order is replaced by the more general notion of scaling of certain level-exceedance probabilities. This generalization is important because for many multifractal processes the local singularity order is not defined.

Certain classes of multifractal processes exist only as random measures $X(dt)$, or equivalently as generalized processes $X(h) = \int_{-\infty}^{\infty} h(t)X(dt)$, where $h(t)$ is a test function.[23] For this reason, the definition of multifractality is given first for such generalized processes. Ordinary processes $X(t)$ are included as they can be viewed as measure densities (as point values of generalized processes), such that $X(dt) = X(t)dt$ and $X(h) = \int_{-\infty}^{\infty} h(t)X(t)dt$, but for clarity we explicitly state the multifractal scale-invariance condition also for them. Self-similarity is a special case of multifractality and is included as well.

Let $X(h)$ be a generalized random process (a positive or signed random measure) on the real line, defined for test functions h in some class Θ that is closed under scaling, meaning that if $h(t)$ belongs to Θ, then also $h_r(t) = rh(rt)$, $0 < r < \infty$, belongs to Θ. Notice that $r > 1$

produces contraction of the time axis and amplification of $h(t)$, whereas $0 < r < 1$ produces dilation of the time axis and contraction of the amplitude. In either case, the integrals of h and $|h|$ remain unchanged.

We say that $X(h)$ is multifractal under contraction in Θ if there exists a set of non-negative random variables A_r such that

$$X(h_r) \stackrel{d}{=} A_r X(h), \quad \text{for any } r \geq 1, \; h \in \Theta \qquad (4.10)$$

If Eq. (4.10) holds for any $0 < r \leq 1$, then $X(h)$ is multifractal under dilation. Equation (4.10) usually holds over a restricted range of scales, in the case of contraction for h functions with characteristic scale s smaller than some outer multifractal limit $s_{\max}$, and in the case of dilation for characteristic scales greater than some lower multifractal limit $s_{\min}$.

Some generalized random processes are defined in a broad function space Θ, but are multifractal only for h in some sub-space. Important cases are when the sub-space includes only functions h whose first $j-1$ moments $\mu_k = \int_{-\infty}^{\infty} t^k h(t) dt$, $k = 0, 1, ..., j-1$, vanish. We denote this sub-space by Θ_j. For example, wavelets of order j have vanishing moments of order up to $j-1$. If $X(h)$ satisfies Eq. (4.10) for $h \in \Theta_j$, then we say that $X(h)$ has multifractal increments of order j. The case $j = 1$ is analogous to a process not being ss but having ss increments (see Sec. 4.3.1). When we just say that $X(h)$ is multifractal without further qualification, we mean that scale invariance includes test functions h whose zero-order moment $\mu_0 = \int_{-\infty}^{\infty} h(t) dt$ is non-zero and, therefore, belongs to Θ_0.

When $X(h)$ has point values $X(t)$, multifractality can be stated in terms of $X(t)$ or its increments $\Delta^j X(t, \tau)$ of various order j. To unify notation, we set $\Delta^0 X(t, \tau) = X(t)$ and recursively define the higher-order increments $\Delta^j X(t, \tau)$ as

$$\Delta^j X(t, \tau) = \Delta^{j-1} X(t + \tau, \tau) - \Delta^{j-1} X(t, \tau), \qquad j > 0 \qquad (4.11)$$

For example, for $j = 1$ and $j = 2$, Eq. (4.11) gives

$$\begin{aligned} \Delta^1 X(t, \tau) &= X(t + \tau) - X(t) \\ \Delta^2 X(t, \tau) &= X(t + 2\tau) - 2X(t + \tau) + X(t) \end{aligned} \qquad (4.12)$$

If Eq. (4.10) holds in Θ_j and $X(t)$ is not degenerate, the jth order increments $\Delta^j X(t, \tau)$ scale as

$$\Delta^j X(\frac{t}{r}, \frac{\tau}{r}) = A_r \Delta^j X(t, \tau) \tag{4.13}$$

For example, for $j = 0$ and $j = 1$, Eq. (4.13) gives

$$X(\frac{t}{r}) = A_r X(t), \qquad\qquad j = 0$$
$$X(\frac{t+\tau}{r}) - X(\frac{t}{r}) = A_r [X(t+\tau) - X(t)], \qquad j = 1 \tag{4.14}$$

Several important properties follow from the iterated application of Eq. (4.10) or Eq. (4.13) n times, with scaling factors all equal to $r^{1/n}$:

(i) For any n, the random factor A_r satisfies $A_r \stackrel{d}{=} A_{r^{1/n},1} \cdot A_{r^{1/n},2} \cdots A_{r^{1/n},n}$, where the $A_{r^{1/n},i}$ are independent copies of a random variable $A_{r^{1/n}}$. Therefore, $\log(A_r)$ must have infinitely divisible (*id*) distribution (on *id* distributions and their characterization, see Feller[24]). The *id* class includes the normal, stable, gamma, Poisson, compound Poisson (*cP*), and Student's *t* distributions, among others. An important characterization of *id* distributions, which can be used to produce processes with specified multifractal scaling,[25] is that any *id* distribution is either *cP* (the sum of a Poisson number N of *iid* variables) or the limit of *cP* distributions as $E[N] \to \infty$;

(ii) The distribution of A_{r_o} for any admissible scaling factor r_o completely defines the distribution of A_r for any other admissible r ($r > 1$ or $0 < r < 1$). It is often convenient to take as reference the value $r_o = e$ for multifractality under contraction and $r_o = 1/e$ for multifractality under dilation and, hence, characterize multifractality through the distribution of A_e or $A_{1/e}$; and

(iii) If it exists, the mean value of A_r must satisfy $E[A_r] = r^{-H}$ for some H.

Now we introduce some nomenclature: A process that satisfies Eq. (4.10) is said to be A_e-*mf*. The process is said to be $(A_e; j)$-*mf* if Eq. (4.10) applies only for $h \in \Theta_j$ and, hence, multifractality is limited to generalized increments of order j (or greater). We change *mf* to *mfs* if the process is stationary and to *mfsi* if the process has stationary increments. Notice the difference between multifractal processes with stationary increments (A_e-*mfsi* processes) and processes with stationary multifractal increments of order 1 [$(A_e;1)$-*mfsi* processes]. In both cases, the first-order increments are stationary and A_e-*mf*, but the process itself is A_e-*mf* only in the first case. A similar notation is used for ordinary random processes $X(t)$.

In addition to being possibly stationary or having stationary increments, generalized A_e-*mf* processes may have zero or non-zero mean and may be non-negative or signed. Important special cases are e^{-H}-*mf* and e^{-H}-*mfsi* processes (another notation for the *H-ss* and *H-sssi* processes of Sec. 4.3.1); $(A_e;1)$-*mfsi* processes (processes with stationary multifractal increments of order 1); and A_e-*mfs* processes (stationary multifractal measures). In the literature, processes with stationary multifractal increments are often called multi-affine.[26,27]

One can show that if Eq. (4.10) holds under both contraction and dilation (for all $r > 0$), then A_r must be deterministic and equal to r^{-H} for some $-\infty < H \leq \infty$. It follows that *ss* processes form a separating class between processes that are *mf* under contraction and under dilation. Contrary to multifractality, self-similarity may apply over all scales from $s_{min} = 0$ to $s_{max} = \infty$.

One can further show that multifractal processes under contraction and under dilation are in one-to-one correspondence. Specifically, $X(h(t))$ is A_e-*mf* under contraction if and only if $X(h(1/t))$ is $A_{1/e}$-*mf* under dilation.[21] Hence, any property of one class of *mf* processes has a counterpart in the other class and methods to construct processes of one type are readily adjusted to construct processes of the other type. Since, for most applications, one is interested in processes that are multifractal under contraction, when we use *mf* without specification we refer to multifractal scaling under contraction, as expressed by Eq. (4.10). Multifractality under dilation is encountered in the spectral (Fourier) representation of stationary *mf* processes (intuitively because higher frequencies in Fourier space correspond to smaller-scale fluctuations in physical space[21]) and has been proposed in certain areas in hydrology, for example to describe the scaling of maximum annual river floods with basin area.[28]

4.3.3. *Multifractal Fields and Vector-valued Processes*

For many applications, such as those involving topography, clouds, rainfall, and soil moisture, one must deal with random fields in spaces of two or higher dimensions. In other cases, for example subsurface flow, one is interested in vector-valued processes (e.g. the flow vector and the hydraulic gradient). The main novelty with multifractal fields is that the scale-change operation under which the field is invariant may involve anisotropic scaling along various coordinate directions, rotation, and even nonlinear geometrical transformations. The conceptual novelty with vector processes is that each scalar component may not be marginally multifractal, but the vector process may be jointly multifractal. Here, we extend the property of scale invariance to include these cases.

For generalized random fields $X(h) = \int_{R^d} h(\underline{t}) X(d\underline{t})$ in R^d, the simplest condition of multifractal scale invariance is given by Eq. (4.10), where the scale-change transformation $h_r(\underline{t}) = r^d h(r\underline{t})$ involves isotropic contraction (for $r > 1$) or isotropic dilation ($0 < r < 1$) of the support of h. To express multifractality of the increments, one constrains $h \in \Theta_j$, where Θ_j includes test functions for which the lowest non-zero moment $\mu_{k_1...k_d} = \int_{R^d} t_1^{k_1} \cdots t_d^{k_d} h(\underline{t}) d\underline{t}$ has order $k_1 + \cdots + k_d = j$.

More in general, a random field $X(h)$ might be invariant under a group of contractive space transformations T_r indexed by the scale-change factor r. In the case of deterministic linear transformations, T_r is a matrix $\underline{T}_r$ that varies with r as

$$\underline{T}_r = e^{\underline{Q} \ln r} = \sum_{i=0,\infty} (\ln r)^i \underline{Q}^i / i! \qquad (4.15)$$

where $\underline{Q}$ is a matrix called the generator of the group. The eigenvalues of $\underline{Q}$ must have positive real parts.[29,30] Then, generalizing Eq. (4.10), we say that a random field $X(h)$ in R^d is $(\underline{T}_e, A_e)$-*mf* (under contraction) if, for some random variables A_r and the matrices $\underline{T}_r$ in Eq. (4.15), $X(h)$ satisfies

$$X(h_{\underline{T}_r}) = A_r X(h), \qquad r \geq 1 \qquad (4.16)$$

where $h_{\underline{T}_r}(\underline{t}) = \det(\underline{T}_r) h(\underline{T}_r \underline{t})$. The field $X(h)$ is said to be $(\underline{T}_e, A_e; j)$-*mf* if Eq. (4.16) holds only for $h \in \Theta_j$. For $\underline{Q} = \underline{I}$, one has $\underline{T}_r = e^{\underline{I} \ln r} = r\underline{I}$ and $\det(\underline{T}_r) = r^d$, recovering the condition of isotropic scaling.

An important special case of anisotropic multifractality is when $\underline{Q} = diag\{q_i\}$ with $q_i > 0$ and $\max\{q_1,...,q_d\} = 1$. Then $\underline{T}_r = diag\{r^{q_i}\}$ with $r^{q_i} > 1$ and maximum diagonal value equal to r. This orthotropic form of scale invariance, in which scaling is different along different coordinate axes, has been used to represent the energy distribution in atmospheric turbulence when smaller eddies have increasing vertical elongation,[31,32] and suggested under the name of dynamic scaling to represent differences in the way rainfall scales in space and time.[33-35] If $\underline{Q}$ is symmetric but non-diagonal, orthotropic scaling applies in a rotated reference, whereas inclusion of a skew-symmetric component adds rotation.

Up to now, we have considered deterministic transformation matrices $\underline{T}_r$. An extension of Eq. (4.16) allows $\underline{T}_r$ to be a (log-infinitely divisible) random matrix, for example to include random rotation. We shall encounter an example of multifractal renormalization under random rotation in Sec. 4.8.4, when we discuss Darcy's flow through media with multifractal hydraulic conductivity. A different extension allows the transformations T_r to be nonlinear. For a general discussion, including the nonlinear and stochastic cases, see Pecknold et al.[36] and Schertzer and Lovejoy.[29]

Finally, we consider multifractality for vector-valued processes $\underline{X}(h)$.[37] The trivial case when the processes are independent reduces to Eq. (4.10) or Eq. (4.16), possibly with distinct parameters A_e and $\underline{T}_e$ for each component of $\underline{X}$. A more interesting case is when the vector process $\underline{X}$ is jointly multifractal. For example, for vector processes on the line, Eq. (4.10) generalizes to

$$\underline{X}(h_r) \stackrel{d}{=} \underline{A}_r \underline{X}(h) \qquad (4.17)$$

where $\underline{A}_r$ is a random positive-definite matrix. Except for the problem of Darcy's flow mentioned above, joint multifractality conditions of this or more general types have not been used much in hydrology. In some cases, a coordinate transformation exists that de-couples the scale-invariance condition of the components of the vector. For example, in the Darcy flow problem, one can use polar coordinates in which the mean flow direction and the flow amplitude scale in a de-coupled way, with significant reduction in the complexity of the scale-invariance condition (see Sec. 4.8.4).

In practice, scale invariance may apply only to certain characteristics of a process, over a finite range of scales, or in some other limited sense. We discuss this topic in Sec. 4.4.5 after describing ways to generate scale-invariant processes.

4.4. Generation of Scale-invariant Random Processes

Having described the ways in which a random process may be scale-invariant, we turn to the problem of providing characterizations of (general ways to generate) such processes. This theoretical problem has applications to modeling and simulation, and is important for understanding the generic mechanisms that may underlie scale-invariant phenomena. Specifically, Secs. 4.4.1, 4.4.2, and 4.4.3 describe three methods to generate self-similar and multifractal processes: via transformation of stationary processes, through limits of renormalized processes, and as sums of scaled processes. In addition to these general methods, we show how certain *H-sssi* and multifractal processes can be obtained through fractional integration and exponentiation of white noise processes called α-stable measures. We conclude the section by mentioning processes with limited scale invariance. While perhaps less theoretically important, deviations from exact scaling invariably occur in physical processes.

4.4.1. *Relationship Between Scale-invariant and Stationary Processes*

It has long been known that, for any given self-similarity index H, there is a one-to-one correspondence between stationary processes and *H-ss* processes on the line.[22] Specifically, $X(t)$, $t > 0$, is *H-ss* if and only if there exists a stationary process $Z(t)$, $-\infty < t \leq \infty$, such that

$$X(t) = t^H Z(\ln t), \qquad t > 0 \qquad (4.18)$$

Notice that the scaling exponent H is controlled by the deterministic factor t^H, whereas the random process $Z(\ln t)$ contributes variability and is responsible for the differences among distinct *H-ss* processes $X(t)$.

The smoothest-possible *H-ss* processes are obtained by setting $Z(t)$ to a deterministic or random constant C. This gives the power-law functions $X(t) = Ct^H$. It is clear from Eq. (4.18) that $X(t)$ must vanish at zero if $H > 0$ and must vanish at infinity if $H < 0$. Based on Eq. (4.18), one may verify that a process $X(t)$ is *H-ss* for some given H by checking that the transformed process $Z(t) = e^{-Ht} X(e^t)$ is stationary. To exemplify, Fig. 4.1 shows three *H-ss* processes with $H = 0.5$ obtained through Eq. (4.18). Two of the processes are simply $X(t) = \pm t^{0.5}$ and result from taking $Z(t) = \pm 1$. In the third case, $Z(t)$ is an autoregressive Gaussian process of order 1, with variance 0.25 and correlation coefficient 0.95 at time lag 0.001. Notice that, in this third case, the amplitude of $X(t)$ decreases and the central frequency of $X(t)$ increases as $t \to 0$.

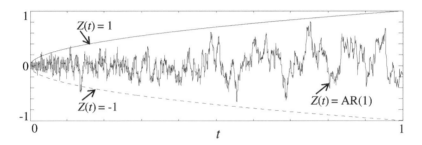

Fig. 4.1. Illustration of the generation of *H-ss* processes with $H = 0.5$ using Eq. (4.18). $Z(t)$ is either a constant or an autoregressive AR(1) Gaussian process with variance 0.25 and correlation coefficient 0.95 at time lag 0.001.

Veneziano[21] extended the stationarity characterization in Eq. (4.18) to *ss* and *mf* processes $X(\underline{t})$ and measures $X(d\underline{t})$ in *d*-dimensional space. Here, we mention only the case of ordinary processes $X(t)$ that are A_e-*mf* under contraction in the unit interval $(0,1]$. This case exemplifies the difference between *ss* and *mf* processes, with a minimum of technical complication. The characterization is similar to Eq. (4.18), except that the deterministic function t^H that confers the *H-ss* property is replaced with a random function that confers the A_e-*mf* property. Specifically, let $\Gamma(\tau), \tau \geq 0$, be the random process with independent stationary increments that satisfies $\Gamma(0) = 0$ and $\Gamma(1) \stackrel{d}{=} \ln(A_e)$. Note that for each infinitely-divisible distribution of $\ln(A_e)$, there is one and only one $\Gamma(\tau)$

process with these properties. Then $X(t)$ is A_e-mf under contraction in $(0,1]$ if and only if there exists a stationary process $Z(t)$, $t \leq 0$, such that

$$X(t) = e^{\Gamma(-\ln t)} Z(\ln t), \qquad 0 < t \leq 1 \qquad (4.19)$$

When A_e is deterministic, it must be $A_e = e^{-H}$ for some real H. Then $\Gamma(t) = -Ht$, $e^{\Gamma(-\ln t)} = t^H$, and Eq. (4.19) reproduces the characterization of H-ss processes in Eq. (4.18). In analogy with the ss case, the A_e-mf scaling property of $X(t)$ is controlled by the factor $e^{\Gamma(-\ln t)}$ in Eq. (4.19), whereas the factor $Z(\ln t)$ is responsible for the difference among identically scaling processes. Setting $Z(t)$ to a deterministic or random constant C produces $X(t) = Ce^{\Gamma(-\ln t)}$, which are the smoothest-possible A_e-mf processes on the line. For the characterization of ss and mf measures $X(dt)$ on the line, replace the stationary process $Z(t)$ in Eqs. (4.18) and (4.19) with a stationary measure $Z(dt)$.

To exemplify Eq. (4.19), we consider the same choices of $Z(t)$ as in Fig. 4.1 and take for $\Gamma(t)$ the Brownian motion with variance $Var[\Gamma(t)] = 0.25t$ and linear drift $E[\Gamma(t)] = -(H+0.125)t$. In this case, $A_e = e^{-H} Y$, where Y is a lognormal variable with $E[Y] = 1$ and $Var[\ln(Y)] = 0.25$. Figures 4.2a and 4.2b show realizations for $H = 0$ and $H = 0.5$, respectively. Also shown in the figures are the corresponding realizations of the factor $e^{\Gamma(-\ln t)}$. Notice that, in both cases, $X(t)$ and its increments are nonstationary and that for $H = 0.5$, $X(0) = 0$. For t close to 1, the processes in Figs. 4.1 and 4.2b are similar, but as t decreases, the process in Fig. 4.2b is increasingly more erratic than that in Fig. 4.1 (notice that the vertical scale is different for the two figures). The increased erraticity is due to multifractality, i.e. to the increased variability of $e^{\Gamma(-\ln t)}$ as $t \to 0$.

4.4.2. *Scale-invariant Processes as Renormalization Limits*

Lamperti[22] gave a second characterization of ss processes on the positive real line, as the renormalization limits of other random processes. In the slightly extended form of Vervaat,[38] the characterization is as follows: $X(t)$ is H-ss if and only if there exists a process $Y(t)$ and a positive function $a(u)$ such that, as $u \to \infty$,

$$a(ru)/a(u) \to r^{-H}$$
$$a(u)Y(ut) \xrightarrow{d} X(t) \tag{4.20}$$

The extension to multifractal processes under contraction replaces the positive scaling function $a(u)_d$ with a positive random process $A(u)$ such that, as $u \to \infty$, $A(ru)/A(u) \to e^{\Gamma(\ln r)}$, where $r \geq 1$ and $\Gamma(t), t \geq 0$, is a process with independent stationary increments with $\Gamma(0) = 0$, as in Eq. (4.19).[21] If $\Gamma(t)$ is deterministic, then $e^{\Gamma(\ln r)} = r^{-H}$ for some H, reproducing the first limit in Eq. (4.20).

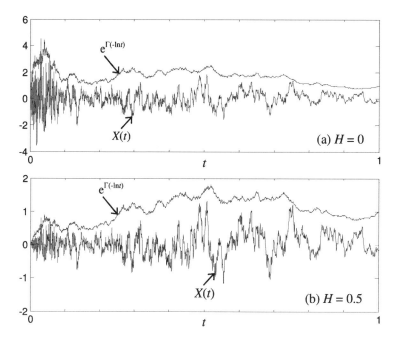

Fig. 4.2. Illustration of A_e-mf processes generated using Eq. (4.19). $Z(t)$ is the same as in Fig. 4.1 and $\Gamma(t)$ is Brownian motion with drift (see text). In all cases $A_e = e^{-H}Y$, where Y is a lognormal variable with unit mean and log-variance $Var[\ln(Y)] = 0.25$. (a) $H = 0$, (b) $H = 0.5$.

To be explicit, a generalized random process $X(h)$ is $e^{\Gamma(1)}$-mf under contraction if and only if there exists a generalized process $Y(h)$ and a process $A(t)$ such that, as $u \to \infty$,

$$A(ru)/A(u) \xrightarrow{d} e^{\Gamma(\ln r)}, \quad \text{for any } r \geq 1$$
$$A(u)Y(h_u) \xrightarrow{d} X(h) \tag{4.21}$$

where $h_u(t) = uh(ut)$. A similar characterization holds for processes that are multifractal under dilation.

4.4.3. *Scale-invariant Processes as Weighted Sums and Products*

The characterizations in Secs. 4.4.1 and 4.4.2 have mainly theoretical interest because it is rare that a scale-invariant process $X(t)$ is physically linked to a stationary process $Z(t)$ through Eq. (4.18) or Eq. (4.19), or is obtained as the renormalization limit of some physical process $Y(t)$ or $Y(h)$ through Eq. (4.20) or Eq. (4.21). Another reason why the characterizations in Eqs. (4.18) and (4.19) are of limited practical interest is that the stationary processes $Z(t)$ associated with important scale-invariant processes, such as ordinary *H-sssi* processes $X(t)$, do not belong to well-known classes.

Most scale-invariant processes are physically and numerically generated as weighted sums or products of fluctuations at different scales. This is why characterizations in terms of additive or multiplicative mechanisms would be of much practical value. Next, we describe characterizations of this type for processes and measures on the line, although the results can be extended to random fields in spaces of higher dimension. First, we give a general characterization of *ss* and *mf* processes. Then, we broaden that characterization and discuss the implications for stationary *mf* processes and processes with stationary *mf* increments.

4.4.3.1. *Self-similar Processes*

A classic way to produce *H-ss* processes is to add independent scaled copies of a given random process.[39] This weighted additive construction has the form

$$X(t) = \sum_{j=-\infty}^{\infty} r_o^{-jH} Z_j(r_o^j t) \tag{4.22}$$

where $r_o > 1$ is a constant scale-change factor, $Z_j(t)$ are independent copies of some random process $Z(t)$, such that $Z(0) = 0$ if $H > 0$ and $Z(-\infty) = 0$ if $H < 0$, and the coefficient r_o^{-jH} is a weight that varies as a power of the resolution r_o^j. If the summation in Eq. (4.22) converges in distribution, then the process $X(t)$ is H-ss under scaling of the support by factors $r = r_o^k$, where k is a positive or negative integer. The transformed process $Y(t) = C + X(t)$, where C is a deterministic or random constant, has H-ss increments (is H-ssi), and if $Z(t)$ has stationary increments, $X(t)$ is H-sssi.

An example application of Eq. (4.22) is the class of pulse processes first developed by Halford,[40] and later rediscovered and applied to rainfall by Lovejoy and Mandelbrot.[41] These processes, called Fractal Sums of Pulses (*FSP*),[41] are generated as

$$X(t) = \sum_i \tau_i^H g_i\left[(t - t_i)/\tau_i\right] \qquad (4.23)$$

where $0 < H < 1$, the $g_i(t)$ are independent pulses equal in probability to a random pulse $g(t)$, the pulse locations t_i form a stationary Poisson point process, and the pulse size parameters τ_i are independent copies of a positive random variable τ. The locations $\{t_i\}$, the pulse sizes $\{\tau_i\}$, and the pulse shapes $\{g_i(t)\}$ are mutually independent.

In order for $X(t)$ to have H-ss increments, the rate of pulses with size between τ and $\tau + d\tau$ must be proportional to $\tau^{-2} d\tau$. Halford[40] and Lovejoy and Mandelbrot[41] considered special cases of Eq. (4.23) in which g is either deterministic or satisfies $g(t) = A g_o(t)$, where A is a random variable and $g_o(t)$ is a deterministic function.

To link the *FSP* processes to Eq. (4.22), we take $Z(t)$ in Eq. (4.22) as

$$Z(t) = \sum_{i : \tau_i \in [1, r_o]} \tau_i^H g_i\left[(t - t_i)/\tau_i\right] \qquad (4.24)$$

Then

$$\sum_{i : \tau_i \in [r_o^{-j}, r_o^{-j+1}]} \tau_i^H g_i\left[(t - t_i)/\tau_i\right] \stackrel{d}{=} r_o^{-jH} Z_j(r_o^j t) \qquad (4.25)$$

which shows that $X(t)$ in Eq. (4.23) has a representation of the type in Eq. (4.22). Since $Z(t)$ in Eq. (4.24) has stationary increments, $X(t)$ is H-sssi.

One should point at an inaccuracy in the argument we have just used to link the pulse processes in Eq. (4.23) to the construction of *ss* processes in Eq. (4.22). The process $Z(t)$ in Eq. (4.24) is stationary and does not vanish at zero. Under these conditions, allowing the pulse sizes τ_i in Eq. (4.23) to range from 0 to ∞ makes $X(t)$ diverge. To prevent this from occurring, one must constrain τ to be in a finite range $[\tau_{min}, \tau_{max}]$. The modified process

$$X'(t) = \sum_{\tau_{min} < \tau_i < \tau_{max}} \tau_i^H g_i \left[(t - t_i) / \tau_i \right] \tag{4.26}$$

is stationary with increments that are approximately *H-ss* over a finite range of scales. Deviations from exact scaling of this type are often accepted in practice, in exchange for the important property of stationarity.

As a general construction of *ss* processes, Eq. (4.22) is unsatisfactory because of the discreteness of the scaling property, the assumption of independence of the processes $Z_j(t)$ when a milder condition might suffice, and the fact that Eq. (4.22) produces but does not give a complete characterization of all *ss* processes. The characterization that follows resolves all three concerns. It does so by setting $r_o = e$ (not essential), replacing the discrete scaling index j with a continuous log-resolution parameter $-\infty < \lambda < \infty$, and replacing the independent random processes $Z_j(t)$ with a random measure $Z(d\lambda, dt)$ that is stationary in λ [meaning that for any given c, $Z(c + d\lambda, dt) = Z(d\lambda, dt)$] and has point values $Z(d\lambda, t)$ in t [meaning that $Z(d\lambda, dt) = Z(d\lambda, t)dt$]. For $H > 0$, $Z(d\lambda, t)$ should vanish for $t = 0$ and for $H < 0$, $Z(d\lambda, t)$ should vanish for $t \to \infty$. Requiring stationarity in λ is analogous to allowing dependence among the $Z_j(t)$ processes in Eq. (4.22) according to the index lag $\Delta j = j_1 - j_2$. The requirement that Z has point values in t is non-essential: relaxation of that requirement leads to the characterization of the broader class of *H-ss* measures $X(dt)$. With Z as just described, a weighted-sum characterization of *H-ss* processes $X(t)$, $t > 0$, is given by

$$X(t) = \int_{-\infty}^{\infty} Z(d\lambda, e^\lambda t) e^{-\lambda H} \tag{4.27}$$

Equation (4.27) expresses $X(t)$ as a continuous weighted sum over λ of scaled random processes $Z_{d\lambda}(t) = Z(d\lambda, e^\lambda t)$, with weights $e^{-\lambda H}$. Before contracting the time support by e^λ, the processes $Z(d\lambda, t)$ for different λ are statistically identical.

Notice that $Z(d\lambda, t)$ must be stationary in λ, but no constraint is imposed on the dependence of Z on t, except for the conditions at zero and infinity and convergence of the integral in Eq. (4.27). If $Z(d\lambda, t)$ has stationary increments in t, then $X(t)$ is H-sssi. This is an appealing characterization of H-sssi processes.

Figure 4.3 exemplifies the generation of H-sssi processes through Eq. (4.27) [using the discrete form of Eq. (4.22)]. The process $Z(t)$ is taken as $Z(t) = Z'(t) - Z'(0)$, where $Z'(t)$ is a 2nd order autoregressive process with unit variance and correlation distance equal to 1. This distance corresponds to a correlation coefficient equal to $1/e$;[42] H is set to 0.2, 0.5, and 0.8 in different simulations. The resulting processes are H-sssi and, due to the choice of $Z(t)$, are very close to fractional Brownian motion.

4.4.3.2. *Multifractal Processes*

Multifractal processes have a similar weighted-sum characterization. The only differences with the *ss* case are that the deterministic weight $e^{-\lambda H}$ in Eq. (4.27) is replaced with a random weight $e^{\Gamma(\lambda)}$, where $\Gamma(\lambda)$ is the process with independent stationary increments that satisfies $\Gamma(0) = 0$ and $\Gamma(1) = \ln(A_e)$, and the integral extends over only the positive values of λ (this limitation is related to the fact that multifractality under contraction applies to t smaller than some finite upper limit t_{max}; see below). With these changes, $X(t)$ is A_e-*mf* under contraction in $0 < t < t_{max}$ if and only if there exists a random measure $Z(d\lambda, dt) = Z(d\lambda, t)dt$ stationary in λ such that

$$X(t) = \int_0^\infty Z(d\lambda, e^\lambda t) e^{\Gamma(\lambda)}, \quad 0 < t < t_{max} \qquad (4.28)$$

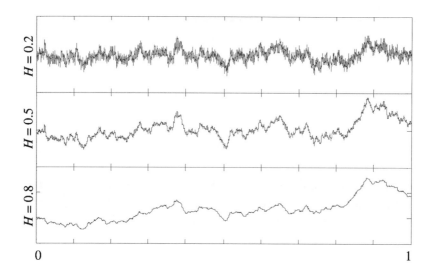

Fig. 4.3. Illustration of *H-sssi* processes as superpositions of scaled *iid* processes with stationary increments; see Eq. (4.27). $Z(t) = Z'(t) - Z'(0)$, where $Z'(t)$ is a 2nd order autoregressive process with unit variance and correlation distance equal to 1. The simulations are for $H = 0.2$, 0.5, and 0.8.

The upper scaling limit t_{max} is such that the contribution to $X(t_{max})$ from log-resolutions λ near zero is negligible relative to the contribution from some higher λ. Like in the *ss* case,

(i) Relaxation of the condition that Z has point values in t produces *mf* measures $X(dt)$;

(ii) Addition of a deterministic or random constant produces processes with A_e-*mf* increments; and

(iii) A_e-*mf* processes with stationary increments (A_e-*mfsi* processes) follow from imposing that Z has stationary increments in t.

Figure 4.4 uses the same format as Fig. 4.3 to show simulations of $X(t)$ from Eq. (4.28), discretized in scale in a way analogous to Eq. (4.22). The only difference with Fig. 4.3 is that $\Gamma(\lambda)$ is now a random process (Brownian motion with linear drift, as in Fig. 4.2b). The processes in Fig. 4.4 are multifractal with stationary increments. They have low-frequency fluctuations similar to those in Fig. 4.3, but higher small-scale variability (due to the fact that in this realization $\Gamma(\lambda)$ tends to be above 1 over a certain range of log-resolutions λ). The increase in the high-frequency content relative to the *ss* case is especially visible for $H = 0.2$. The scaling of $X(t)$ for $H = 0.5$ is the same as in Fig. 4.2b, but here $X(t)$ has stationary increments.

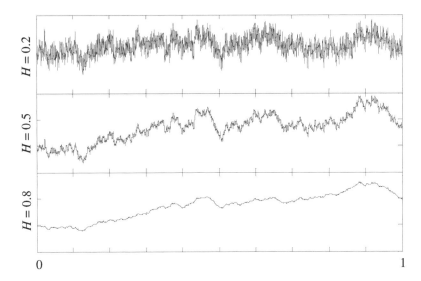

Fig. 4.4. Simulation of A_e-*mfsi* processes using Eq. (4.28), discretized in scale in analogy with Eq. (4.22). Z is the same process as in Fig. 4.3 and Γ is the same process as in Fig. 4.2b, except that here the simulations are for three different values of H and the realizations are different.

Alternative Representation of Processes with Stationary Multifractal Increments

Equation (4.28) views $X(t)$ as contributed additively by processes $Z_{d\lambda}(t) = Z(d\lambda, e^\lambda t)$, with random weights $e^{\Gamma(\lambda)}$. For certain classes of processes, notably processes with stationary multifractal increments and stationary multifractal measures, it is convenient to modify Eq. (4.28) by letting the factor $e^{\Gamma(\lambda)}$ vary with time as $e^{\Gamma([0,\lambda],t)}$, where $\Gamma([0,\lambda],t)$ is obtained as follows (readers who are familiar with multiplicative cascades will recognize $e^{\Gamma([0,\lambda],t)}$ as the bare measure density of a stationary multifractal measure at log-resolution λ; see end of this section).

Let $A(t), -\infty < t < \infty$, be a stationary infinitely divisible process, whose marginal distribution is the same as the distribution of $\ln(A_e)$. We first associate with $A(t)$ a stationary random measure $\Gamma'(d\lambda, dt) = \Gamma'(d\lambda, t)dt$ in the (λ, t) plane, as follows. Here, Γ' is independently scattered in λ, meaning that for any disjoint sets $\Lambda_1, \Lambda_2, ...$ on the λ axis the processes $\Gamma'(\Lambda_1, t), \Gamma'(\Lambda_2, t), ...$ are independent. Moreover, for any λ, Γ' satisfies

$$\Gamma'([\lambda, \lambda+1], t) \stackrel{d}{=} A(t) \qquad (4.29)$$

To obtain the point values $\Gamma(d\lambda, t)$ of the measure Γ of interest, we scale Γ' in t, as

$$\Gamma(d\lambda, t) = \Gamma'(d\lambda, e^\lambda t) \qquad (4.30)$$

In the special case when $A(t)$ is random but constant in time, $\Gamma([0, \lambda], t)$ reduces to $\Gamma(\lambda)$ in Eq. (4.28).

Using the random measures $\Gamma(d\lambda, t)$ in Eq. (4.30), one can construct processes $X(t)$ with stationary A_e-mf increments as

$$X(t) = C + \int_0^\infty Z(d\lambda, e^\lambda t)e^{\Gamma([0,\lambda], t)}, \quad 0 \le t \le t_{max} \qquad (4.31)$$

where C is an arbitrary deterministic or random constant and $Z(d\lambda, t)$ is stationary in λ and has stationary increments in t.

There is an important difference between the *mfsi* processes generated by Eq. (4.31) and those obtained from Eq. (4.28) using the same Z processes. If, in a certain realization, $\Gamma(\lambda)$ in Eq. (4.28) happens to be small (large) in a log-frequency range $[\lambda_1, \lambda_2]$, then, for all t, the resulting $X(t)$ process will be poor (rich) in the corresponding frequencies. For example, simple visual inspection reveals a dominance of certain high frequencies in the realizations of Fig. 4.4 relative to Fig. 4.3. Hence, the processes $X(t)$ in Eq. (4.28) are not ergodic. By contrast, if $\Gamma(d\lambda, t)$ is ergodic, then the processes $X(t)$ in Eq. (4.31) display variability in frequency content in different time periods and may themselves be ergodic.

The wavelet-based construction of processes with stationary A_e-mf increments (i.e. of "multi-affine" processes) suggested by Benzi et al.[27] may be regarded as a discrete version of Eq. (4.31). In the Benzi et al.[27] construction, the random process $Z(d\lambda, e^\lambda t)$ is replaced by a deterministic periodic function (obtained by concatenating in time the Haar wavelet functions at log-resolution λ) and $e^{\Gamma([0,\lambda],t)}$ is the bare measure density of a discrete cascade at the same log-resolution, whose values serve as wavelet coefficients.

For $A_e = e^{-H}$ deterministic, Eq. (4.31) becomes

$$X(t) = C + \int_0^\infty Z(d\lambda, e^\lambda t) e^{-\lambda H} \qquad (4.32)$$

Equation (4.32) is the same as Eq. (4.27) except for the lower limit of integration, the constant C, and the restriction that Z has stationary increments in t. Equation (4.32) characterizes processes with stationary H-ss increments.

Stationary Multifractal Measures and Discrete Multiplicative Cascades

Consider now the special case of Eq. (4.31) when $E[A_e] = e^{-H} = 1$ (hence $H = 0$). Then $E[e^{\Gamma(d\lambda,t)}] = 1$ and, if in particular one takes $Z(d\lambda, e^\lambda t) = C(e^{\Gamma(d\lambda,t)} - 1)$, Eq. (4.31) becomes

$$X(t) = C[1 + \int_0^\infty (e^{\Gamma(d\lambda,t)} - 1) e^{\Gamma([0,\lambda],t)}]$$

$$= C[1 + \int_0^\infty (e^{\Gamma([0,\lambda+d\lambda],t)} - e^{\Gamma([0,\lambda],t)})] \qquad (4.33)$$

$$= C e^{\int_0^\infty \Gamma(d\lambda,t)}, \qquad 0 \le t \le t_{max}$$

i.e. $X(t)$ is the product of independent non-negative scaled *iid* processes $e^{\Gamma(d\lambda,t)}$ with mean value 1. This special case of Eq. (4.31) produces stationary A_e-mf measures. Technically, the integral in Eq. (4.33) does not converge, but X often exists as a stationary generalized process $X(h)$, obtained as

$$X(h) = C \lim_{\lambda_{max} \to \infty} \left(\int_{-\infty}^{\infty} h(t) e^{\Gamma([0,\lambda_{max}],t)} \, dt \right) \quad (4.34)$$

As an example, one may take $\Gamma(d\lambda,t)$ to be a stationary normal process with mean value $-1/2\,\sigma^2 d\lambda$ and variance $\sigma^2 d\lambda$, where σ^2 is the variance of $\ln(A_e)$. This generates the class of stationary lognormal multifractal measures.

Discrete multiplicative cascades may be viewed as approximations to the stationary multifractal measures generated through Eq. (4.33). The importance and usefulness of discrete cascades stem from their simplicity, the fact that the so-called "bare" and "dressed" measure densities (see later) are unambiguously defined, and the possibility to derive important results on existence, moments, distributions, and extremes (see Secs. 4.5.3 and 4.5.4). The same results are generally assumed to hold for continuous stationary multifractal measures, although this is true in approximation.

The construction of discrete cascades follows from a discretization of the log-resolution λ in Eq. (4.33) (whereby $d\lambda$ is replaced by a finite log-resolution increment $\Delta\lambda$), and the use of particular random functions $\Gamma(\Delta\lambda,t) = \Gamma'(\Delta\lambda, e^\lambda t)$. However, it is simpler to explain the construction in terms of products of *iid* variables, as follows.

The canonical construction considers a measure with mean value 1 inside the unit d-dimensional cube S_d. One starts at level 0 with a single d-dimensional cubic tile $\Omega_{0_1} = S_d$ and a uniform unit measure density inside Ω_{0_1}. At subsequent levels $n = 1, 2, \ldots$, each tile at the previous level $n-1$ is partitioned into m^d cubic tiles, where $m > 1$ is the integer linear multiplicity of the cascade. The measure density inside each cascade tile Ω_{n_i} ($i=1,\ldots,m^{nd}$) is obtained by multiplying the measure density in the parent tile at level $n-1$ by a random variable W_{n_i} with unit mean value. Although, in general, the random variables W_{n_i} could be dependent both within and among cascade levels, the standard cascade model assumes that all such variables are independent copies of a non-negative random variable W, called the generator of the cascade.

Linkage of this construction to Eqs. (4.31) and (4.33) is obtained by taking $\Delta\lambda = \ln(m)$ and $\Gamma'([0,\Delta\lambda],t)$ as a process with value $\ln(W_i)$ for $i-1 < t \leq i$, where the variables W_i are independent copies of W.

However, in the discrete cascade construction one does not require W to be infinitely divisible (*id*). The *id* requirement originates from densification of the cascade construction in scale, as $\Delta\lambda \to d\lambda$.

A fundamental quantity in a discrete cascade is $\varepsilon_{m,n}$, the average measure density inside a generic tile Ω_{n_i} at level n. Note that the distribution of $\varepsilon_{m,n}$ does not depend on i and depends on m and n through the resolution $r = m^n$. Hence, alternative notations for $\varepsilon_{m,n}$ are ε_{m^n} and ε_r. One should distinguish between two average densities: the "bare" density $\varepsilon_{m^n;b}$ and the "dressed" density ε_{m^n}. The former does not include fluctuations at scales smaller than m^{-n}. Hence, $\varepsilon_{m^n;b}$ is the measure density in Ω_{n_i} when the multiplicative cascade construction is terminated at level n. By contrast, the dressed measure density ε_{m^n} is the average density in Ω_{n_i} for the fully developed cascade. The two measure densities satisfy

$$\varepsilon_{m^n;b} = W_1 W_2 \cdots W_n$$
$$\varepsilon_{m^n} = \varepsilon_{m^n;b} Z \quad (4.35)$$

where the variables W_i are independent with the distribution of W, and Z is the so-called dressing factor. The dressing factor is independent of $\varepsilon_{m^n;b}$ and has the same distribution as ε_{m^0}, the dressed measure density in the unit cube.

Figure 4.5 shows a simulation of a binary cascade ($m = 2$) at resolution levels $n = 1, 2, 3$, and 6. The top four plots give the bare cascade for these values of n, whereas the plot at the bottom shows the dressed cascade at level 6. The cascade uses a generator W with lognormal distribution, $E[W] = 1$ and $Var[\ln(W)] = 0.2\ln(2)$.

Measures $X(S)$, $S \subset R^d$, generated by discrete cascades have stationary marginal distribution if one restricts S to be a cascade tile at some level n. Stationarity of the higher-dimensional (joint) distributions holds under tighter constraints. For example, the joint distribution of the measure densities in two tiles at level n depends on the level N of the closest common ancestor tile. The level N is a function not only of n but also of the separating distance between the two level-n tiles. Therefore, discrete cascades are fundamentally nonstationary.

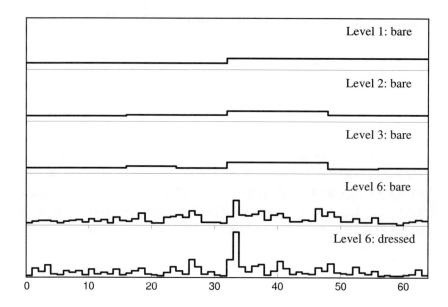

Fig. 4.5. Simulation of a binary cascade. The top four plots show the bare density at resolution levels 1, 2, 3, and 6. The plot at the bottom is the dressed realization at level 6. The generator W has lognormal distribution with unit mean value and log-variance $\sigma^2_{\ln(W)} = 0.2\ln(2)$.

4.4.4. *Scale-invariant Processes from Fractional Integration of α-stable Measures*

While covered by the general characterizations given earlier, important classes of *H-sssi* and multifractal processes can be constructed by other methods, most notably the filtering of certain white noise processes called α-stable measures, which we introduce first. For a more detailed presentation limited to *H-sssi* processes, see Samorodnitsky and Taqqu.[43] Within the *H-sssi* class, we consider first processes with independent increments and then Gaussian and stable non-Gaussian processes. These include fractional Brownian motion (*fBm*) and stable extensions of *fBm*. Then, we turn to stationary multifractal measures and processes with stationary multifractal increments. In addition to linear filtering, the generation of these multifractal processes requires exponentiation.

α-stable Random Measures and their Integrals

An important class of random measures for the construction of scale-invariant processes are the so-called α-stable measures $M_\alpha(S)$.[43] Here, we consider a sub-class of such measures, which is sufficient for our purpose.

Recall that α-stable distributions $S_\alpha(\sigma,\beta,\mu)$ have four parameters: the stability index $0 < \alpha \leq 2$ (with $\alpha = 2$ corresponding to normal distributions), the location $-\infty < \mu < \infty$, the dispersion $\sigma \geq 0$, and the skewness $-1 \leq \beta \leq 1$. The α-stable measures $M_\alpha(S)$ are independently-scattered, meaning that for disjoint sets $S_1,...,S_n$, the measures $M_\alpha(S_1),...,M_\alpha(S_n)$ are mutually independent. In addition, for the class of α-stable measures considered here, $M_\alpha(S)$ has distribution $S_\alpha(|S|^{1/\alpha},\beta,0)$. Different choices of α and β produce distinct α-stable measures.

For generating *H-sssi* processes, one needs to consider integrals of the type

$$I(f) = \int_{-\infty}^{\infty} f(t) M_\alpha(dt) \qquad (4.36)$$

where $f(t)$ is a deterministic function that satisfies $\int_{-\infty}^{\infty} |f(t)|^\alpha dt < \infty$. Using the definition of such integrals in Samorodnitsky and Taqqu,[43] $I(f)$ has α-stable distribution $S_\alpha(\sigma_f,\beta_f,0)$ with parameters

$$\sigma_f = \left[\int_{-\infty}^{\infty} |f(t)|^\alpha\, dt\right]^{1/\alpha}, \qquad \beta_f = \beta\, \frac{\int_{-\infty}^{\infty} |f(t)|^\alpha\, [\text{sign}\, f(t)]\, dt}{\int_{-\infty}^{\infty} |f(t)|^\alpha\, dt} \qquad (4.37)$$

4.4.4.1. H-sssi Processes

With the above as background, we consider specific classes of *H-sssi* processes generated through linear filtering of α-stable measures. We start with *H-sssi* processes with independent increments, which must necessarily be of the α-stable type and have self-similarity index $H = 1/\alpha$. The case $\alpha = 2$ corresponds to Brownian motion (*Bm*), which is the only Gaussian *H-sssi* process with independent increments and $H = 0.5$. Then, we consider fractional Brownian motion (*fBm*), which generalizes *Bm* for the case of dependent increments and $H \neq 0.5$. Fractional Brownian motion has several representations as linearly filtered Gaussian white noise. When applied to white Levy stable noise, these representations produce α-stable *H-sssi* processes with dependent increments and $H \neq 1/\alpha$. While all the representations are equivalent in the Gaussian case ($\alpha = 2$), this is not true for $\alpha \neq 2$. Here, we describe only the α-stable *H-sssi* processes called Linear Fractional Stable Motion (LFSM). Other α-stable *H-sssi* processes are described in Samorodnitsky and Taqqu.[43]

H-sssi Processes with Independent Increments (α-stable Motion)

Processes with independent stationary increments must have infinitely divisible distribution, but the additional constraint of self-similarity limits the distribution to be of the stable type $S_\alpha(\sigma,\beta,\mu)$. This is why this class of *H-sssi* processes is called α-stable Levy motion. The self-similarity index $H = 1/\alpha$ follows from properties of sums of *iid* α-stable variables.

A simple way to generate α-stable Levy motion is as

$$X(t) = \begin{cases} M_\alpha([0,t]) = \int_0^t M_\alpha(d\tau), & t \geq 0 \\ M_\alpha([t,0]) = \int_t^0 M_\alpha(d\tau), & t < 0 \end{cases} \quad (4.38)$$

In the case $\alpha = 2$, $X(t)$ is Brownian motion. As α decreases, the trajectories of $X(t)$ are increasingly dominated by large discontinuities.

Fractional Brownian Motion

Fractional Brownian motion (*fBm*) is the only Gaussian *H-sssi* process; it has zero mean and covariance function given in Sec. 4.5.1. Fractional Brownian motion with index H is often denoted by $B_H(t)$. There are several representations of $B_H(t)$ of the integral type

$$B_H(t) = \int_{-\infty}^{\infty} f_t(\tau) M_2(d\tau), \quad t \geq 0 \quad (4.39)$$

which differ in the form of the filter function $f_t(\tau)$. Two simple cases are the symmetric ("well-balanced") filter

$$f_t(\tau) = |t - \tau|^{H-0.5} - |\tau|^{H-0.5} \tag{4.40}$$

and the causal ("non-anticipative") filter

$$f_t(\tau) = (t - \tau)_+^{H-0.5} - (-\tau)_+^{H-0.5} \tag{4.41}$$

where $(t)_+ = t$ for $t > 0$ and $(t)_+ = 0$ for $t < 0$. Notice that for both filter functions in Eqs. (4.40) and (4.41), $f_t(\tau)$ vanishes for $t = 0$; hence $B_H(0) = 0$. In the case of Eq. (4.41), $f_t(\tau)$ vanishes for $\tau > t$ and the integral in Eq. (4.39) extends over only the interval $(-\infty, t]$. Filtering representations of this non-anticipative type are especially useful for predicting the future evolution of a process from observation of its infinite past (see Sec. 4.6.1).

Linear Fractional Stable Motions

There are several extensions of *fBm* to the α-stable case. Those that are most commonly used are obtained by extending the integral representations of *fBm* in Eq. (4.39) and are called linear fractional stable motions (*LFSM*). For each (α, H) combination, there are many *LFSM* processes $L_{\alpha,H}(t)$, because the various integral representations that are equivalent for *fBm* are not equivalent for *LFSM*. Here, we mention only the extension of the well-balanced representation in Eq. (4.39) with $f_t(\tau)$ in Eq. (4.40), which for *LFSM* has the form

$$L_{\alpha,H}(t) = \int_{-\infty}^{\infty} \left[|t - \tau|^{H-1/\alpha} - |\tau|^{H-1/\alpha} \right] M_\alpha(d\tau) \tag{4.42}$$

For other *H-sssi* processes obtained through fractional integration of α-stable measures, see Chapter 7 in Samorodnitsky and Taqqu,[43] Vervaat,[38] and Maejima.[44]

4.4.4.2. *Multifractal Processes*

An important class of stationary multifractal measures and processes with stationary multifractal increments is obtained by appropriately filtering and exponentiating the α-stable (Gaussian or Levy-stable) measures introduced above. The resulting measures and processes are sometimes labeled "universal," based on the argument that, by virtue of the attraction of sums of *iid* variables to Levy-stable distributions, many generating mechanisms in nature should produce scale-invariant processes of this type.[45] While the "attraction argument" may be refuted, universal multifractal processes retain significant theoretical and practical interest, as they are multifractal generalizations of fractional Brownian and Levy motions. Here, we prefer the terminology lognormal and log-Levy multifractal processes, with reference to the distribution of the random factor in the multifractal scale-invariance property.

Let $M_\alpha(S)$ be an α-stable measure (see above) in R^d with Levy stable distribution $S_\alpha(\sigma_S, \beta, \mu)$, where $0 < \alpha \le 2$, $\sigma_S = |S|^{1/\alpha}$, $\beta = -1$, and $\mu = 0$. The constraint of maximum negative skewness $\beta = -1$ is needed to avoid degeneracy of the derived multifractal measure. Notice that for $\alpha = 2$ the distribution is normal and, therefore, is symmetrical irrespective of the skewness parameter β. Log-Levy multifractal measures are obtained through the following sequence of operations:[46-48]

(i) Simulate $M_\alpha(S)$ on a fine computational grid in *d*-dimensional space. This operation requires simulation of *iid* Gaussian or Levy-stable variables $S_\alpha(\sigma, \beta = -1, \mu = 0)$, where σ is a positive constant. This is followed by a linear transformation of the simulated variables (see below);

(ii) Multiply the Fourier transform of the above independent field by $|\omega|^{-d/\alpha'}$, where *d* is the space dimension and α' is such that $1/\alpha + 1/\alpha' = 1$. Then transform back to physical space;

(iii) Exponentiate the back-transformed field to obtain a stationary lognormal or log-Levy multifractal measure density ($H = 0$);

(iv) To simulate processes with multifractal increments and $H > 0$, multiply the Fourier transform of the exponentiated process by $|\omega|^{-H}$. Then transform back to physical space.

The result at the end of Step (iii) is a stationary measure. This measure is sometimes said to be "conservative" because, without further filtering by $|\omega|^{-H}$, it can be obtained by a multiplicative cascade process that at each cascade step preserves the mean. Step (iv) essentially accomplishes fractional integration of order H in Fourier space. In analogy with the techniques to generate H-sssi processes, described in Sec. 4.4.4.1, fractional integration could be performed through different filtering operations in physical space, producing other classes of multifractal log-Levy processes. However, the procedure described above is the only one in use for universal multifractal processes.

The moment scaling function $K(q)$ of universal multifractal processes [for $H = 0$, this is the moment scaling function of the average measure density (see Sec. 4.5.2), whereas for $H > 0$, $K(q)$ is the moment scaling function of the increments of the process] has the form

$$K(q) = \begin{cases} -Hq + \dfrac{C_1}{\alpha-1}(q^\alpha - q), & \alpha \neq 1 \text{ (only for } q \geq 0 \text{ when } \alpha < 2) \\ -Hq + C_1 q \ln(q), & \alpha = 1 \end{cases} \quad (4.43)$$

where $0 < C_1 < d$ is called the co-dimension parameter and controls the erraticity of the field.

The linear transformation in Step (i) of the simulation procedure aims at producing fields with the $K(q)$ function in Eq. (4.43), and in particular controls the value of C_1. For details on this transformation and the simulation of maximally-skewed Levy stable variables, see Appendix B of Wilson et al.[46]

Figure 4.6(a) illustrates the above procedure in the lognormal case, for $C_1 = 0.1$ and different values of H. For comparison, Fig. 4.6(b) shows realizations of fractional Brownian motion $B_H(t)$ for the same positive values of H. All plots start from the simulation of a normal process $Y(t)$ with spectral density $S(\omega) \propto \omega^{-1}$ (over a broad but finite range of frequencies); see top plot in Fig. 4.6(b). The top plot of Fig. 4.6(a) is obtained through exponentiation of $Y(t)$ and the plots for $H > 0$ in Figs. 4.6(a) and 4.6(b) are generated by filtering $Y(t)$ or $\exp\{Y(t)\}$ with ω^{-H} in Fourier space.

4.4.5. Processes with Limited Scale Invariance

In many cases, one must deal with processes that display systematic deviations from scale invariance or are scale-invariant in some limited sense. For example, a random function may not be strictly scale-invariant, but its first two moments may be (this condition is sometimes referred to as weak scale invariance, as in the distinction between strict and weak stationarity).

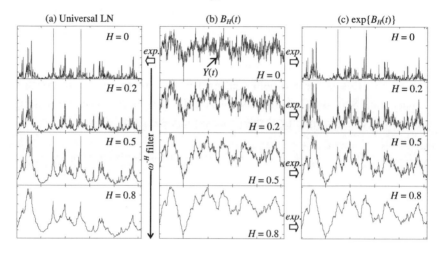

Fig. 4.6. (a) Simulated realizations of lognormal universal processes with multifractal increments, co-dimension parameter $C_1 = 0.1$ and different values of H. For $H = 0$, the process becomes the degenerate density of a stationary multifractal measure. (b) Realizations of fractional Brownian motion for the same positive values of H. (c) Exponentiated fractional Brownian motion $\exp\{B_H(t)\}$ for different values of H. The simulations of $B_H(t)$ are those shown in (b).

In other cases, the mechanism that generates a process is scale-invariant, but the process itself is not. For example, in generating scale-invariant processes one often starts with a smooth function (such as a constant) and adds details in a scale-invariant way at progressively higher resolutions. If this iterative procedure is terminated at some finite resolution, the resulting process is not exactly scale-invariant, not even over a finite range of scales. Rather, what is scale-invariant (over a finite range of resolutions) is the generating mechanism. Along similar lines, one often encounters processes with different scaling exponents in different resolution ranges. What is actually meant is that the generating mechanism has distinct scaling properties in different ranges, whereas the process itself is not scaling.

In yet other cases, a process may not be scaling, but a simple transformation of it may be scale-invariant. One example is the whole class of stationary processes, which, as we saw in Sec. 4.4.1, become self-similar under certain transformations. Another example is the class of exponentiated self-similar processes, such as exponentiated fractional Brownian motion. Except for special cases, exponentiation destroys scale invariance, but the fact that the log of a process is scale-invariant may still be of interest.

In what follows, we first describe four different deviations from multifractal scaling that are of particular interest in rainfall and turbulence. We start with bounded cascades and extended self-similarity (ESS). Bounded cascades are models of stationary measures $X(dt)$ whose construction generalizes that of multifractal cascades, whereas ESS processes are extensions of processes $X(t)$ with multifractal increments. As we shall see, these apparently different generalizations are, however, quite similar. We then talk about processes in which the distribution of the generator at any given scale depends on the bare intensity at some larger scale, and finally mention processes with segmented multifractality.

Bounded Cascades

The generator W of a stationary multifractal cascade (the random factor by which the bare densities at level n are multiplied to produce the bare densities at level $n + 1$) has the same distribution at all levels n (see Sec. 4.4.3.2). Bounded cascade models[49,50] allow the distribution of W to depend on n. If $\log(W)$ has infinitely-divisible distribution, then the dependence of W_n on n is typically as follows. Let $\Gamma(t)$ be the process with independent stationary increments such that $\Gamma(0) = 0$ and $\Gamma(m) = \ln(W)$, where m is the multiplicity of the cascade. Then, in bounded cascades, one often takes W_n as

$$W_n \stackrel{d}{=} e^{\Gamma(m\gamma^n)} \quad (4.44)$$

where $0 < \gamma < 1$ is a given constant. For example, if $\ln(W)$ has normal distribution with mean value $-0.5\sigma^2$ and variance σ^2, then $\Gamma(m\gamma^n)$ in Eq. (4.44) has normal distribution with mean value $-0.5\gamma^n\sigma^2$ and variance $\gamma^n \sigma^2$. If one densifies this bounded cascade construction in scale, the result is exponentiated fractional Brownian motion $\exp\{B_H(t)\}$ with H that satisfies $m^{-2H} = \gamma$, i.e. $H = -0.5\log_m(\gamma)$.

Much has been said (but not always correctly) about the scaling properties of $\exp\{B_H(t)\}$. One can show that this process does not scale, although in the high-resolution limit it has the same H-ss properties of the increments as $B_H(t)$. In hydrology, bounded cascades have been widely used as models of rainfall[49,51] (see also Sec. 4.8.1) and exponentiated *fBm* has been used as a model of hydraulic conductivity in natural aquifers (see Sec. 4.8.4). Examples of exponentiated *fBm* are shown in Fig. 4.6(c). The case $H = 0$ is identical to the plot at the top of Fig. 4.6(a) and has multifractal scaling properties, whereas for $H > 0$, $\exp\{B_H(t)\}$ does not scale, except for the very local increments, which scale like $B_H(t)$. This small-scale property is especially visible for large H (compare bottom plots in Fig. 4.6(b) and 4.6(c)).

Extended Self-Similarity

A second important departure from multifractality is known as extended self-similarity (ESS). Extended self-similarity was proposed by Benzi and collaborators[52-55] to represent the scaling of turbulent velocity fields. If $\Delta v(\tau)$ is the longitudinal velocity increment at distance τ in the mean flow direction, then a standard multifractal assumption is that, for any $r > 1$ there exists some random variable A_r such that

$$\Delta v(\tau/r) \stackrel{d}{=} A_r \Delta v(\tau)$$
$$\Rightarrow E[|\Delta v(\tau)|^q] \propto \tau^{-\xi(q)} \qquad (4.45)$$

where $\xi(q) = \log_r E[A_r^q]$. Equation (4.45) holds reasonably well at high Reynolds numbers, but breaks down at the small scales where viscosity becomes significant. To extend validity to the latter regime, Benzi and collaborators suggested a weaker condition on the moments of $\Delta v(\tau)$, of the type

$$E[|\Delta v(\tau)|^{q_1}] \propto E[|\Delta v(\tau)|^{q_2}]^{\xi(q_1)/\xi(q_2)} \qquad (4.46)$$

While Eq. (4.45) implies Eq. (4.46), the reverse is not true, because Eq. (4.46) does not require that the moments be power functions of τ. A simple way to verify whether ESS applies is to see whether the plot of $\log E[|\Delta v(\tau)|^{q_1}]$ against $\log E[|\Delta v(\tau)|^{q_2}]$ for different τ is linear. If it is, then the slope equals the ratio $\xi(q_1)/\xi(q_2)$.

A condition for stationary cascades analogous to Eq. (4.46) would be

$$E[\varepsilon^{q_1}(S)] \propto E[\varepsilon^{q_2}(S)]^{K(q_1)/K(q_2)} \quad (4.47)$$

where $\varepsilon(S)$ is the average measure density in S. Equation (4.47) is satisfied if at level n of the multiplicative cascade one takes a generator W_n which, in the notation of Eq. (4.44), is given by

$$W_n \stackrel{d}{=} e^{\Gamma(t_n)} \quad (4.48)$$

where $t_n, n = 1, 2, ...$ is a sequence of non-negative numbers. Comparison of Eqs. (4.44) and (4.48) shows that the ESS and bounded-cascade constructions are similar, as both correspond to allowing the generator to depend on the resolution level (although in bounded cascades the dependence is constrained to be of a particular type). The *ESS* construction could be further generalized by taking $t_n, n = 1, 2, ...$ in Eq. (4.48) to be a random sequence.

Dependence of the Generator W_n on the State at Level n

The previous extensions of multifractality allow the generators W_n to have different distributions at different cascade levels n. In studies of rainfall, Olsson,[56] Güntner et al.,[57] and Veneziano et al.[58] found it necessary to make a further generalization, in which the distribution of W_n depends also on the bare rainfall intensity in the parent cascade tile at the previous level $n - 1$, $\overline{X}_b(n-1)$. This is a special case of randomness of the distribution of W_n. We emphasize that in this extension W_n depends on the bare rainfall intensity at level $n - 1$ (dependence on the dressed intensity at level $n - 1$ is present also in standard multifractal cascades).

Segmented Multifractality

A special case of dependence of W on scale is when the cascade generator has piece-wise constant distribution in different scale regimes, i.e. when W is distributed like W_i for $n_i < n \le n_{i+1}$, where the n_i satisfy

$1 = n_1 < n_2 < ... < n_s = \infty$. We call this scale-segmented multifractality. Measures with segmented multifractality are actually not multifractal at any scale, but if a range $[n_i, n_{i+1}]$ is wide, the measure will display approximate scale invariance inside that range. For example, if it exists, the spectral density $S(\omega)$ would display approximate power-law behavior in the frequency range associated with $[n_i, n_{i+1}]$. The existence of distinct scaling regimes in rainfall time series has long been recognized; see for example Fraedrich and Larnder[59] and Sec. 4.8.1.

4.5. Properties of Scale-invariant Processes

This section gives basic properties of scale-invariant processes. We first consider self-similar processes with stationary increments and then deal with multifractal processes. For the latter, special attention is given to stationary multifractal measures.

4.5.1. H-sssi Processes

Among the self-similar processes, those with stationary increments (*H-sssi* processes) are of particular theoretical and practical importance. These processes have several interesting properties, including those on second moments, the spectral density, and the fractal dimension of graphs listed below. For a more detailed account, see Samorodnitsky and Taqqu.[43]

First, we note that the existence of moments of certain orders has an implication on the possible range of the self-similarity index H, and vice-versa. Let $X(t)$ be *H-sssi* with $H > 0$. If $X(t)$ is non-degenerate, meaning that $X(t) \neq 0$ with positive probability for all $t > 0$, then the following is true:

- If $E[|X(t)|^q] < \infty$ for some $q < 1$, then $H < 1/q$
- If $E[|X(t)|] < \infty$, then $H \leq 1$ (4.49)
- If $0 < H < 1$, then $E[X(t)] = 0$

Notice that if $E[X(t)^2] < \infty$, then $E[|X(t)|] < \infty$ and by the second property above it must be $H \leq 1$. Next, we give several second-moment properties of *H-sssi* processes when $E[X(t)^2]<\infty$, followed by the fractal-dimension of $X(t)$ graphs when $E[|X(t)|]<\infty$.

Second Moments of X(t) and its Increments

Under the condition that the second moment $E[X(t)^2]$ is finite, one can derive the second moments $E[X(t)X(s)]$ of the process and $E[\xi(s)\xi(s+n)]$ of the increments $\xi(k)=X(k+1)-X(k)$. Using the identity $X_1 X_2 = 0.5\{X_1^2 + X_2^2 - (X_1-X_2)^2\}$ and the fact that $E[X(t)^2]=t^{2H}E[X(1)^2]$, one obtains

$$\begin{aligned} E[X(t)X(s)] &= \frac{1}{2}\left\{E[X(t)^2]+E[X(s)^2]-E[|X(t)-X(s)|^2]\right\} \\ &= \frac{1}{2}\left\{E[X(t)^2]+E[X(s)^2]-E[X(t-s)^2]\right\} \quad (4.50) \\ &= \frac{1}{2}\left\{t^{2H}+s^{2H}-(t-s)^{2H}\right\}E[X(1)^2] \end{aligned}$$

where the second equality uses the property of stationarity of the increments, and the last equality uses the *H-ss* property. The second moment function in Eq. (4.50) is characterized completely by two parameters, the self-similarity index H and the second moment $E[X(1)^2]$. Note from Eq. (4.49) that for $H < 1$, it must be $E[X(t)] = 0$; hence, in this case $E[X(t)X(s)]$ in Eq. (4.50) is also the covariance function of $X(t)$. Since Gaussian processes are completely described by their first- and second-moment properties, the only Gaussian *H-sssi* process is the Gaussian process with zero mean and the function in Eq. (4.50) as covariance function. This process is known as fractional Brownian motion (*fBm*).

Next consider the increments $\xi(k)$. Since these increments are stationary, $B(n)=E[\xi(s)\xi(s+n)]$ depends only on the lag n. This function is given by

$$\begin{aligned} B(n) &= E[\xi(0)\xi(n)] = E\{X(1)[X(n+1)-X(n)]\} \\ &= E\{X(1)X(n+1)-X(1)X(n)\} \quad (4.51) \\ &= \frac{1}{2}\left\{(n+1)^{2H}-2n^{2H}+(n-1)^{2H}\right\}E[X(1)^2] \end{aligned}$$

From Eq. (4.49), if H is strictly less than 1, then $E[\xi(k)] = 0$ and $r(n)$ is also the covariance function of $\xi(k)$. As $n \to \infty$, $B(n)$ approaches the form

$$\begin{cases} B(n) \approx H(2H-1)n^{2H-2}E[X(1)^2], & \text{if } H \neq 0.5 \\ B(n) = 0, \quad n \geq 1, & \text{if } H = 0.5 \end{cases} \quad (4.52)$$

Therefore,

- If $0 < H \leq 0.5$, the increments of $X(t)$ have *negative correlation* and $\sum_{n=0}^{\infty} |B(n)| < \infty$. This means that $X(t)$ has *weak dependence*.

- If $H = 0.5$, the increments of $X(t)$ are *uncorrelated*. (4.53)

- If $0.5 < H \leq 1$, the increments of $X(t)$ have *positive correlation* and $\sum_{n=0}^{\infty} |B(n)| = \infty$. This means that $X(t)$ has *long-range dependence*.

On weak and long-range dependence, see Beran.[60]

Spectral Density

For $H > 0$, the process $X(t)$ is nonstationary. For example, $X(0) = 0$ and the covariance function in Eq. (4.50) depends on s and t, not just through $(s - t)$. However, the increments of $X(t)$ are stationary and, if one relaxes the condition $X(0) = 0$, one may regard $X(t)$ as a stationary process with infinite-variance and a well-defined spectral density. The spectral density $S(\omega)$ can be obtained as follows. For $s = t$, Eq. (4.50) gives

$$E[X(\tau)^2] = E[|X(\tau) - X(0)|^2] \propto \tau^{2H} \quad (4.54)$$

The second moment $E[X(\tau)^2]$ is also the variance of the process $X_\tau(t) = X(t+\tau) - X(t)$, obtained by passing $X(t)$ through a linear filter $h_\tau(t) = \delta(t - \tau) - \delta(t)$ whose transfer function is $e^{i\omega\tau} - 1$. Therefore, $S(\omega)$ must satisfy

$$\int_0^\infty |e^{i\omega\tau} - 1|^2 S(\omega) \, d\omega = \frac{1}{\tau} \int_0^\infty |e^{i\omega'} - 1|^2 S(\omega'/\tau) \, d\omega' \propto \tau^{2H} \quad (4.55)$$

where we have put $\omega' = \omega\tau$. In order for Eq. (4.55) to hold, $S(\omega)$ must have the form

$$S(\omega) \propto \omega^{-(1+2H)} \quad (4.56)$$

This is the spectral density of fractional Brownian motion. However, the result in Eq. (4.56) holds, in general, for all *H-sssi* processes with finite second moments.

Fractal Dimension of Graphs

Suppose that $E[|X(t)|] < \infty$. From the *H-sssi* property, the process $X(t)$ is statistically equivalent to $X'(t) \stackrel{d}{=} r^{-H} X(rt)$. Therefore, the numbers of tiles of area A_δ that cover $X(t)$ and $X'(t)$ in the interval [0,1], N_δ and N'_δ respectively, have the same distribution. Also note that $E[N'_{\delta d}] = E[N_\delta] = m_\delta$ is finite due to the condition $E[|X(t)|] < \infty$. From $X'(t) = r^{-H} X(rt)$, one further concludes that, in the interval [0, 1/r], $X'(t)$ must be covered by N_δ tiles of area $A_\delta r^{-(1+H)}$ (tile size contraction by r in the horizontal direction and by r^H in the vertical direction), again with the same distribution of N_δ. Then, using the definition of Hausdorff fractal dimension in Sec. 4.2.1,

$$M_D = \lim_{r \to \infty} r \, E[N_\delta] \frac{A_\delta r^{-(1+H)}}{(\delta/r)^2} \left(\frac{\delta}{r}\right)^D \quad (4.57)$$

$$\propto \lim_{r \to \infty} r^{2-H} \, r^{-D}$$

Since M_D is zero for $D < 2 - H$ and diverges for $D > 2 - H$, we calculate that the Hausdorff fractal dimension of graphs of $X(t)$ is

$$D_H = 2 - H \quad (4.58)$$

For example, graphs of Brownian motion have fractal dimension 1.5.

4.5.2. *Moment Scaling of Multifractal Processes*

The multifractal property in Eqs. (4.10) and (4.13) (see Sec. 4.3.2) has important implications on the distribution of $X(h_r)$ and the increments $\Delta^j X(t/r, \tau/r)$. Here, we focus on the scaling of the moments for different resolutions r. Other implications concern mainly stationary multifractal measures, and are reviewed in Secs. 4.5.3 and 4.5.4.

It follows immediately from Eqs. (4.10) and (4.13) that the non-diverging moments of $X(h)$ and $\Delta^j X$ scale as

$$E[X(h_r)^q] = r^{K(q)} E[X(h)^q]$$

$$E[\Delta^j X(\frac{t}{r}, \frac{\tau}{r})^q] = r^{K(q)} E[\Delta^j X(t,\tau)^q] \qquad (4.59)$$

where $K(q) = \log_r E[A_r^q] = \ln E[A_e^q]$. In the *ss* case, $A_r = r^{-H}$ and $K(q) = -qH$ is proportional to q. Many authors take a non-proportional (or more often a nonlinear) dependence of $K(q)$ on q as the definition of multifractality.

Using an argument similar to that for *H-sssi* processes in Sec. 4.5.1, one can show that, if $X(h)$ is isotropic multifractal or has isotropic multifractal increments in R^d and its spectral density function $S(\underline{\omega})$ exists, then

$$S(\underline{\omega}) \propto |\underline{\omega}|^{-d+K(2)} \qquad (4.60)$$

Equation (4.59) applies for $q < q^*$, where q^* is the order of moment divergence ($q^* = \infty$ if all the positive moments are finite). To understand the source of moment divergence, we denote by $h_{\max}$ a test function h at the outer scale of multifractality and take r to be the resolution relative to this outer scale. Then Eq. (4.10) gives

$$X(h_r) = A_r X(h_{\max}) \qquad (4.61)$$

In the context of discrete multifractal cascades (a particularly simple class of quasi-stationary multifractal measures; see Sec. 4.4.3.2), one associates the factor A_r with the so-called "bare component" of the process, $X_b(h_r)$, and $X(h_{\max})$ with the dressing factor Z. The factor Z accounts for fluctuations of the process at resolutions higher than r, and $X(h_r) = X_b(h_r) Z$ is called the "dressed" process. However, Eq. (4.61) is more general and applies also to nonstationary multifractal processes for which $X(h_{\max})$ does not have this dressing-factor interpretation.

It follows from Eq. (4.61) that the order of moment divergence for $X(h_r)$ satisfies $q^* = \min\{q_A^*, q_{\max}^*\}$, where q_A^* and $q_{\max}^*$ are the orders of moment divergence of the random variables A_r and $X(h_{\max})$ in Eq. (4.61). In the case of stationary multifractal processes in d-dimensional space, $q_{\max}^*$ is the value of $q > 1$, such that $K(q) = d(q-1)$ and $q^* = q_{\max}^*$ (see Sec. 4.5.3), but for nonstationary multifractal processes one may have $q^* = q_A^* < q_{\max}^*$. The value of q^* is important for what follows.

4.5.3. *Existence, Moments, and Distributions of Stationary Multifractal Measures*

Stationary multifractal measures, in particular discrete cascades, have high theoretical and practical interest. Theoretically, they can be analyzed to produce important results, while the condition of stationarity (for discrete cascades, quasi-stationarity) often has a strong practical appeal. Here, we review results related to the existence, moments, and distributions of discrete cascades. In good approximation, the same properties apply to continuous stationary multifractal measures. We start from conditions under which the dressed measure exists and moments of certain order diverge. We also show how one can numerically calculate the dressed moments and dressed marginal distributions of these measures.

Existence and Moment Divergence

Recall from Sec. 4.4.3.2 the construction of discrete cascades and the distinction between bare measure densities $\varepsilon_{m^n;b}$ and dressed measure densities ε_{m^n}, where m is the linear multiplicity of the cascade and n is the cascade level. The bare densities are simply the product of n independent variables W_i, all distributed like the generator W, and the dressed densities are obtained by multiplying the bare densities by the dressing factor Z (see Eq. (4.35)).

Clearly, the bare densities always exist and their moments are the n powers of the corresponding moments of W. Therefore, the qth moment of $\varepsilon_{m^n;b}$ exists if and only if the qth moment of W exists. Matters are more complicated for the dressed densities, because ε_{m^n} is non-degenerate (has non-zero probability of being finite non-zero) if and only if Z is non-degenerate. Similarly, the qth moment of ε_{m^n} exists if and only if the qth moment of Z exists. Therefore, one is led to examine the degeneracy and moment divergence problems for the dressing factor Z.

Kahane and Peyriere[61] studied these problems and proved the following fundamental results, which involve the function $K(q) = \log_m E[W^q]$ and its derivative $K'(q)$:

- Z is non-degenerate if and only if $K'(1) < d$

- $E[Z^q]$, $q > 1$, is finite if and only if $K(q) < d(q-1)$ (4.62)

These are also the conditions for non-degeneracy and finiteness of the q-moment of the dressed density ε_{m^n}. The lowest order of moment divergence $q^* > 1$ is found from the condition $K(q^*) = d(q^*-1)$. Next, we discuss how to calculate the moments and distribution of Z under the constraints of Eq. (4.62) and then use these results to obtain the moments and marginal distribution of ε_{m^n}.

Dressed Distributions

The distribution of Z depends exclusively on the distribution of the generator W and the "volumetric multiplicity" m^d, where d is the space dimension. Specifically, Z satisfies the fundamental consistency relation

$$Z \stackrel{d}{=} \frac{1}{m^d} \sum_{i=1}^{m^d} W_i Z_i \qquad (4.63)$$

where the variables W_i and Z_i are independent copies of W and Z, respectively. Equation (4.63) follows from considering the dressed density in the unit cube of the cascade, which must be the average of the dressed densities in the m^d descendant tiles at level 1. The dressed density in the unit cube has the distribution of Z and the dressed density in the ith descendant tile has the same distribution as $W_i Z_i$. Equation (4.63) does not uniquely determine the distribution of Z. For example, $Z = 0$ is always a solution of Eq. (4.63). The solution that corresponds to the dressing factor can be obtained by first setting $Z_{(0)} = 1$ and then iteratively updating the distribution of $Z_{(j)}$ at steps $j = 1, 2, \ldots$ using

$$Z_{(j)} = \frac{1}{m^d} \sum_{i=1}^{m^d} W_i Z_{i(j-1)} \qquad (4.64)$$

where the variables $Z_{i(j-1)}$ are independent copies of $Z_{(j-1)}$. As $j \to \infty$, $Z_{(j)} \stackrel{d}{\to} Z$. For the numerical implementation of this iterative procedure, see Veneziano and Furcolo.[62] When the critical order q^* is finite, the distribution of Z has an algebraic upper tail of the type $P[Z > z] \sim z^{-q^*}$. An illustration of the convergence of $Z_{(j)}$ to Z is shown in Fig. 4.7 where j ranges from 1 to 27 ($\approx \infty$). In this case, the cascade is binary and one-dimensional ($m = 2$, $d = 1$), and the generator W has lognormal distribution with $E[W] = 1$ and $Var[\log_2(W)] = 0.2$; hence $C_1 = 0.1$.

The inset in Fig. 4.7 displays the same results in log-log space. For $j = 1$, $Z_{(1)}$ has the same (lognormal) distribution as W, but as j increases, $Z_{(j)}$ becomes more skewed to the right and eventually develops an algebraic upper tail with probability density $f_Z(z) \propto z^{-(q^*+1)}$, in this case with $q^* = 1/C_1 = 10$.

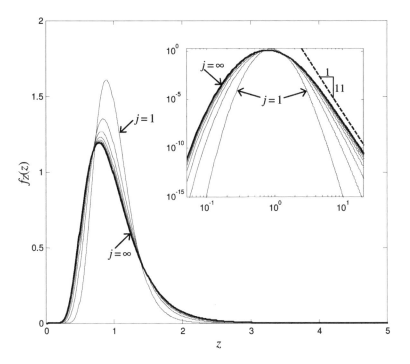

Fig. 4.7. Convergence of the probability density function of $Z_{(j)}$ in Eq. (4.64) to that of the dressing factor Z for a binary lognormal cascade on the line with co-dimension parameter $C_1 = 0.1$. Results are shown in both arithmetic and log-log (inset) scales.

The distribution of the dressed density ε_r, $r = m^n$, can be found numerically as the convolution product between the distributions of the bare density $\varepsilon_{r;b} = \prod_{i=1}^{n} W_i$ and Z. In the (frequent) case, when Z has a power-law upper tail $P[Z > z] \sim z^{-q^*}$, the dressed density $\varepsilon_r = \varepsilon_{r;b} Z$ has a similar algebraic upper tail, $P[\varepsilon_r > \varepsilon] \sim \varepsilon^{-q^*}$. However, this power behavior occurs for values of ε that rapidly increase as the resolution r

increases. As a consequence, at high resolutions, the power-law tail may not be observable, even in large samples.

Veneziano and Furcolo[62] studied the asymptotics of the distribution of ε_r as $r \to \infty$. They found that, in the high-resolution limit, the power-law tail of Z becomes unimportant and the distribution of ε_r is attracted to one of three types: lognormal, log-Levy, or a third type with a lognormal or log-Levy distribution of $\varepsilon_r^+ = (\varepsilon_r | \varepsilon_r > 0)$. The last distribution type arises when W has a probability atom at zero. For the cascade in Fig. 4.7, Fig. 4.8 shows the probability density function of the dressed density ε_r at resolutions $r = 2^n$ for $n = 1, 2, 4, 8$, and 16. The inset shows log-log plots of the exceedance probability $P[\varepsilon_r > \varepsilon]$. In this log-log representation, the asymptotic algebraic upper tail corresponds to a straight line with slope $-q^* = -10$ (see dashed lines). For $r = 2$, this algebraic behavior is attained at about $\varepsilon = 5$ (exceedance probability about 10^{-3}). However, as r increases, the algebraic tail moves rapidly to the right. For example, for $r = 2^4$, the algebraic tail is reached at about $\varepsilon = 1000$ (for an exceedance probability around 10^{-18}). At this or higher resolutions, the distribution of ε_r is practically lognormal. This behavior of the marginal distribution of ε_r for different r is critical to understanding the extremes of multifractal cascades (see Sec. 4.5.4).

Dressed Moments

Calculation of the non-diverging moments of Z is easier than calculation of the distribution, especially in the case of low volumetric multiplicity m^d.[62] Here, we consider the simplest case of binary cascades on the line, for which $m^d = 2$. Then Eq. (4.63) reduces to $Z \stackrel{d}{=} 1/2(W_1 Z_1 + W_2 Z_2)$. Raising both sides to a positive integer power q, using the binomial expansion for the right hand side and taking expectations, gives

$$E[Z^q] = \frac{1}{2^q} \sum_{q'=0}^{q} \binom{q}{q'} \{E[W^{q'}]E[Z^{q'}]E[W^{q-q'}]E[Z^{q-q'}]\} \quad (4.65)$$

where

$$\binom{q}{q'} = \frac{q!}{q'!(q-q')!} \quad (4.66)$$

is the binomial coefficient. Finally, using $E[W^q] = 2^{K(q)}$ and solving for $E[Z^q]$ [there is a term in $E[Z^q]$ on the right hand side of Eq. (4.65)] gives

$$E[Z^q] = \frac{2^{-q}}{1 - 2^{K(q)-(q-1)}} \sum_{q'=1}^{q-1} \binom{q}{q'} 2^{K(q')+K(q-q')} E[Z^{q'}] E[Z^{q-q'}] \quad (4.67)$$

This is a recursive relation, which, starting from $E[Z] = 1$, sequentially gives $E[Z^q]$ for $q = 2, 3, \ldots < q^*$.

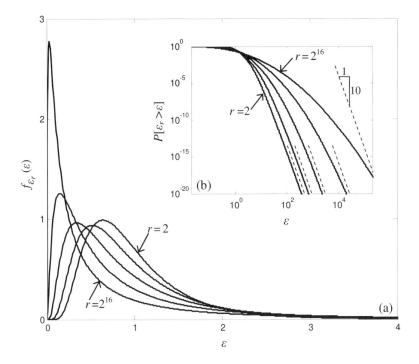

Fig. 4.8. Marginal distribution of the dressed density ε_r at resolutions $r = 2, 2^2, 2^4, 2^8$, and 2^{16} for the lognormal cascade in Fig. 4.7: (a) probability density functions, (b) exceedance probability in log-log scale.

Once the moments $E[Z^q]$ are known, the dressed moments $E[\varepsilon_r^q]$ are found simply as

$$E[\varepsilon_r^q] = r^{K(q)} E[Z^q] \quad (4.68)$$

Equation (4.68) is a special case of more general moment relations involving two or more cascade tiles.[33,63,64] To exemplify, consider a continuous stationary multifractal measure on the unit interval, with unit mean value. For that measure, one can basically follow Marsan et al.[33] to derive the two-point moments $B_{q_1,q_2,l,\Delta} = E[\varepsilon^{q_1}(l_1)\varepsilon^{q_2}(l_2)]$, where $\varepsilon(l_1)$ and $\varepsilon(l_2)$ are the average dressed densities in two segments of common length l, obtained one from the other through translation by $\Delta \geq l$. The analysis is approximate, as it combines characteristics of discrete cascades with features of continuous multifractal processes. The discrete cascade paradigm is convenient due to its hierarchical geometrical organization and the clear distinction between bare densities and dressing factor, but has the drawback of not producing strictly stationary processes; hence, for cascades, the moment function $B_{q_1,q_2,l,\Delta}$ depends on the actual location of the tiles l_1 and l_2, not just the distance Δ between the centerpoints. Reference to continuous processes eliminates this complexity and further allows one to consider any positive value of l and Δ, not just certain negative powers of the discrete cascade multiplicity.

In a discrete cascade idealization, one may regard $L = l + \Delta$ as the size of the nearest common ancestor tile to l_1 and l_2. It follows that $\varepsilon(l_1)$ and $\varepsilon(l_2)$ may be expressed as

$$\varepsilon(l_1) = \varepsilon_b(L) W_{L \to l_1} Z_1$$
$$\varepsilon(l_2) = \varepsilon_b(L) W_{L \to l_2} Z_2 \qquad (4.69)$$

where $\varepsilon_b(L)$ is the common bare density at resolution L, $W_{L \to l_1}$ and $W_{L \to l_2}$ are independent with the distribution of $W_{L \to l}$ (the additional bare components to go from resolution L to resolution l), and Z_1 and Z_2 are independent copies of the dressing factor Z. All five variables on the right hand side of Eq. (4.69) are mutually independent. Further, considering that $E[\varepsilon_b^q(L)] = L^{-K(q)}$ and $E[W_{L \to l}^q(L)] = (L/l)^{K(q)}$, one obtains

$$B_{q_1,q_2,l,\Delta} = L^{-K(q_1+q_2)+K(q_1)+K(q_2)} l^{-[K(q_1)+K(q_2)]} E[Z^{q_1}] E[Z^{q_2}] \qquad (4.70)$$

This derivation can be extended to stationary multifractal measures in spaces of higher dimension, including the case of non-isotropic scaling.[33]

4.5.4. *Extremes of Stationary Multifractal Measures and Origin of "Multifractality"*

Next, we discuss properties of extremes of discrete multiplicative cascades. We start by considering a fundamental large-deviation result by Cramer[65] and its implications for the bare densities. We then extend the results to the dressed densities, and finally consider the problem of cascade extremes (the values exceeded by bare and dressed measure densities on average once every n cascade realizations). These results are connected with the origin of the term "multifractality," of which we give a brief account at the end of the section.

4.5.4.1. *A Large Deviation Result*

There is a duality between the moment-scaling property in Eq. (4.59) and the behavior of certain tail probabilities in the small-scale limit.[66] This duality stems from a large deviation property of the random variable A_r for different r. Let γ be a given positive number. It follows from Cramer's large deviation theorem that in the high-resolution limit $r \to \infty$ the exceedance probability $P[A_r > r^\gamma]$ behaves like[67]

$$P[A_r > r^\gamma] = g(\gamma, r) r^{-C(\gamma)} \tag{4.71}$$

where $g(\gamma, r)$ is a function that varies slowly with r as $r \to \infty$ and $C(\gamma)$ is the Legendre transform of $K(q)$

$$C(\gamma) = \max_q \{q\gamma - K(q)\} \tag{4.72}$$

Equation (4.71) is often written in the form of a "rough limit," as

$$P[A_r > r^\gamma] \sim r^{-C(\gamma)} \tag{4.73}$$

Cramer's Theorem actually refers to the sum of a large number n of independent and identically distributed variables. To obtain Eqs. (4.71) and (4.72), one sets $r = r_o^n$, where $r_o > 1$ is fixed, and applies Cramer's Theorem to $\log(A_r) = \sum_{i=1}^{n} \log(A_{r_o, i})$.

Under certain conditions on the distribution of A_r, one can further show that the prefactor $g(\gamma, r)$ behaves asymptotically as[65,68,69]

$$g(\gamma,r) = \left(2\pi \ln(r) \frac{[C'(\gamma)]^2}{C''(\gamma)}\right)^{-1/2} (1+o(1)) \quad (4.74)$$

where C' and C'' are the first and second derivatives of $C(\gamma)$ and $o(1)$ is a term that vanishes asymptotically as $r \to \infty$.

4.5.4.2. Implications on the Bare Cascades and Extension to the Dressed Cascades

Consider now a discrete cascade of the type described in Sec. 4.4.3.2. The cascade has linear multiplicity m, unit mean value, and A_e-mf property inside the unit cube of R^d. Since the bare densities at resolution $r = m^n$, $\varepsilon_{r;b}$, have the same distribution as A_r, the previous results for A_r apply directly also to $\varepsilon_{r;b}$.

What happens for the dressed densities $\varepsilon_r = \varepsilon_{r;b} Z$ is less obvious, because Z does not have a distribution of the same type as A_r, for any r. The problem has been studied by Schertzer and Lovejoy,[45] and in greater detail by Veneziano.[70] The following is proven in the latter reference. Define the dressed moment scaling function $K_d(q)$ as

$$K_d(q) = \lim_{r \to \infty} \frac{\log E[\varepsilon_r^q]}{\log r} = \begin{cases} K(q), & q < q^* \\ \infty, & q \geq q^* \end{cases} \quad (4.75)$$

where q^* is the order of moment divergence (see Sec. 4.5.3). Also let $C_d(\gamma)$ be the Legendre transform of $K_d(q)$ given as

$$C_d(\gamma) = \max_q \{q\gamma - K_d(q)\} = \begin{cases} C(\gamma), & \gamma \leq \gamma^* \\ C(\gamma^*) + q^*(\gamma - \gamma^*), & \gamma > \gamma^* \end{cases} \quad (4.76)$$

where $\gamma^* = K'(q^*)$ is the slope of $K(q)$ at q^*. Then one can show that a relationship similar to Eq. (4.71) holds for the dressed densities, of the type

$$P[\varepsilon_r > r^\gamma] = g_d(\gamma, r) r^{-C_d(\gamma)} \quad (4.77)$$

where $g_d(\gamma, r)$ is a function that varies slowly with r at infinity. The asymptotic form of this function is similar to that of $g(\gamma, r)$ in Eq. (4.74).[70] Again ignoring the prefactor $g_d(\gamma, r)$ gives the rough approximation

$$P[\varepsilon_r > r^\gamma] \sim r^{-C_d(\gamma)} \tag{4.78}$$

Note that for $\gamma > \gamma^*$, Eqs. (4.76) and (4.78) give $P[\varepsilon_r > r^\gamma] \sim r^{-q^*\gamma}$, which by putting $\varepsilon = r^\gamma$ becomes $P[\varepsilon_r > \varepsilon] \sim \varepsilon^{-q^*}$. This power law is a reflection of the algebraic upper tail of the dressed density ε_r when the order of moment divergence q^* is finite.

Equation (4.78) is fundamental to the study of stationary multifractal extremes, and is the source of the term "multifractal". Next, we comment on these two implications. In addition, Eq. (4.78) could be used to infer $C_d(\gamma)$ from data and then to obtain $K_d(q)$ through the Legendre transform $K_d(q) = \max_\gamma \{q\gamma - C_d(\gamma)\}$.[71,72] However, estimates of $K_d(q)$ obtained by this procedure are generally less accurate than direct estimates from scaling of the moments; therefore, this approach is seldom used.

4.5.4.3. *Cascade Extremes*

Let $P_{r,d}(\varepsilon) = P[\varepsilon_r > \varepsilon]$ be the probability with which, in a d-dimensional cascade, the dressed density at resolution r exceeds a given level ε. Since there are r^d cascade tiles at resolution r, the product $r^d P_{r,d}(\varepsilon)$ is the expected number of exceedances of ε by ε_r in a single cascade realization. One may then take the reciprocal of this quantity

$$T_{r,d}(\varepsilon) = \frac{1}{r^d P_{r,d}(\varepsilon)} \tag{4.79}$$

as the return period of ε expressed in units of cascade realizations. Calculating this return period or equivalently finding how the "T-cascade value" $\varepsilon(r,d,T)$ from Eq. (4.79) depends on the linear resolution r, the spatial dimension d, and the return period T is a problem of much importance in the study of extremes.

The accuracy with which $T_{r,d}(\varepsilon)$ or $\varepsilon(r,d,T)$ are calculated depends, of course, on how accurately the probability $P_{r,d}(\varepsilon)$ in Eq. (4.79) is estimated. Three possibilities, each with strengths and weaknesses, are: (i) use the rough approximation in Eq. (4.78); (ii) use more accurate analytic approximations; and (iii) use the exact value calculated numerically from the marginal distribution of ε_r (see Sec. 4.5.3). The first approach is the least accurate one, but suffices to derive important asymptotic scaling properties of $\varepsilon(r,d,T)$. The second alternative is more complicated, but provides valuable numerical estimates of $\varepsilon(r,d,T)$ and information on the ranges of r and T for which the asymptotic scaling relations reasonably apply. The last alternative is, of course, exact, but is computationally tedious and provides little insight.

Other possibilities to determine return-period values arise from using different definitions of the return period, such as taking T as the reciprocal of the probability with which ε is exceeded in a single cascade realization. We shall consider this alternative at the end of the section.

(i) Asymptotic Scaling Properties of the Return-period Values

The prefactor $g_d(\gamma,r)$ is immaterial to the asymptotic behavior of $\varepsilon(r,d,T)$ for ($r\to\infty$, T finite) and (r finite, $T\to\infty$). Therefore, for these asymptotic analyses, one may use the rough approximation in Eq. (4.78) with $C_d(\gamma)$ in Eq. (4.76). The objective is to find γ such that $P[\varepsilon_r > r^\gamma] \sim r^{-C_d(\gamma)} = 1/(r^d T)$, i.e. such that $C_d(\gamma) = d + \log_r(T)$.

Consider first the high-resolution limit ($r\to\infty$, T finite). As shown below, in this case $\gamma < \gamma^*$. Therefore $C_d(\gamma) = C(\gamma)$ and γ must satisfy $C(\gamma) = d + \log_r(T)$. Also note that, since T is finite, $\log_r(T)\to 0$ as $r\to\infty$. Hence γ is infinitesimally close to the value γ_d such that $C(\gamma_d) = d$, and one may use linear Taylor series expansion of $C(\gamma)$ around γ_d to study the effect of T on γ. The expansion is $C(\gamma) = d + q_d(\gamma - \gamma_d)$, where $q_d = C'(\gamma_d)$ is the moment order associated with γ_d. Solving $d + q_d(\gamma - \gamma_d) = d + \log_r(T)$ for γ gives $\gamma = \gamma_d + (1/q_d)\log_r(T)$. We conclude that, under ($r\to\infty$, T finite), the return-period value $\varepsilon(r,d,T) = r^\gamma$ scales with r and T as $\varepsilon(r,d,T) \sim r^{\gamma_d} T^{1/q_d}$. Since $\gamma < \gamma^*$, our previous use of the first expression in Eq. (4.76) is justified.

Now, consider the case (r finite, $T\to\infty$). Since r is finite and $r^\gamma\to\infty$, it must be that $\gamma\to\infty$, and one must use the expression of $C_d(\gamma)$ in Eq. (4.76) for $\gamma > \gamma^*$. The condition $q^*\gamma - d(q^*-1) = d + \log_r(T)$ gives $\gamma = d + (1/q^*)\log_r(T)$. Therefore, $\varepsilon(r,d,T) \sim r^d T^{1/q^*}$. To summarize, the two asymptotic scaling results are

$$\varepsilon(r,d,T) \sim \begin{cases} r^{\gamma_d} T^{1/q_d}, & r \to \infty, T \text{ finite} \\ r^d T^{1/q^*}, & r \text{ finite}, T \to \infty \end{cases} \quad (4.80)$$

The parameters γ_d, q_d, and q^* are illustrated in Fig. 4.9. The result in Eq. (4.80) for (r finite, $T \to \infty$) was first derived by Hubert et al.,[73] and that for ($r \to \infty$, T finite) was obtained by Veneziano and Furcolo,[74] both in the context of intensity-duration-frequency (IDF) estimation of rainfall extremes. The latter reference also shows that these asymptotic scaling properties do not depend on the precise definition of the return period T and, for example, hold also when T is defined using $\varepsilon_{\max}(r,d)$, the maximum of ε_r inside the unit cube in R^d. This result and the first limit in Eq. (4.80) imply that, at high resolutions, the maximum $\varepsilon_{\max}(r,d)$ in a cascade realization scales with the linear resolution r as

$$\varepsilon_{\max}(rr_o, d) \stackrel{d}{=} r_o^{\gamma_d} \varepsilon_{\max}(r,d) \quad (4.81)$$

Equation (4.81) holds in good approximation also at moderate resolutions, except in the far upper tail, which in Eq. (4.80) corresponds to the limit for very high return periods T.

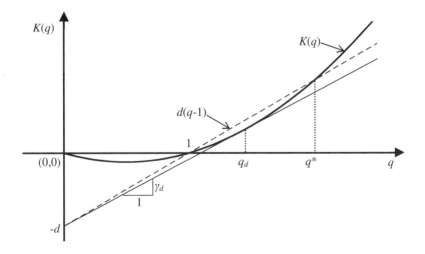

Fig. 4.9. Illustration of the moment-scaling function $K(q)$. The moment orders q_d and q^* and the "singularity order" γ_d control the asymptotic scaling of multifractal extremes; see Eq. (4.80).

When dealing with measures in spaces of dimension $d > 1$, for example in $(d - 1)$ spatial dimensions plus time, interest is often in the maximum of the moving average in a d-dimensional region A as A translates in a sub-space of dimension $d' < d$. For example, in the case of rainfall in two spatial dimensions plus time, one may be interested in the annual maximum of the rainfall intensity averaged in space over a river catchment and in time over the response time of the basin. In this case, averaging is in 3D, whereas maximization is along time, i.e. $d' = 1$.

Denote by $\varepsilon_{\max}(r,d,d')$ the maximum in a single cascade realization under these more general conditions. Using results in Veneziano and Langousis,[75] one can show that at high resolutions

$$\varepsilon_{\max}(rr_o, d, d') \stackrel{d}{=} r_o^{\gamma_{d'}} \varepsilon_{\max}(r, d, d') \qquad (4.82)$$

This is the same as Eq. (4.81), with $\gamma_{d'}$ in place of γ_d. Therefore, the scaling of the maximum depends on the dimension d' of the space over which the maximum is taken, and not the dimension d of the space of averaging. In Sec. 4.8.3, we use Eq. (4.82) to interpret the scaling of floods from the scaling properties of rainfall and the width function.

(ii) Analytical Approximation of Non-asymptotic Return-period Values

Equation (4.80) shows that, as $r \to \infty$ or $T \to \infty$, the effects of the resolution r and the return period T on $\varepsilon(r,d,T)$ become separable and the dependence on each parameter is of the power-law type. These asymptotic scaling properties are not sufficient when one needs the actual return period value $\varepsilon(r,d,T)$ for finite r and T. As indicated above, one possibility is to calculate $\varepsilon(r,d,T)$ from the marginal distribution of ε_r, but doing so requires tedious multiple convolutions (see Sec. 4.5.3).

As an alternative, Langousis et al.[76] suggested approximations to the marginal distribution of ε_r that produce analytical results. This was done for a one-dimensional cascade of the beta-lognormal type, used to represent temporal rainfall at a fixed geographic location. For such a cascade, the amplitude scaling factor A_r has a probability mass $1 - r^{-C_\beta}$ at 0 and $A_r^+ = (A_r | A_r > 0)$ has lognormal distribution with log-mean $m_{\ln(A_r^+)} = (C_\beta - C_{LN})\ln(r)$ and log-variance $\sigma^2_{\ln(A_r^+)} = 2C_{LN}\ln(r)$, where

C_β and C_{LN} are non-negative parameters such that $C_\beta + C_{LN} < 1$. The function $K(q)$ has the form $K(q) = C_\beta(q-1) + C_{LN}(q^2 - q)$.

The bare density at resolution r has the same distribution as A_r. To approximate the distribution of the dressed density, Langousis et al.[76] developed a two-step procedure. First they approximated the dressing factor Z by A_{r_Z}, where r_Z is such that A_{r_Z} matches some integer moment of Z. Calculation of the integer moments of Z is rather straightforward (see Eq. (4.67)), while the moments of A_{r_Z} are given by $E[(A_{r_Z}^q)] = r_Z^{K(q)}$. With Z replaced by A_{r_Z}, the dressed density at resolution r has the same distribution as A_{rr_Z}.

The second step improves the above approximation in the upper tail region, by "grafting" a power-law tail to the distribution of A_{rr_Z} above the point where the log-log slope of the exceedance probability equals $-q*$. Using this approximating distribution, one obtains an analytical expression for $T_{r,d=1}(\varepsilon)$, the return period of the event $[\varepsilon_r > \varepsilon]$. A further approximation allows one to obtain the return period value $\varepsilon(r,1,T)$ in analytical form[76]

$$\varepsilon(r,1,T) = \begin{cases} (rr_Z)^{C_\beta - C_{LN} + 2\sqrt{C_{LN}[\log_{rr_Z}(rT/\delta) - C_\beta]}}, & T \leq T_r^* \\ (rr_Z)^{1 + \frac{C_{LN}}{1-C_\beta}[\log_{rr_Z}(rT/\delta) - 1]}, & T > T_r^* \end{cases} \quad (4.83)$$

where

$$T_r^* = \frac{\delta}{r}(rr_Z)^{[\frac{(1-C_\beta)^2}{C_{LN}} + C_\beta]} \quad (4.84)$$

and δ is a constant close to 5. Over a wide range of r and T, $\varepsilon(r,1,T)$ in Eq. (4.83) is close to the exact value calculated numerically from the marginal distribution of ε_r.

One may use either Eq. (4.83) or the exact (numerically calculated) distribution of ε_r to investigate the range of resolutions and return periods for which the asymptotic scaling relations in Eq. (4.80) hold in good approximation. Figure 4.10 shows log-log plots of $\varepsilon(r,1,T)$ against r and T for a lognormal cascade with $C_1 = 0.1$. The values were obtained numerically using the procedure of Sec. 4.5.3 (see for example Fig. 4.8). The regions where the slopes of the plots are within 10% of the asymptotic values from Eq. (4.80) are delimited by dashed-dotted lines.

As one can see, there is a wide region of the (r,T)-plane where $\varepsilon(r,1,T)$ is not separable in r and T, and the asymptotic scaling relations cannot be assumed to hold.

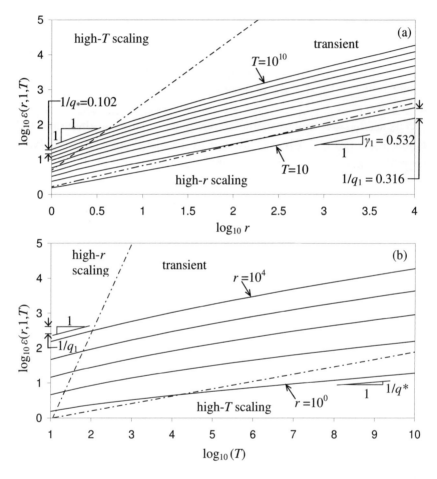

Fig. 4.10. Return-period values $\varepsilon(r,1,T)$ for a one-dimensional lognormal cascade with co-dimension parameter $C_1 = 0.1$: (a) $\varepsilon(r,1,T)$ as a function of r for $\log_{10}(T) = 0(1)10$; (b) $\varepsilon(r,1,T)$ as a function of T for $\log_{10}(r) = 0(1)4$.

In addition, one finds that, already for moderate resolutions, say r on the order of 5, the asymptotic behavior under $T \to \infty$ is attained for very long return periods, which are typically beyond the range of practical interest. This is consistent with the marginal distribution plots in Fig 4.8b.

(iii) Extremes from the Maximum in One Cascade Realization

The previous results are based solely on the marginal distribution of ε_{m^n}. A more appropriate approach to the extremes is to calculate the distribution of $\varepsilon_{m^n,\max}$, the maximum dressed density ε_{m^n} among the m^{nd} cubic tiles of the cascade at level n. In this case, $T_{r,d}(\varepsilon)$ is defined as the reciprocal of the exceedance probability $P[\varepsilon_{m^n,\max} > \varepsilon]$. Next, we give a numerical method from Veneziano and Langousis[77] to calculate the distribution of $\varepsilon_{m^n,\max}$.

The cumulative distribution function $F_{\varepsilon_{m^n,\max}}$ can be found recursively for $n = 0, 1, \ldots$ by noting that $\varepsilon_{m^0,\max}$ has the distribution of the dressing factor Z and may be found by the method of Sec. 4.5.3, and for any $n > 0$, $\varepsilon_{m^n,\max}$ is the maximum of m^d iid variables distributed like $W\varepsilon_{m^{n-1},\max}$, where W is the generator of the cascade. Therefore, working with logs,

$$F_{\log(\varepsilon_{m^n,\max})} = \begin{cases} F_{\log(Z)}, & n=0 \\ [F_{\log(\varepsilon_{m^{n-1},\max})} * f_{\log(W)}]^{m^d}, & n=1,2,\ldots \end{cases} \quad (4.85)$$

where f_X is the probability density function of X and $F*f$ is the convolution $\int_{-\infty}^{\infty} F(x)f(s-x)dx$.

While generally feasible, this numerical procedure is computationally demanding; it requires repeated convolutions, first to find the distribution of Z and then to implement Eq. (4.85). To avoid such complications, one may want to use approximations. For example, as described earlier in this section, one might replace Z with a random variable of the W type. One might also neglect dependence among the dressed densities in different tiles and replace Eq. (4.85) with

$$F_{\log(\varepsilon_{m^n,\max})} = [F_{\log(Z)} * f_{\log(W_1 W_2 \ldots W_n)}]^{m^{nd}}, \quad n=1,2,\ldots \quad (4.86)$$

Veneziano and Langousis[77] evaluated these approximations and found that replacing Z by a W-type variable without a "grafted" Pareto tail generally produces accurate results, except in the extreme upper tail where the power-law behavior of the exact distribution of $\varepsilon_{m^n,\max}$ is lost. If a Pareto tail is grafted, the approximation becomes accurate also in the extreme upper region of the distribution. By contrast, ignoring dependence among the cascade tiles produces larger errors in the body of the distribution of $\varepsilon_{m^n,\max}$, but has little effect on the extreme quantiles. The reason why Eq. (4.86) is accurate in the extreme upper tail of $\varepsilon_{m^n,\max}$ is that, as $\varepsilon \to \infty$, the exceedance events $[\varepsilon_{m^n} > \varepsilon]$ in different tiles become independent.

4.5.4.4. Origin of the Term "Multifractality"

Consider the rough limit in Eq. (4.78) for a measure in d-dimensional space. Since there are r^d tiles at resolution r, the expected number of tiles where the level r^γ is exceeded is $r^{d-C_d(\gamma)}$. Therefore, $D_\gamma = d - C_d(\gamma)$ has formally the meaning of a box fractal dimension (see Sec. 4.2.1) and $C_d(\gamma)$ has formally the meaning of box fractal co-dimension. Notice, however, that these are not the fractal dimension and co-dimension of an individual set; rather, they are fractal-like exponents for the sequence of exceedance sets (what changes with r is not only the size of the covering tiles, as in the case of individual fractal sets, but also the set itself!). Unless $K(q)$ is linear in q, the "fractal dimension" D_γ varies with the threshold exponent γ and, for this reason, the measure is said to be multifractal. If $K(q)$ is linear in q, D_γ is constant with γ and the measure is "monofractal".

More often, the term multifractality is associated with the distribution of singularities of different orders. A measure is said to be singular of order γ at $\underline{t}$ if the average measure density in a ball of diameter $1/r$ centered at $\underline{t}$, $\varepsilon(r,\underline{t})$, satisfies

$$\lim_{r \to \infty} \frac{\log(\varepsilon(r,\underline{t}))}{\log r} = \gamma \qquad (4.87)$$

If the exceedance events in Eq. (4.78) are caused by singularities of order γ or greater, then Eq. (4.78) may be interpreted as saying that the set of points with singularity of order γ is fractal, with fractal dimension $D_\gamma = d - C_d(\gamma)$. However, this loose interpretation is unsatisfactory on two accounts: (i) As noted above, D_γ is not the fractal dimension of any particular set; and (ii) more fundamentally, for realizations of stationary multifractal measures, the limit in Eq. (4.87) generally does not exist;

hence, at any given location $\underline{t}$ the singularity exponent γ is usually not defined. We conclude that it is more accurate and general to interpret $C_d(\gamma)$ as the exceedance-probability scaling function in Eq. (4.78) than seeing it as a "co-dimension of singularities," as is often done in the context of stochastic multifractal processes.

4.6. Forecasting and Downscaling of Stationary Multifractal Measures

In this section, we discuss techniques to forecast stationary multifractal measures from observations of their infinite past and to produce estimates or simulations of fine-scale values given observations at coarser scales. The latter problem is often referred to as downscaling.

4.6.1. *Forecasting*

Let $\varepsilon(d\underline{x}, dt)$ be a stationary multifractal measure density in d spatial dimensions plus time. The special case $d = 0$ corresponds to a measure density $\varepsilon(dt)$ in time only. A frequently encountered problem is to forecast the future evolution of ε from observation of its past values.

Standard linear prediction methods based on second-moments are not well-suited for this purpose due to non-negativity and, in general, strong non-normality of ε. Two main ideas have been proposed for forecasting, which often require some approximation of the original process. One is to seek a representation of $\varepsilon' = \log(\varepsilon)$ as causally filtered zero-mean white noise (CFWN). If such a representation can be found, then forecasting is accomplished by recovering the past noise from past observations of ε' and then estimating future values of ε' by setting the future noise to its zero mean. Marsan *et al.*[33] developed this approach for universal multifractal measures in d spatial dimensions plus time and applied the procedure to space-time rainfall.

The second idea, which has been used with processes in time only, is to approximate ε as a discrete multifractal cascade with hidden Markov properties. Specifically, the log of the measure density, ε', is expressed as the sum of independent processes $W_j'(t)$, $j = 1, 2, \ldots, k$, where j is an index of scale and k is the number of cascade levels retained in the representation, plus possibly a random noise $n(t)$ to account for higher-frequency fluctuations. If the processes $W'_j(t)$ are Markov, then ε' can be expressed in terms of a hidden Markov process whose state at time t is the vector $\underline{W}'(t) = [W'_1(t),...,W'_k(t)]$ and one may use Bayesian updating and forecasting algorithms similar in concept to the Kalman filter to recursively estimate the future state $\underline{\hat{W}}'(t+\eta)$ from information up to t and thus forecast ε' as $\hat{\varepsilon}'(t+\eta) = \sum_1^k \hat{W}'_j(t+\eta)$. Calvet and Fisher[78] developed a method of this type for what they call Poisson Multifractal Measures (PMM) with applications to financial time series, and Chou[79] used a similar approach with lognormal cascades in time for temporal rainfall prediction. Next, we describe these forecasting strategies in greater detail.

Causally Filtered White Noise Method

We start by reviewing the classic problem of predicting future values of a stationary, but not necessarily multifractal, process $X(t)$ using its second-moment properties and exact or noisy observations of its infinite past.[80-82] Then, we describe how Marsan et al.[33] adapted this classical theory to deal with universal multifractal processes in space and time.

Suppose that $X(t)$ is a stationary process with zero mean and covariance function $B(\tau) = E[X(t)X(t+\tau)]$. The optimal linear predictor of $X(t+\eta)$ based on the exact (non-noisy) observation of X up to t has the form

$$\hat{X}(t+\eta) = \int_0^\infty h_\eta(\tau) X(t-\tau) d\tau \qquad (4.88)$$

where the filter function $h_\eta(\tau)$ satisfies

$$\int_0^\infty h_\eta(\tau) B(t-\tau) d\tau = B(t+\eta) \qquad (4.89)$$

As an alternative to Eqs. (4.88) and (4.89), one may seek a representation of $X(t)$ as linearly filtered zero-mean white noise $n(t)$ with a filter function $g(\tau)$ that vanishes for $\tau < 0$ (causal or non-anticipative representation). The process $X(t)$ needs not have a spectral density $S(\omega)$; if it does, then the Fourier transform of $g(\tau)$, $G(\omega)$, must satisfy $|G(\omega)|^2 \propto S(\omega)$.

Scaling and Fractals in Hydrology

If the causal filter function $g(\tau)$ is known and the past noise $n(\tau)$, $\tau \leq t$, has been recovered from the observation of X up to t, then $X(t + \eta)$ is optimally predicted as

$$\hat{X}(t+\eta) = \int_{-\infty}^{t} g(t+\eta-\tau)n(\tau)d\tau \qquad (4.90)$$

The recovery of $n(\tau)$ up to t from $X(\xi)$, $\xi \leq t$, follows from noting that, if $g^{(-1)}(\tau)$ is the Fourier transform of $G^{-1}(\omega)$, then the transformation $n(\tau) \to X(t)$ can be inverted to give

$$n(\tau) = \int_{-\infty}^{t} g^{(-1)}(\tau-\xi)X(\xi)d\xi \qquad (4.91)$$

Equation (4.91) gives the exact noise for $\tau \leq t$. For $\tau > t$, the same equation gives values of n such that X vanishes after t. These values are, of course, incorrect, but they are not used for prediction; see Eq. (4.90).

Marsan et al.[33] proposed a method of this type to predict the log of universal multifractal fields in space and time. The fields considered are isotropic in space, but may have different scaling properties in space and time. Here, we review the approach for the simpler case of lognormal fields with isotropic scaling in d spatial dimension plus time.

Marsan et al.[33] notice that, if a filter function $g(\underline{x},t)$ in $(d+1)$-space has the form

$$g(\underline{x},t) \sim \|\underline{x},t\|^{-(d+1)/2} \qquad (4.92)$$

then its Fourier transform satisfies

$$G(\underline{k},\omega) \sim \|\underline{k},\omega\|^{-(d+1)/2} \qquad (4.93)$$

This holds for $\|\cdot\|$ any norm such that $\|r\cdot\| = r\|\cdot\|$ (for details, see their Appendix B). For example, if a filter of the type in Eq. (4.92) with $\|\cdot\| = |\cdot|$ is applied to Gaussian white noise, the resulting field is isotropic, with spectral density

$$S(\underline{k},\omega) \propto |G(\underline{k},\omega)|^2 \sim |\underline{k},\omega|^{-(d+1)} \qquad (4.94)$$

The function $S(\underline{k},\omega)$ in Eq. (4.94) corresponds to the spectral density of the log of a lognormal multifractal measure. However, this filtering representation is not causal and, therefore, is not suitable for prediction. The challenge is to find a causal filter that produces spectral densities similar (if not identical) to Eq. (4.94). Marsan et al.[33] consider filters of the type in Eq. (4.92) with norm

$$\| \underline{x}, t \| = \left(|\underline{x}|^{(d+1)/2} + |t|^{(d+1)/2} \right)^{2/(d+1)} \tag{4.95}$$

In this case, the causal filter function

$$g^+(\underline{x},t) \sim H(t) \left(|\underline{x}|^{(d+1)/2} + |t|^{(d+1)/2} \right)^{-1} \tag{4.96}$$

where $H(t)$ is the Heaviside function [$H(t) = 0$ for $t < 0$, $H(t) = 1$ for $t \geq 0$] gives spectral density functions $S(\underline{k},\omega)$ of the filtered process of the type

$$S(\underline{k},\omega) \sim \frac{1}{|\underline{k}|^{d+1} + |\omega|^{d+1}} \tag{4.97}$$

While for $d \neq 1$ this spectral density differs from the target function in Eq. (4.94), Marsan et al.[33] consider it an acceptable approximation.

Practical implementation of the predictor is not discussed in Marsan et al.,[33] but one could proceed as follows:

- Take the Fourier transform of the original data;
- Apply the inverse filter $G^{+^{-1}}(\underline{k},\omega)$;
- Transform back to physical space to recover the noise $n(\underline{x},t)$; and
- Set $n(\underline{x},t) = 0$ after the present time and filter forward in either physical or Fourier space.

Hidden Markov Process Method

Calvet and Fisher[78] developed a different forecasting strategy for what they call Poisson Multifractal Measures (PMM). A PMM is a stationary variant of a discrete cascade in time, with random exponentially distributed tile sizes defined by a Poisson point process. One starts from a uniform measure in the unit interval and proceeds in discrete levels, like in the standard cascade construction. At the first level, a Poisson point process with intensity λ produces a partition of the unit interval. Inside each level-1 tile, the measure density is multiplied by an independent realization of a random variable W with mean value 1. At level 2, each level-1 tile is further partitioned using a Poisson point process with rate $b\lambda$, $b > 1$, and the measure density in each sub-tile is multiplied by an independent realization of W. This construction procedure continues to higher levels. At the generic level j, each tile at level $j - 1$ is partitioned by a Poisson process with intensity $b^{j-1}\lambda$ and again the measure density in each sub-tile is multiplied by an independent realization of W. At small scales, the resulting measure has the same multifractal scaling properties as a standard cascade with multiplicity b and generator W.

What makes the PMM model attractive for prediction is that its value depends on a hidden continuous-time Markov process whose state at time t is the collection of all the W multipliers that contribute to the process at that time. For the fully developed measure, the number of these W variables is infinite. However, that number becomes finite if, in approximation, one terminates the cascade model at some finite level k. This is what Calvet and Fisher[78] do for prediction.

The hidden Markov property and the Poisson structure of the PMM model make it conceptually easy (under simplifying assumptions) to find the distribution of the future Markov state and future value of the process, given the current state. In essence, one uses a Bayesian updating approach conceptually similar to the Kalman filter to recursively find the distribution of the state at time t, given the same distribution at time $t - 1$ and the observation of the measure density ε in $[t - 1, t]$. A limitation of the uncertainty updating algorithm is that the dimension of the state vector is very large (the W variables are discretized and any combination of discrete values of $W_1,...,W_k$ constitutes a possible state).

As Calvet and Fisher[78] point out, knowledge of the latent Markov state and knowledge of only the measure in $[t - 1, t]$ produce different forecasts. For example, the same value of $\varepsilon[t - 1, t]$ can result as the product of low small-scale W's and high large-scale W's or vice-versa, but these two latent Markov states would produce quite different forecasts. This is consistent with the observations by Marsan et al.[33]

The PMM model of Calvet and Fisher[78] makes sense for financial time series in which changes in market conditions occur rapidly, marking the beginning and end of relatively stable trading periods. For other applications, where the multiplier W at a given level fluctuates in a continuous manner, continuous models or mixed models with a probability mass at zero make better sense. A simple example is the lognormal cascade used by Chou[79] to forecast temporal rainfall.

Chou[79] considered a cascade in which at each level of construction j the measure density at level $j-1$ is multiplied by $W_j(t) = \exp\{W_j'(b^j t)\}$, $b > 1$, where $W_j'(t)$ is an independent copy of a Gauss-Markov process $W'(t)$ with variance σ^2, mean value $m = -0.5\sigma^2$, and correlation distance r_o. At small scales, the resulting measure has the same multifractal properties as a standard lognormal cascade with generator $W = e^{W'}$ and multiplicity b. We call this the continuous lognormal cascade (CLC) model (the model is continuous in time but still discrete in scale).

Like the PMM, the CLC model has a latent Markov state, which at time t comprises all the log multipliers $W_j'(t)$. Again, like in the PMM, one can approximate the latent state by the set of multipliers at levels up to k and absorb the rest of the variability of the log measure into an independent additive noise term $n(t)$. If $n(t)$ is approximately normal, then one can use the Kalman filter to update the mean vector and covariance matrix of the log state vector at a series of discrete times. This results in a significant simplification of the Calvet-Fisher updating procedure. The simplification is especially critical if one wants to extend the prediction problem to space-time processes, for which the latent state varies with spatial location (hence the number of W' state variables equals the number of discrete spatial locations times the number of cascade levels k).

From an application of the CLC model to temporal rainfall, Chou[79] concluded, like Marsan et al.[33] and Calvet and Fisher,[78] that the forecasting error increases rapidly with the forecast lead time η and that, as η increases, forecasting relies more heavily on the state variables with higher correlation in time.

In summary, two fundamentally different methods exist to forecast stationary multifractal measures. One method assumes that the measure can be represented (or approximated) as causally filtered white noise, whereas the other approximates the measure as the product of a small number of processes at different scales, with Markov dependence in time. While these methods offer effective forecasting tools, the fact remains that the dominant role of the high-frequency components in stationary multifractal measures causes the forecast errors to be large, unless the interest is in temporal averages over long future time periods.

4.6.2. *Downscaling*

A common problem in geophysics and hydrology is the limited availability of measurements at high temporal and spatial resolutions with which to run detailed models. This calls for methods to "downscale" the available low-resolution data in a way that is consistent with both the probability laws of the phenomenon and the observations. A prime example is the generation of detailed rainfall time series or space-time rainfall distributions consistent with results from global circulation models or large-scale remote sensing observations.[83,84] Several methods have been developed specifically for this application. Here we discuss the problem of downscaling with primary focus on rainfall.

One should distinguish between two approaches to downscaling. One is conditional analysis, which aims at calculating the distribution or at least the first two moments of the fine-scale values conditioned on coarser-scale observations. This operation is also referred to as data assimilation. More often, what one needs is conditional simulation, i.e. the generation of an ensemble of high-resolution realizations that are consistent with the observations. While, in principle, conditional analysis could be followed by simulation and ensemble simulation could be used to infer conditional distributions, these operations may be difficult to implement (especially the former, which, for example, is impossible if conditional analysis is limited to the first two moments). Therefore, it is useful to keep the distinction between conditional analysis and conditional simulation.

Another useful distinction concerns the format of the data, for which we consider three representative cases: First, the simplest case is when one observes the average rain rate in a single large region S of R^d and downscaling is needed within S. A second case is when S is partitioned into coarse sub-regions $S_1,...,S_n$, the average rainfall intensity $\varepsilon(S_i)$ is observed in each sub-region, and downscaling is needed in S. The third, and most general, case is when S is partitioned into $S_{j,1},...,S_{j,n_j}$ at different resolutions j, and observations are available for some subregions at each resolution. This is a reasonable idealization when a rainfall field is observed using different instruments, for example satellites, radar, and raingages. In all three cases, the observations may be exact or contain measurement or model errors. These errors should be accounted for when downscaling. Hereafter, we refer to the above data availability cases as data of Types 1, 2, and 3.

Finally, concerning the rainfall model used, one may distinguish between models that are strictly scaling (multifractal models of various kinds) or embody the notion of scale but are not scale-invariant (e.g. bounded cascades), and models that are not built around scale relations (such as traditional clustered pulse models). Our primary interest is in the former. One would expect advantages from the use of scaling or scale-based models due to the fact that these models explicitly link the values of the process at different resolutions, which is what downscaling essentially requires. We start with a brief review of early developments in rainfall downscaling and then focus on techniques based on scaling concepts.

Downscaling emerged as an important issue long before scaling models of rainfall were developed. There is now a large body of literature on conditional analysis using linear prediction models and first- and second-order moments of the rain rate. This approach was initially proposed by Valencia and Schaake[85,86] and developed further by many authors. For reviews of the original method and the variants that followed, see Salas[87] and Koutsoyiannis et al.[88] Due to exclusive consideration of first and second moments, these methods cannot resolve the nonlinearities and highly intermittent structure that are characteristic of rainfall. Application is usually to data of Type 2. As we shall see later, conditional second-moment methods have been recently adapted to data of Type 3 using scale-based rainfall models. When rainfall is represented as a lognormal field and one works with logs,[89] second-moment analysis becomes sufficient to provide full-distribution conditional information and, at least conceptually, can be used for subsequent conditional simulation. However, this is not possible if there are deviations from lognormality, for example due to the presence of dry periods.

For conditional simulation, full-distribution models are more appropriate than second-moment models. An especially attractive class of full-distribution rainfall models are the clustered-pulse representations pioneered by Le Cam.[90-96] The models embody some notion of scaling (pulses at different resolutions have an essentially self-similar branching arrangement) and their construction proceeds from lower to higher resolutions, but it is difficult to specify the initial coarse-scale conditions implied by the large-scale observations, even for the simpler case of Type 1 data. One approach is to make an educated guess at those large-scale conditions (for example, the location and intensity of large mesoscale precipitation areas), simulate the finer-scale fluctuations using the model, and finally scale the simulated rainfall field to match the observations.

Conditional Simulation with Data of Types 1 and 2

Next, we discuss the use of scale-based models with data of Type 1 or 2. In addition to multifractals, these models include wavelet representations,[97] and bounded cascades.[49,51,98,99] An advantage of scale-based models, especially of the discrete cascade or wavelet type, is that they have a Markovian structure in scale, with the bare average rain rates defining the state at each given scale. This Markov property allows one to de-couple the use of information from the coarse-scale observations (which provide initial conditions at large scales) from knowledge of the stochastic rainfall process (which is used to generate conditional high-resolution values). This strategy is especially attractive for data of Types 1 and 2 and has been applied to rainfall downscaling in both space[97,98,100,101] and time.[49,51,56,57,99]

While greatly simplified by the use of scale-based representations, conditional simulation from Type 1 or Type 2 data is not without problems. Consider first the case of Type 1 data, which consist of a single average-rainfall value $\varepsilon(S)$. The standard procedure is to simulate rainfall starting from a uniform density in S and then scale the simulated rain rates such that the average in S matches the observation $\varepsilon(S)$, or an appropriately simulated value in the case of noisy measurements. However, technically, the conditional simulation should use generators that depend on $\varepsilon(S)$. For example, if $\varepsilon(S)$ exceeds the expected rain rate μ, the small-scale field should tend to exceed $\varepsilon(S)$ near the center of S and be below $\varepsilon(S)$ near the boundary of S. The opposite is true if $\varepsilon(S) < \mu$. Another effect that is not captured by standard simulation methods is the tendency of the fraction of dry area to increase with decreasing $\varepsilon(S)$. This is again caused by the fact that $\varepsilon(S)$ is informative on the generators of the cascade that have produced that average rainfall intensity. A simple way to account for these effects is to generate more simulations than the number needed and use only those with $\varepsilon(S)$ close to the observed value. One should avoid using discrete cascade algorithms, as these models produce "boxy" and nonstationary realizations.

The case of Type 2 data is more complicated. Applying Type 1 downscaling procedures independently in each sub-region S_i is sub-optimal and generates unwanted discontinuities at the sub-region boundaries. The approach of over-simulating and retaining only realizations whose large-scale averages are close to the observations would reduce such discontinuities, but rapidly becomes impractical as the number of observations exceeds 2 or 3. The development of alternative methods would be an objective worth pursuing.

Conditional Analysis with Data of Type 3

Conditional simulation using Type 3 data is even more complicated and, to our knowledge, has not been pursued using scale-based models. However, conditional distribution methods for data of Type 3 exist, and are reviewed below. The methods take advantage of the Markovian dependence between the (bare) rainfall intensities at consecutive scales to update uncertainty of the field at the resolutions of interest. This is done through Kalman-smoothing algorithms. In some cases, the algorithm is applied to the original field, and in others to log-transformed rain rates after setting the zeros to a positive threshold.

A model of this type, called scale recursive assimilation or SRA, was developed by Primus et al.[89] The procedure uses concepts from multiplicative cascades to find the conditional second-moments of log rainfall averages at some desired spatial scale from noisy multi-resolution measurements. The essence of the method is as follows. Consider a two-dimensional cascade with multiplicity $m = 2$, and denote by ε_n the average rainfall intensity at level n, normalized to have unit mean value. Each parent tile at level $n-1$ with average measure density ε_{n-1} is partitioned into four descendant tiles with measure densities $\varepsilon_{n,i}$ ($i = 1,...,4$) given by

$$\varepsilon_{n,i} = W_{n,i}\varepsilon_{n-1} \qquad (4.98)$$

where the $W_{n,i}$ are independent copies of a unit-mean positive variable W_n whose distribution is allowed to depend on n. Note that Eq. (4.98) applies to bare, not dressed, measure densities (see Sec. 4.4.3.2), but in this formulation the distinction between these two densities is ignored. Primus et al.[89] assume ε_n and W_n to be lognormal, and write

$$\varepsilon_n = \exp\{\varepsilon'_n - 0.5\sigma^2_{\varepsilon'_n}\}, \qquad W_n = \exp\{W'_n - 0.5\sigma^2_{W'_n}\} \qquad (4.99)$$

where ε'_n and W'_n are zero mean normal variables with variances $\sigma^2_{\varepsilon'_n}$ and $\sigma^2_{W'_n}$, respectively. Under Eq. (4.99), Eq. (4.98) has the additive form

$$\varepsilon'_{n,i} = \varepsilon'_{n-1} + W'_{n,i} \qquad (4.100)$$

Primus et al.[89] further assume that, when normalized to mean-1, the generic measurement at level n, Z_n, is related to ε_n as $Z_n = \varepsilon_n V_n$, where V_n is a mean-1 lognormal noise variable. In analogy with ε_n and W_n, the variables Z_n and V_n are written as $Z_n = \exp\{Z'_n - 0.5\sigma^2_{Z'_n}\}$ and $V_n = \exp\{V'_n - 0.5\sigma^2_{V'_n}\}$, where Z'_n and V'_n are normal variables with zero mean. Consequently, the observation model becomes

$$Z'_n = \varepsilon'_n + V'_n \qquad (4.101)$$

The objective of the analysis is to obtain the mean value and variance of the log-transformed rain-rate $\varepsilon'_{n,i}$ for each tile i at any given level n, conditional on all measurements at finer and coarser scales. This is achieved in two steps. In the first step (upward sweep), the algorithm recursively moves from the finest scale to coarser scales and updates the unconditional mean and variance of the $\varepsilon'_{n,i}$ variables to account for observations at resolution levels $\geq n$. In the second step (downward sweep), the algorithm moves from coarser to finer scales to include information from measurements at levels $< n$. The results depend on the error variances $\sigma^2_{V'_n}$ and the generator variances $\sigma^2_{W'_n}$. The former are assumed known, whereas the latter are estimated recursively in such a way that the measurements and measurement residuals (estimates minus measurements) are uncorrelated and the variance of the residuals equals the value from the Kalman estimator.

Tustison et al.[102] followed the same methodology as Primus et al.,[89] but estimated the generator variances $\sigma^2_{W_n'}$ by *a priori* fitting a cascade model with lognormal generator to the rainfall data. The cascade is of the bounded type, with variance $\sigma^2_{W_n'}$ that depends on the cascade level n. A limitation of the Primus et al.[89] method is that conditional analysis for the rainfall intensity in tile $T_{n,i}$ considers exclusively measurements in tiles that either include $T_{n,i}$ or are entirely inside $T_{n,i}$, with no "lateral" transfer of information from other tiles. Other limitations (boxiness and nonstationarity of the field) are intrinsic to the use of discrete cascades. Finally, the algorithm operates in log space and, therefore, cannot explicitly account for zero values (dry regions). Primus et al.[89] exclude dry regions from their analysis, whereas Tustison et al.[102] set the zero values to a small positive quantity to enable log transformation. Both operations are *ad hoc* and affect the calculated conditional moments.

To address the issue of zeros, Gupta et al.[103] suggested a variant of the Primus et al.[89] method in which one works in the original space and applies Eqs. (4.100) and (4.101) directly to the rainfall values without log transformation. Hence, in this formulation, ε_n' is the average measure density at level n when the mean of the field is subtracted and W_n' is a zero-mean random variable with variance that depends on the level n. A limitation of this methodology is that the variance $\sigma^2_{W_n'}$ is not made to depend on the rainfall intensity ε_{n-1}' at the immediately larger scale. This assumption is tenable only for log-transformed variables.

In summary, the Markov structure of multifractal and other scale-based models reduces the complexity of downscaling and multi-resolution data assimilation. However, the present state-of-the-art is not entirely satisfactory, mainly due to the difficulty of enforcing the observations in conditional simulation and, in general, of using continuous models rather than simpler but unrealistic discrete representations.

4.7. Inference of Scaling from Data

This section discusses methods to estimate the scale-invariance properties of stationary multifractal measures $X(dt)$ and processes $X(t)$ with stationary multifractal increments (the so-called multi-affine processes). In both cases, scale invariance is characterized by the probability distribution F_{A_r} of the random factor A_r in Eq. (4.10) or Eq. (4.13) (see Sec. 4.3.2). Therefore, the objective is to estimate F_{A_r} or any of several alternative characterizations of F_{A_r}, such as the moment scaling function $K(q) = \log_r(E[A_r^q])$ or the co-dimension function $C(\gamma)$ in Eq. (4.72) (see Sec. 4.5.4.1). Estimation methods depend on the random process they consider [$X(dt)$ or $X(t)$], the characterization they seek [through F_{A_r}, $K(q)$, $C(\gamma)$, etc.], and the specific procedure they use.

We do not aim to systematically cover all methods. Rather, we consider popular procedures for four inference problems, point at their limitations, and propose corrections or improvements. The problems we consider are: (i) the estimation of $K(q)$, $q > 0$, for stationary measures $X(dt)$; (ii) the estimation of $K(q)$, $q < 0$, for the same measures in the presence of noise or quantization; (iii) the estimation of $K(q)$, $q > 0$, for multi-affine processes $X(t)$ by the gradient amplitude (GA) method; and (iv) the estimation of universal multifractal parameters by the double trace moment (DTM) method. In all four cases, estimation is based on how the moments of quantities extracted from the signal scale with resolution.

Some preliminary remarks should be made on methods that are based on moment-scaling relations. First, in spite of the significant attention it has attracted, the importance of estimating $K(q)$ for negative moments (problem (ii) above) is somewhat overstated. One reason is that the negative moments of A_r often diverge; for example, this is the case when $P[A_r = 0] > 0$. When negative moments exist, their estimation is sensitive to data inaccuracies, for example due to additive noise or quantization. Consequently, if the moment scaling function $K(q)$ is known except for a few parameters, it is generally better to fit the parameters to the positive moments and then obtain $K(q)$ for $q < 0$ through parametric extrapolation, rather than using direct estimation. It should also be noted that, in many applications, the interest is on the large values of the process, which are controlled by the positive moments.

Another general observation is that the factors A_r are not directly accessible and $K(q)$ must be inferred from the scaling of moments of quantities extracted from the signal. The range of finite moments depends on which specific quantity is analyzed (average densities, wavelet coefficients, etc.) and never exceeds the range of finite moments of A_r. Differences may be especially large for the negative moments. For example, in a stationary lognormal multifractal process, all positive and negative moments of A_r are finite, but the moments of the average densities and wavelet coefficients exist for a finite range of q and for wavelets of order 1 or higher, the moments of order $q \leq -1$ diverge. Hence, the range of estimable moments of A_r, especially the negative moments, depends on which quantity is used in the moment-scaling analysis.

4.7.1. *Estimation of $K(q)$, $q > 0$, for Stationary Multifractal Measures*

The standard method to estimate $K(q)$ for a stationary multifractal measure, say in the unit interval, is to determine how the empirical moments of the measure or average measure density scale with resolution. For this purpose, one typically partitions the unit interval into sub-intervals of length 2^{-n}, where $n = 1, 2, \ldots$ is the resolution level, and for each n calculates the empirical moments of ε_n, the average measure density at level n. Multifractality implies that the non-diverging moments $\mu_n^q = E[\varepsilon_n^q]$ vary with n and q as $\mu_n^q \propto 2^{nK(q)}$, where $K(q) = \log_r(E[A_r^q])$.

The function $K(q)$ is usually estimated in two steps.[104] In the first step, one calculates the empirical moments $\hat{\mu}_n^q = \langle \varepsilon_n^q \rangle$ of ε_n and finds $\hat{K}(q)$ for each q as the slope of the linear regression of $\log_2(\hat{\mu}_n^q)$ against n over a range $[n_{\min}, n_{\max}]$ of resolution levels. This operation is performed without assuming any parametric form of $K(q)$; hence, we refer to $\hat{K}(q)$ as a nonparametric estimator. One may generalize this basic estimator by replacing the local average densities ε_n with the absolute wavelet coefficients $2^n |\int \phi(2^n t) X(dt)|$, where ϕ is a wavelet of order 0 or higher. However, this extension is not fundamental and has modest effects on $\hat{K}(q)$. In the second step, one uses least-squares (LS) or other criteria to fit a parametric model $\hat{K}_{par}(q)$ to $\hat{K}(q)$ over a range $[q_{min}, q_{max}]$ of moment orders.

The bias and variance of the nonparametric estimator $\hat{K}(q)$ depend on the actual function $K(q)$ and the range of resolution levels $[n_{min}, n_{max}]$ used in the analysis. The bias and variance of the parametric estimator $\hat{K}_{par}(q)$ depend, in addition, on the range of moment orders $[q_{min}, q_{max}]$ used for parameter fitting. Next, we review what is currently known about these performance statistics and evaluate a broader class of estimators aimed at reducing the error variance. The material in this section is based largely on Furcolo and Veneziano.[105]

Performance of the Nonparametric Estimator $\hat{K}(q)$

Not much is known theoretically about the performance of $\hat{K}(q)$, except that, as the resolution level $n_{max} \to \infty$, $\hat{K}(q)$ becomes linear with some positive slope γ^+ for q above some $q^+ > 1$ and becomes linear with some negative slope γ^- for q below some negative value q^-.[106,107] In this, γ^+ and γ^- are the slopes of the tangents to $K(q)$ with Y-intercept equal to -1 (equal to $-d$ in the general case of measures in R^d) and q^+ and q^- are the values of q at the points of tangency (see Fig. 4.11). Next, we elaborate on the "straightening" of $\hat{K}(q)$ for $q > q^+$, but similar considerations apply also to the negative moments.

The straightening of $\hat{K}(q)$ above q^+ is explained as follows. As the resolution level $n \to \infty$, the ensemble moments $\mu_n^q = E[\varepsilon_n^q]$ are dominated by values of ε_n in excess of $2^{nK'(q)}$, where the "$'$" sign denotes differentiation, but in a single cascade realization these values occur with probability 1 for $q < q^+$ and probability 0 for $q > q^+$.[106-109] It follows that the sample maximum of ε_n, $\varepsilon_{n,max}$, satisfies

$$\lim_{n \to \infty} \frac{\log_2(\varepsilon_{n,max})}{n} = \gamma^+ \qquad (4.102)$$

For $n \to \infty$, the sample maxima $\varepsilon_{n,max}$ control the empirical moments $\hat{\mu}_n^q$ for $q > q^+$, making them behave like $\hat{\mu}_n^q \sim 2^{n(q\gamma^+ - 1)}$. This causes $\hat{K}(q)$ to become linear for $q > q^+$, with slope γ^+.

The straightening of $\hat{K}(q)$ described above applies as the upper limit of the resolution range $[n_{min}, n_{max}]$ goes to infinity. In practice, n_{max} is finite and $\hat{K}(q)$ displays gradual straightening as q increases, becoming asymptotically linear as $q \to \infty$. The asymptotic slope in this finite-resolution case, say $\hat{\gamma}^+$, varies from sample to sample, and is obtained as the slope of the linear regression of $\log_2(\varepsilon_{n,\max})$ against n in the range $[n_{\min}, n_{\max}]$. Transition to linearity might be considered to occur at the moment order $\hat{q}^+$ for which the sample maxima $\varepsilon_{n,\max}$ contribute more than some fixed fraction of the empirical moments $\hat{\mu}_n^q$ (for alternative, but essentially similar criteria, see Harris et al.[104]). As noted already, both the transition order $\hat{q}^+$ and the asymptotic slope $\hat{\gamma}^+$ are realization-dependent, in addition to varying with the resolution range $[n_{\min}, n_{\max}]$ used in the analysis.[104]

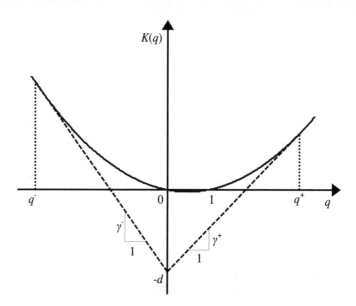

Fig. 4.11. Moment orders q^- and q^+ beyond which the empirical $K(q)$ function becomes linear, also for a record with infinite resolution.

For $n_{\max}$ finite, it is difficult to theoretically quantify the bias and variance of $\hat{K}(q)$. Harris et al.[104] used numerical simulation of discrete cascades to show how the bias and statistical variability of $\hat{K}(q)$ depend on the resolution level n_{dat} of the data (in that study, $n_{\min}$ is close to zero and $n_{\max}$ is close to n_{dat}; hence, what is effectively varied is $n_{\max}$). The results confirm that, for $q > 1$, the estimator $\hat{K}(q)$ is biased towards low values, and that both bias and variance increase as q increases or the resolution level n_{max} decreases.

Alternative Nonparametric Estimators

Furcolo and Veneziano[105] introduced a wider class of nonparametric estimators, with the objective of reducing the error variance of $\hat{K}(q)$. While focus is on the positive moments, the estimators apply equally well to the negative moments if the data are not corrupted by noise or other artifacts and, of course, if those moments exist.

Suppose that the available data consist of a single cascade realization at some finite resolution level n_{dat}, which is also the case considered by Harris et al.[104] The rather large variance of $\hat{K}(q)$ is due to the strong correlation of ε_n in time and across resolution levels n, and could be reduced by increasing the number of independent cascade samples. To artificially increase the number of samples when just one cascade realization is available, we partition the record into sub-cascades, each rooted in a different tile at some resolution level n_o, find $\hat{K}_j(q)$ from each sub-cascade $j = 1,...,2^{n_o}$ using a resolution range $[n_{min}, n_{max}]$ with $n_{min} \geq n_o$, and average the results to obtain

$$\overline{K}(q) = 2^{-n_o} \sum_{j=1}^{2^{n_o}} \hat{K}_j(q) \tag{4.103}$$

The "average nonparametric estimator" $\overline{K}(q)$ depends on (n_o, n_{min}, n_{max}) and for $n_o = 0$ reduces to the standard nonparametric estimator $\hat{K}(q)$. Notice that the sub-cascade estimators $\hat{K}_j(q)$ in Eq. (4.103) are independent. Therefore, the bias of $\overline{K}(q)$ is the same as the bias of $\hat{K}_j(q)$, but the variance of $\overline{K}(q)$ is 2^{n_o} times smaller. The price one pays for increasing the number of independent samples by using a value of n_o larger than zero is that the sub-cascades have a reduced effective resolution $n'_{dat} = n_{dat} - n_o$ compared with the resolution n_{dat} of the original data, causing the bias of the average nonparametric estimator to increase somewhat relative to that of the standard nonparametric ($n_o = 0$) (see below).

The sub-sampling and averaging operations in Eq. (4.103) are formally similar to a variance-reduction technique frequently used in estimating the power spectrum of stationary signals.[110] For power spectrum estimation, one may use also other variance-reduction methods, such as smoothing the empirical spectral density over a range of frequencies, but these alternative methods do not work well for $K(q)$ due to the high correlation of $\hat{K}(q)$ for different q.

Following the approach of Harris et al.[104] for $\hat{K}(q)$, Furcolo and Veneziano[105] estimated the bias and statistical variability of the estimator $\overline{K}(q)$ in Eq. (4.103) by simulating binary multifractal cascades. The cascade generator $W = A_2$ was taken to have lognormal distribution with mean value 1 and log-variance $\sigma_{\ln W}^2 = 2C_1 \ln(2)$, where $0 < C_1 < 1$ is a parameter that controls the variability of $X(dt)$. In this case, $K(q) = \log_2(E[W^q]) = C_1(q^2 - q)$ is finite for all positive and negative moments and $q^{\pm} = \pm\sqrt{1/C_1}$. The bias and variance of $\overline{K}(q)$ depend on C_1, q, and the parameters $(n_o, n_{\min}, n_{\max})$ of the estimators $\hat{K}_j(q)$ in Eq. (4.103). Figure 4.12 shows the bias, standard deviation, and root mean square error (RMSE) of $\overline{K}(q)$ for $C_1 = 0.1$, $q = -2, 2, 3$, and 4, for all possible combinations of $n_o \leq n_{\min} < n_{\max} = 14$. Notice that for $C_1 = 0.1$, $q = 4$ exceeds the moment order $q^+ = \sqrt{10} = 3.162$, beyond which $\overline{K}(q)$ is expected to become linear in q; hence, the bias for $q = 4$ is expected to be high. All results are based on 1000 dressed cascade simulations.

The standard deviation and RMSE in Fig. 4.12 refer to a single cascade realization. If s independent cascade realizations are available and the realization-specific estimates of $K(q)$ are averaged, the bias remains the same but the standard deviation is reduced by a factor $\sqrt{s}$. For $q > 0$, both bias and standard deviation tend to increase with increasing q. In particular, the bias increases rapidly for $q > q^+$ (here $q^+ = 3.162$) due to the linearization discussed above. For $q = -2$, the bias is comparable to that for $q = 2$, but the standard deviation is larger.

In general, the bias decreases as n_o decreases and $n_{\min}$ increases. The standard deviation has an opposite dependence on n_o (due to the number of independent sub-cascades as n_o increases) and is insensitive to $n_{\min} - n_o$, provided that $n_{\min}$ is not too close to $n_{\max}$ ($n_{\min} = n_{\max} - 1$ at the right terminal point of each constant-n_o curve). The standard estimator $\hat{K}(q)$

corresponds to $n_o = n_{min} = 0$, which is the point marked in Fig. 4.12 by an asterix.

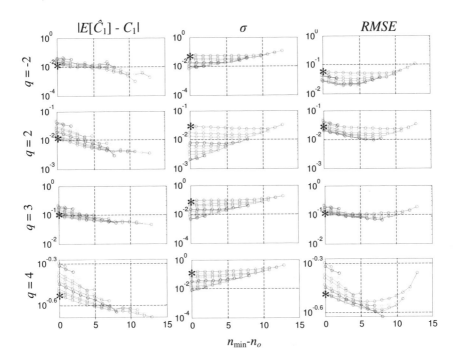

Fig. 4.12. Bias, standard deviation, and RMSE of the estimator $\bar{K}(q)$ in Eq. (4.103) for a lognormal cascade with $C_1 = 0.1$ and $q = -2, 2, 3,$ and 4. Results are for a single cascade at resolution level $n_{dat} = 14$. The parameters n_o and n_{min} of the estimator are varied in each panel: n_o is constant along each line ($n_o = 0$ for the longest line and $n_o = 13$ for the shortest) and $n_{min} - n_o$ varies along the horizontal axis. In all cases, the third parameter n_{max} is set to 14.

The different sensitivities of the bias and standard deviation noted above make for an interesting behavior of the RMSE. For a single cascade realization, which is the case considered in Fig. 4.12, RMSE is dominated by the bias for (high – n_o, low $n_{min} - n_o$) combinations; otherwise, the variance dominates. In general, there is a well-defined optimality region for $n_o \approx 5$ and $n_{min} - n_o$ that varies from 3 to 8 as the moment order q increases from –2 to 4. A rule of thumb for q around 2–3 is to take $n_o = n_{dat}/3$ and $n_{min} - n_o \approx n_{dat}/2$. If many realizations are available, the error variance is lower, and one should use smaller values of n_o and values of n_{min} closer to n_{max}.

Parametric Estimation and Bias Correction

Harris et al.[104] discuss at length the selection of the range $[q_{min}, q_{max}]$ of positive moment orders, so that the straightening of $\hat{K}(q)$ at large q has a limited biasing effect on the parametric estimator $\hat{K}_{par}(q)$. For this purpose, they set $q_{max} = \hat{q}^+$ and suggest two specific ways to obtain $\hat{q}^+$, the moment order above which $\hat{K}(q)$ is essentially linear. The statistical variability of $\hat{q}^+$ and, therefore, of q_{max} is large. Here, we consider the same problem for the average estimator $\overline{K}(q)$, and use a different method to set q_{max} for each sub-cascade j in Eq. (4.103). First, we find the asymptotic slope $\hat{\gamma}^+$ of $\hat{K}_j(q)$ using the maximum densities $\varepsilon_{n,max}$ for the jth sub-cascade, as explained earlier in this section. Then, we find q_{max} as the moment order for which the slope of $\hat{K}_j(q)$ equals $\eta\hat{\gamma}^+$, where η is a parameter between 0 and 1, with recommended values around 0.5. Since for lognormal cascades the function $K(q)$ is a parabola with vertex at $q = 0.5$, the slope of $K(q)$ is proportional to $q - 0.5$. Further, note that the slope $\hat{\gamma}^+$ is attained approximately for $q = q^+ = \sqrt{1/C_1}$. Therefore, typical values of q_{max} are

$$q_{max} \approx 0.5 + \eta(\sqrt{1/C_1} - 0.5) \quad (4.104)$$

For example, for $C_1 = 0.1$ and $\eta = 0.5$, Eq. (4.104) gives $q_{max} \approx 1.83$. Although the value of q_{max} varies randomly from one sub-cascade to the other, we have found, through simulation, that the statistical variability around the value in Eq. (4.104) is small.

Fitting $K(q) = C_1(q^2 - q)$ by least squares to $\overline{K}(q)$ between $q_{min} = 0$ and q_{max}, one obtains an estimate $\hat{C}_1$ of C_1. Figure 4.13 shows the bias, standard deviation, and RMSE of $\hat{C}_1$ as a function of the parameters (n_o, n_{min}) used for $\hat{K}_j(q)$, when $C_1 = 0.1$, $\eta = 0.5$ and $n_{max} = 14$.

As expected, the variance of $\hat{C}_1$ decreases as n_o increases and, in this case of high-resolution data ($n_{dat} = 14$), can be made very small by choosing a large value of n_o. By contrast, the bias increases (slightly) with increasing n_o. The optimal combination of n_o and n_{min} for RMSE is similar to that of the nonparametric estimator $\overline{K}(q)$ for q around 2 (see

Fig. 4.12). This is because, C_1 is sensitive to the values of $\overline{K}(q)$ near the upper limit of the moment range $[0, q_{max}]$ used for parametric estimation and, in the present case, q_{max} is close to 2.

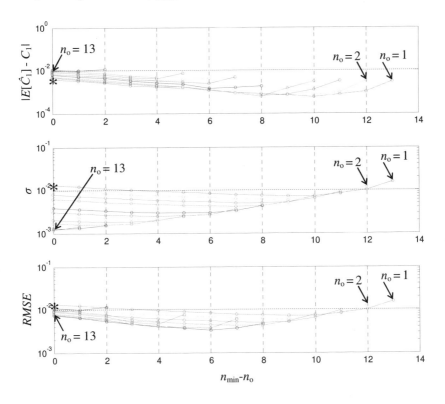

Fig. 4.13. Lognormal cascade as in Fig. 4.12. Bias, standard deviation, and RMSE of the estimator $\hat{C}_1$ as a function of the parameters n_o and n_{min}, for $n_{max} = 14$ and $\eta = 0.5$.

Since, here, $K(q)$ has a known parametric form, one may use simulation to evaluate the bias of $\hat{C}_1$ as a function of the true C_1. This is shown in Fig. 4.14 for the case $n_o = n_{min} = n_{max} - 1 = 13$ and $\eta = 0.5$. With this bias correction, one becomes free to select the parameters (n_o, n_{min}) to minimize the error variance. The minimum is typically attained for $n_o = n_{min} = n_{max} - 1$ and $n_{max} = n_{dat}$ (this is why we use this parameter setting in Fig. 4.14). Bias-corrected minimum-variance estimators are much more accurate than conventional parametric estimators. For example, in the present case, the conventional estimator (with the present selection of q_{max}) has standard deviation and RMSE values around 10^{-2} (see * symbols in Fig. 4.13), whereas the optimum bias-corrected estimator has $\sigma = \text{RMSE} \approx 10^{-3}$ ($n_o = 13$ in Fig. 4.13).

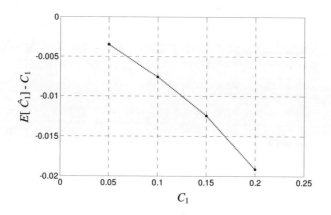

Fig. 4.14. Lognormal cascade as in Fig. 4.12. Bias of $\hat{C}_1$ as a function of C_1 for $n_o = n_{min} = n_{max} - 1$ and $\eta = 0.5$.

4.7.2. *Estimation of K(q), q < 0, for Stationary Multifractal Measures in the Presence of Noise or Quantization*

While there are many similarities between the nonparametric estimation of $K(q)$ for positive and negative moments, there are also important differences. As noted already, the negative moments may diverge and, in any case, are sensitive to additive noise, data quantization, and other measurement artifacts.

Wavelet Tranform Modulus Maxima (WTMM) Estimators

The problem of estimating the scaling of the negative moments when the data are affected by imperfections has been extensively discussed in the literature.[104,111,112] The basic strategy is to calculate the negative moments using values of ε_n away from zero, being careful not to alter the moment-scaling relation. A popular technique is to first calculate the continuous wavelet transform

$$\varepsilon_n(\tau) = 2^n \int \phi[2^n(t-\tau)]X(dt) \qquad (4.105)$$

where ϕ is a wavelet of order zero or higher, and then calculate the absolute moments using only the local maxima of $|\varepsilon_n(\tau)|$.[111-114] [Note that when ϕ has value one inside the unit interval and is zero elsewhere, $\varepsilon_n(\tau)$ is just the average measure density in an interval of duration 2^{-n}.] This is called the wavelet transform modulus maxima (WTMM) method. Several variants of this procedure have been proposed, mainly to avoid local maxima that are still too close to zero. The most popular variant, known as "*WTMM-with-sup*," connects the local maxima at different resolution levels n to form modulus maxima lines and then replaces each local maximum of $|\varepsilon_n(\tau)|$ with

$$\sup_{n \leq n' \leq n_{dat}} |\varepsilon_{n'}(\tau)| \qquad (4.106)$$

where the supremum is along the maxima line that passes through the point.[113,115,116]

The WTMM estimates $\hat{K}_{WTMM}(q)$ have two bias components. One component is similar to the bias of the standard estimator $\hat{K}(q)$, and is due to dependence among the local maxima of $|\varepsilon_n(\tau)|$ for different n. The other component is due to a symmetry breaking when, as is done in practice, the wavelet transform in Eq. (4.105) is evaluated at constant intervals $\Delta\tau = 2^{-n_{dat}}$ corresponding to the resolution of the data. Not to bias the scaling of the moments, the number of $\varepsilon_n(\tau)$ values from which the local maxima are extracted should vary with n as 2^n, whereas in the WTMM method that number is kept constant and equal to $2^{n_{dat}}$ for all n. For $q > 1$, this tends to increase the moments of the local maxima at low resolutions, relative to those calculated at high resolutions, and therefore the estimate of $K(q)$ decreases.

Another problem is that the WTMM procedure is inefficient in filtering out values of $\varepsilon_n(\tau)$ that are close to zero. Alternatives, like *WTMM-with-sup*, are more effective in this regard, but suffer from other problems, such as the symmetry break from the fact that the supremum is over a range of resolution levels $[n, n_{dat}]$ that becomes narrower as n increases. Moreover, the *WTMM-with-sup* method is ineffective when the values of ε along the maxima lines tend to increase with increasing resolution, is complicated to implement, lacks transparency, etc.

Alternative Methods

Furcolo and Veneziano[105] suggested alternatives to *WTMM*-based methods, with the objectives of avoiding bias-inducing operations, retaining simplicity, and including tunable parameters to focus on values of $|\varepsilon_n(\tau)|$ that are at different "distances" from zero. This last feature allows one to obtain accurate estimates of $K(q)$ under different levels of data corruption, for example different additive noise or quantization levels.

To avoid biases, one should subject the wavelet transform to a filtering operation that preserves the scaling of the moments. For example, one could use only the coefficients $\varepsilon_n(\tau_j)$ for $\tau_j = j2^{-n}$ and $j = 1, 2, ..., 2^n$. One could then use the maximum of $|\varepsilon_n(\tau_j)|$ in each non-overlapping interval of length $2^{-n+\Delta}$, where Δ is a given positive integer (larger values of Δ let one focus on increasingly higher values of $|\varepsilon_n(\tau_j)|$, although the number of maxima retained for moment analysis decreases). More generally, one could retain the largest m values of $|\varepsilon_n(\tau_j)|$ in each $2^{-n+\Delta}$ interval, with m a given number between 1 and 2^Δ. We refer to these as (m, Δ) estimators.

Figure 4.15 illustrates the performance of (m, Δ) estimators for all the combinations of $m = 4, 8$ and $\Delta = 4, 5, 6$. The estimators are applied to simulated continuous lognormal multifractal processes of the type shown at the top of Fig. 4.6(a). The C_1 parameter is set to 0.1 and the simulation is dressed, at resolution 2^{14}. The moment scaling function $K(q)$ is estimated for each of 1000 simulations and for q in the range [−4, 2]. For each simulation, four cases are considered: noiseless data, data with additive noise, data with multiplicative noise, and data quantization.

In the additive noise case, the noise has normal distribution with a standard deviation equal to 20% of the standard deviation of the noiseless process at the highest resolution. A minimum is also imposed to avoid negative or zero values. In the multiplicative noise case, the log of the noise has normal distribution with a standard deviation equal to 30% of the standard deviation of the log of the noiseless process at the highest resolution. Quantization assumes an "increment reduction" equal to the 10th percentile of the process at the highest resolution.

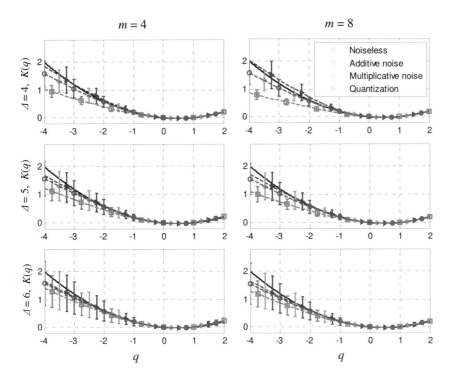

Fig. 4.15. Mean ± one-sigma intervals of different (m, Δ) estimators applied to continuous lognormal multifractal processes with $C_1 = 0.1$. See text for a description of the data, the additive and multiplicative noise, and the level of data quantization.

The range of scales over which power laws are fitted to the empirical moments is $[16, D/4]$, where $D = 2^{14}$ is the total length of the series and 1 is the smallest interval. This choice is based on the fact that, in the cases of noisy or quantized data, the log-moments display reasonable scaling within this range of scales.

For each combination of data and (m, Δ), Fig. 4.15 shows the mean value and ± one-sigma ranges of the estimator. For clarity, the ± one-sigma error ranges of the different estimates are shown at alternating values of q. As one would expect, the width of the error ranges is smallest when the largest number of data values are used ($m = 8$, $\Delta = 4$) and is largest when only a few data values are retained ($m = 4$, $\Delta = 8$). On the other hand, the bias has an opposite trend. The estimation with $m = 4$

and $\Delta = 5$ perhaps has the best overall performance. Also notice that, all the estimators perform well with noiseless data and multiplicative errors, slightly less well for additive noise, and worst in the case of quantization.

For the same cases of noiseless data, additive noise, multiplicative noise, and data quantization and the same simulations, Fig. 4.16 compares the performance of the $(m = 4, \Delta = 5)$ estimator with the standard WTMM estimator and a scale-invariant version of the WTMM estimator. The scale-invariant version uses the discrete wavelet transform in which only the coefficients for non-overlaping Haar wavelets are retained in the WTMM analysis. In the noiseless and multiplicative-noise cases, all three estimators perform rather well, with the scale-invariant WTMM estimator having somewhat smaller bias for the high negative moments. However, in the more critical case of additive noise and quantization, the (m, Δ) estimator is clearly superior.

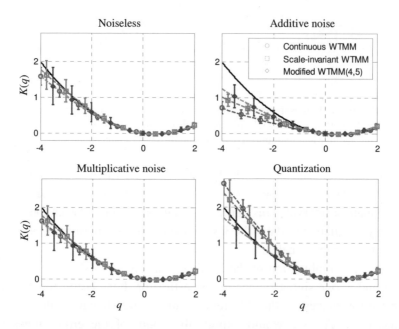

Fig. 4.16. Comparison of $(m = 4, \Delta = 5)$ estimators of $K(q)$ with the standard *WTMM* estimator and its scale-invariant version based on the discrete wavelet transform. Same data as in Fig. 4.15.

Another context in which the above techniques for negative moments are relevant is the analysis of multifractal measures with additive polynomial trends.[117] The non-scaling trend can be filtered out by choosing for ϕ in Eq. (4.105) a wavelet of suitable order. However, in this case, the absolute wavelet coefficients $|\varepsilon_n(\tau)|$ have non-zero probability density at 0 and, hence, have divergent moments of order $q \leq -1$. In order to estimate $K(q)$ for higher negative moments, one must avoid using coefficients $\varepsilon_n(\tau)$ close to zero and, for this purpose, one may use the methods discussed above.

A similar problem is encountered in the analysis of multi-affine processes $X(t)$, for which one must use wavelets of order at least 1. Again, the moments of the increments of order $q \leq -1$ generally diverge, and one may use the same techniques to extend the estimation of $K(q)$ to moment of lower order.

4.7.3. *Estimation of K(q), q > 0, for Multi-affine Processes by the Gradient Amplitude Method*

Tessier *et al.*[118] observed that multi-affine processes $X(t)$ (processes with stationary multifractal increments) can be obtained through integration of some fractional order H of a stationary multifractal measure (see the log-Levy "universal" multifractal processes in Sec. 4.4.4.2). The scaling of the increments of X depends on H and the multifractal characteristics of the parent stationary measure, for example its moment scaling function $K(q)$. Hence, a possible data analysis strategy is to first estimate H, then fractionally differentiate (order H or higher) $X(t)$, and finally estimate $K(q)$ from the differentiated process. Tessier *et al.*[118] have suggested a shortcut procedure to obtain $K(q)$, by calculating the amplitude of the gradient of $X(t)$, $|\nabla X|$, at some high resolution (typically the resolution of the data) and using this gradient amplitude in place of the fractionally differentiated signal. The claim is that, since the gradient transformation resembles differentiation, the small-scale gradient amplitude measure $|\nabla X|$ should be multifractal, with moment-scaling function $K(q)$.

This gradient amplitude (*GA*) method can be applied in spaces of any dimension and has acquired high popularity. For example, three of the four multifractal estimation methods evaluated by Turiel et al.[119] use local gradient amplitudes. Theoretical arguments have been advanced in support of the claim that $|\nabla X|$ is stationary multifractal.[49,120] These arguments further suggest that the moment scaling function of $|\nabla X|$, $K_{|\nabla X|}(q)$, is related to the moment scaling function of the increments of X, $K_{\Delta X}(q)$, as

$$K_{|\nabla X|}(q) = qK_{\Delta X}(1) - K_{\Delta X}(q) \qquad (4.107)$$

However, not only Eq. (4.107) is incorrect, but, more fundamentally, the measure $|\nabla X|$ does not scale. The basic reason for lack of scaling is that $|\nabla X|$ is the amplitude of the gradient (for a process on the line, the absolute value of the increment), not the (signed) increment of X.

This criticism of the gradient amplitude method was raised by Veneziano and Iacobellis[121] in the context of topographic surfaces, for which they found strong evidence of scale invariance of the self-similar type. Previous studies that had concluded in favor of multifractality were based on GA analysis. To show that the empirical moments of $|\nabla X|$ do not scale and that, when power functions are fitted by least squares to the empirical moments, the resulting $K(q)$ function is curved, spuriously indicating multifractality, Veneziano and Iacobellis[121] considered a Brownian surface, which is known to be ss with $H = 0.5$. For this surface, one can make simplifying, but good, approximations to theoretically calculate the moments of averages of $|\nabla X|$ over regions of different size. These approximate theoretical moments (i) do not scale with the size of the averaging region; (ii) behave similarly to the moments of $|\nabla X|$ from actual topographic surfaces; and (iii) when fitted by power laws, produce multifractal-looking $K_{|\nabla X|}(q)$ functions.

The scaling properties of multi-affine processes and fields can be easily determined using increments or generalized increments (wavelet coefficients).[113,115,116,121,122] The gradient-amplitude transformation distorts these scaling properties and has no significant practical advantage over the correct analysis methods. In our opinion, the use of gradient amplitudes should be discontinued.

4.7.4. Estimation of Universal Parameters by the Double Trace Moment Method

Log-Levy ("universal") multifractal measures have been widely used to model many geophysical fields, including atmospheric turbulence and temperature, cloud reflectivity, topographic surfaces, ocean surface radiance, rainfall and river flow, and other geophysical fields.[32,45,71,118,123-132] In the conservative case $H = 0$, the moment-scaling function $K(q)$ has the form

$$K(q) = \begin{cases} \dfrac{C_1}{\alpha-1}(q^\alpha - q), & \text{for } \alpha \neq 1 \\ C_1 q \ln(q), & \text{for } \alpha = 1 \end{cases} \qquad (4.108)$$

where $0 < \alpha \leq 2$ is the index of stability of the Levy distribution and C_1 is a positive constant that controls the dispersion of $\ln(A_e)$.[45] For $\alpha = 2$, $\ln(A_e)$ has normal distribution with mean value $m = -C_1$ and variance $\sigma^2 = 2C_1$ (this corresponds to the sub-class of lognormal multifractal measures), whereas for $\alpha \to 0$ the distribution of A_e develops a mass e^{-C_1} at e^{C_1} and a mass $(1 - e^{-C_1})$ at zero (the so-called beta multifractal measures). The double trace moment (DTM) method of Lavallee et al.[71,123] is the preferred procedure to estimate the scaling parameters α and C_1.[118,125-127] The method works as follows.

Recall from Sec. 4.4.3.2 the important distinction between bare and dressed measures. Bare measures at resolution r are obtained by developing the multiplicative cascade down to resolution r and no further, whereas dressed measures are obtained by continuing the cascade process to infinite resolution and then degrading the process by averaging to a finite resolution. We use the notation $\varepsilon(S)$ for dressed densities and $\varepsilon_b(S)$ for bare densities. Under multifractality, bare and dressed densities scale in the same way: for any $r \geq 1$,

$$\varepsilon(S) \stackrel{d}{=} A_r\, \varepsilon(rS)$$
$$\varepsilon_b(S) \stackrel{d}{=} A_r\, \varepsilon_b(rS) \qquad (4.109)$$

where A_r is a non-negative random variable independent of $\varepsilon(rS)$ or $\varepsilon_b(rS)$. It follows from Eq. (4.109) that, if they exist, the indicated q-moments scale as

$$E[\varepsilon(rS)]^q \approx r^{-K(q)}$$
$$E[\varepsilon_b(rS)]^q \approx r^{-K(q)} \qquad (4.110)$$

where $K(q) = \log_r E[(A_r)^q]$. When A_r has log-Levy distribution, $K(q)$ is given by Eq. (4.108).

Suppose that the process is observed at some finite resolution, through the dressed measure densities $\varepsilon(S)$, where the tiles S form a partition of the region of observation. For what follows, it is important to distinguish between two sets of quantities derived from the original densities $\varepsilon(S)$ through different sequences of exponentiation, normalization, and averaging operations. For any given $r \geq 1$, let:

- $\varepsilon^\eta(rS)$ = result from first raising $\varepsilon(S)$ to the power η, then dividing by $\langle \varepsilon^\eta(S) \rangle$ (' sign), and finally averaging over all S in rS; and (4.111)

- $[\varepsilon(rS)]^{\eta'}$ = result from first averaging $\varepsilon(S)$ in rS, then raising $\varepsilon(rS)$ to the power η, and finally dividing by $\langle \varepsilon^\eta(rS) \rangle$. (4.112)

The DTM method uses the quantities $\varepsilon^{\eta'}(rS)$ in Eq. (4.111), whose moments are claimed to scale with r as

$$E\{[\varepsilon^{\eta'}(rS)]^q\} \propto r^{-K(\eta,q)} \qquad (4.113)$$

where

$$K(\eta, q) = K(\eta q) - qK(\eta) \qquad (4.114)$$

For universal multifractal measures, substitution of Eq. (4.108) into Eq. (4.114) gives $K(\eta, q) = \eta^\alpha K(q)$. Therefore, the DTM method estimates α as the slope of $\log[K(\eta, q)]$ against $\log(\eta)$. Once α is known, C_1 can be estimated from the same plots, using the relation $K(\eta, q) = \eta^\alpha K(q)$ and $K(q)$ in Eq. (4.108).

The problem with this method is that the scaling relation in Eqs. (4.113) and (4.114) holds for only the bare densities, not the dressed densities.[133] As an alternative, Veneziano and Furcolo[133] suggested using the quantities in Eq. (4.112), for which they show that the moments scale as

$$E\{([\varepsilon(rS)]^{\eta'})^q\} \propto r^{-K(\eta q) + qK(\eta)} \qquad (4.115)$$

This is the type of scaling needed by the DTM method.

Veneziano and Furcolo[133] studied the conditions under which the original DTM method gives accurate results. In spite of the fact that the method makes erroneous scaling assumptions, its estimates of α are generally accurate. Higher bias is found in the estimator of C_1, especially for small α. The modified DTM method is unbiased for both α and C_1.

The general conclusion one draws from the previous discussion of inference methods for multifractal processes is that not enough has been done to establish the accuracy, and in some cases even the validity, of the currently popular procedures. This is an issue of much importance to ensure the credibility and accuracy of scaling results.

4.8. Selected Applications in Hydrology

In reviewing the applications of scale invariance in hydrology, we choose to focus on four specific areas: rainfall, river networks and fluvial erosion topography, floods, and flow through saturated porous media. These areas certainly do not exhaust the range of "fractal interests" in hydrology. A more comprehensive coverage would have included clouds and turbulence, soil moisture, processes in the oceans, problems of transport, issues at the interface between hydrology and ecology, and percolation and flow through fractured rocks, among other topics. However, one may argue that the areas we have selected are central to hydrology and, to some extent, to the development of fractal theory from its inception. They also exemplify a wide range of models and methods that are relevant to other applications.

4.8.1. *Rainfall*

There is not yet a clear explanation for the origin of scale invariance in rainfall, starting from first principles. Its possible connection to atmospheric turbulence, which in the inertial range is known to display multifractality in the energy dissipation and velocity increments, has been repeatedly mentioned,[33,45,118] but a detailed connection involving the microphysics of droplet formation and coalescence, rainfall advection, and precipitation is still missing. Therefore, most of what is known about rainfall scaling comes from empirical evidence. Such evidence shows that multifractal models of various types can reproduce many statistical properties of rainfall at scales from below one hour to several days in time and from a few kilometers to more than 100 km in space. These properties include the scaling of the moments,[45,49,100,134,135] the power spectral density,[35,128,136,137] the alternation of wet and dry periods,[56,57,138,139] and the distribution of extremes.[73-76,140] However, significant deviations from multifractal scaling have also been observed,[49,56-58,139] motivating the development of some of the extensions of scale-invariant models mentioned in Sec. 4.4.5. Here, we review this evidence for and against scaling. For the application of scaling principles to rainfall extremes, forecasting, and downscaling, see Secs. 4.5.4, 4.6.1, and 4.6.2.

Scaling in Time

Denote by $I(t)$ the time series of rainfall intensity at a given site and by I_D the average of $I(t)$ in a generic interval of duration D. A property that has been used to verify the multifractality of $I(t)$ is that the spectral density $S(\omega)$ behaves like $\omega^{-\beta}$ with $0 < \beta < 1$ (see Sec. 4.5.2). More comprehensive checks involve the dependence of the moments $E[I_D^q]$ on D (see Secs. 4.7.1 and 4.7.2). Similar concepts apply to rainfall in space and space-time, for which however the additional issue of isotropic or anisotropic scaling arises (see below).

Olsson[136] and Fraedrich and Larnder[59] are among the studies of rainfall scaling in time based on the spectral density function. Olsson[136] examined a 2-year long, 8-min resolution time series from Sweden and found that, between 8 minutes and 11 days, $S(\omega)$ has power-law behavior with exponent $\beta = 0.66$. At larger temporal scales, β is essentially zero, indicating independence. In a more extensive study, Fraedrich and Larnder[59] examined 90-year rainfall records at 5-min resolution from 13 European stations and identified different scaling regimes: for durations below about 2.4 hours, the spectrum is of the power-law type with $\beta \approx 0.9$; for durations between 2.4 hours and about 10 days, β decreases to about 0.5; a transition regime follows between 10 days and about 1 month, and above 1 month the spectrum is essentially flat. The existence of multiple scaling regimes has been confirmed by many other studies.[130,141-143]

More detailed evaluations of multifractality use the scaling of the moments of I_D with the averaging duration D. For example, Tessier et al.[128] used this scaling to fit a universal multifractal model to 30 records of daily rainfall in France. The model has log-Levy generator with three scaling parameters (the index of stability α, the intermittency parameter C_1, and the order of fractional integration H; see Sec. 4.4.4.2) and a power-law spectral density with exponent

$$\beta = 1 + 2H - \frac{C_1}{\alpha - 1}(2^\alpha - 2) \qquad (4.116)$$

Tessier et al.[128] fitted the parameters α and C_1 using the double-trace-moment method (see Sec. 4.7.4) and calculated H to match the empirical value of β through Eq. (4.116). They found two scaling regimes, one with ($\alpha = 1.6 \pm 0.2$, $C_1 = 0.1 \pm 0.05$, $H = -0.35 \pm 0.2$) for $D > 16$ days, and the other with ($\alpha = 0.7 \pm 0.2$, $C_1 = 0.4 \pm 0.1$, $H = -0.1 \pm 0.1$) for D between 1 and 16 days. The corresponding spectral exponents β vary from record to record, around 0.20 for $D > 16$ days and around 0.3 for shorter durations. The first value of β indicates near independence and is consistent with other studies mentioned above. However, the second value is smaller than what is typically observed. The fact that H is on the average negative implies fractional differentiation of a stationary multifractal field (see Sec. 4.4.4.2) and is somewhat suspicious. This might be the consequence of bias in the estimation of α and C_1 (see also Sec. 4.7.4).

Values of α, C_1, and $H < 0$ that result in low spectral exponents have been obtained also by Hubert et al.[142] and Pathirana et al.[72] The latter study fitted a universal multifractal model to 18 temporal rainfall series from Japan, obtaining universal multifractal parameters that for $D < 2$ days imply $\beta \approx 0.3$. Interestingly, Pathirana et al.[72] also calculated the spectral density exponent directly from the data, but did not use the results to constrain the multifractal parameters. For durations below one day, direct analysis gives $\beta = 1.03 \pm 0.1$. This estimate is consistent with that of Fraedrich and Larnder[59] and is more credible than 0.3. Hence, one concludes that, like for Tessier et al.,[128] estimation of the multifractal parameters is likely biased.

In a more recent study, Langousis and Veneziano[140] fitted a so-called beta-lognormal multifractal model to a 24-year, 5-min resolution rainfall record from Florence, Italy. The model has a generator W_r ($r > 1$ is the scale-change factor) with probability mass $P_0 = 1 - r^{-C_\beta}$ at 0. The positive variable $W_r^+ = (W_r | W_r > 0)$ has lognormal distribution with log-mean $m_{\log_r(W_r^+)} = C_\beta - C_{LN}$ and log-variance $\sigma^2_{\log_r(W_r^+)} = 2C_{LN}$, where C_β and C_{LN} are non-negative parameters. The parameters $C_\beta = 0.45$ and $C_{LN} = 0.05$ estimated for durations between 20 minutes and 15 days imply a spectral slope $\beta = 1 - C_\beta - 2C_{LN} = 0.45$, which corresponds to the spectral slope obtained directly from the record.[144]

Scaling in Space

Since the advent of meteorologic radar and earth-observing satellites, there have been many studies of rainfall scaling in space. In most cases, one uses the scaling of the moments $E[I_L^q] \propto L^{-K(q)}$, where I_L is the average rainfall intensity inside a square region of side length L. Of special interest are the exponents $K(0) \leq 0$, which corresponds to the scaling of the wet area fraction [$2 + K(0)$ is the fractal dimension of the rainy set], and $K(2) > 0$ for the scaling of the second moment. Some studies report a scaling exponent γ for the variance of I_L. If I is a stationary multifractal measure, its variance does not strictly scale. However, if the coefficient of variation of I_L is much larger than 1, then the variance displays approximate scaling with $\gamma \approx K(2)$.

In an application to spatial rainfall, Kundu and Bell[145] used a sequence of 2-km radar scans from the TOGA–COARE (Tropical Ocean and Global Atmosphere program – Coupled Ocean Atmosphere Response Experiment) dataset to study how various statistics depend on the scale of spatial averaging, L. They found that for L in the range 2–16 km, the exponents $-K(0)$ for the rainy fraction and γ for the variance are in the range 0.4–0.6 and 0.5–0.7, respectively. Similar results were obtained by Gebremichael and Krajewski,[146] who used 2-km ground radar data from two TRMM (Tropical Rainfall Measuring Mission) campaigns. This study further compared scaling results from radar and relatively dense raingage networks (about 1 gage/4 km^2). While the radar scans show good moment scaling down to the 2-km resolution, the raingage measurements indicate lack of scaling for $L < 16$ km. As shown in Veneziano and Langousis,[75] this lack of scaling may be due in part to biases caused by the finite and spatially uneven density of the raingage instruments: in the limit as L approaches the inter-station distance, only one station is used to calculate the spatial average, with significant distortion of the statistical moments.

In more extensive studies, Deidda et al.[35,137] used radar scans at 4-km resolution extracted from the GATE (GARP Atlantic Tropical Experiment), TOGA–COARE, and TRMM–LBA (Large-scale Biosphere Atmosphere) experiments. For L between 4 and 100 km, rainfall was found to scale in a multifractal way, with $K(2)$ in the range 0.5–1.2. The value of $K(2)$ was found to increase as the large-scale rainfall intensity over the entire frame, $\bar{I}$, decreases. In a similar study, Gebremichael et al.[147] used satellite and ground radar rainfall data from TRMM. Both datasets show that for $L < 64$ km, rainfall scales in an approximately multifractal way with exponents $-K(0)$ and $K(2)$ in the ranges 0.5–1.5 and 0.8–1.5, respectively. Similarly to Deidda et al.,[35,137] Gebremichael et al.[147] found that, as the large-scale average rainfall intensity $\bar{I}$ decreases, $-K(0)$ and $K(2)$ increase. A strong dependence of the wet fraction exponent $K(0)$ on $\bar{I}$ was found earlier by Over and Gupta.[138] However, this dependence is somewhat spurious, since frames with a smaller wet fraction necessarily tend to have lower values of $\bar{I}$.

Scaling in Space-Time

When rainfall is studied in more than one dimension, in particular when averages are taken over space-time regions, scale invariance may hold under non-isotropic transformations (see Sec. 4.3.3). A simple (orthotropic) case is when the field in (x, y, t)-space renormalizes under transformations of the type $(x \to x/r_x, y \to y/r_y, t \to t/r_t)$, where $r_x, r_y, r_t > 1$ are possibly different contraction factors. This orthotropic scaling is also known as "dynamic scaling."[34,148-150] In the notation of dynamic scaling, one typically sets $r_t = r_x^{z_x} = r_y^{z_y}$, where $z_x, z_y > 0$ are the "dynamic scaling exponents." In what follows, we use this notation.

Estimates of z_x and z_y can be obtained by comparing the directional spectra of the rainfall intensity along the x, y, and t coordinate directions. In the case of dynamic scaling, these directional spectra have the form

$$S_x(\omega) \propto \omega^{-f_x}, \quad S_y(\omega) \propto \omega^{-f_y}, \quad S_t(\omega) \propto \omega^{-f_t} \quad (4.117)$$

where f_x, f_y, and f_t are positive constants that satisfy $f_x = z_x f_t$ and $f_y = z_y f_t$. Unless preferential directions in space exist, the rainfall field should be spatially isotropic with $z_x = z_y = z$. This condition is usually satisfied.[33,35,97,151] Hence, in what follows, we focus on the spatially isotropic case when $r_x = r_y = r$ and $r_t = r^z$ and discuss whether z differs from 1.

Under the assumption that rainfall is a passive tracer advected by fully developed turbulent flow, one obtains the theoretical estimate $z = 2/3$.[33,45,118] However, this assumption is only approximate and empirical analyses of space-time rainfall fields generally support isotropic rainfall scaling in space and time.[33-35,75,100,137] For example, Marsan et al.[33] used a composite of rainfall datasets from the US National Weather Service (NWS) radars to estimate z. By comparing the scaling behavior of one-dimensional energy spectra in space and time as indicated above, they found $z \approx 1.1$, which is close to the unit value under isotropic scaling. A similar estimate ($z \approx 1.13$) was obtained by Deidda[100] from 22 GATE radar sequences. In a more extensive study, Deidda et al.[35] investigated rainfall scaling in space and time using 102 radar sequences from the TOGA–COARE campaign. The average value $z = 0.99$ strongly supports isotropic scaling.

Using three convective storms from Darwin, Australia, Venugopal et al.[34] reached different conclusions. Their analysis is based on the assumption that under dynamic scaling the variance

$$\mathrm{Var}\left[\log\left(\frac{I_{rL,D}(t+r^z\tau)}{I_{L,D}(t)}\right)\right] \quad (4.118)$$

does not depend on $r > 0$. In Eq. (4.118), $I_{L,D}(t)$ is the rainfall intensity averaged over an interval of duration D centered at t and over a geographic region of linear size L, z is the dynamic scaling exponent, and (τ, D, L) are fixed positive constants. Venugopal et al.[34] obtained estimates of z in the range [0.6–1.2] with most values close to 0.6.

However, for stationary multifractal measures, the variance in Eq. (4.118) generally depends on r.[152] Veneziano et al.[153] calculated the bias when z is estimated using the above assumption and concluded that the estimates of Venugopal et al.[34] are not accurate.

Additional indirect evidence of isotropic scaling in space and time comes from charts of the areal reduction factor (ARF), which is the ratio between spatially averaged rainfall and rainfall at a point. In the (log$L$, log$D$)-plane, where L and D are the space and time scales of rainfall averaging, the ARFs are approximately constant along 45° lines.[154-156] In the case of anisotropic space-time scaling, the contour lines would have a different slope.[75]

Dependence of Scaling on Rainstorm Characteristics

In an effort to link the scale invariance of rainfall to physical parameters, a number of studies have tried to relate rainfall statistics, such as the moment-scaling exponents $K(q)$, to prevailing atmospheric conditions,[97,134,151,157] or the orography.[137,158-162]

For example, Olsson and Niemczynowicz[134] used daily rainfall records from a dense raingage network (230 raingages covering an area of about 10000 km^2) in Sweden to study how rainfall scaling in space depends on the characteristics of the rainfall generating mechanisms at the mesoscale: i.e. warm fronts, cold fronts, and mesoscale convective systems (MCS). They found different scaling in the range 10–100 km depending on the storm type. Note, however, that their database is limited and that temporal averaging over one day may affect the scaling results.

Perica and Foufoula-Georgiou[97] used the Haar-wavelet decomposition of 47 storms to study how the variability of the rainfall field depends on the characteristics of the pre-storm environment. Among the 21 thermodynamic and kinematic parameters analyzed, only the convective available potential energy (CAPE) was found to be significantly related (with positive correlation) to the variance of the rainfall intensity. This makes physical sense, as higher values of CAPE are associated with increased convection and a more lacunar structure of the rain field.

In a more recent study, Carvalho et al.[157] studied the scaling properties of mesoscale convective systems in tropical South America. They fitted a stationary universal multifractal model with parameters (α, C_1, $H = 0$) to images from the LBA campaign with a maximum resolution of 4 km. Multifractality was found to hold at spatial scales L ranging from 4 to 100 km. The estimated stability index α is close to 1.7, irrespective of the storm characteristics. By contrast, in convective areas, the intermittency parameter C_1, which for fixed α controls the variability of rainfall, increases as the fraction of deep updrafts increases. This is in general agreement with the findings of Perica and Foufoula-Georgiou.[97]

The effect of orography on the scaling of rainfall is not entirely clear. From a space-time analysis of 187 rainfall sequences from GATE, TOGA–COARE and TRMM–LBA, Deidda et al.[137] concluded that topographic elevation does not influence the scaling of rainfall. This agrees with earlier findings by Pathirana and Herath[160] and Badas et al.[162] The latter study shows that one can account for orographic influences by multiplying a spatially homogeneous multifractal field by a smooth function of elevation. This operation does not affect the scaling properties. However, different conclusions were reached by Harris et al.,[158] Purdy et al.,[159] and Nykanen and Harris.[161] The first two studies used temporal rainfall series, whereas the third study used radar scans from three (same-day) storms. Harris et al.[158] and Purdy et al.[159] concluded that rainfall variability increases with increasing elevation, whereas Nykanen and Harris[161] reported opposite effects. Note, however, that all three studies are based on a small number of events; therefore, differences in the results may be just a consequence of statistical storm-to-storm variability.

Deviations from Multifractal Scaling

A systematic deviation of rainfall from multifractality occurs at small scales. As described in Fraedrich and Larnder,[59] at the high frequencies, there is a change in the algebraic exponent of the spectral density and a power deficit relative to a multifractal spectrum. This observation has been subsequently confirmed by many studies, including Menabde et al.,[49] Menabde and Sivapalan,[51] and Paulson and Baxter[99] for rainfall in time, and Veneziano et al.,[58] Perica and Foufoula-Georgiou,[97,151] Deidda,[100] and Paulson[163] for rainfall in space. In a multiplicative-cascade construction, this behavior can be modeled by reducing the dispersion of the generator W at the small scales, as for example is done in bounded cascades (see Sec. 4.4.5).

Another deviation from multifractality concerns the alternation of dry and wet periods, as first documented by Schmitt et al.[139] for the temporal record of Uccle, Belgium. Similar findings have been reported by Veneziano and Iacobellis[144] and Langousis and Veneziano[140] for Florence, Italy. What is also found is that the dry periods inside storms are not the same as rescaled versions of the dry periods between storms. Therefore, in estimating the probability that the cascade generator W is 0, there are significant differences if one uses the continuous rainfall record or only the portion of the record inside storms. For example, Langousis and Veneziano[140] found that, irrespective of the duration of storms, storm interiors are almost compact with negligible lacunarity: for the storm interiors, the value of $K(0)$ is about -0.02, whereas when one uses the whole record including the inter-storm periods one finds $K(0) \approx -0.45$.

Finally, there are statistical differences in the lacunarity of rainfall during periods of intense and light precipitation, or near the beginning and end of a storm compared with the central portion.[56,57] Olsson[56] used a 2-yr rainfall record at 8-min resolution from Sweden, and Güntner et al.[57] used four hourly records from Brazil and the United Kingdom with a length of about 4 years. Both studies found that the probability P_0 that the cascade generator W is zero increases with decreasing rainfall intensity of the parent tile and with increasing number of dry tiles flanking the parent tile.

More generally, Veneziano et al.[58] noted that the entire distribution of the generator W at cascade level n (not just the probability mass at zero) depends on the (bare) rainfall intensity $\overline{I}_{n-1}$ at the immediately coarser scale during the cascade construction. Multiplicative models with this type of self-exciting structure were mentioned at the end of Sec. 4.4.5. This is generally consistent with the finding by Venugopal et al.[122] and Over and Gupta[138] that the scaling of rainfall is different in high- and low-intensity rainy regions, although Veneziano et al.[58] make W depend on the rainfall intensity at the immediately coarser scale, not at a large fixed scale.

Specifically, Veneziano et al.[58] assumed W_n to have lognormal distribution except for a probability mass $P_{0,n} \geq 0$ at zero (beta-lognormal model) and found that $P_{0,n}$ is smaller at larger scales and for higher bare rainfall intensities. This means that rainfall displays a more compact support at larger scales and in regions where rainfall intensity is higher. Also, the dispersion of the lognormal distribution of $W^+ = (W \mid W > 0)$ was found to depend on n and $\bar{I}_{n-1}$, generally decreasing at smaller scales and with increasing rainfall intensity R_n (more tamed local multiplicative fluctuations and in regions of intense rainfall).

4.8.2. Fluvial Topography and River Networks

The simulation of rough terrain is one of the earlier applications of fractal and self-similarity theories,[5,164] and many classical laws governing the planar organization of river basins are expressions of what we now call scale invariance.[165-168] Yet, the mathematical description of topographic regularities, their relation to physical processes, and the simulation of realistic-looking landscapes remain among the most challenging problems of theoretical geomorphology and computer graphics.[1,169,170]

Our main focus is fluvial erosion topography, by which we mean landforms that are shaped predominantly by the erosive action of rivers and streams. This focus poses limitations on the geology and climate as well as the range of scales considered, since at very large and very small scales processes other than fluvial erosion tend to dominate the landscape. These include large-scale tectonic movements, floodplain deposition, and diffusive phenomena, like landslides, eolian transport, and soil creep, to name a few. In addition, the spatial variation of geology, precipitation, and vegetation induces differences in soil erodibility and channelization thresholds, resulting in river networks that are developed to varying degrees and, therefore, in spatially non-uniform drainage density. These factors need to be kept in mind when interpreting the scaling of observed river networks and topographic surfaces.

Broadly speaking, the scaling properties of topographic surfaces have been investigated using two distinct approaches. One approach is to study topography over extended regions or long transects using tools like spectral analysis and variograms.[171-177] Under the condition of self-similarity, both the spectral density and the variogram (a plot of the variance of topographic elevation increments against distance) have a power-law form, from which the self-similarity index H can be recovered. There have also been attempts at detecting multifractality using moment-scaling methods.[36,124,131,177] Most of these studies have aimed at validating or rejecting scale invariance of the self-similar or multifractal type and at assessing the dependence of the scaling exponents on geoclimatic characteristics and the scale range of the observations.

The second approach has been developed mainly by hydrologists, with focus on river basins. Most of this work has been on the fractal properties of two-dimensional river networks,[165,178-182] but significant attention has been given also to the scaling of river profiles and three-dimensional topography.[18,121,183-188] The main tools of these investigations have been the numerical analysis of mapped river courses and drainage networks and the simulation of realistic-looking topographies using dynamic equations of topographic evolution. Relative to the first line of inquiry, this approach emphasizes the role of physical processes and possibly principles like self-organized criticality,[189,190] and the minimization of energy functionals associated with erosional processes.[191-195]

Results from the two approaches are generally different and may appear inconsistent. One reason is that the scale-invariance properties of topography within river basins are not the same as those possibly satisfied along transects or over geographic regions that strand multiple basins. For example, inside a basin, the terrain must satisfy drainage requirements and flow must accrete from the source areas towards the outlet, constraining the type of scale invariance that can be satisfied. A second reason is that, different scaling properties hold on the hillslopes and the drainage network. This differentiation is not made in studies of the first kind, where one generally seeks an overall assessment of scaling. Finally, there are differences in the methods used to investigate scaling (in particular multifractality), which affect the results. In reviewing the literature, we rely mainly on the work of Veneziano and Niemann[187,188] and Veneziano and Iacobellis[121] on fluvial topography within river basins.

General Scale-invariance Condition for River Basin Topography

While it may be unrealistic to try to explain all observations about scale invariance and fractality of river basins through a single invariance principle tied to fluvial erosion, it is worth making an attempt in this direction. The potential benefits are many: connecting seemingly disparate observations, understanding what parameters control or destroy scaling, and identifying needed refinements and extensions. The unifying view presented here is based on Veneziano and Niemann.[187,188]

One must first determine the type of scaling that topographic relief $h(x,y)$ inside a river basin Ω might have (the relief is the elevation relative to the basin outlet). Considering the symmetry-breaking effect of gravity, the non-homogeneity of the increments due to the general steepening of the topography as one moves upstream from the outlet and other factors, Veneziano and Niemann[187] concluded that scale invariance could be expressed as follows. Consider the main sub-basins Ω_i of Ω. These are sub-basins of different areas A_i with outlets at different points O_i along the main channel of Ω. The line that connects the main-stream source S (which is common to all main sub-basins) to O_i defines the general orientation of Ω_i. Further, let $h_i(\underline{x})$ be the relief of Ω_i relative to its outlet O_i, in a Cartesian reference translated and rotated such that the origin is at O_i and the x_i axis passes through S (see Fig. 4.17 for an illustration). A possible steady state (i.e. time-invariant) self-similarity condition is that for some H and any i and j,

$$h_i(\underline{x}) \stackrel{d}{=} r_{ij}^{-H} h_j(r_{ij}\underline{x}) \qquad (4.119)$$

where $r_{ij} = \sqrt{A_j/A_i}$. Equation (4.119) may hold exactly for the main sub-basins and in approximation for all other sub-basins, for example the sub-basins drained by side tributaries, provided that the origin of the axes is translated to the main source of each sub-basin. In the multifractal case, the deterministic amplitude-scaling factor r_{ij}^{-H} in Eq. (4.119) is replaced by a non-negative random variable $W_{r_{ij}}$, and r_{ij} is constrained to be either greater than 1 (multifractality under contraction) or between 0 and 1 (multifractality under dilation) (see Sec. 4.3.2).

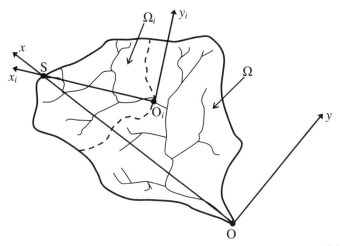

Fig. 4.17. Illustration of reference frames used in the scale-invariance condition of Eq. (4.119); (x,y) are the coordinates for the entire basin (outlet O) and (x_i,y_i) are the coordinates for the sub-basin with outlet O_i.

To investigate the plausibility of Eq. (4.119), Veneziano and Niemann[187] considered a model of fluvial topography evolution in which the relief $h(x,y,t)$ relative to the outlet of Ω evolves in time according to

$$\frac{\partial h}{\partial t} = U - f(\underline{\beta}, \tau) \qquad (4.120)$$

$$\tau \propto A^m S^n$$

where $U > 0$ is the tectonic uplift rate relative to the outlet and f is the rate of fluvial erosion, assumed to depend on a vector $\underline{\beta}$ of local erodibility-related parameters and the hydraulic shear stress τ. The shear stress processes $\tau_{x,y}(t)$ at different (x,y) locations are not related in any particular way. However, here τ refers not to the entire temporal process, but to some characteristic shear stress for erosion, for example the maximum annual value. This characteristic value is expected to depend on location mainly through the local slope of the channel S and a characteristic flow Q, say the annual maximum flow. Such characteristic flow Q is approximately a power-law function of the drained area A (see Sec. 4.8.3). This is why Eq. (4.120) assumes that τ is a power function of A and S. The exponents m and n in that function are fixed, while the uplift U may vary randomly in time and the erodibility β may, in general, vary randomly in three-dimensional space. Under these assumptions, Veneziano and Niemann[187] considered whether the relief h might reach and maintain a scale-invariant state in the sense of Eq. (4.119) or its multifractal extension, and what specific scale-invariance conditions it would satisfy. This was done by studying the renormalization properties of Eq. (4.120), i.e. by looking for scale-invariant topographies h that under Eq. (4.120) are in static or dynamic equilibrium.

An important special case is when β is constant in space and U is constant in time. Then Eq. (4.120) admits solutions $h(x,y)$ that satisfy Eq. (4.119) with $H = 1 - 2m/n$. Realistic-looking topographies and drainage networks have H between 0 and 0.5, which is also the range obtained from plausible exponents m and n. If β varies with geologic depth z and U varies as a stationary process in time, the relief h evolves randomly in time and may asymptotically reach a dynamic self-similar state. In other cases (when β is constant with depth z but varies geographically as a stationary multifractal measure), the topography may reach time-invariant states $h(x,y)$ with multifractal scale invariance. As remarked by Veneziano and Niemann,[187] the existence of scale-invariant solutions does not necessarily imply that such solutions are attractive (that they are attained under a wide spectrum of initial and boundary conditions[19]). However, numerical experiments reported by Veneziano and Niemann[187] and others[1] indicate that indeed Eq. (4.120) tends to produce scale-invariant topographies.

One can use Eq. (4.119) to make several predictions and compare them with observations. Next, we examine a few of them.

River Courses

Under the renormalization property in Eq. (4.119), the sinuosity of the main river changes scale with the square root of the drained area, as one moves from the outlet towards the source. An idealized example of river course with this *ss* property is shown in Fig. 4.18. The simulation was obtained using Eq. (4.18) with $H = 1$ and $Z(t)$ a band-limited Gaussian process with zero mean. The details of the Z process do not matter for what follows. Also, the fact that our idealization is in the form of a function, rather than a general curve on the plane, is unimportant. Suppose that one measures the length L of the river with a divider of small length l. The result is L_l. If the river course is locally smooth (non-fractal, as is the case in Fig. 4.18), then there are upstream reaches where the divider resolves well the sinuosity and downstream reaches where it does not. As $l \to 0$, L_l approaches the actual finite length L of the river. Viewed as a function of l, L_l is a random process whose mean value behaves like $E[L_l] = L - c_1 l^{c_2}$, where c_1 and c_2 are positive constants. If, on the other hand, the river course is fractal, then no finite l can capture the detailed sinuosity of any reach and $E[L_l]$ behaves like l^{1-D_s}, where $D_s > 1$ is the fractal sinuosity dimension of the river (see Sec. 4.2.1). In this case, $E[L_l]$ diverges as $l \to 0$.

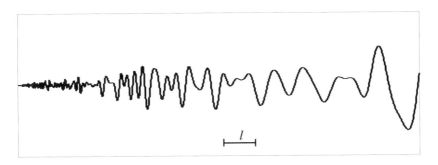

Fig. 4.18. Schematic example of a river course that satisfies the self-similarity condition of Eq. (4.119) and is non-fractal.

The difference in the behavior of $E[L_l]$ in the fractal and non-fractal cases should make it possible to distinguish between the two hypotheses. However, actual rivers cannot be fractal down to infinitesimal scales (or else it would take an infinite amount of time for a water particle to be transported), making the distinction difficult in practice. In our opinion, if fractality of the river courses exists, it is limited to a narrow range of scales and is, therefore, relatively unimportant, whereas what may be more important is the *ss* condition.

The present discussion is relevant to Hack's law.[168] When the length of the main river is measured using a divider of constant length l and L_l is plotted against drained area A, one typically observes a relation of the type

$$L_l \propto A^\alpha \qquad (4.121)$$

with an exponent α around 0.55–0.60. Much has been said about the reason why α differs from the "dimensionally correct" value 0.5.[182] The leading explanations are that the deviation is due to either elongation of the larger river basins (allometric hypothesis, as originally suggested by Hack[168]) or fractality of the river courses (fractal hypothesis, as advocated by Mandelbrot[5]), or a combination of both. There is evidence that elongation alone could not explain the deviation from 0.5[182] and, therefore, the fractal interpretation is currently favored, but in the light of what was said above also this hypothesis may be challenged. As a third possibility, we suggest that self-similarity may play a role. Although self-similarity does not produce power-law relations of the type in Eq. (4.121), an exponent $\alpha > 0.5$ could result from power-law fitting data from *ss* river courses over a finite range of drained areas.

River Networks

Most of the work on the scaling of river networks has focused on their topological (branching) structure, although in some cases metric properties (link lengths, drained areas) have also been considered. Fractal dimension analysis requires branching properties and at least one metric consideration (stream length L or drained area A).[182] Traditionally, these studies have used the Horton-Strahler classification of streams[165,196,197] and Horton's ratios

$$R_B = n_{\omega-1} / n_\omega$$
$$R_L = L_\omega / L_{\omega-1} \qquad (4.122)$$
$$R_A = A_\omega / A_{\omega-1}$$

where ω is the Strahler order of a stream, increasing in the direction of flow, and R_B, R_L, and R_A are the so-called bifurcation, length, and area ratios. The fact that these ratios are approximately independent of ω is an expression of network self-similarity, although it has been shown that this is a general property of most rooted networks.[198,199]

As Peckham,[181] Tarboton,[182] Turcotte and Newman[200] and others have pointed out, a more appropriate characterization of topologic scale invariance is that proposed by Tokunaga.[178,201,202] According to Tokunaga, (i) the branching numbers n_{ij} = number of order-j tributaries of each stream of order i depend only on $k = i - j$; and (ii) n_k has a geometric dependence on k, of the type $n_k \propto n_1 \gamma^k$. The combination of these two properties is generally referred to as Tokunaga's cyclicity. Peckham[181] further observed that, like Horton's ratios, Tokunaga's rules should be interpreted probabilistically, meaning that the distribution of n_{ij} depends only on $k = i - j$, and that $n_k \propto n_1 \gamma^k$ holds in expectation.

The *ss* condition in Eq. (4.119) implies Tokunaga's first property and admits, but does not require, Tokunaga's second property. It would thus be of interest to understand the origin of the latter. We suggest that the geometric dependence of n_k on k might originate from the competition for drainage area of tributaries that sequentially form in the channelization process: tributaries of order $j = i - 1$ form first and tend to drain most of the area far away from the order-i stream. This leaves smaller disconnected regions closer to the i-stream that generate tributaries of order $j = i - 2$. This process continues in a scale-independent way, causing n_k to depend geometrically on k.

Recently, a body of work has focused on the multifractal properties of river networks extracted from blue lines.[203-205] A variety of data analysis techniques are used, but the main idea is to partition the basin into tiles of size l, measure the total length of channels in each tile, and treat this as a random measure on the plane. One can then characterize multifractality through the scaling of the moments with l or in some other equivalent way, e.g. through the so-called singularity spectrum

$\{\alpha, f(\alpha)\}$. In essence, what one considers as multifractal in these analyses is the drainage density. Since the drainage density is higher in regions where the geologic material is more erodible, one way in which the drainage density can be multifractal in the context of topographic evolution models of the type in Eq. (4.120) is if the soil erodibility varies spatially in a multifractal way. However, depending on the geology and climate, the extent of channelization may be controlled by other factors, for example soil instability.[206-208]

Width Function

The width function $F(l)$ gives the area density at flow distance l from a basin outlet. Interest in this function is due to the fact that, if one neglects water losses and the residence time on the hillslopes and further assumes constant flow speed v in the channels, then the normalized function $F(vt)/A$ is the unit hydrograph for a rainfall pulse in time with uniform spatial distribution over the basin. Hence, to a first approximation, $F(vt)$ characterizes the linear dynamic response of the basin and, together with temporal rainfall, determines the characteristics of discharge at the basin outlet.

Two aspects of the width function have received significant attention: the dependence of $F(l)$ on basin area A and the scale-invariance properties of $F(l)$ for any given basin. Both issues are related to the scaling of the river network with drained area, and in particular to Hack's exponent α in Eq. (4.121). Under Eq. (4.119), the deviations of α from 0.5 are due to self-similarity (and possibly fractality) of the river courses and the width function scales in the obvious way with contributing area, i.e.

$$F_A(l) \stackrel{d}{=} A^{1/2} F_1(A^{-1/2} l) \quad (4.123)$$

where $F_A(l)$ is the width function for a basin of area A. In a given geologic and climatic environment, this self-similarity relation is largely satisfied. The deviations from Eq. (4.123) are small compared to the basin-to-basin fluctuations in the shape of $F(l)$ and, when the width function is used to approximate the unit hydrograph, to the effects of hillslope processes, hydrodynamic dispersion, the variation in celerity from stream to stream and other factors.[1,209-214] Therefore, Eq. (4.123) may be accepted as accurate.

The other aspect of interest is the possible scaling of the function $F(l)$ for any given basin. To investigate this issue, Veneziano et al.[215] constructed a model of the width function that is consistent with the self-similarity property in Eq. (4.119) and Tokunaga's trees. The model idealizes the contribution to $F(l)$ from a sub-basin as a deterministic or random pulse whose location, shape, and size reflect the location, shape, and size of the sub-basin. Sub-basins are classified according to whether they discharge directly into the main stream (sub-basins of Order 1) or into the main stream of Order 1 (these are sub-basins of Order 2), and so on for higher orders. The model of $F(l)$ is then built recursively as follows: Initially $F(l)$ is idealized as a single pulse $F_0(l)$. In the next step, $F_0(l)$ is replaced with the sum of a random number of randomly located rescaled pulses, each representing a sub-basin of Order 1. At subsequent stages of construction, each Order-n pulse is replaced with the sum of a random number of randomly located rescaled pulses, each representing a sub-basin of Order $n + 1$.

While this model is simplistic in various ways (for example, it ignores the territorial competition among sub-basins), it produces realistic simulations and reflects fundamental features of real basins, including their hierarchical organization and the fact that small and large sub-basins have essentially the same shape.[193,206] The model allows one to analytically derive interesting results. One is that the spectral density of $F(l)$ behaves like $S(\omega) \propto \omega^{-\beta}$, where $\beta = 1/\alpha$ is the reciprocal of Hack's exponent α. This is consistent with much empirical evidence that the spectral density has a power-law decay with exponent β around 1.8.[1,216] If one considers that the deviation of α from 0.5 is due, at least in part, to bias from numerical discretization, then the true spectrum of $F(l)$ should behave like ω^{-2} (the same discretization issues that affect the estimate of α also affect the estimate of β).

Notice that, while $\sim \omega^{-2}$ is also the spectrum of Brownian motion, the width function is far from Brownian motion (e.g. it is non-negative, does not have stationary increments, and has finite duration). It is possible that the increments of $F(l)$ have local self-similarity or multifractal properties (on local scaling, see Sec. 4.3.1), but at large scales no such property seems plausible due to the constraints $F(0) = F(L) = 0$, where L is the length of the main river.

There have been also claims of multifractality. Lashermes and Foufoula-Georgiou[117] analyzed the fluctuations of $F(l)$ using a wavelet decomposition method. Their main finding is that, after filtering out low-order polynomial trends, the local increments are multifractal. The origin and hydrologic significance of this local multifractality are, however, unclear. Multifractality of $F(l)$ or its increments have also been claimed by Marani et al.[216] and Richards-Pecou.[217] However, the methods used in these studies are inappropriate and produce spurious scaling results. The method and results of Marani et al.[216] have been discussed in Veneziano et al.[218] For the method of Richards-Pecou,[217] which uses the absolute gradient of the increments of $F(l)$, see Veneziano and Iacobellis[121] and Sec. 4.7.3.

River Profiles and Topography

Next, we consider scaling properties that involve the relief inside basins, not just their planar organization. A fundamental geomorphologic relation in river basins links the local slope S of riverbeds to the drained area A. The self-similarity condition in Eq. (4.119) specifically implies

$$S_{A^{0.5}\Delta L}(\underline{x}_A) \stackrel{d}{=} A^{-(1+H)/2} S_{\Delta L}(\underline{x}_1) \qquad (4.124)$$

where $\underline{x}_A$ is the section of the main river that drains area A and $S_{\Delta L}(\underline{x})$ is the average slope between the main channel sections at $\underline{x}$ and at an upstream flow distance ΔL from $\underline{x}$. This is similar to the slope-area relation $S \propto A^{-\theta}$ often reported in the literature.[219-222] However, Eq. (4.124) says that scaling [with exponent $\theta = (1+H)/2$] applies if S is measured as the average slope over channel segments whose length varies with the locally drained area as $A^{0.5}\Delta L$.

In common practice,[1,183,219] one calculates the slope S over some fixed distance ΔL, irrespective of A. Doing so affects the slope-area exponent. The less common alternative of using the average slope of links is appropriate, if the length of the links varies proportionally to the square root of the contributing area. It is interesting that the use of a constant distance ΔL not only makes the empirical value of θ deviate from

$(1+H)/2$, but also introduces nonlinearity in how the moments of S scale with area.[188] Sometimes, this nonlinearity has been incorrectly interpreted as an indication of multifractality.[219]

Going beyond the slope-area relationship in Eq. (4.124), it is of interest to develop stochastic models of the river profile $h(l)$, where h is elevation relative to the outlet O and l is flow distance from O. Early models treated the river flow Q as a given function of l and derived $h(l)$ from physical principles of erosion/deposition, equilibrium, and energy. More recently, numerical models have been used to study the evolution of geomorphology inside two-dimensional basins, making it possible to analyze river profiles in more realistic settings.[18,223] The main new feature of these models is that the distribution of flow along the river course is not specified in advance, but is determined by the complex process of drainage network formation and evolution. This is important because the mean flow affects the slope of river channels and, therefore, their profile. Hence, there are two aspects to the stochasticity of river profiles: one is that the area function $A(l)$ is stochastic, and the other is that, for given $A(l)$, the slope and elevation along the river are random.

Stochastic river profile models usually represent $h(l)$ directly, as opposed to first characterizing the random area process $A(l)$ and then modeling $h(l)$ given $A(l)$. One example of direct model is that of Gupta and Waymire,[183] further developed by Tarboton et al.[219] The latter study presents that model as multifractal, but in the context of the self-similarity definition in Eq. (4.119), it is self-similar. Only preliminary work has been done on modeling $A(l)$ and $h[l|A(l)]$ as random scale-invariant processes.[224]

We conclude this section by returning to the initial quest of the type of scale invariance displayed by topographic surfaces. The main issues are: whether the topographic surface is scale-invariant; if so, whether scale invariance is of the *ss* or *mf* type; and, in either case, with what particular scaling exponents. This is a broad topic, due to the different ways topography can be analyzed and the different geologic settings, climates, and ranges of scale that may be considered. Here, we mention some important results and refer the reader to Veneziano and Iacobellis[121] for a detailed analysis.

The analysis of topographic increments inside large geographic regions and along topographic transects generally support self-similarity with H close to 0.5, with little evidence of multifractality.[177] Significant exceptions are the studies of Lavallee et al.[124] and Lovejoy et al.,[131] who specifically considered the possibility of multifractality. The conclusion that topography is indeed multifractal was reached in these studies by working with gradient amplitudes rather than the more conventional topographic increments. As explained in Sec. 4.7.3, gradient amplitudes do not scale, and their use produces moments with multifractal-like dependence on scale also when the surface is self-similar. Therefore, these claims of multifractality should be largely discarded.

Veneziano and Iacobellis[121] analyzed both simulated and actual river basin topographies, using for the latter elevation readings either on a regular grid or along the river network. The distinction is important when dealing with actual topographies because topographic increments on the hillslopes are locally self-similar with H close to 1, whereas those on the channel network are self-similar with H between 0 and 0.4. Veneziano and Iacobellis[121] concluded that the often-reported value $H = 0.5$ from regional analyses is a "composite" index that mediates between the much higher values of H on the hillslopes and the lower values on the channel network. Veneziano and Iacobellis[121] found no evidence of multifractality. For points on the drainage network, the study provides strong support for renormalization conditions of the type in Eq. (4.119).

4.8.3. Floods

Considering the multifractality of rainfall (Sec. 4.8.1) and the self-similar scaling of the drainage network with basin area (Sec. 4.8.2), one might expect some scale invariance to apply also to river floods. Two types of scale invariance are of particular interest for floods: (i) scaling of the discharge time series $Q(t)$ at a given river station; and (ii) scaling of the extreme floods across basins, specifically scaling of the annual maximum $Q_{\max}(A)$ with drained area A. The latter might hold under some homogeneity constraints on the rainfall climate, the soil type, the general relief, etc. These issues have been addressed in some cases theoretically,[225] and more often using recorded river flows[28,226] or synthetic records obtained by filtering simulated rainfall through basin response models.[227]

Temporal Scaling of Q(t)

To qualitatively frame the issue of scaling in time, we start by considering a simple model for river flow generation. Let $I(dx,dy,dt)$ be the space-time rainfall volume in (dx,dy,dt) and $Q(t)$ be the associated flow at the outlet of a basin Ω. Assuming, for simplicity, that the basin responds linearly, Q is related to I as

$$Q(t) = \int_{-\infty}^{t} \int_{\Omega} I(dx,dy,dt')g(x,y,t-t') \qquad (4.125)$$

where $g(x,y,\tau)$ is the response function of the catchment. In general, g is a nonlinear function that for any $(x,y) \in \Omega$ integrates to 1 over time and has its mass concentrated inside an interval of duration t_r, called the response or concentration time of the basin. The duration t_r depends on the size of the basin, its average slope, the degree of channelization, and other variables, and typically ranges from one hour or less to several days.[228] The other characteristic dimension is the linear size of the basin, say $L \sim \sqrt{A}$, but if $t_r \propto L$, then one characteristic dimension suffices.

The filtering operation in Eq. (4.125) affects minimally the rainfall frequencies ω below about $\omega_r = 2\pi/t_r$ (if such frequencies are associated with wavenumbers $k_x, k_y < 2\pi/L$), whereas the higher frequencies are attenuated approximately by $\sim \omega^{-1}$. Consequently, one would expect the low-frequency properties of the flow to reflect the low-frequency properties of rainfall, whereas the high-frequency fluctuations of Q are significantly reduced relative to the rainfall input. The frequency range that separates these two regimes is controlled by t_r.

These observations agree well with what has been observed on the temporal scaling of floods.[128,229,230] Specifically, Labat *et al.*[230] studied the temporal scaling of rainfall and discharge time series using daily data from three karstic springs in France. The techniques used include spectral analysis and examination of how the moments scale with the duration D of temporal averaging. They found that for D below about 15 days, rainfall at a point exhibits multifractal scaling with a spectral density $S(\omega) \sim \omega^{-\beta}$ and $\beta = 0.65$. At larger temporal scales, $S(\omega)$ is virtually constant, indicating independence. From the discharge series, Labat et al.[230] found that, at temporal scales longer than about a week, the spectral density has an exponent $\beta \approx 0.8$, which is not far from that of rainfall (although the range with comparable spectral scaling, from 1–2 weeks, is very modest). At temporal scales D shorter than about one week, the spectral exponent β for flow ranges from 1.13 to about 2, with the larger values associated with less karstified basins and longer response times. This steep slope of the spectrum is consistent with the smoothing effect of the rainfall-flow transformation.

Similar findings on river flow have been reported by Tessier et al.[128] and Pandey et al.[229] Tessier et al.[128] made a spectral analysis of discharge time series from 30 small catchments (40–200 km^2) in France. At temporal scales that exceed 7–10 days, the spectral exponent is about 0.7, whereas at shorter scales $\beta \approx 1.6$. Pandey et al.[229] used spectral techniques to investigate the scaling regimes of daily discharges from 19 river basins in the United States, ranging in size from 5 to 1.8 x 10^6 km^2. They found $\beta \approx 0.7$ at temporal scales longer than about one week and $\beta \approx 2.5$ at shorter scales. The difference with the small-scale exponents of Tessier et al.[128] is probably due to the different size of the basins considered in the two studies.

These results on the power spectrum of $Q(t)$ hardly support any conclusion of scale invariance: at the larger temporal scales, the range of possible scaling is modest, as it is bounded from above by the whitening of the rainfall input and from below by the response time of the basin. At the shorter scales, the rainfall-discharge transformation in Eq. (4.125) destroys any multifractality that rainfall might have. What might conceivably scale at short temporal scales are the flow increments. For such increments, one would have to distinguish between rainy periods when the increments reflect characteristics of the rainfall input and dry periods when the flow is controlled by the response function g. We are not aware of studies of this type.

Scaling of the Annual Floods with Area

Flood frequency estimation at ungaged locations has always been a pressing problem in hydrology. An attractive approach is to use scaling arguments to link the distribution of peak floods at ungaged and gaged locations of a river network or inside a statistically homogeneous geographic region.[28,225-227,231-237] Scaling is usually verified by examining how the moments of Q_{max} or of the normalized peak flows $Q'_{max} = Q_{max}/A$ depend on A. If Q_{max} varies in a self-similar way with A, then for some positive or negative H' and any $r > 0$,

$$Q'_{max}(rA) \stackrel{d}{=} r^{H'} Q'_{max}(A) \qquad (4.126)$$

The implication on the moments is

$$E[\{Q_{max}(A)\}^q] \propto A^{q+H'q} = A^{Hq} \qquad (4.127)$$

where $H = 1 + H'$. Alternatively, Q'_{max} might vary in a multifractal way with A under contraction or under dilation (see Sec. 4.3.2). In this case, there exist random variables B_r such that, for any $r \geq 1$,

$$Q'_{max}(A) \stackrel{d}{=} B_r Q'_{max}(rA) \quad \text{(mf under contraction), or}$$
$$Q'_{max}(A) \stackrel{d}{=} B_r Q'_{max}(A/r) \quad \text{(mf under dilation)} \qquad (4.128)$$

Consequently the moments of Q_{max} scale as

$$E[(Q_{max}(A))^q] \propto \begin{cases} A^{q-K(q)} & \text{(mf under contraction)} \\ A^{q+K(q)} & \text{(mf under dilation)} \end{cases} \qquad (4.129)$$

where $K(q) = \log_r E[B_r^q]$ is a convex function. One may include all *ss* or *mf* moment-scaling relations in Eqs. (4.127) and (4.129) by writing

$$E[(Q_{max}(A))^q] \propto A^{q-\hat{K}(q)} \qquad (4.130)$$

where $\hat{K}(q)$ is proportional to q [$\hat{K}(q) = -H'q$] in the *ss* case, is a convex function $\hat{K}(q) = K(q)$ for multifractality under contraction, and is a concave function $\hat{K}(q) = -K(q)$ for multifractality under dilation.[28] A simple way to distinguish among these three different scaling conditions is to consider CV, the coefficient of variation of Q_{max}. In the *ss* case, CV is independent of A, whereas for multifractality under contraction (dilation) CV decreases (increases) with increasing area.

In an early study, Cadavid[238] analyzed instantaneous streamflow records from different regions of the United States and basin areas in the range 10–1000 km². He found that the moments of Q_{max} behave like $A^{q-K(q)}$ with $K(q)$ convex, which is an indication of multifractality under contraction. However, his $K(q)$ function is nearly linear (see also Fig. 2 of Gupta and Waymire[232]), making self-similarity a plausible alternative.

In a more recent study, Pandey[231] used records from 180 stations in Canada to study how the distribution of Q_{max} varies with catchment area A. The size of the basins ranges from 0.8 to nearly 10^5 km². The analysis was conducted using both ordinary and probability weighted moments. The qualitative finding is that the ordinary moments exhibit self-similar scaling, whereas the probability weighted moments indicate multifractality. Note, however, that under multifractality the probability weighted moments do not scale; hence, one should discard the latter results. The conclusion that $Q_{max}(A)$ is approximately self-similar is further supported by the fact that the coefficient of variation is nearly constant with A.

Gupta et al.[28] studied the scaling properties of $Q_{max}(A)$ using daily discharges from 270 basins in central Appalachia, with sizes in the range 2–26000 km². They argue that Eq. (4.130) holds with $\hat{K}(q)$ convex for $A > A_c$ and $\hat{K}(q)$ concave for $A < A_c$, where A_c is about 50 km². However, the moments for given A show significant dispersion from basin to basin, making it difficult to distinguish between multifractality and simple scaling. Robinson and Sivapalan[234] revisited the dataset of Gupta et al.,[28] focusing on the coefficient of variation CV. The calculated values of CV show significant dispersion with a slight ascending trend with A for $A < 100$ km² and a slight descending trend for larger catchments. They conclude that these trends are, at least in part, due to biases induced by the different record lengths and the non-uniformity of the sample with respect to basin size, and ultimately consider them to be statistically not significant.

Bhatti[226] made a more extensive moment-scaling analysis of the annual maximum floods using 2150 unregulated basins from the USGS database. The basins belong to 15 hydrologic units and their sizes vary from 2 to 26000 km². The main finding is that Q_{max} varies in a self-similar way with area, with $H = 0.8 \pm 0.15$ for $A < 10^3$ km² and

$H = 0.5 \pm 0.2$ for larger areas (the fluctuations of H refer to differences among hydrologic units). Bhatti[226] also studied the dependence of the coefficient of variation CV with basin area A. As observed also by others, the empirical values of CV for different basins are highly dispersed and, if one excludes a few very large basins with low CV, do not exhibit significant trends with area.

Further evidence of self-similar scaling of daily and peak annual discharges comes from the analysis of 99 daily flow series in the US by Dodov and Foufoula-Georgiou.[239] The main finding is that both daily and peak annual discharges exhibit self-similar scaling with area, with two distinct scaling regimes. For the daily fluxes, $H \approx 0.8$ for $A < 700$ km^2 and $H \approx 0.5$ for $A > 700$ km^2, whereas for the peak annual discharges $H \approx 0.7$ for $A < 700$ km^2 and $H \approx 0.3$ for $A > 700$ km^2. This is in general agreement with Bhatti's conclusions.[226] Dodov and Foufoula-Georgiou[239] attribute the different values of H for A above and below 700 km^2 to the prevalence of below- or over-bank discharge conditions (for basins with $A > 700$ km^2, almost half of the peak annual discharges correspond to above-bank full flows, whereas for $A < 700$ km^2 only 15% are associated with such conditions).

The scaling of Q_{max} with A has been investigated also using simple theoretical models of runoff generation and routing. In a comprehensive study of this type, Menabde and Sivapalan[227] used a multifractal cascade model to generate rainfall, a self-similar model to represent the river network, and a simple flow model with connected reservoirs to model runoff and routing. When rainfall is uniform in space, the peak floods exhibit self-similar scaling with $H = 1$ for storms of duration $D > t_r$ and H that depends in a nonlinear way on D/t_r for shorter storm durations. This finding suggests that the observed self-similar scaling of peak floods is likely controlled by storms with duration D longer than the response time of the basin.

Menabde and Sivapalan[227] also performed a 100-year simulation using storms with durations drawn at random from an exponential distribution with mean value equal to 7 hours. Also, the inter-storm periods were taken to be exponentially distributed, with a mean value of 150 hours. For such well-separated storms, no interaction of flood waves is expected in the channel network. The mean storm duration is close to the response time $t_r \approx 10$ hours of the largest basin. Within storms, rainfall intensity was taken to be constant in time and multifractal in space, with parameters representative of mesoscale convective systems. When using an infiltration rate of 5 mm/hr, the peak annual flood scales in a self-similar way with area, with $H = 0.68$. This value of H is within the range from empirical studies. When the infiltration rate is increased to 25 mm/h, the same value of H holds for areas smaller than about 200 km^2, whereas for larger areas H is about 0.6. Notice that, by increasing the infiltration rate, one creates non-scaling dry areas in regions of the basin with divergent topography. This affects the calculated value of H.

In essence, available evidence points at two main conclusions: (i) peak annual fluxes may be taken to scale in a self-similar way with catchment area A; and (ii) the self-similarity index H is different for areas below and above about 700–1000 km^2, with smaller H for the larger basins. Next, we show how these findings can be explained using results from multifractal theory, and multifractal extremes in particular.

Multifractal Interpretation of the Scaling of Annual Floods with Area

There are two results that are relevant to the scaling of $Q_{\max}(A)$, one for large areas A and the other for small areas. In both cases, it is convenient to idealize the flow $Q(t)$ as resulting from direct integration of the rain rate over the river catchment in space and a suitable duration in time, such as the response time of the basin, t_r. This is the basis of the classical "rational method" of flood frequency estimation.[228] Specifically, let $S(t)$ be the above space-time region of integration centered at time t and denote by $\varepsilon(t)$ the average rainfall intensity in $S(t)$. We assume that the p-quantiles of $Q(t)$, $Q_p(t)$, satisfy

$$Q_p(t) \stackrel{d}{=} c_p A \varepsilon_p(t) \qquad (4.131)$$

where $\varepsilon_p(t)$ is the p-quantile of $\varepsilon(t)$ and c_p is a constant, usually referred to as runoff coefficient, that may depend on p and accounts for water losses and other effects.

For large basins, the response time t_r is dominated by channel transport and may be taken as proportional to the linear size of the basin.

This means that, as A varies, the region $S(t)$ scales by the same factor $\propto \sqrt{A}$ in space and time. For space-time regions $S(t)$ of sizes below the outer limits of multifractality of rainfall, one may apply the theory of multifractal extremes to ε_{max}, the annual maximum of $\varepsilon(t)$ (see Sec. 4.5.4.3, and in particular Eq. (4.82)). For the present problem ($r \propto A^{-1/2}$, $d = 3$ (immaterial), and $d' = 1$), Eq. (4.82) gives

$$\varepsilon_{max}(A) \stackrel{d}{=} A^{-\gamma_1/2} \varepsilon_{max}(1) \tag{4.132}$$

where $\gamma_1 > 0$ satisfies $C(\gamma_1) = 1$ (i.e. γ_1 is the slope of the tangent to the moment-scaling function $K(q)$ with Y-intercept equal to -1). Equations (4.131) and (4.132) imply the following self-similar scaling of the annual floods:

$$Q_{max}(A) \stackrel{d}{=} A^{1-\gamma_1/2} Q_{max}(1) \tag{4.133}$$

Values of γ_1 for rainfall are typically in the range 0.7–0.8,[76] implying a scaling of Q_{max} with area of the type $\sim A^H$ with H around 0.60–0.65. Considering that the present analysis is based on crude approximations, this result is consistent with the scaling of Q_{max} at relatively large areas observed empirically. A feature of interest is that, although rainfall models are multifractal, the implied flow maxima are self-similar, as generally observed from flow records. Note that, for extremely large basins, the size of S may exceed the limits of multifractal scaling of rainfall, producing some reduction in H.

In very small basins, the response time is dominated by hillslope processes, which do not scale with basin area. Consequently, as A approaches the size of a single hillslope, t_r approaches a non-zero limit t_o. When rainfall is averaged over space-time regions of different sizes, if the area A in space is small and the averaging duration t_o is relatively long, the distribution of $\varepsilon(t)$ becomes insensitive to A,[75,152] and, following the arguments above for large basins, one concludes that Q_{max} scales proportionally to A. This is again consistent with the observation that the scaling exponent increases towards 1 as the basin size decreases.

4.8.4. Flow Through Porous Media

At the macroscale, the flow of a liquid through a saturated porous medium is accurately described by Darcy's law

$$\underline{q} = -K \cdot \nabla H \qquad (4.134)$$

where $\underline{q}(\underline{x})$ is the specific discharge, $K(\underline{x})$ is the saturated hydraulic conductivity, and $H(\underline{x})$ is the hydraulic head. For a zero-divergence flow field, the hydraulic head H and the log hydraulic conductivity $F = \ln K$ must satisfy

$$\nabla^2 H + \nabla F \cdot \nabla H = 0 \qquad (4.135)$$

Over the past three decades, much work has been done on the statistical properties of the hydraulic head ∇H and the flow $\underline{q}$ that result from Eqs. (4.134) and (4.135) when K is a random field. In particular, the first-order spectral theory developed by Gelhar and co-workers through the 1990s has led to closed form relations between the power spectrum of F and the spectral density tensors of ∇H and $\underline{q}$.[240-242] The spectral tensor of $\underline{q}$ has then been used to explain how the macroscopic dispersivities A_{ij} vary with travel distance or time.[243-247]

First-order analysis reduces the complexity of the problem by replacing Eqs. (4.134) and (4.135) with linear approximations to the fluctuations $f = F - E[F]$ and $h = H - E[H]$. Using mainly perturbation methods, several authors have examined the effects of higher-order terms in f and h on the spectral density tensors of ∇H and $\underline{q}$.[243,248-251] In general, these studies have found that the inclusion of second-order terms has significant effects on the spectra of ∇H and $\underline{q}$ when the variance of $\ln K$ exceeds unity. However, for large log-conductivity variances, higher-order perturbation theories are not necessarily more accurate than first-order solutions.

More recently, there has been interest in the case when the hydraulic conductivity is a broad-band field with some type of scale invariance.[242,247,252-256] The models proposed typically assume that $F = \ln K$ is a homogeneous Gaussian field with spectral density S_F that decays like a power law along any given direction in Fourier space. In the isotropic case, this means that $S_F(\underline{k}) \propto k^{-\alpha}$, where $k = |\underline{k}|$ and α is some constant. For example, when α is between $d + 1$ and $d + 3$, d being the dimension of the space, the log-conductivity is a Brownian surface. The tools used to derive properties of the hydraulic gradient and flow are again those of first-order or second-order perturbation analysis.

Values of α close to the space dimension d have also been reported.[242] This is an interesting special case, since for $\alpha = d$, the hydraulic conductivity K is a stationary lognormal multifractal measure. Veneziano and Essiam[257,258] have studied this case, with special concern for the scale-invariance properties of the resulting hydraulic gradient and flow fields (see also Kozlov[259]). It is convenient to write the spectral density of the log-conductivity as

$$S_F(\underline{k}) = \begin{cases} \dfrac{2}{\Omega_d} C_K\, k^{-d}, & k_{\min} \leq k \leq k_{\max} \\ 0, & \text{otherwise} \end{cases} \qquad (4.136)$$

where $k = |\underline{k}|$, Ω_d is the area of the unit d-dimensional sphere (hence, $\Omega_1 = 2$, $\Omega_2 = 2\pi$, and $\Omega_3 = 4\pi$), C_K is the so-called co-dimension parameter of K, and $k_{\min}$ and $k_{\max}$ are wavenumbers that define the limits of multifractal scaling of K. The variability of the K field increases as C_K increases. The reason why the spectral range in Eq. (4.136) is assumed finite is that, for $k_{\max}/k_{\min} \to \infty$, the variance of F diverges and the medium becomes non-conductive.

The approach used by Veneziano and Essiam[257,258] is different from first- or higher-order perturbation analysis; it is nonlinear and exploits the fact that F has a Markov property in scale. The essence of the method is as follows. Let Ω be the region of R^d of interest, say a ball of unit diameter. The fluctuations of F at scales much larger than 1, say for k less than some low cutoff value k_1, are assumed to influence the flow in Ω through only the large-scale hydraulic gradient applied to Ω (simulation has confirmed the accuracy of this assumption). Let F_1 be the log-conductivity field low-passed below k_1 and let ∇H_1 and $\underline{q}_1$ be the resulting constant hydraulic gradient and flow inside Ω. Veneziano and Essiam[257] considered a cascade of problems of this type at different resolutions $r \geq 1$, in which ∇H_r and $\underline{q}_r$ are the hydraulic gradient and flow in the contracted region Ω/r under F_r, the log-conductivity field low-passed below $k_r = rk_1$. Hence, as r increases, F_r includes more of the high-frequency components, while ∇H_r and $\underline{q}_r$ are the resulting hydraulic gradient and flow in regions of decreasing size. The scaling properties of ∇H and $\underline{q}$ can be obtained by considering what happens at resolutions r and $r + dr$. The result is that ∇H_r and $\underline{q}_r$ have Markovian properties in r. Specifically, for any $r, r_1 \geq 1$, the hydraulic gradient ∇H satisfies

$$\nabla H_{rr_1}(\underline{x}) \stackrel{d}{=} (A_r \underline{R}_r) \cdot \nabla H_{r_1}(r\underline{R}_r^T \underline{x}), \qquad \underline{x} \in \Omega_r \qquad (4.137)$$

where A_r is a non-negative random variable, $\underline{R}_r$ is a random rotation matrix, and A_r, $\underline{R}_r$, and the random field ∇H_{r_1} on the right hand side are mutually independent. Equation (4.137) corresponds to a multifractal scaling condition that under contraction involves random rotation of the field ΔH in addition to random amplitude scaling. The variable A_r has the same distribution as the amplitude of the hydraulic gradient $|\nabla H_r(\underline{x})|$ at a generic location $\underline{x}$. This distribution is lognormal, with parameters

$$\begin{cases} E[\ln(A_r)] = \dfrac{d-4}{d(d+2)} C_K \ln(r) \\ \\ Var[\ln(A_r)] = \dfrac{6}{d(d+2)} C_K \ln(r) \end{cases} \qquad (4.138)$$

where C_K is the parameter in Eq. (4.136). As for the matrix $\underline{R}_r$, it is sufficient to characterize its first column. This is a unit random vector $\underline{e}_r$ that gives the rotation of the first coordinate axis and has the distribution of Brownian motion on the unit d-dimensional sphere, at "time" $t = 2/(d^2+2d) C_K \ln r$. The motion starts at the point $[1,0,\ldots,0]$ at time $t = 0$.

From Eq. (4.138) and well-known properties of the lognormal distribution, one can obtain the moments of A_r and, therefore, of $|\nabla H_r(\underline{x})|$. These are

$$E[|\nabla H_r(\underline{x})|^s] = E[A_r^s] = r^{C_K \left[\frac{d-1}{d(d+2)} s + \frac{3}{d(d+2)} (s^2 - s) \right]} \qquad (4.139)$$

Equation (4.139) shows that, in spaces of dimension $d > 1$, the mean hydraulic gradient amplitude diverges as the resolution $r \to \infty$; hence, the amplitude of the hydraulic gradient $|\nabla H|$ is non-conservative.

Divergence is due to the rotation in the renormalization condition in Eq. (4.137). By contrast, the hydraulic gradient itself has the same mean value for all r and is conservative.

Similarly, the specific flow $\underline{q}_r$ satisfies

$$\underline{q}_{rr_1}(\underline{x}) \stackrel{d}{=} (B_r \underline{R}_r) \cdot \underline{q}_{r_1}(r\underline{R}_r^T \underline{x}), \qquad \underline{x} \in \Omega_r \qquad (4.140)$$

where $\underline{R}_r$ is the same matrix as in Eq. (4.137) and B_r is a lognormal variable with parameters

$$\begin{cases} E[\ln(B_r)] = \left[\dfrac{d-4}{d(d+2)} - 1\right] C_K \ln(r) \\ \mathrm{Var}[\ln(B_r)] = \dfrac{2(d^2-1)}{d(d+2)} C_K \ln(r) \end{cases} \qquad (4.141)$$

The amplitude $q_r = |\underline{q}_r|$ has the same marginal distribution as B_r. Its moments are

$$E[q_r^s] = E[B_r^s] = r^{C_K\left[-\frac{d+5}{d(d+2)}s + \frac{d^2-1}{d(d+2)}(s^2-s)\right]} \qquad (4.142)$$

Contrary to the mean hydraulic gradient amplitude, the mean flow amplitude decreases as the resolution of the conductivity field r increases, and for $r \to \infty$ the medium becomes non-conductive. Another difference is that the mean flow vector depends on the resolution r. Specifically,

$$E[\underline{q}_r(\underline{x})] = r^{-\frac{2}{d}C_K}[1, 0, \ldots, 0] \qquad (4.143)$$

Hence, the flow is non-conservative and vanishes asymptotically as $r \to \infty$. This behavior results from the negative correlation ρ between the log conductivity F_r and the log hydraulic gradient amplitude $\ln|\nabla H_r|$ at any given location $\underline{x}$. The correlation coefficient is independent of the resolution r, but varies with the space dimension d as $\rho = -[(d+2)/3d]^{0.5}$. The amplitude $r^{-2C_K/d}$ of $E[\underline{q}_r(\underline{x})]$ in Eq. (4.143) may be interpreted as the effective conductivity K_{eff} of the medium and coincides with a formula conjectured by Matheron.[260] To our knowledge, this is the only case for which multifractal scaling of the generalized type (with random rotation and random amplitude scaling in the renormalization conditions) has been derived analytically.

Using the scaling properties in Eqs. (4.137) and (4.140), Veneziano and Essiam[257,258] obtained the spectral density tensors of the ∇H and q fields. Results differ from conventional linear perturbation theory[261] in that the spectral densities from linear theory decay at a faster rate and, as the wavenumber k increases, the contour lines from nonlinear analysis become circular (reflecting the isotropy of the hydraulic head field at small scales), whereas those from linear theory remain non-circular.

Using the spectral density tensor of q, Veneziano and Essiam[257,258] obtained the evolution of the ensemble longitudinal and transversal macrodispersivities with mean traveled distance. Reflecting the local isotropy of the flow field, at small distances, the longitudinal and transversal macrodispersivities coincide. This feature is not captured by the linear perturbation approach. Another difference with perturbation analysis is that the macrodispersivities are larger at small distances, due to the increased high-frequency content of the flow predicted by the nonlinear analysis.

In many natural porous formations, the hydraulic conductivity cannot be represented as an isotropic lognormal multifractal field. In some cases, the spectral density of $F = \ln K$ decays approximately like $k^{-\alpha}$ with $\alpha > d$. In other cases, the spectral density of F does not have a power-law form, indicating that the K field lacks scaling invariance of any kind. Veneziano and Tabaei[262] extended the analysis summarized above to the case when K is any isotropic lognormal field, i.e. when S_F is any integrable radially-symmetric spectral density. This includes the case of bounded-cascade representations of the hydraulic conductivity (see Sec. 4.4.5). Of course, the ∇H and q fields are no longer scale-invariant, but other properties like the marginal distributions and spectral density tensors can still be found analytically. The marginal distribution of q determines the effective conductivity of large blocks (through the mean value $E[q]$) and the spectral density tensors characterize the macrodispersivities. Further extension to cases involving anisotropy, for example in the form of differences in hydraulic conductivity in the vertical and horizontal directions, would be very challenging.

4.9. Concluding Remarks

This chapter has attempted to provide a concise, yet comprehensive, view of scale invariance in hydrology. Since no systematic treatment of the important class of multifractal processes exists, we have devoted much space to their definition and properties. We have stressed from the beginning that mono- and multi-fractality are best seen as statistical renormalization properties of the object under study (typically a set, a function or a measure). This notion is widely accepted for self-similarity but not for multifractality. The alternative, and more popular, view of multifractality, based on the fractal dimension of the support of singularities of the object, is less general and further lacks the appeal of a renormalization property.

Methods to generate (quasi-)stationary multifractal measures through discrete cascades and processes with stationary self-similar or multifractal increments through the additive superposition of certain scaled processes are well-known, but general methods to produce scale-invariant processes are not. In Sec. 4.4, we have dealt with this topic in some detail. Some of the material in the same section is from our unpublished works.

The review of methods to infer scaling properties from data is non-systematic. In our opinion, inference procedures for multifractality still await in-depth study and assessment. To alert the reader of possible pitfalls, we have critically discussed some of the more standard methods, pointing out deficiencies and suggesting improvements. In practice, the limited performance of the methods often compounds with an excessive readiness by the analyst to claim scale invariance (for example, using narrow scaling ranges or fitting power laws when the data do not display algebraic behavior) to produce questionable results. The development of statistical tests for scale invariance would be of significant help, but so far only modest progress has been made in this direction.

Our coverage of hydrologic applications is also limited, as we have provided only a summary of developments in four areas: rainfall, fluvial geomorphology, floods, and subsurface flow. The choice and coverage are further influenced by our own interests and past work. However, even from this selected coverage, it is clear that the area of hydrology, which has always been the source of inspiration for fractal theory, continues to be a fertile ground for application. In nature, nothing is exactly fractal or scale-invariant. At best, these properties apply within limits. The ability to use scaling approaches while also recognizing their limits will be a clear sign that scale invariance in hydrology has reached its mature stage.

Acknowledgments

This chapter was made possible in part by the generous support of the Government of Portugal through the Portuguese Foundation for International Cooperation in Science, Technology and Higher Education and was undertaken in the MIT-Portugal Program. The second author was also partially supported by the Alexander S. Onassis Public Benefit Foundation under Scholarship No. F-ZA 054/2005-2006. We thank Pierluigi Furcolo for useful discussions and for generating many of the figures, and Bellie Sivakumar for his help with editing the manuscript. The constructive comments of two anonymous reviewers are also acknowledged.

References

1. I. Rodriguez-Iturbe and A. Rinaldo, Eds., *Fractal River Basins: Chance and Self-Organization* (Cambridge University Press, Cambridge, UK, 1997).
2. M. F. Barnsley, Ed., *Fractals Everywhere* (Academic Press, 1988).
3. C. E. Puente and N. Obregón, *Water Resour. Res.*, 2825 (1996).
4. C. E. Puente and B. Sivakumar, *Nonlinear Proc. Geophys.*, 525 (2003).
5. Mandelbrot, Ed., *The Fractal Geometry of Nature* (W. H. Freedman and Company, New York, 1983).
6. K. J. Falconer, Ed., *Fractal Geometry: Mathematical Foundations and Applications* (John Wiley & Sons, New York, 1990).
7. J. E. Hutchinson, *Indiana Univ. Math. J.*, 713 (1981).
8. K. J. Falconer, *Math. Proc. Camb. Phil. Soc.*, 559 (1986).
9. R. D. Mauldin and S. C. Williams, *Trans. Am. Math. Soc.*, 325 (1986).
10. S. Graf, *Probab. Th. Rel. Fields*, 357 (1987).
11. U. Zahle, *Prob. Th. Rel. Fields*, 79 (1988).
12. C. Tricot, Ed., *Curves and Fractal Dimension* (Springer-Verlag, New York, 1993).
13. A. Fournier, D. Fussell and L. Carpenter, *Comm. ACM*, 371 (1982).

14. S. Lu, F. Molz and H. H. Liu, *Computers and Geosciences*, 15 (2003).
15. B. B. Mandelbrot and T. Vicsek, *J. Phys.* **A**, L377 (1989).
16. H. O. Peitgen, H. Jurgens and D. Saupe, Eds., *Fractals for the Classroom* (Springer-Verlag, New York, 1992).
17. J. S. Lee, D. Veneziano and H. H. Einstein, *Proceedings: 31st U.S. Symposium on Rock Mechanics* (Golden, Colorado, Balkema, 1990).
18. G. Willgoose, R. L. Bras and I. Rodriguez-Iturbe, *Water Resour. Res.*, 1671 (1991).
19. B. Dubrulle, in *Scale Invariance and Beyond*, Eds. B. Dubrulle, F. Graner and D. Sornette (Springer-Verlag, Berlin, 1997), p. 275.
20. I. Yekutieli, B. B. Mandelbrot and H. Kaufman, *J. Phys. A: Math. Gen.*, 275 (1994).
21. D. Veneziano, *Fractals*, 59 (1999).
22. J. Lamperti, *Trans. Amer. Math. Soc.*, 62 (1962).
23. A. M. Yaglom, Ed., *Correlation Theory of Stationary and Related Random Functions, Vol. 1: Basic Results* (Springer-Verlag, New York, 1986).
24. W. Feller, Ed., *An Introduction to Probability Theory and Its Apllications*, Vol II (John Wyley & Sons, New York, 1966).
25. E. Bacry and J. F. Muzy, *Comm. Math. Phys.*, 449 (2003).
26. T. Vicsek and A. L. Barabasi, *J. Phys. A: Math. Gen.*, L845 (1991).
27. R. Benzi, L. Biferale, A. Crisanti, G. Paladin, M. Vergassola and A. Vulpiani, *Physica D*, 352 (1993).
28. V. K. Gupta, O. J. Mesa and D. R. Dawdy, *Water Resour. Res.*, 3405 (1994).
29. D. Schertzer and S. Lovejoy, in *Non-linear Variability in Geophysics*, Eds. D. Schertzer and S. Lovejoy (Kluwer Academic Publishers, The Netherlands, 1991), p. 41.
30. K. Pflug, S. Lovejoy and D. Schertzer, *J. Atmos. Sci.*, 538 (1993).
31. D. Schertzer and S. Lovejoy, in *Turbulent Shear Flows 4*, Ed. B. Launder (Springer, New York., 1985).
32. S. Lovejoy and D. Schertzer, *Phys. Can.*, 62 (1990).
33. D. Marsan, D. Schertzer and S. Lovejoy, *J. Geophys. Res.*, **D21**, 26333 (1996).
34. V. Venugopal, E. Foufoula-Georgiou and V. Sapoznhnikov, *J. Geophys. Res.*, **D24**, 31599 (1999).
35. R. Deidda, M. G. Badas and E. Piga, *Water Resour. Res.*, doi: 10.1029/2003WR002574 (2004).
36. S. Pecknold, S. Lovejoy, D. Schertzer and C. Hooge, in *Scale in Remote Sensing and GIS*, Eds. D. Quattrocchi and M.F. Goodchild (CRC Press, New York., 1997), Chapter 16.
37. C. Meneveau, K. R. Sreenivasan, P. Kailasnath and M.S. Fan, *Phys. Rev. A*, 894 (1990).
38. W. Vervaat, *Bull. Int. Statist. Inst.*, 199 (1987).
39. R.F. Voss, in *The Science of Fractal Images*, Eds. H. O. Peitgen and D. Saupe (Springer-Verlag, New York, 1988), Chapter 1.
40. D. Halford, *Proceedings IEEE*, 251 (1968).

41. S. Lovejoy and B. B. Mandelbrot, *Tellus*, **A**, 209 (1985).
42. A. Gelb, Ed., *Applied Optimal Estimation* (MIT Press, Cambridge, Mass, 1974), p. 44.
43. G. Samorodnitsky and M. S. Taqqu, Eds., *Stable Non-Gaussian Random Processes* (Chapman & Hall, New York, 1994), Chapters 7 and 8.
44. M. Maejima, *Sugaku Expositions*, 103 (1989).
45. D. Schertzer and S. Lovejoy, *J. Geophys. Res.*, 9693 (1987).
46. J. Wilson, D. Schertzer and S. Lovejoy, in *Non-linear Variability in Geophysics*, Eds. D. Schertzer and S. Lovejoy (Kluwer Academic Publishers, 1991), p. 185.
47. S. Pecknold, S. Lovejoy, D. Scherzer, C. Hooge and J. F. Malouin, in *Cellular Automata: Prospects in Astrophysical Applications*, Eds. J. M. Perdang and A. Lejeme (World Scientific, 1993), p. 228.
48. D. Schertzer and S. Lovejoy, Eds., *Resolution Dependence and Multifractals in Remote Sensing and Geophysical Information Systems* (Lecture Notes, McGill University, Montreal, 1996).
49. M. Menabde, D. Harris, A. W. Seed, G. L. Austin and C. D. Snow, *Water Resour. Res.*, 2823 (1997).
50. M. Menabde, *Nonlinear Proc. Geophys.*, 63 (1998).
51. M. Menabde and M. Sivapalan, *Water Resour. Res.*, 3293 (2000).
52. R. Benzi, S. Ciliberto, C. Baudet and S. Succi, *Phys. Rev. E*, R29 (1993).
53. R. Benzi, S. Ciliberto, C. Baudet and G. Ruiz Chavarria, *Physica D*, 385 (1995).
54. R. Benzi, L. Biferale, S. Ciliberto, M. V. Struglia and R. Tripiccione, *Physica D*, 162 (1996).
55. M. Briscolini, P. Santangelo, S. Succi and R. Benzi, *Phys. Rev. E*, 1745 (1994).
56. J. Olsson, *Hydrol. Earth Syst. Sci.*, 19 (1998).
57. A. Güntner, J. Olsson, A. Calver and B. Gannon, *Hydrol. Earth Syst. Sci.*, 145 (2001).
58. D. Veneziano, P. Furcolo and V. Iacobellis, *J. Hydrol.*, 105 (2006).
59. K. Fraedrich and C. Larnder, *Tellus*, **A**, 289 (1993).
60. J. Beran, Ed. *Statistics of Long Memory Processes* (Chapman & Hall, New York, 1994).
61. J. P. Kahane and J. Peyriere, *Adv. Math.*, 131 (1976).
62. D. Veneziano and P. Furcolo, *Fractals*, 253 (2003).
63. M. E. Cates and J. M. Deutsch, *Phys. Rev. A*, 4907 (1987).
64. J. O'Neil and C. Meneveau, *Phys. Fluid A*, 158 (1993).
65. H. Cramer, *Actualites Scientifiques et Industrielles*, No. 736 of Colloque consacre a la theorie des probabilities (Herrman, Paris, 1938), p. 5.
66. C. J. G. Evertsz and B. B. Mandelbrot, in *Chaos and Fractals: New Frontiers of Science*, Eds. H. O. Peitgen, H. Jurgens and D. Saupe (Springer-Verlag, New York, 1992), p. 921.
67. D. W. Stroock, Ed. *Probability Theory: An Analytic View* (Cambridge University Press, USA, 1994).

68. S. R. S. Varadhan, Ed. *Large Deviations and Applications* (Society for Industrial and Applied Mathematics, Philadelphia, 1984).
69. A. Dembo and O. Zeitouni, Eds. *Large Deviation Techniques and Applications* (Jones and Barlett, Boston, 1993).
70. D. Veneziano, *Fractals*, 117 (2002); Erratum, *Fractals*, 1 (2005).
71. D. Lavallee, D. Schertzer and S. Lovejoy, in *Non-linear Variability in Geophysics*, Eds. D. Schertzer and S. Lovejoy (Kluwer Acad. Publishers, The Netherlands, 1991), p. 99.
72. A. Pathirana, S. Herath and T. Yamada, *Hydrol. Earth Syst. Sci.*, 668 (2003).
73. P. Hubert, H. Bendjoudi, D. Schertzer and S. Lovejoy, *Proceedings: Int. Conf. On Heavy Rains and Flash Floods* (Istanbul, Turkey, 1998).
74. D. Veneziano and P. Furcolo, *Water Resour. Res.*, 1306 (2002).
75. D. Veneziano and A. Langousis, *Water Resour. Res.*, W07008, doi:10.1029/2004WR003765 (2005).
76. A. Langousis, D. Veneziano, P. Furcolo and C. Lepore, *Chaos Soliton Fract.*, doi:10.1016/j.chaos.2007.06.004 (2007).
77. D. Veneziano and A. Langousis, *Fractals*, 311 (2005).
78. L. Calvet and A. Fisher, *J. Econometrics*, 27 (2001).
79. Y. J. Chou, Ed. *Short-Term Rainfall Prediction Using a Multifractal Model* (MS Thesis, Department of Civil and Environmental Engineering, MIT, Cambridge, MA, 2003).
80. W. B. Davenport and W. L. Root, Eds. *An Introduction to the Theory of Random Signals and Noise* (McGraw-Hill, New York, 1958), Chapter 11.
81. L. A. Wainstein and V. D. Zubakov, Eds. *Extraction of Signals from Noise* (Dover Publications, New York, 1962), Chapter 2.
82. G. Kallianpur, Ed. *Stochastic Filtering Theory* (Springer-Verlag, New York, 1980).
83. N. G. Mackay, R. E. Chandler, C. Onof and H. S. Wheater, *Hydrol. Earth Syst. Sci.*, 165 (2001).
84. N. Rebora, L. Ferraris, J. von Hardenberg and A. Provenzale, *J. Hydrometeor.*, 724 (2006).
85. D. Valencia and J. C. Schaake, Report no. 149 (Ralph M. Parsons Laboratory of Water Resources and Hydrodynamics, MIT, Cambridge, MA, 1972).
86. D. Valencia and J. C. Schaake, *Water Resour. Res.*, 211 (1973).
87. J. D. Salas, in *Handbook of Hydrology*, Ed. D. Maidment (McGraw-Hill, New York, 1993), Chapter 19.
88. D. Koutsoyiannis, C. Onof and H. S. Wheater, *Water Resour. Res.*, 1173, doi:10.1029/2002WR001600 (2003).
89. I. Primus, D. Mclaughlin and D. Entekhabi, *Adv. Water Resour.*, 941 (2001).
90. L. LeCam, in *Proceedings of Fourth Berkeley Symposium on Mathematical Statistics and Probability*, Vol. 3, Ed. J. Neyman (University of Califorinia Press, Berkeley, 1961), p. 165.
91. E. C. Waymire and V. K. Gupta, *Water Resour. Res.*, 1261 (1981).

92. E. C. Waymire and V. K. Gupta, *Water Resour. Res.*, 1273 (1981).
93. E. C. Waymire and V. K. Gupta, *Water Resour. Res.*, 1287 (1981).
94. I. Rodriguez-Iturbe, D. R. Cox and V. Isham, *Proc. Royal Soc. London,* **Ser. A**, 269 (1987).
95. P. S. P. Cowpertwait, *Proc. Royal Soc. London,* **Ser. A**, 163 (1995).
96. P. Willems, *J. Hydrol.*, 126 (2001).
97. S. Perica and E. Foufoula-Georgiou, *J. Geophys. Res.*, **D21**, 26347 (1996).
98. B. Ahrens, *J. Geophys. Res.*, **D8**, 8388, doi:10.1029/2001JD001485 (2003).
99. K. S. Paulson and P. D. Baxter, *J. Geophys. Res.*, doi:10.1029/2006JD007333 (2007).
100. R. Deidda, *Water Resour. Res.*, 1779 (2000).
101. D. D. Hodges, R. J. Watson and K. S. Paulson, *IEEE*, 55, doi:10.1109/APS.2005.1552580 (2005).
102. B. Tustison, E. Foufoula-Georgiou and D. Harris, *J. Geophys. Res.*, **D8**, 8377, doi:10.1029/2001JD001073 (2003).
103. R. Gupta, V. Venugopal and E. Foufoula-Georgiou, *J. Geophys. Res.*, D02102, doi:10.1029/2004JD005568 (2006).
104. D. Harris, A. Seed, M. Menabde and G. Austin, *Nonlinear Proc. Geophys.*, 137 (1997).
105. P. Furcolo and D. Veneziano, *Symposium on New Statistical Tools in Hydrology* (Capri, Italy, October 13-14, 2008).
106. M. Ossiander and E. C. Waymire, *Ann. Statist.*, 1533 (2000).
107. M. Ossiander and E. C. Waymire, *Indian J. Stat.*, 323 (2002).
108. B. Lashermes, P. Abry and P. Chainais, *Int. J. Wavelets Multir. Inf. Proc.*, 497 (2004).
109. D. Veneziano, A. Langousis and P. Furcolo, *Water Resour. Res.*, W06D15, doi: 10.1029/2005WR004716 (2006).
110. W. H. Press, S. A. Teukolsky, W. T. Vetterling and B. P. Flannery, Eds., *Numerical Recepies in C: The Art of Scientific Computing,* 2nd Edition (Cambridge University Press, 1992).
111. J. F. Muzy, E. Bacry and A. Arneodo, *Phys. Rev. Lett.*, 3515 (1991).
112. E. Bacry, J. F. Muzy and A. Arneodo, *J. Stat. Phys.*, 635 (1993).
113. J. F. Muzy, E. Bacry and A. Arneodo, *Int. J. Bifurcation Chaos*, 245 (1994).
114. Z. R. Struzik, *Fractals*, 163 (2000).
115. J. F. Muzy, E. Bacry and A. Arneodo, *Phys. Rev. E*, 875 (1993).
116. A. Arneodo, E. Bacry and J. F. Muzy, *Physica A*, 232 (1995).
117. B. Lashermes and E. Foufoula-Georgiou, *Water Resour. Res.*, W09405, doi:10.1029/2006WR005329 (2007).
118. Y. Tessier, S. Lovejoy and D. Schertzer, *J. Appl. Meteor.*, 223 (1993).
119. A. Turiel, C. J. Perez-Vicente and J. Grazzini, *J. Comput. Phys.*, 362 (2006).
120. S. I. Vainshtein, K. R. Sreenivasan, R. T. Pierrehumbert, V. Kashyap and A. Juneja, *Phys. Rev. E*, 1823 (1994).

121. D. Veneziano and V. Iacobellis, *J. Geophys. Res.*, **B6**, 12797 (1999).
122. V. Venugopal, S. G. Roux, E. Foufoula-Georgiou and A. Arneodo, *Water Resour. Res.*, W06D14, doi:10.1029/2005WR004489 (2006).
123. D. Lavallee, S. Lovejoy and D. Schertzer, in *Proceedings 1558, Society of Photo-Optical Instrumentation Engineers (SPIE)* (San Diego, 1991), p. 60.
124. D. Lavallee, S. Lovejoy, D. Schertzer and P. Ladoy, in *Fractals in Geography*, Eds. L. De Cola and N. Lam (Prentice-Hall, 1993), p. 171.
125. F. Schmitt, D. Lavallee, D. Schertzer and S. Lovejoy, *Phys. Rev. Lett.*, 305 (1992).
126. F. Schmitt, S. Lovejoy, D. Schertzer, D. Lavallee and C. Hooge, *Acad. Sci. Paris, Series II*, 749 (1992).
127. F. Schmitt, D. Schertzer, S. Lovejoy and Y. Brunet, *Fractals*, 569 (1993).
128. Y. Tessier, S. Lovejoy, P. Hubert, D. Schertzer and S. Pecknold, *J. Geophys. Res.*, 26427 (1996).
129. S. Lovejoy and D. Schertzer, *J. Appl. Meteor.*, 1167 (1990).
130. J. Olsson, J. Niemczynowicz and R. Berndtsson, *J. Geophys. Res.*, 23265 (1993).
131. S. Lovejoy, D. Lavallee, D. Schertzer and P. Ladoy, *Nonlinear Proc. Geophys.*, 16 (1995).
132. E. M. Douglas and A. P. Barros, *J. Hydrometeor.*, 1012 (2003).
133. D. Veneziano and P. Furcolo, *Fractals*, 181 (1999).
134. J. Olsson and J. Niemczynowicz, *J. Hydrol.*, 29 (1996).
135. R. Deidda, R. Benzi and F. Siccardi, *Water Resour. Res.*, 1853 (1999).
136. J. Olsson, *Nonlinear Proc. Geophys.*, 23 (1995).
137. R. Deidda, M. Grazia-Badas and E. Piga, *J. Hydrol.*, 2, doi: 10.1016/j.jhydrol.2005.02.036 (2006).
138. T. M. Over and V. K. Gupta, *J. Geophys. Res.*, **D21**, 26319 (1996).
139. F. Schmitt, S. Vannitsem and A. Barbosa, *J. Geophys. Res.*, **D18**, 23181 (1998).
140. A. Langousis and D. Veneziano, *Water Resour. Res.*, doi:10.1029/2006WR005245 (2007).
141. S. Lovejoy and D. Schertzer, in *New Uncertainty Concepts in Hydrology and Water Resources*, Ed. Z. W. Kundzewicz, (Cambridge Press, 1995).
142. P. Hubert, Y. Tessier, P. Ladoy, S. Lovejoy, D. Schertzer, J. P. Carbonnel, S. Violette, I. Desurosne and F. Schmitt, *Geophys. Res. Lett.*, 931 (1993).
143. D. Veneziano, R. L. Bras and J. D. Niemann, *J. Geophys. Res.*, **D21**, 26371 (1996).
144. D. Veneziano and V. Iacobellis, *Water Resour. Res.*, 13.1 (2002).
145. P. K. Kundu and T. L. Bell, *Water Resour. Res.*, doi:10.1029/2002WR001802 (2003).
146. M. Gebremichael and W. F. Krajewski, *J. Appl. Meteor.*, 1180 (2004).
147. M. Gebremichael, T. M. Over and W. F. Krajewski, *J. Hydrometeor.*, 1277 (2006).
148. M. Kardar, G. Parisi and Y. Zhang, *Phys. Rev. Lett.*, 889 (1986).
149. A. Czirok, E. Somfai and T. Vicsek, *Phys. Rev. Lett.*, 2127 (1993).
150. V. Venugopal, E. Foufoula-Georgiou and V. Sapozhnikov, *J. Geophys. Res.*, **D16**, 19705 (1999).

151. S. Perica and E. Foufoula-Georgiou, *J. Geophys. Res.*, **D3**, 7431 (1996).
152. D. Veneziano and P. Furcolo, *Fractals*, 147 (2002).
153. D. Veneziano, P. Furcolo and V. Iacobellis, *EGU General Assembly* (Nice, France, April 2002).
154. G. Leclerc and J. C. Schaake, Eds., *Derivation of Hydrologic Frequency Curves* (Report 142, Department of Civil Engineering, MIT, Cambridge, MA, 1972).
155. Natural Environmental Research Council (NERC), *Flood Studies Report*, Vol. 2 (Institute of Hydrology, Wallingford, UK, 1975).
156. D. Koutsoyiannis and Th. Xanthopoulos, Eds., *Engineering Hydrology*, 3rd Edition (National Technical University of Athens, Athens, 1996).
157. L. M. V. Carvalho, D. Lavallée and C. Jones, *Geophys. Res. Lett.*, doi:10.1029/2001GL014276 (2002).
158. D. Harris, M. Menabde, A. Seed and G. Austin, *J. Geophys. Res.*, **D21**, 26405 (1996).
159. J. C. Purdy, D. Harris, G. L. Austin, A. W. Seed and W. Gray, *J. Geophys. Res.*, **D8**, 7837 (2001).
160. A. Pathirana and S. Herath, *Hydrol. Earth Syst. Sci.*, 659 (2002).
161. D. K. Nykanen and D. Harris, *J. Geophys. Res.*, **D8**, doi:10.1029/2001JD001518 (2003).
162. M. G. Badas, R. Deidda and E. Piga, *Adv. Geosci*, 285 (2005).
163. K. S. Paulson, *Radio Sci.*, doi:10.1029/2001RS002527 (2002).
164. R. F. Voss, in *Fundamental Algorithms for Computer Graphics*, Ed. R. A. Earnshaw (Springer-Verlag, Berlin, 1985).
165. R. E. Horton, *Bull. Geol. Soc. Am.*, 275 (1945).
166. R. L. Shreve, *J. Geol.*, 17 (1966).
167. R. L. Shreve, *Geology*, 527 (1975).
168. J. T. Hack, *USGS Professional Paper* (USGS, 1957), pp. 45.
169. B. B. Mandelbrot, in *The Science of Fractal Images*, Eds. H. O. Peitgen and D. Saupe (Springer-Verlag, New York, 1988).
170. K. Musgrave, *Chapman Conference on Fractal Scaling, Nonlinear Dynamics, and Chaos in Hydrologic Systems* (Clemson, South Carolina, 1998).
171. P. A. Burrough, *Nature*, 240 (1981).
172. S. A. Brown, *Geophys. Res. Lett.*, 1095 (1987).
173. G. Dietler and Y. C. Zhang, *Physica A*, 213 (1992).
174. B. Klinkenberg and M. F. Goodchild, *Earth Surf. Process. Landf.*, 217 (1992).
175. A. Malinverno, *PAGEOPH*, 139 (1989).
176. D. M. Mark and P. B. Aronson, *Math. Geol.*, 671 (1984).
177. J. K. Weissel, L. F. Pratson and A. Malinverno, *J. Geophys. Res.*, **B7**, 13997 (1994).
178. E. Tokunaga, *Geogr. Rep. Tokyo Metrop. Univ.*, 1 (1978).
179. P. La Barbera and R. Rosso, *Water Resour. Res.*, 735 (1989).
180. V. I. Nikora, *Water Resour. Res.*, 1327 (1991).
181. S. D. Peckham, *Water Resour. Res.*, 1023 (1995).

182. D. G. Tarboton, *J. Hydrol.*, 105 (1996).
183. V. K. Gupta and E. C. Waymire, *Water Resour. Res.*, 463 (1989).
184. A. D. Howard, *Water Resour. Res.*, 2261 (1994).
185. G. E. Moglen, *Simulation of Observed Topography Using a Physically Based Basin Evolution Model* (PhD Thesis, Department of Civil and Environmental Engineering, MIT, Cambridge, MA, 1994).
186. G. E. Tucker, *Modeling the Large-Scale Interaction of Climate, Tectonics, and Topography* (PhD thesis, Pennsylvania State University, 1996).
187. D. Veneziano and J. Niemann, *Water Resour. Res.*, 1923 (2000).
188. D. Veneziano and J. Niemann, *Water Resour. Res.*, 1937 (2000).
189. E. J. Ijjasz-Vasquez, R. L. Bras and I. Rodriguez-Iturbe, *Eos*, 202 (1991).
190. A. Rinaldo, I. Rodriguez-Iturbe, R. Rigon, E. Ijjasz-Vasquez and R. L. Bras, *Phys. Rev. Lett.*, 822 (1993).
191. A. Rinaldo, I. Rodriguez-Iturbe, R. Rigon, R. L. Bras, E. Ijjasz-Vasquez and A. Marani, *Water Resour. Res.*, 2183 (1992).
192. A. Rinaldo, A. Maritan, F. Colaiori, A. Flammini, R. Rigon, I. Rodriguez-Iturbe and J. R. Banavar, *Phys. Rev. Lett.*, 3364 (1996).
193. R. Rigon, A. Rinaldo, I. Rodriguez-Iturbe, R. L. Bras and E. Ijjasz-Vasquez, *Water Resour. Res.*, 1635 (1993).
194. T. Sun, P. Meakin, and T. Jossang, *Phys. Rev. E*, 5353 (1995).
195. K. Sinclair and R. Ball, *Phys. Rev. Lett.*, 3360 (1996).
196. R. E. Horton, *Trans. Am. Geophys. Union*, 350 (1932).
197. A. N. Strahler, *Geolog. Soc. Am. Bull.*, 1117 (1952).
198. J. W. Kirchner, *Geology*, 591 (1993).
199. K. Paik and P. Kumar, *Eur. Phys. J.*, **B**, 247 (2007).
200. D. L. Turcotte and W. J. Newman, *Proc. Natl. Acad. Sci.*, 14295 (1996).
201. E. Tokunaga, *Jap. Geomorph. Un.*, 71 (1984).
202. E. Tokunaga, in *Research of Pattern Formation*, Ed. R. Takaki (KTK Scientific Publishers, Tokyo, 1994), p. 445.
203. S. G. De Bartolo, R. Gaudio and S. Gabriele, *Water Resour. Res.*, W02201, doi:10.1029/2003WR002760 (2004).
204. S. G. De Bartolo, M. Veltri and L. Primavera, *J. Hydrol.*, 181 (2006).
205. R. Gaudio, S. G. De Bartolo, L. Primavera, S. Gabriele and M. Veltri, *J. Hydrol.*, 365 (2006).
206. D. R. Montgomery and W. E. Dietrich, *Science*, 826 (1992).
207. P. Tailing and M. J. Sowter, *Earth Surf. Process. Landf.*, 809 (1999).
208. Z. Lin and T. Oguchi, *Geomorphol.*, 159 (2004).
209. O. J. Mesa and E. R. Mifflin, in *Scale Problems in Hydrology*, Eds. V. K. Gupta, I. Rodriguez-Iturbe and E. F. Wood (D. Reidel, Norwell, MA, 1986), Chapter 1.
210. J. S. Robinson, M. Sivapalan and J. D. Snell, *Water Resour. Res.*, 3089 (1995).
211. P. S. Naden, *Hydrol. Sci. J.*, 53 (1992).

212. V. K. Gupta and E. C. Waymire, in *Scale Dependence and Invariance in Hydrology*, Ed. G. Sposito (Cambridge University Press, New York, 1998), p. 88.
213. M. Sivapalan, C. Jothityangkoon and M. Menabde, *Water Resour. Res.*, 4.1, doi:10.1029/2001WR000482 (2002).
214. P. M. Saco and P. Kumar, *Water Resour. Res.*, 1244, doi:10.1029/2001WR000694 (2002).
215. D. Veneziano, G. Moglen, P. Furcolo and V. Iacobellis, *Water Resour. Res.*, 1143 (2000).
216. M. Marani, A. Rinaldo, R. Rigon and I. Rodriguez-Iturbe, *Geophys. Res. Lett.*, 2123 (1994).
217. B. Richards-Pecou, *Hydrol. Sci. J.*, 387 (2002).
218. D. Veneziano, G. Moglen and R. L. Bras, *Phys. Rev. E*, 1387 (1995).
219. D. G. Tarboton, R. L. Bras and I. Rodriguez-Iturbe, *Water Resour. Res.*, 2037 (1989).
220. G. Willgoose, R. L. Bras and I. Rodriguez-Iturbe, *Water Resour. Res.*, 1697 (1991).
221. A. D. Howard, W. E. Dietrich and M. A. Seidl, *J. Geophys. Res.-Solid Earth*, 13971 (1994).
222. J. R. Banavar, F. Colaiori, A. Flammini, A. Giacometti, A. Maritan and A. Rinaldo, *Phys. Rev. Lett.*, 4522 (1997).
223. G. E. Moglen and R. L. Bras, *Water Resour. Res.*, 2613 (1995).
224. D. Veneziano and J. D. Niemann, *EGS General Assembly* (Nice, France, 1998).
225. V. K. Gupta, S. L. Castro and T. M. Over, *J. Hydrol.*, 81 (1996).
226. M. B. Bhatti, *Extreme Rainfall, Flood Scaling and Flood Policy Options in the United States* (MS Thesis, Department of Civil and Environmental Engineering and Engineering Systems Division, MIT, Cambridge, MA, 2000), pp. 227.
227. M. Menabde and M. Sivapalan, *Adv. Wat. Resour.*, 1001 (2001).
228. V. P. Singh, Ed., *Elementary Hydrology* (Prentice-Hall, New Jersey, USA, 1992).
229. R. Pandey, S. Lovejoy and D. Schertzer, *J. Hydrol.*, 62 (1998).
230. D. Labat, A. Mangin and R. Ababou, *J. Hydrol.*, 176 (2002).
231. R. Pandey, *J. Hydrol. Eng.*, 169 (1998).
232. V. K. Gupta and E. Waymire, *J. Geophys. Res.*, **D3**, 1999 (1990).
233. V. K. Gupta and D. Dawdy, *Hydrol. Process.*, 347 (1995).
234. J. S. Robinson and M. Sivapalan, *Water Resour. Res.*, 1045 (1997).
235. J. S. Robinson and M. Sivapalan, *Water Resour. Res.*, 2981 (1997).
236. S. A. Veitzer and V. K. Gupta, *Adv. Water Resour.*, 955 (2001).
237. B. M. Troutman and T. M. Over, *Adv. Water Resour.*, 967 (2001).
238. E. E. Cadavid, *Hydraulic Geometry of Channel Networks* (Department of Civil Engineering, University of Mississippi, USA, 1988).
239. B. Dodov and E. Foufoula-Georgiou, *Water Resour. Res.*, W05005, doi:10.1029/2004WR003408 (2005).
240. L. W. Gelhar and C. L. Axness, *Water Resour. Res.*, 161 (1983).

241. L. W. Gelhar, in *Fundamentals of Transport in Porous Media*, Eds. J. Bear and M. Y. Corapcioglu (Dordrecht, The Netherlands: Martinus Nijhoff, 1987), p. 657.
242. R. Ababou and L. W. Gelhar, in *Dynamics of Fluids in Hierarchical Porous Media*, Ed. J. H. Cushman (Academic, San Diego, California, 1990), p. 393.
243. G. Dagan, *Water Resour. Res.*, 573 (1985).
244. D. L. Koch and J. F. Brady, *Phys. Fluids*, 965 (1988).
245. G. Dagan and S. P. Neuman, *Water Resour. Res.*, 3249 (1991).
246. J. Glimm, W. B. Lindquist, F. Pereira and Q. Zhang, *Trans. Porous Media*, 97 (1993).
247. H. Zhan and S. W. Wheatcraft, *Water Resour. Res.*, 3461 (1996).
248. F. W. Deng and J. H. Cushman, *Water Resour. Res.*, 103 (1998).
249. K. C. Hsu, D. Zhang and S. P. Neuman, *Water Resour. Res.*, 571 (1996).
250. T. V. Lent and P. K. Kitanidis, *Water Resour. Res.*, 1197 (1996).
251. K. C. Hsu and S. P. Neuman, *Water Resour. Res.*, 625 (1997).
252. A. Arya, T. A. Hewett, R. Larson and L. W. Lake, *SPE Reservoir Eng.*, 139 (1988).
253. S. W. Wheatcraft and S. W. Tyler, *Water Resour. Res.*, 566 (1988).
254. S. P. Neuman, *Water Resour. Res.*, 1749 (1990).
255. H. Rajaram and L. W. Gelhar, *Water Resour. Res.*, 2469 (1995).
256. V. Di Federico and S. P. Neuman, *Water Resour. Res.*, 1075 (1997).
257. D. Veneziano and A. K. Essiam, *Water Resour. Res.*, 1166, doi:10.1029/2001WR001018 (2003).
258. D. Veneziano and A. K. Essiam, *Chaos Solitons Fract.*, 293 (2004).
259. S. M. Kozlov, in *Proceedings, Second Workshop on Composite Media and Homogenization Theory* (ICTP, Trieste, Italy, World Scientific, 1993), p. 217.
260. G. Matheron, Ed., *Elements pour une theorie des millieux poreaux* (Masson et Cie, Paris, 1967).
261. L. W. Gelhar, *Stochastic Subsurface Hydrology* (Prentice Hall, Englewood Cliffs, New Jersey, 1993).
262. D. Veneziano and A. Tabaei, *J. Hydrol.*, 4 (2004).

CHAPTER 5

REMOTE SENSING FOR PRECIPITATION AND HYDROLOGIC APPLICATIONS

Emmanouil N. Anagnostou

Department of Civil and Environmental Engineering, University of Connecticut, Storrs, CT 06269, USA
E-mail: manos@engr.uconn.edu
and
Institute of Inland Waters, Hellenic Center for Marine Research, Anavissos, Attica, Greece 19013
E-mail: manos@ath.hcmr.gr

This chapter presents the state-of-the-art on rainfall estimation from satellite sensor observations and applications in hydrology and water resources management. The observations considered are from active (precipitation radar) and passive microwave sensors from earth orbiting platforms, as well as Visible (VIS) and Infrared (IR) sensors onboard geostationary satellites. We present the physical basis of combined radar and radiometer retrieval algorithms from the Tropical Rainfall Measuring Mission (TRMM) satellite and its applications on overland rain estimation. We discuss the current status of research on overland rain estimation from passive microwave observations, and outstanding issues associated with those techniques. The significance of lightning information and cloud life stage in advancing high-frequency rainfall estimation is also discussed. Current approaches to merging the infrequent passive microwave-based rainfall estimates with the high-frequency, but lower-accuracy, rainfall fields derived from proxy parameters (e.g. lightning and IR) are presented. The use of remotely-sensed data in precipitation forecasting on the basis of various data assimilation schemes is discussed. We close this chapter with a section on the use of satellite precipitation retrievals in flood forecasting and water management applications.

5.1. Introduction

Understanding the fate of precipitation as soil moisture, snowpack, evapotranspiration, and runoff requires capability to accurately predict energy and water cycle processes. This is also a key underpinning issue for characterization of memories, pathways, and feedbacks between the water cycle system's various components. The current (and anticipated) global availability of high-resolution remote sensing datasets on precipitation and land surface parameters (e.g. soil moisture, vegetation state, surface temperature, snowpack), coupled with physically-based land surface models (LSM), offers a unique capability to advance water- and energy-cycle predictions.[1] Two widely used systems in the United States that rely on off-line land surface models and remotely-sensed data, namely, the Global Land Data Assimilation System (GLDAS)[2] and NASA's Land Information System (LIS),[3] represent a characteristic example of our current capability to provide regional to global estimates of the land surface hydrologic state. However, uncertainties and sampling constraints in space-based observations (particularly over the hydrologically-active tropical regions) combined with model representativeness issues limit the fidelity of those systems to predict water and energy cycle variability.

In terms of improving the prediction accuracy, a critical aspect is the need to force land surface models at high temporal (< 3 hours) and spatial (< 10 km) resolutions. This is important because physical processes that characterize land-vegetation-atmosphere interactions evolve at these fine space-time scales. Although land surface models are physically capable of modeling land-atmosphere interactions at resolutions down to 1 km, most current attempts to estimate rainfall near those scales rely on proxy data (e.g. Infrared-IR, from Geostationary satellites) merged with that of infrequent, but more physically-based, passive microwave estimates.[4-8] Recent work indicates that the desired progression to finer scales in satellite rain estimation is actually counter-balanced by an increasing dimensionality of the retrieval error, which has a consequentially complex effect on the propagation through land surface-atmosphere interaction simulations.[9-13] In essence, this scale incongruity between meteorologic data and its hydrologic application represents a competing trade-off for lowering the satellite retrieval error versus modeling land-vegetation-atmosphere processes at the finest scale possible. Fortunately, recent studies have also shown that the inclusion of continuous lightning observations (proxy to convective precipitation) in satellite observations improves the physical basis of the retrieval, facilitating higher-resolution rain estimates (< 0.1 deg) and improved accuracy in the estimation of high (convective type) rain rates.[14,15] This advancement in high-frequency satellite rain estimation was further corroborated[16] to have consequential improvements on the accuracy of land surface model simulations.

Another important aspect of land surface model prediction accuracy is the need to integrate satellite retrieved rain rates with model-predicted surface forcing variables (i.e. near-surface meteorologic and radiation budget fields). It is critical that those satellite retrieved rain rates and model-predicted meteorologic variables are physically consistent to properly force the land surface model. At coarse resolutions (> 5–10 km), which are the scales associated with regional meteorologic model applications, the moist processes that occur in convective systems are represented by convective parameterization schemes.[17] However, numerical weather prediction (NWP) models, despite on-going developments, continue to exhibit low skill in forecasting convective precipitation at regional scales. The two error sources that have long been recognized as responsible culprits are: (i) the lack of sufficient data and the coarse-grid resolutions that limit our ability to correctly specify those features acting to trigger convection in model initializations; and (ii) the weak assumptions used in developing the convective cumulus parameterization schemes. Clearly, the high nonlinearity in modeling atmospheric processes causes sensitivities to even small perturbations in the atmospheric initial state, commonly termed as 'chaos.'[18] Improved initialization of the local environment in mesoscale models (particularly those of moisture and temperature) is, therefore, an avenue for potential improvements in the prediction of convection. In this regard, one aspect that has shown convincing signs of improving numerical weather prediction is data assimilation. Advanced assimilation frameworks, such as three- (or four-)dimensional variational and/or ensemble Kalman filter (EnKF), driven by observations,[19-20] are computationally demanding and difficult to execute at regional/global scales. Continuous lightning observations, on the other hand, offer an alternative way to rapidly provide information about the growth, location, life cycle, and ice microphysics of convection in a variety of continental regimes. Charge separation leading to lightning is a physical process that takes place in regions of a thunderstorm associated with rigorous vertical forcing, which make the continuous measurement of this meteorologic parameter potentially useful for improving the prediction of a storm's evolution and intensification.[21] Making use of the physically sound link between lightning occurrence and convection, a technique was recently developed[22] to amend the initialization data inadequacies and limitations in formulating sub-grid-scale processes using continuous lightning observations. It was shown that nudging model humidity profiles to convective profiles related to flash rates leads to more realistic model soundings and consequential improvements in convective precipitation forecasts. Our collective expectation from these studies is that the use of lightning data to consistently integrate satellite retrieved rain rates with model-predicted variables could maintain an adequate level of consistency in land surface model forcing.

Another avenue for improving the consistency of surface forcing variables is by dynamically updating the meteorologic model land surface boundary conditions through soil moisture and sensible heat flux fields derived from the off-line land surface model. In a recent work on this subject, Papadopoulos et al.[23] have investigated a technique for improving convective precipitation forecasts through ingesting radar rainfall data in the land surface scheme of a mesoscale model. The results indicate that using observed rainfall data as forcing term in the land surface parameterization scheme of the mesoscale model leads to better characterization of the soil moisture variability and improved atmospheric predictability, as confirmed by a comparison between radar-based rainfall observations and model quantitative precipitation forecasts. Some limitations are imposed on the proposed technique by the uncertainties and errors associated with the rainfall estimates.[24] Moreover, the coupling procedure between NWP models and land surface models is subject to substantial model biases and errors that may negatively impact the quality of their output due to the positive feedbacks caused by the nonlinear land-atmosphere interactions. Careful testing is also necessary in terms of soil moisture representation[25] and initialization procedures.[26] Furthermore, recent studies[27,28] have shown the value of assimilating satellite observations of surface soil moisture in land surface models. Consequently, future work may include the implementation of integrative data-modeling systems, using state-of-the-art land surface models in combination with a weather forecasting system and advanced data assimilation schemes; such a system would possibly result in an even better representation of the modeled precipitation fields and significant improvements in quantitative precipitation forecasting.

In terms of quantification of predictive uncertainty of energy and water cycle processes, there is a clear need to improve our diagnosis of the degree of uncertainty related to the complex issue of high-resolution (1–5 km) forcing of land surface models at regional-to-global scales. This is primarily due to the inherent complexity in the error structure of precipitation fields derived from space-based observations that has an intimate relationship with the structural/parametric uncertainty of land surface models related to model identifiability. One approach to address this fine-scale limitation is through explicit characterization of the complex stochastic nature of the estimated space-borne rainfall fields and its nonlinear propagation (interaction with model uncertainty) in land surface model simulations. This can be achieved using ensemble rainfall fields representing equiprobable scenarios of satellite sensor retrievals. The ensembles could be generated from a state-of-the-art stochastic model that is mathematically cognizant of the complex limitations of satellite rainfall estimation.[9,12,29] The latest advancement in the diagnosis of the implications of satellite rainfall error for land surface modeling is the two-dimensional error model formulated by Hossain and Anagnostou.[10] Studies with this model[12] have revealed that its multi-dimensional stochastic error structure conceptualization renders it more amenable to capturing the spatio-temporal characteristics of soil moisture uncertainty than the simpler error modeling strategies commonly used in the literature. An additional versatility offered by this modeling framework is the derivation of ensemble realizations of satellite-retrieved rain fields (independent of the retrieval technique) that preserve well the spatio-temporal error characteristics across scales of aggregation.

Finally, as shown in recent studies, the response of land surface model-predicted hydrologic variables to the uncertainty in atmospheric forcing variables (i.e. error propagation) is complex and nonlinearly dependent on scale.[13,30,31] Characterizing this uncertainty in land surface simulations and understanding its dependence on scale would facilitate the development of an efficient integrative observationally-based modeling framework to diagnose predictability of continental water- and energy-cycle processes. Hossain and Anagnostou[9] have exemplified the need for detailed understanding of the interaction of rainfall retrieval with model state error as a way for achieving optimal integration of satellite rain remote sensing in land data assimilation systems. A comprehensive assessment of the nonlinear error propagation in land surface simulations requires accounting for the fact that parameters describing the model state may not be representative. Hossain and Anagnostou[9] have shown that a statistical characterization of land surface model prediction error associated with simultaneous sources of uncertainty (rainfall and model parameter) could potentially advance the integration of remote sensing data in the high-resolution prediction of global land surface hydrologic processes.

In this chapter, current rainfall estimation techniques developed for microwave and infrared satellite sensors are reviewed. The techniques include those applied to TRMM (Tropical Rainfall Measuring Mission) precipitation radar profiling retrievals, overland passive microwave rain estimation, high-frequency rainfall estimation from geostationary satellite infrared observations, and multi-sensor rain retrieval merging approaches. The chapter ends with a discussion on the limitations of these techniques and future research directions related to error modeling and hydrologic applications.

5.2. Precipitation Nowcasting from Space-based Platforms

The estimation and quantification of precipitation over large areas is only truly viable from space-borne instrumentation due to deficiencies in conventional surface networks. A large number of satellite observations now exist from which precipitation/rainfall information can be extracted.

These cover a range of wavelengths within the Visible (VIS) and Infrared (IR) spectrum gathered from satellites in geostationary orbits (GEO) and low Earth orbits (LEO), and both passive and active measurements gathered at microwave (MW) frequencies obtained from LEO satellite sensors.

Visible/Infrared techniques have a long history and generally relate the cloud top brightness temperatures (Tbs) to surface and/or other satellite observations. Such techniques range from simple time-integral techniques, such as the GOES Precipitation Index (GPI)[32] through to artificial neural network (NN) techniques.[8] Newer multi-spectral sensors, such as the MODIS (in LEO) and SEVIRI (in GEO), now permit the microphysical properties of clouds to be identified,[33] leading to the prospect of improving rainfall retrievals. The GEO multi-spectral VIS/IR images can be acquired at a nominal 15 min/3 km resolution (1 km VIS), permitting the monitoring of cloud systems and their evolution over time. However, VIS/IR techniques cannot directly retrieve precipitation from the observations of cloud tops.

Radiation in the MW region of the spectrum is affected by hydrometeor-sized particles (i.e. precipitation-sized ice and water particles): the signal received by the satellite sensor being physically linked to the size and phase of the hydrometeors present within the observed atmospheric column. Algorithms used to retrieve the precipitation can be broadly divided into empirical and physical techniques.[34] Empirical techniques are computationally simple and include many of the inherent errors (beam-filling effects and others) associated with the retrievals through their calibration. Purely physical retrievals are somewhat impractical due to the complexity of the inverse radiative transfer modeling required. Instead, cloud resolving models (CRMs) generate a population of hydrometeor profiles, which are then used to generate a set of Tbs for each profile through radiative transfer modeling. The observed Tbs can then be related to this library of profiles.[35-39]

A critical aspect of satellite data uses in hydrometeorologic applications is the need to resolve the precipitation variability at high temporal (< 3 hrs) and spatial (< 25 km) scales. Observations from LEO

MW sensors, while more physical in relating to rainfall, offer only intermittent coverage of a given region of interest (although the Global Precipitation Measurement era in 2013 will significantly reduce this sampling uncertainty[40]), resulting in significant errors due to their poor temporal sampling. There has been an effort to bridge these sampling gaps by estimating precipitation from proxy parameters (e.g. cloud-top temperature), which can be inferred from GEO observations of VIS/IR radiances. There are several global rainfall products that are based on combined LEO MW and GEO VIS/IR observations for high-resolution rainfall estimation. The most widely available satellite blended techniques that are global in scope include the MW-calibrated IR techniques of Turk and Miller[6] (named NRL-Blend) and Kidd et al.,[41] the neural network-based fusion technique of Sorooshian et al.[42] (named PERSIANN), and the combined MW/IR technique of Huffman et al.[4] (named TMPA). These techniques commonly use the IR observations for high-frequency rain estimation, and the MW rain rates as a mean for calibrating their IR rain retrieval algorithm. For example, Turk and Miller developed a calibration scheme based upon the most recent VIS/IR–MW collocated observations within a 5 × 5 deg region, while Kidd et al. used a center-weighted 5 × 5 deg region using weighted observations over the previous five days. These techniques are capable of providing precipitation estimates at 30 min/5 km resolution. However, they still rely upon the cloud-top temperatures to provide an assessment of rain extent and intensity. In the mean time, Joyce et al.[43] developed the Climate Prediction Center Morphing (CMORPH) scheme, whereby the MW observations are seen as the 'true' rainfall, whilst the IR observations provide information on the movement of the precipitation system: in this way, the MW rainfall field is moved from one MW observation to the next by IR-derived cloud motion vectors. On this basis, the technique retrieves high-frequency (half-hourly) precipitation fields by performing time-weighted linear interpolations between sequential passive microwave (PMW) rain estimates. The most commonly available satellite global rain products are summarized in Table 5.1.

Table 5.1. Summary of global satellite rainfall products.

Product name	Agency/Country	Scale
TMPA (Huffman et al.)[4]	NASA-GSFC/USA	25 km/3-hr
CMORPH (Joyce et al.)[43]	NOAA-Climate Prediction Center/USA	8 km/½-hr
PERSIANN (Sorooshian et al.)[42]	University of Arizona/USA	25 km/6-hr
PERSIANN–CCS (Hong et al.)[44]	University of California Irvine/USA	4 km/½-hr
NRL-Blend (Turk and Miller)[6]	Naval Research Lab/USA	10 km/3-hr
GSMAP (sharaku.eor.jaxa.jp)	JAXA/Japan	10 km/1-hr
UBham (Kidd et al.)[41]	University of Birmingham/UK	4 km/½-hr

Results of inter-comparison and validation studies summarized in Adler et al.[45] and Ebert et al.[46] show that IR techniques perform reasonably well at timescales from a few hours through to the monthly scale, but also often suffer from regional biases due to different climatologic rain processes. Passive microwave (PMW) sensor techniques can produce extremely good estimates of instantaneous surface precipitation, although this quickly degrades with accumulated estimates of an hour or more. Combined schemes based upon the calibration of the IR by the PMW are, in general, better than the IR-alone schemes, although the CMORPH scheme is significantly better. Overall, however, the validation studies have shown that the accuracy of the final precipitation product is very much dependent upon the precipitation processes: different precipitation systems result in different relationships, such as convective vs. stratiform rainfall and the hydrometeor size, density, and phase within the cloud system.

Improvements to the estimation and application of rainfall from satellite observations, therefore, rely upon advances in three main areas. First, improving our knowledge of the radiation-hydrometeor interactions; second, improved combinations of satellite observations to reduce sampling-induced errors; and, third, knowledge of the accuracy and associated errors of the precipitation products. Recent research has shown that using physical information from cloud storm dynamics[47,48] and continuous cloud-to-ground lightning measurements[14,15] could help moderate some of the shortcomings in satellite rainfall estimation. In particular: (i) the microphysical characterization based on lightning information and the storm life stage category can be applied to assign a category of cloud and precipitation type; and (ii) the different cloud types/storm maturity stages, calibrated for the different satellite rain products, can be used to establish cloud-type/ground-rainfall relationships.

Some recent studies have combined satellite IR data with lightning observations that provide surrogate information on intense thunderstorms[14,15,49] for rainfall estimation. These studies have shown that lightning measurements can provide reliable delineation of the convective cores in a storm, which can lead to significant improvement in determining the rainfall variability. Morales and Anagnostou[14] and Chronis et al.[15] developed comprehensive schemes for continuous thunderstorm monitoring over very large regions with the use of a long-range lightning detection network data. They demonstrated, using lightning information in IR estimation of convective rain area, that there is significant (~30%) bias-reduction in reference to radar rainfall data from the TRMM satellite. In regards to the correlation with gages, the increase was shown to be as high as 20% in hourly estimates at 0.1-deg resolution.

5.3. Data Uses in Precipitation Forecasting

Due to the rapid progress in computing performance and atmospheric modeling, high-resolution numerical meteorologic models can run nowadays with grid resolutions down to a few kilometers and can be used to predict weather operationally at local scales. The refinement in the model resolution, however, did not prevent the quantitative forecasting of precipitation from exhibiting large uncertainties at these scales, and thus the improvement of the rainfall prediction remains one of the most difficult tasks for meteorologic modeling.

Apart from general predictability issues, the main sources of uncertainty are the errors associated with the initial conditions used in the forecasting models and errors associated with the highly nonlinear modeling of small-scale physical processes. Accurate observations and analyses of variables related to the formation of precipitation and the water cycle, such as cloud coverage, liquid and ice water content, atmospheric humidity, and temperature may contribute to a better definition of latent heating, divergence, and moisture, and may well improve the short-range precipitation forecast.[50]

In the last few years, the inadequacy of traditional analysis methods, based essentially on radio soundings and surface observations, has been partially overcome. The availability of meteorologic satellites improved data assimilation quality and was one of the major incentives for the development of variational data assimilation systems used in global and large-scale regional models. On global scales, recent developments have successfully proven the benefit of assimilating cloud and rain-affected satellite observations in an operational context.[51-53] However, these systems are complicated to develop and difficult to maintain.[54] For limited area applications, it is more convenient to develop simpler data assimilation procedures that do not involve the integration of the entire model and, therefore, tend to greatly simplify the link between observed precipitation and prognostic model variables. The combination of an operational global weather forecasting system, such as those at ECMWF (European Center for Medium Range Weather Forecasts) and NCEP (National Center for Environmental Prediction), with regional modeling frameworks running more detailed physical parameterizations, and whose spatial resolution is more suited for hydrologic modeling, therefore, permits detailed analysis of forecast performance as a function of constraints (e.g. data, process parameterizations) introduced along the modeling chain.

Generally, cloud and precipitation data assimilation involves a large and very diverse number of observational and background data. One important development to connect observations and model state (control) variables is, therefore, the 'observation operator' that maps geophysical quantities into the observation space. This operator may comprise physical parameterization schemes, radiative transfer models, and interpolation schemes. Due to the large dynamic range of scales covered by the spectrum of the available models, the scale-dependent representation of moist physical processes is of great importance. The nonlinear character of these processes requires the assessment of analysis stability and forecast sensitivity with regard to the initial conditions using different configurations of data assimilation and initialization schemes, global model data, and potentially through ensemble forecasting. The issue of nonlinearity greatly affects the assimilation performance schemes through non-convergence in the analysis.

5.4. Data Uses in Hydrology

5.4.1. *Soil Moisture*

Estimates of soil moisture can be made through different approaches: (i) direct ground measurements; (ii) integration of a land surface model forced with meteorologic data derived from observations; (iii) retrievals from low-frequency active and passive microwave data; and (iv) through land data assimilation, combining the complementary information from measurements and models of the land surface into a superior estimate of soil moisture. These are described next.

A common approach to estimate soil moisture is to run a land surface model forced with meteorologic observations. The physical formulation of a land surface model integrates the forcing and produces estimates of soil moisture. Some errors affect these model products: errors in the meteorologic forcing, faulty estimates of the model parameters, and deficient model formulations.[28]

Indirect measurements of surface soil moisture can be obtained from satellite sensors that observe microwave emission by the land surface.[55] However, satellite data coverage is spatially and temporally incomplete and retrieval errors are present, because of limitations in the instruments, difficulties in the parameterization of the physical processes that relate brightness temperature with soil moisture, and difficulty in obtaining a global distribution of the parameters of the retrieval algorithm. Moreover, it is impossible to retrieve soil moisture in areas where the fraction of water is significant (coastal areas) and when the soil is frozen. The current satellite data typically infer soil moisture from its impact on the C-band or X-band passive microwave signal. This adds errors to the retrieval because modest amounts of vegetation obscure the soil moisture signal in those frequencies. The future NASA SMAP (Soil Moisture Active Passive) and European SMOS (Soil Moisture and Ocean Salinity) satellite missions are based on lower-frequency (L-band) passive microwave measurements that are better suited for the penetration of dense vegetation canopies.

Data assimilation systems merge satellite retrieval information with the spatially and temporally complete information given by the land surface models to provide a superior product.[28] This is achieved by correcting the model-generated values of soil moisture towards the observational estimates depending on the level of error associated with each. Constraining the model with observations using data assimilation methods has been demonstrated to be an effective way to moderate model errors and improve the estimates. Recently, data assimilation techniques have been developed to exploit the high availability of remotely-sensed land surface variables,[56,57] and recent advancements in developing sequential land data assimilation systems have demonstrated that retrieved soil moisture may improve the dynamic representation of soil moisture in hydrologic models.[58] Most assimilation systems consider only the covariance information of the model, and the observational errors that are assumed Gaussian. For example, sequential data assimilation algorithms (such as the Kalman filter) propagate error covariance information from one update to the following, subject to model uncertainties, to provide estimates of the uncertainty in the assimilation products. The most common ensemble filter is the Ensemble Kalman filter (EnKF), a Monte Carlo variant of the Kalman filter that derives state forecast error covariance information through the introduction of random noise in the model states, fluxes, and forcing data. The EnKF is based on the idea that a small ensemble of model trajectories captures the relevant parts of the error structure.

A key point in data assimilation techniques is that the model and the observational uncertainties are poorly known: methods driven by poor estimates may produce poor estimates of land surface variables.[59] For this reason, the calibration of the input error parameters is a delicate point in a data assimilation system; however, it is possible to calibrate the input error parameters adaptively during the assimilation integration.[57,58] The quality of the assimilation estimates depends critically on the realism of the error estimates for the model and the observations. Arguably, the way model errors are handled in standard land data assimilation systems is still very simplistic and can use improvement, which should lead to improvements in the data assimilation estimates themselves. These considerations point out the necessity of investigating the impact of different model error characterizations on the assimilation of remotely-sensed soil moisture observations in a land surface model. Crow and Van Loon[59] explored the consequences of making incorrect assumptions about the source and magnitude of model error on the efficiency of assimilating surface soil moisture observations in a land surface model. They showed that inappropriate assumptions may lead to cases in which the assimilation of soil moisture observations degrades the performance of the model. Thus, improved error modeling strategies are needed to characterize the uncertainty in simulation of soil moisture fields from a land surface model forced with satellite rainfall data in order to enhance the efficiency of data assimilation system. The current system for describing model error is a simple scaling of the input precipitation forcing with a multiplicative perturbation at each time and space location and for each ensemble member.[28] This implies, for example, that all ensemble members have zero precipitation whenever the input precipitation is zero. The perturbations are temporally correlated through an AR-1 (autoregressive order-1) process, and a spatial correlation structure based on two-dimensional Gaussian correlation functions is used. These spatial and temporal error structures are numerically convenient, but they do not describe how precipitation errors behave, particularly in the cases of satellite retrievals that are susceptible to rain detection and false alarms. Recent studies have developed more complex satellite rainfall error models for generating ensembles of satellite rain fields on the basis of high accuracy 'reference' rain fields. Hossain and Anagnostou[11] showed that more complex error schemes are capable of conserving the satellite retrieval error structure, while simpler approaches of error modeling revealed biases exceeding 100%. These results are encouraging toward an optimal integration of remotely-sensed rainfall data in land data assimilation system. Furthermore, Hossain and Anagnostou[9,12] investigated the implication of using complex error schemes to describe the uncertainty in soil moisture predictions from a land surface model. The spatial and temporal characteristics of soil moisture uncertainty were well captured that present a higher consistency than simpler two-dimensional error structures. These considerations lead to the following questions that are open to scientific investigations: (i) how do different satellite rainfall error modeling techniques affect the efficiency of a land data assimilation system? and (ii) particularly, how would an improved precipitation forcing error model impact the performance of an LDAS in soil moisture prediction?

5.4.2. *Flood Forecasting and Water Management*

A major challenge faced by many lowermost riparian nations in flood-prone international river basins today is in regards to the availability of in-situ hydrologic data across geopolitical boundaries for issuing early flood warnings and for water management. It is estimated that about 40% of the world's population lives in such international basins accounting for about 60% of the global freshwater flows.[60] About 33 countries have more than 95% of their territory within these basins,[61] and many of these nations are thus forced to cope with a large proportion of the flood mass generated beyond their borders. Some examples of such flood-prone nations are: (i) Bangladesh in the Ganges-Brahmaputra-Meghna (GBM) basin comprising three other nations (Bhutan, Nepal, and India);[62] (ii) Cambodia in the Mekong River Basin (other riparian countries are: Myanmar, Laos, Vietnam, Thailand, and China); and (iii) Senegal in the Senegal River Basin (comprising Senegal, Mali, and Mauritania), among others.

Floods can be forecasted at a point downstream knowing the river flow at some point upstream in conjunction with a hydrologic/land use model. Based on this information, simple regression forecasts can give fairly accurate short-term estimates of river discharges. However, for many of these flood-prone nations situated within international basins, the challenge of issuing effective flood forecasts for water management applications can be particularly difficult to overcome under two conditions: (i) when surface measurements of rainfall and other land surface parameters are largely absent due to inadequate resources or complex terrain;[29,63] and (ii) when there is lack of cooperative agreement among the riparian nations to share hydrologic information in real-time for proactive flood management.[61,62,64] The first condition is quite commonly observed within tropical basins of Asia, Africa, and South America that lack financial resources for real-time monitoring of hydrologic parameters. The second condition, barring a few exceptions (such as the Mekong River Commission — MRC), is endemic in most flood-prone developing nations and limits severely the lead-time in skillful flood forecasting beyond a few days. For example, Bangladesh does not receive any upstream river flow and rainfall information in real-time from India during the critical Monsoon season. Bangladeshi authorities, therefore, measure river flow at staging points where the two major rivers (Ganges and Brahmaputra) enter Bangladesh, and at other locations downstream. On the basis of these data, it is possible to forecast flood levels in the interior and the south of Bangladesh with only 2–3 days lead-time (Flood Forecasting and Warning Center, Bangladesh: www.ffwc.net).[64]

Since rainfall is the single most important determinant of the state of surface runoff leading to large-scale floods, it is logical to expect that satellite remote sensing of rainfall and numerical weather forecasting, along with other satellite-derived surface parameters (such as elevation, vegetation, soils, and river network) can potentially address the current challenge of improving the accuracy and range of flood forecasting for many flood-prone nations constrained within international river basins. A longer-term forecasting range would have a consequentially beneficial impact of enhancing the utility of a decision support system that ingests these warnings. For example, 7–10-day forecasts are much more useful than daily forecasts in agricultural decision support, as they inform farmers on the potential benefits of delayed sowing or reaping of crops, while it is considered that a 21-day forecast is the most ideal.[65] Extended forecasts also assist in economic decision-making through early disbursement of rehabilitatory loans to regions anticipated to be affected by floods.[66] Most importantly, the longer the warning, the better the disaster preparedness and the lesser are the chances of damage to property and loss of lives.

Given the unique vantage that is possessed by satellites in space (unlike ground-based systems), satellite rainfall data has the potential to: (i) extend the accuracy and range of forecasted flood levels in the lowermost riparian states through early assessment of the surface runoff evolution in the upstream nations; and (ii) minimize the negative impact of unavailable data and/or high operational costs of in-situ networks. Recently, the Mekong River Commision's River Monitoring Network (http://www.mrcmekong.org/info_resources/ffw/overview.htm) has demonstrated capability for 7–10-day flood forecasting in downstream Cambodia using simple ingestion of satellite imagery from upstream regions in its forecasting system. An additional aspect that makes satellite rainfall data an ideal candidate for flood forecasting over international river basins is the anticipated abundance of high-resolution global rainfall measurements (0.05–0.1 degree, 3–6 hourly) from the upcoming constellation of PM satellites known as the Global Precipitation Measurement (GRM) mission.[40] As mentioned earlier in this chapter, the higher sampling frequency of PM data from GPM mission, combined with high-frequency rainfall observations available from Geostationary Infrared sensors,[4,5] auxiliary data (such as lightning),[15] and multiple sensor estimation techniques can be expected to yield global rainfall products of various levels of utility and accuracy. However, these satellite retrievals are subject to errors caused by various factors, ranging from sampling to the high complexity and variability in the relationship of the measurements to precipitation parameters, which consequentially have a highly nonlinear effect on the quality of flood prediction.[13,29,31]

A critical aspect associated with the use of satellite rainfall data in an operational flood forecasting setting in flood-prone international river basins is, therefore, the need to assess the performance (of satellite data) at spatio-temporal resolutions pertinent for rainfall-runoff modeling. Recently, Hossain and Anagnostou[29] showed that the use of the more frequent IR rainfall used to bridge between PM overpasses over a medium-sized (< 200 km^2) saturation-excess watershed can yield significant improvements in the estimation of water budget and the magnitude of flood wave, while improvements are elusive for predicting the arrival time of peak runoff. Although studies have recently begun to address how errors in satellite rainfall retrieval manifest themselves as flood prediction uncertainty,[29,31,63] it is currently unknown as to how this error propagation affects the accuracy of forecasted flow levels and the consequential impact on decision support applications (e.g. water management, disaster management, agriculture).

Further, climate-based approaches have also been recently initiated for addressing the limitations of flood forecast over Monsoon-affected nations, such as India and Bangladesh.[67] Although based on physically sound principles of early detection of weather patterns and intra-seasonal variability, these approaches rely heavily on coupled cloud model observations. Unfortunately, the rainfall derived from these coupled global climate models is not always necessarily of sufficient accuracy due to several sources of uncertainty, including coarse resolutions, weak model relationships, and sensitivity to initial conditions.[68] The currently available large array of higher-resolution earth science data (from satellites) and anticipated global availability of higher-accuracy PM rainfall data are largely unaddressed in these approaches. Additionally, climate-based approaches, due to their theoretical complexity, represent a far greater challenge in rapid prototype enhancement of operational flood warning systems compared to satellite data, which are considered much more readily amenable for ingestion in any existing forecasting system for an enhanced decision support system.

5.5. Conclusions and Outlook

This chapter focused on the use of satellite remote sensing from various space-based platforms for precipitation nowcasting (i.e. quantitative precipitation estimation), forecasting, and applications in hydrology and the management of water resources. Collective evidence shows that our current ability to predict floods and droughts and our general knowledge of the water-cycle mechanisms are limited. We believe that part of this limitation is caused by the insufficient use of observing systems that are available at global and local scales and the still poor capability to accurately represent cloud and precipitation processes in numerical weather prediction (NWP) and General Circulation Models (GCMs).[69,70] Improving the observation of clouds and precipitation, the understanding of the involved physical processes, and the integration of observations in atmospheric and hydrologic models will greatly advance our capability to make better quantitative precipitation forecasts with consequential improvements on flood forecasting and water resources management.

It is pertinent, at this point, to note that remotely-sensed data used in hydrology is founded on two hypotheses: (i) the vantage of space to view the Earth makes satellite rainfall and other satellite-derived land surface parameters ideal to address transboundary limitations of flood forecasting and water resources management across international river basins; and (ii) satellite retrievals are not perfect and, hence, the uncertainty associated with satellite-based measurements (in particular, rainfall, which is the primary determinant of the water cycle) have a consequentially nonlinear and deteriorating impact on the accuracy of the forecasts. These two hypotheses jointly represent a competing trade-off for using remote sensing earth observation data and numerical weather forecasts in hydrologic and water resources applications. Future research should seek to establish a clear understanding of the scales at which these products can be proficiently used and the implications of rainfall error propagation on the prediction and forecast accuracy of water cycle parameters, which is vital towards achieving better decision support for water and disaster management through earth observations.

Acknowledgments

This work was supported by EU Marie Curie Excellence Grant project PreWEC (MEXT-CT-2006-038331). Material used in this article was derived from contributions by Faisal Hossain, Themis Chronis, Vincenzo Levizzani, Chris Kidd, and Peter Bauer.

References

1. K. E. Mitchell, D. Lohmann, P. R. Houser, E. F. Wood, J. C. Schaake, A. Robock, B. A. Cosgrove, J. Sheffield, Q. Duan, L. Luo, R. W. Higgins, R. T. Pinker, J. D. Tarpley, D. P. Lettenmaier, C. H. Marshall, J. K. Entin, M. Pan, W. Shi, V. Koren, J. Meng, B. H. Ramsay and A. A. Bailey, *J. Geophys. Res.*, D07S90, doi:10.1029/2003JD003823 (2004).

2. M. Rodell, P. R. Houser, U. Jambor, and J. Gottschalck, K. E. Mitchell, C.-J. Meng, K. Arsenault, B. A. Cosgrove, J. Radakovich, M. Bosilovich, J. K. Entin, J. P. Walker, D. Lohmann and D. Toll, *Bull. Amer. Meteor. Soc.*, 381 (2004).
3. S. V. Kumar, C. D. Peters-Lidard, J. L. Eastman and W.-K. Tao, *Environ. Model. Softw.*, 169 (2008).
4. G. J. Huffman, R. F. Adler, D. T. Bolvin, G. Gu, E. J. Nelkin, K. P. Bowman, Y. Hong, E. F. Stocker and D. B. Wolff, *J. Hydrometeor.*, 38 (2007).
5. G. J. Huffman, R. F. Adler, E. F. Stocker, D. T. Bolvin and E. J. Nelkin, *12th Conference on Satellite Meteorology and Oceanography* (Long Beach, California, 2003).
6. F. J. Turk and S. D. Miller, *IEEE Trans. Geosci. Remote Sens.*, 1059 (2005).
7. F. S. Marzano, M. Palmacci, D. Cimini, G. Giuliani and F. J. Turk, *IEEE Trans. Geosci. Remote Sens.*, 1018 (2004).
8. F. J. Tapiador, C. Kidd, V. Levizzani and F. S. Marzano, *J. Appl. Meteor.*, 576 (2004).
9. F. Hossain and E. N. Anagnostou, *Adv. Water Resour.*, 1336 (2005).
10. F. Hossain and E. N. Anagnostou, *IEEE Trans. Geosci. Remote Sens.*, 1511 (2006).
11. F. Hossain and E. N. Anagnostou, *IEEE Geosci. Remote Sens. Lett.*, 419 (2006).
12. F. Hossain and E. N. Anagnostou, *Geophys. Res. Lett.*, 1 (2006).
13. E. N. Anagnostou, in *Measuring Precipitation from Space: EURAINSAT and the Future*, Eds. V. Levizzani, P. Bauer and F. J. Turk (Kluwer Academic Publishers, 2007), p. 357.
14. C. Morales and E. N. Anagnostou, *J. Hydrometeor.*, 141 (2003).
15. T. G. Chronis, E. N. Anagnostou and T. Dinku, *Quart. J. Roy. Meteor. Soc.*, 1555 (2004).
16. E. N. Anagnostou, K.-H. Lee, T. Dinku, D. Wang and A. Tadese, *18th Conference on Hydrology, 84th AMS Annual Meeting* (Seattle, WA, USA, 2004).
17. C. Cohen, *Mon. Wea. Rev.*, 1722 (2002).
18. E. N. Lorenz, *J. Atmos. Sci.*, 130 (1963).
19. J. L. Anderson, H. M. van den Dool, *Mon. Wea. Rev.*, 507 (2004).
20. F. Weng and Q. Liu, *J. Atmos. Sci.*, 2633 (2003).
21. G. D. Alexander, J. A. Weinman, V. M. Karyampudi, W. S. Olson and A. C. L. Lee, *Mon. Wea. Rev.*, 1433 (1999).
22. A. Papadopoulos, T. G. Chronis and E. N. Anagnostou, *Mon. Wea. Rev.*, 1961 (2005).
23. A. Papadopoulos, E. Serpetzoglou, E. N. Anagnostou, *Adv. Water Resour.*, 1456 (2008).
24. G. J. Ciach, W. F. Krajewski and G. Villarini, *J. Hydrometeor.*, 1325 (2007).
25. J. C. Schaake, Q. Duan, V. Koren, K. E. Mitchell, P. R. Houser, E. F. Wood, A. Robock, D. P. Lettenmaier, D. Lohmann, B. A. Cosgrove, J. Sheffield, L. Luo, R. W. Higgins, R. T. Pinker, J. D. Tarpley, *J. Geophys. Res.*, D01S90, doi:10.1029/2002JD003309 (2004).

26. M. Rodell, P. R. Houser, A. A. Berg and J. S. Famiglietti, *J. Hydrometeor.*, 146 (2005).
27. M. Drusch, *J. Geophys. Res.*, D03102, doi:10.1029/2006JD007478 (2007).
28. R. H. Reichle, R. D. Koster, P. Liu, S. P. P. Mahanama, E. G. Njoku and M. Owe, *J. Geophys. Res.*, D09108, doi:10.1029/2006JD008033 (2007).
29. F. Hossain and E. N. Anagnostou, *J. Geophys. Res.-Atmos.*, D07102 (2004).
30. K.-H. Lee and E. N. Anagnostou, *Canadian J. Remote Sensing*, 706 (2004).
31. B. Nijssen and D. P. Lettenmaier, *J. Geophys. Res.-Atmos.*, D02103 (2004).
32. P. A. Arkin, R. Joyce and J. E. Janowiak, *Rem. Sens. Rev.*, 107 (1994).
33. D. Rosenfeld, in *Measuring Precipitation from Space, EURAINSAT and the Future*, Eds. V. Levizzani, P. Bauer, F. J. Turk (Kluwer Academic Publishers, 2007), p. 61.
34. C. Kidd, D. Kniveton and E. C. Barrett, *J. Atmos. Sci.*, 1576 (1998).
35. E. A. Smith, A. Mugnai, H. J. Cooper, G. J. Tripoli and X. Xiang, *J. Appl. Meteor.*, 506 (1992).
36. A. Mugnai, E. A. Smith and G. J. Tripoli, *J. Appl. Meteor.*, 17 (1993).
37. C. D. Kummerow and L. Giglio, *J. Appl. Meteor.*, 3 (1994).
38. C. D. Kummerow and L. Giglio, *J. Appl. Meteor.*, 19 (1994).
39. C. D. Kummerow, Y. Hong, W. S. Olson, S. Yang, R. F. Adler, J. McCollum, R. Ferraro, G. Petty, D.-B. Shin and T. T. Wilheit, *J. Appl. Meteor.*, 1801 (2001).
40. E. A. Smith, G. Asrar, Y. Furuhama, A. Ginati, C. Kummerow, V. Levizzani, A. Mugnai, K. Nakamura, R. F. Adler, V. Casse, M. Cleave, M. Debois, J. Durning, J. Entin, P. R. Houser, T. Iguchi, R. Kakar, J. Kaye, M. Kojima, D. P. Lettenmaier, M. Luther, A. Mehta, P. Morel, T. Nakazawa, S. Neeck, K. Okamoto, R. Oki, G. Raju, M. Shepherd, E. F. Stocker, J. Testud, and E. F. Wood, in *Measuring Precipitation from Space – EURAINSAT and the Future*, Eds. V. Levizzani, P. Bauer, F. J. Turk (Kluwer Academic Publishers, 2007), p. 611.
41. C. Kidd, D. R. Kniveton, M. C. Todd and T. J. Bellerby, *J. Hydrometeor.*, 1088 (2003).
42. S. Sorooshian, K.-L. Hsu, X. Gao, H. V. Gupta, B. Imam and D. Braithwaite, *Bull. Amer. Meteor. Soc.*, 2035 (2000).
43. R. J. Joyce, J. E. Janowiak, P. A. Arkin and P. Xie, *J. Hydrometeor.*, 487 (2004).
44. Y. Hong, K.-L. Hsu, S. Sorooshian and X. Gao, *J. Appl. Meteor.*, 1834 (2004).
45. R. F. Adler, C. Kidd, G. Petty, M. Morissey and H. M. Goodman, *Bull. Amer. Meteor. Soc.*, 1377 (2001).
46. E. E. Ebert, J. E. Janowiak and C. Kidd, *Bull. Amer. Meteor. Soc.*, 47 (2007).
47. A. Tadesse and E. N. Anagnostou, *Atmos. Res.* (2008).
48. A. Tadesse and E. N. Anagnostou, *J. Atmos. Ocean. Tech.* (2008).
49. M. Grecu, E. N. Anagnostou and R. F. Adler, *J. Hydrometeor.*, 211 (2000).
50. J. Manobianco, S. Koch, V. M. Karyampudi and A. J. Negri, *Mon. Wea. Rev.*, 341 (1994).
51. P. Bauer, P. Lopez, A. Benedetti, D. Salmond and E. Moreau, *Quart. J. Roy. Meteor. Soc.*, 2277 (2006).

52. P. Bauer, P. Lopez, A. Benedetti, D. Salmond, S. Saarinen and M. Bonazzola, *Quart. J. Roy. Meteor. Soc.*, 2307 (2006).
53. G. Kelly, P. Bauer, A. Geer, P. Lopez, and J.-N. Thépaut, *Mon. Wea. Rev.*, 2713 (2007).
54. R. M. Errico, G. Ohring, P. Bauer, B. Ferrier, J.-F. Mahfouf, J. Turk and F. Weng, *J. Atmos. Sci.*, 3737 (2007).
55. E. G. Njoku, T. J. Jackson, V. Lakshmi, T. K. Chan and S. V. Nghiem, *IEEE Trans. Geosci. Remote Sens.*, 215 (2003).
56. J. P. Walker, P. R. Houser and R. H. Reichle, *EOS*, 545 (2003).
57. R. H. Reichle, R. D. Koster, J. Dong and A. A. Berg, *J. Hydrometeor.*, 430 (2004).
58. R. H. Reichle, W. T. Crow, R. D. Koster, H. O. Sharif, S. P. P. Mahanama, *Geophys. Res. Lett.*, L01404, doi:10.1029/2007GL031986 (2008).
59. W. T. Crow and E. Van Loon, *J. Hydrometeor.*, 421 (2006).
60. A. T. Wolf, J. Nathrius, J. Danielson, B. Ward and J. Pender, *Int. J. Water Resour. Develop.*, 387 (1999).
61. M. A. Giordano and A. T. Wolf, *Natural Resources Forum*, 163 (2003).
62. G. N. Paudyal, *Hydrol. Sci. J.*, S5 (2002).
63. F. Hossain, E. N. Anagnostou, T. Dinku and M. Borga, *Hydrol. Process.* (2004).
64. P. J. Webster, R. Grossman, C. Hoyos, T. Hopson, K. Sahami and T. N. Palmer, (*Exchanges-Scientific Contributions, Climate Variability and Predictatbility*, CLIVAR, 2004).
65. ADPC, *Application of Climate Forecasts in the Agriculture Sector. Climate Forecasting Applications in Bangladesh Project, Report 3* (Asian Disaster Preparedness Center (ADPC), Bangkok, 2002).
66. C. del Ninno, P. A. Dorosh, L. C. Smith and D. K. Roy, *International Food Policy and Research Institute, Research Report 122* (Washington, DC, 2001).
67. P. J. Webster and C. Hoyos, *Bull. Amer. Meteorol. Soc.*, 1745 (2004).
68. S.-J. Chen, Y.-H. Kuo, W. Wang, Z.-Y. Tao and B. Cui, *Mon. Wea. Rev.*, 2330 (1998).
69. D. Randall, S. Krueger, C. Bretherton, J. Curry, P. Duynkerke, M. Moncrieff, B. Ryan, D. Starr, M. Miller, W. Rossow, G. Tselioudis and B. Wielicki, *Bull. Amer. Meteor. Soc.*, 455 (2003).
70. G. L. Potter and R. D. Cess, *J. Geophys. Res.*, D02106, doi:10.1029/2003JD004018 (2004).

CHAPTER 6

NEARLY TWO DECADES OF NEURAL NETWORK HYDROLOGIC MODELING

Robert J. Abrahart

*School of Geography, University of Nottingham,
Nottingham NG7 2RD, United Kingdom
E-mail: bob.abrahart@nottingham.ac.uk*

Linda M. See

*School of Geography, University of Leeds,
Leeds LS2 9JT, United Kingdom
E-mail: l.m.see@leeds.ac.uk*

Christian W. Dawson

*Department of Computer Science, Loughborough University,
Loughborough LE11 3TU, United Kingdom
E-mail: c.w.dawson1@lboro.ac.uk*

Asaad Y. Shamseldin

*Department of Civil and Environmental Engineering,
University of Auckland, Private Bag 92019, Auckland, New Zealand
E-mail: a.shamseldin@auckland.ac.nz*

Robert L. Wilby

*Department of Geography, Loughborough University,
Loughborough LE11 3TU, United Kingdom
E-mail: r.l.wilby@lboro.ac.uk*

This chapter provides an overview of the field of neural network hydrologic modeling. The recognition and approval of neural networks as useful tools in this field is considered in terms of the Revised

Technological Adoption Life Cycle (RTALC). This market adoption and penetration model provides a useful construct for reviewing developments in neural network hydrologic modeling from the initial explorations that were performed in the early 1990s to current state-of-the-art: developments that have greatly expanded the field in terms of solutions involving other artificial intelligence technologies, modular and ensemble modeling, embedded solutions, and neuroemulation. The review reveals that neural network hydrologic modeling has not yet left the first stage of that model — referred to as Innovators. Bridging the chasm that exists between Innovators and Early Adopters — the next group in that model — presents a major challenge to researchers in this field. Future trajectories in terms of ongoing research directions that might serve to bridge this gap are highlighted including: the establishment of a more open research community, for better sharing of methodologies, datasets and findings, internal inspection of neural solutions to provide a better understanding of their components and the relationship of such components to observed physical and mechanical behaviors, and the estimation of uncertainty. The closing section identifies a pair of 'killer apps': two potential solutions that could deliver a quantum leap forward in terms of greater acceptance by the wider hydrometerologic community.

6.1. The Long and Winding Road

The history of neural networking can be traced back to at least the late 1800s in the form of scientific attempts to study brain activity patterns and the workings of the human mind.[1] McCulloch and Pitts,[2] in due course, produced a two-part mathematical model of the neuron that is still used today in artificial neural networks comprising: (i) a summation of weighted inputs; and (ii) an input-output transfer function that is applied to the summation of weighted inputs. These initial concepts started to be applied in computational tools, starting with Turing's *B-type machines*[3] and Rosenblatt's *Perceptron*[4] — the simplest type of neural network (NN: used in the rest of this text to cover both singular and plural cases). Unfortunately, single-layer perceptrons can only learn patterns that can be separated in a linear manner (i.e. a linear classifier), and Minsky and Papert's later book on the subject[5] argued that this type of network had the fundamental limitation of not being able to learn a simple XOR (eXlusive OR) function. It was, at that point, conjectured, based on 'intuitive' judgment, that a similar fundamental limitation would hold for an analogous mechanism that had additional layers of units and connections (p. 232) — the so called Multi-Layered Perceptron (MLP). Minsky and Papert's 'intuitive judgment' was, however, a bit too rash, since more powerful learning procedures that enabled MLP networks to be trained in a superior manner were soon (re)discovered and thereafter popularized, e.g. 'backpropagation of error algorithm' (BP) — often attributed to Rumelhart et al.,[6] rather than to Amari.[7-11] The solution to such problems was nevertheless too slow in coming forward to help put matters right. The damage had been done! The often-cited Minsky and Papert text,[5] with its incorrect conclusions, caused a significant decline in interest and funding — NN modeling was no longer seen as a fashionable topic to be researching. It took ten or more years for interest in this field to recover from such a damaging blow, with fresh entrants and a major resurgence of research undertakings occurring during the 1980s and 1990s commensurate with the arrival of desktop computing.

The past two decades have indeed witnessed an increased trialing and testing of neural modeling procedures, spanning a broad range of disciplines. Moreover, computer science innovations and investigations into the development of improved neurocomputing methodologies are pressing forward at a fast pace. This is all well and good, but associated progress in transferring the resultant innovation products into the hydrologic sciences, or seeking to exploit the potential gains on offer across a broad spectrum of relevant application domains, has not experienced equal levels of 'renaissance' or 'acceptance.' The wider use of such tools has instead encountered a chequered pattern of historical approval/disapproval. This chapter first highlights some important milestones in the reported use of such tools for hydrologic modeling purposes — a process that can be traced to the start of the 1990s and a field of investigation that is sometimes referred to as *neurohydrology*: a term that originated at the *First International Conference on GeoComputation* (http://www.geocomputation.org/).[12] The discussion thereafter considers the overall sequence of import and take-up of neural solutions within the hydrologic sciences using the *Revised Technological Adoption Life Cycle* (RTALC: Fig. 6.1).[13]

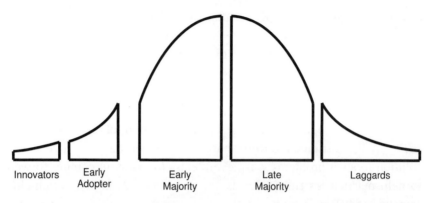

Fig. 6.1. Revised Technological Adoption Life Cycle. Modified from Moore, G. A., *Crossing the Chasm: Marketing and Selling High-Tech Products to Mainstream Customers*, HarperCollins Publishers, New York, 1991.

This psychographic model of adoptive behavior is intended to describe the market penetration of a new technological product in terms of progression with respect to the type of consumers that it attracts throughout its useful life. Five different stages are identified in the model: (i) *Innovators/Technology Enthusiasts*, who aggressively seek out new technological developments; (ii) *Early Adopters/Visionaries*, who are not technologists but nonetheless find it easy to understand and appreciate the potential rewards; (iii) *Early Majority/Pragmatists*, who accept that most new-fangled inventions are passing fads and are content to adopt a wait and see strategy — until strong proof of usefulness exists — but who would not want to be disadvantaged and are driven by a strong sense of practicality; (iv) *Late Majority/Conservatives*, who will wait until a particular technological development has become an established standard with lots of support; and (v) *Laggards/Skeptics*, who, for some reason or other, simply don't want anything to do with the latest technological innovations.

The difference between the original life cycle model and the *revised* life cycle model is important and has a particular bearing in terms of neural network hydrologic modeling. Figure 6.1 depicts a set of four natural 'separation gaps' in the revised model that are used to disconnect the five individual life cycle groups. Each gap represents the potential opportunity for a particular innovation or adoption to lose its forward momentum and, as a result, technological products that do not achieve a successful crossing of such gaps will fail to accomplish a full and proper transition from one segment to the next and never gain leadership or the central ground. The extent to which such gaps in the model reflect real crisp segregations in terms of distinct groups on the ground, or are perhaps indicative of some gradual sequence of overlapping transitional states, or fuzzy membership groupings, could, of course, be debated and will differ according to the situation that is being modeled. The specified model that is to be applied does, however, possess such gaps, and the size of the gap that exists between different groups in that model is significant. Most gaps are small, but there is one much larger gap or 'chasm' that separates the *Early Adopters* from the *Early Majority*. These two groups may seem similar but differences exist in their attitudes towards a particular innovation. The *Early Adopters* are the first to implement a change in their field, so they are prepared for the social and institutional challenges that will emerge arising from the major resistance to change that is so often encountered. This group is also prepared to work through any problems with the innovation, creating several versions and refining the product over time. The *Early Majority*, on the other hand, are looking for an improvement that can be slotted into their existing working practices, and would want to minimize any change between their current operations and the adoption of a particular innovation. The *Early Majority* also want a product that is largely finished and can be implemented as part of their existing technological *modus operandi*.

Neural networks are an important technological development and the initial promise of such tools was anticipated to be of great interest to *Innovators*. The list of 'sales pitch' advantages is extensive. Neural network models are observed over and over again to produce excellent results under appropriate circumstances: such solutions are fast to develop and run, more tolerant of imperfect and incomplete datasets than other methodologies, and do not require a prescribed set of predetermined formulas or rules to describe their input-output mapping transformation(s). Neural products are good at pattern recognition, generalization, and trend prediction. Neural networks can learn, if they are presented with an appropriate range of training examples, deduce their own rules for solving problems, and produce valid answers from noisy data. Nevertheless, important concerns have been expressed in certain hydrologic quarters that this might be an academic cul-de-sac:[14] no objective guidelines have been developed for model construction and/or application (a criticism that is equally valid for conventional hydrologic modeling); little or no effort has been made in terms of traditional support activities, such as sensitivity analysis, or to develop confidence intervals on forecasts (a shortcoming which is also shared by many traditional hydrologic models[15]); and research to date has done little or nothing to build upon existing knowledge or to provide greater understanding of hydrologic processes. Total reliance, moreover, is placed upon empirical testing for model development and subsequent legitimization of the final product in terms of 'fitness-for-purpose.' Thus, each solution is dependent on the selected inputs and datasets that are used to build it and on the method of assessment that is used to select a favored construct. Editorial reservations also exist with regard to the abuse of such tools and the related use of 'numerous technical details,' 'exotic vocabulary,' and 'anthropomorphisms.'[16]

Neural network enthusiasts, nevertheless, claim that neural products, in most cases, provide superior numerical accuracies compared to traditional constructs, such as statistical or conceptual models. Some studies refute this claim: for example, in a single case study investigation of autoregressive monthly discharge forecasting for the River Nile, a stochastic model produced higher levels of statistical fit than a NN model, and analog modeling was found to be better in terms of simplicity and ease of application.[17] However, to be conclusive, the scientific method demands that such experiments must be applied to a wide range of datasets and the reported findings reproduced. Neural network approaches are, nevertheless, considered to deliver faster, cheaper, and more robust operational management solutions — in contrast to complex bespoke software — but such qualities are seldom, if ever, tested or reported. The way forward is to publish sound research papers that deliver comprehensive and authoritative assessments of their good and bad points, expressed in terms of usefulness and/or limitations, including even negative conclusions that the proposed approach "does not offer too much" in a particular field of modeling.[16]

It is important, in such arguments, to establish the exact reason(s) for building a particular model. It is axiomatic that hydrologic models can be developed to serve a number of different purposes and that such matters will inform decisions on the nature of the route that is to be followed in the selection of an appropriate methodology for producing a particular modeling solution. Two different hydrologic objectives that should not be in direct competition with one another can be identified in this respect: (i) the provision of reliable solutions for operational management purposes; and (ii) the development of models for scientific reasons, such as hypothesis testing operations, or to assist in the process of knowledge acquisition through the mechanisms of model construction and development. Hydrologic models are tools that should be tested in terms of their operational 'fitness-for-purpose,'[18] e.g. based on items, such as the timing of flood events,[19,20] abstraction control thresholds,[21] impact of flooding on river protection works, throughput volumes for sediment budgets, or drought estimation capabilities. Functional qualities and constraints, such as robustness (impervious to perturbing influences) or graceful degradation (limited error occurs when individual components malfunction) that do not have an optimal relationship with model output accuracies, should also be considered. It is also important to perform experiments that can be tested against other procedures in an open and unencumbered manner by sharing datasets and raw model outputs. This seldom happens; but for a good example of hydrologic modelers who were willing to put their datasets on the internet for other scientists to model, see Gaume and Gosset.[22] However, posting such datasets on the internet is not always possible due to ownership issues and copyright restrictions. The end result of such matters is to create a pressing need for an easily accessible, quality-controlled database, containing observed datasets for a range of international catchments that possess different hydrologic, meteorologic, and climatic properties, but housed on a site that can also store raw model outputs submitted by different users. Such a database, if widely used, would support good opportunities for large-scale modeling inter-comparison exercises and produce a major step forward in the identification of conclusive pros and cons for different model types. This point is discussed in more detail in Sec. 6.6.

In summation, the pros and cons of different arguments and endorsements, plus a mounting opposition to its poor scientific foundations, have created a stifled environment in which neural networking software is still struggling in the hydrologic sciences to successfully span the initial gap that exists between *Innovators* and *Early Adopters*. Most reported applications are still restricted to the research environment and form part of the learning process. Indeed, for both practical and socio-theoretical reasons, the number of operational hydrologic implementations is limited to a handful of specialist situations — producing what amounts to a knowledge transfer gap. The substantial 'chasm' that exists between *Early Adopters* and the *Early Majority* has not yet been encountered — such that crossing it presents a far-off challenge for neurohydrologists!

The rest of this chapter provides a contextualization of relevant hydrologic developments in terms of the RTALC. It is, of course, recognized that scientific approval and commercial acceptance are somewhat different, with testing and argument published in scientific papers providing a stronger and more transparent part of the former process. The commercial model, nevertheless, offers a good descriptive instrument for the purposes of understanding and communicating a series of historical events and/or practical restraints. The field of hydrologic application has also expanded to the extent that it is far too big to be covered in detail within the confines of a single chapter. The greatest amount of published hydrologic material relates to rainfall-runoff modeling and flood forecasting applications. Thus, most referenced examples are taken from allied fields, but with supplemental illustrative material being used, as required. Important historical steps are first highlighted. The next four sections present a categorization of subject advancement in the manner of a phased transitional process. This reported summation of events covers the beginnings of NN hydrologic modeling, in the early 1990s, and engages with numerous twists and turns in the development and recognition of this field during the period that followed. The analysis culminates in pertinent issues that 'next generation modelers' are attempting to address — interested parties who should still be rejoicing at the arrival of multi-functional, multi-representational, interoperable, intelligent software products that can be run on shared multi-user platforms — but are, in fact, struggling to move forward in a meaningful or determined manner. Finally, the closing section advocates the development of two potential applications for rapid bridging of the divide that exists between *Innovators* and *Early Adopters*.

6.2. From Little Acorns Grow Mighty Oaks

The spread of NN modeling within the hydrologic sciences has, to a large extent, mirrored the microcomputer and desktop computing revolution(s), providing an ever-increasing supply of cheaper, more sophisticated and more user-friendly software packages. Initial studies seldom provided explicit details on the NN simulation software that was used; it was not considered to be a significant factor in the overall picture at that point. It is clear that some pioneering neurohydrologists wrote their own software programs developed in a high-level language.[23,24] This permitted maximum customization potential and greater flexibility; but, unless the code was rigorously tested, such programs could also be prone to software bugs and algorithmic errors that might produce spurious outputs. Several hydrologic pioneers opted to use popular freely downloadable software packages, such as the *Stuttgart Neural Network Simulator* (SNNS: http://www.ra.cs.uni-tuebingen.de/SNNS/). This particular tool was used in 18% of the studies reported in Dawson & Wilby.[25] Early versions of that software required access to a UNIX environment, but it was later ported to 'windows machines' as the platform independent JavaNNS. However, its successor has never achieved the same level of popularity. Later hydrologic studies have instead opted to switch towards using Commercial Off-The-Shelf (COTS) software products, such as *NeuroSolutions* (http://www.neurosolutions.com/) that was introduced in 1994;[26,27] or the *Trajan Neural Network Simulator* (http://www.trajan-software.demon.co.uk/) that also originated in the mid-1990s.[28,29] These two commercial packages support an advanced set of functions, including 'code generation.' Others have opted to conduct their hydrologic investigations using some of the dedicated software toolboxes that are provided as part of an extended computational environment in MATLAB (http://www.mathworks.com/). The *MATLAB Neural Network Toolbox* offers a comprehensive set of algorithms and supporting functions that can be compared and contrasted; e.g. Feed Forward Neural Network (FFNN), General Regression Neural Network (GRNN), Radial Basis Neural Network (RBNN). The outputs can also be compared against pertinent functions that are provided in the *MATLAB Fuzzy Logic Toolbox*; e.g. Adaptive-Network-based Fuzzy Inference Systems (ANFIS).[30,31]

Important milestones can be identified in the start-up period. Daniell[32] provides what could be regarded as the first ever paper on the use of NN tools for hydrologic modeling: that paper received an award for the best presentation of a paper at the International Hydrology and Water Resources Symposium, in Perth, Western Australia. The paper itself listed ten potential applications in hydrology and water resources. It also offered two illustrative examples: a 'security of supply' investigation that predicted per capita water consumption demands based on climate datasets; and a 'regional flood estimation procedure' that used catchment parameters to estimate flood frequency distributions, i.e. Log-Pearson Type 3. Later pioneering papers broadened the range of different hydrologic topics addressed, including emulation of a space-time mathematical model of rainfall patterns;[33] prediction of flood hydrographs and peak flows using 5-min storm hyetograph datasets for five storm events at Bellevue, in King County, Washington State, USA;[34] one-step-ahead rainfall-runoff modeling using daily and hourly datasets for the Pyung Chang River Basin in South Korea;[35] and replication of simulated model output runoff volumes for recorded single burst storm events in Singapore.[36]

The scientific press has thereafter experienced an incremental growth in hydrologic publications related to the development and application of experimental solutions. Seminal papers that helped spread the use of NN were: Minns and Hall[37] on the emulation of a conceptual model that had a nonlinear reservoir and produced simulated discharge sequences; plus four other papers that involved the use of real-world datasets: Hsu *et al.*[38] developed viable hydrologic alternatives to linear time series models and conceptual models using daily datasets; Maier and Dandy[39] provided

14-day forecasts of water quality parameters using daily records for the River Murray, South Australia; Shamseldin[23] compared a set of different rainfall-runoff models developed on similar inputs for six international catchments using daily datasets; Dawson and Wilby[24] developed robust models of 15-min flows with 6-hr lead times for two flood-prone rivers in England.

Time series forecasting has continued to be a particular focus of interest with NN models having been developed in several different areas of water resources science and engineering; however, in most cases, either embracing or replicating the existing challenges of traditional water resources applications, such as the development of conventional autoregressive functions[31,40,41] or replicating conventional rainfall-runoff modeling (including two static feed forward solutions calibrated on different algorithms compared against a Recurrent NN;[42] a conceptual model compared against a Tapped Delayed Line NN and a Recurrent NN;[43] and NN models compared against wavelet analysis approaches[44]). There are also examples of research supporting advances in related disciplines, such as the downscaling of climate change predictions.[45-47]

To help put matters in perspective, a systematic search of published papers was undertaken to establish the growth in popularity of NN hydrologic modeling since the early 1990s. This search was performed on the ISI Web of Knowledge (http://isiwebofknowledge.com/), a premier research platform that claims to be the "world's leading citation database with multidisciplinary coverage of over 10,000 high-impact journals in the sciences, social sciences, and arts and humanities, as well as international proceedings coverage for over 120,000 conferences." The search was performed using a two-item 'keyword query' of the 'publication title' field. Text inputs to the search engine comprised:

Drought	Inflow
Evaporation	River
Evapotranspiration	Runoff
Flood	Stage
Groundwater	Stream or Streamflow
Hydrology	Watershed
Infiltration	

used in combination with either 'neural,' 'neuro' or 'ANN.' The returned list was filtered so as to ensure that only those papers that were relevant to NN hydrologic modeling were included in the final count. For example, 'stage' identified a number of papers that were not related to hydrologic science, that included the term 'multi-stage' or 'two-stage' neural network. Figure 6.2 depicts the individual popularity of each hydrologic keyword that was employed in the two-item search procedure, showing the total number of publications returned in respect of specific words for the period in question. It should be noted that some keywords appeared alongside others in the same paper.

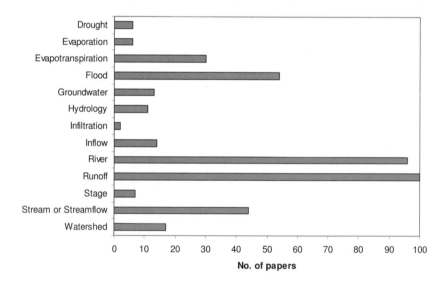

Fig. 6.2. Keyword count in relevant publications for 1992–2008.

Figure 6.3 shows the overall trend in total publications for the field from 1992 to 2008. These numbers were calculated from the keyword search results for each year. In this case, care was taken to remove duplicate copies from the total, i.e. publications that included more than one keyword. In addition, short comment letters, critiques of existing papers or responses to discussions were excluded from the final count to ensure that the statistics were not unduly affected by long, drawn out

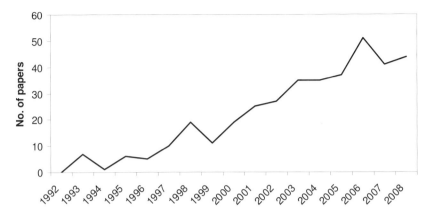

Fig. 6.3. Total number of relevant publications for 1992–2008.

dialog that sometimes occur between authors, reviews, and critics in the press.

In total, around 400 scientific papers related to rainfall-runoff modeling and flood forecasting have now been published, and also rising in recent times to an acceptance rate of some 45–50 papers per year. This topic is still the main subject of interest and a good marker of the technological dissemination process; however, wider appreciation and testing of predictive models in related fields of hydrologic interest, such as water pollution and sediment budgeting, continues to grow.[48-52] Moreover, it must be stressed that papers related to precipitation, rainfall, sediment, and water quality were not included in Figs. 6.2 and 6.3, although, from a similar set of queries, such topics would appear to account for at least another 200 publications.

It is accepted that a number of papers will have been missed in this initial search, since querying was limited to a restricted consideration of published titles, instead of content, and because papers continue to be spread out over numerous fields, ranging from technical subjects, such as artificial intelligence and computer science, to the natural sciences, including environmental planning and process modeling. The figures do, nevertheless, provide a general feel of increasing momentum in the field. That momentum can also be partitioned into a set of distinct phases, and such phases will be used in the following sections to provide a temporal structure for more detailed analysis.

6.3. Establishing the Field (1990–2000)

The start-up period comprised an increasing number of pioneering proof-of-concept studies that occurred throughout the 1990s. This period had steep learning curves and was, to a large extent, 'problem driven.' The pioneering studies sought to explore what could be done: focusing on past hydrologic modeling issues and the need for operational solutions, that were simple to develop and implement, and could support suitable representations of the nonlinear dynamics that are inherent in physical watershed processes. Neural network solutions that could resolve the associated difficulties of structure identification and parameter estimation were also sought — related to what were at that point termed 'system theoretic models,' investigations concerned with the solution of technological problems within the constraints imposed by the available datasets.[53] No physical hydrology was intended other than in the selection of meaningful or logical inputs; no effort was made to describe the behavior and interdependence of the different hydrologic processes involved; it was not intended to provide a complete synthesis of the hydrologic cycle. The numerical equivalents, at that point, comprised linear/piecewise-linear mathematical solutions or simplified conceptual models. Neural networks were found to provide an effective nonlinear mapping, without the need for an explicit internal hydrologic structure of the watershed, and required only limited conceptual understanding of hydrologic dynamics related to the particular watershed, or water resources issue, that was being addressed. Each paper started with a general explanation of NN concepts and provided a detailed account of the mechanisms and procedures involved in developing a standard MLP. Of particular interest is the fact that such papers contained a number of good ideas about the prospective use of NN that have not really been followed up: such as their capabilities to operate on different or mixed types of inputs and outputs (e.g. nominal data; fractal dimension), use of additional outputs that preserve or monitor global features, perhaps related to the overall structure of what is being modeled, or to particular aspects of a dataset;[33] potential use of binaries to mark the start and end of rainfall;[54] potential use of alternative representations that produce better modeling solutions, e.g. Fourier series parameters instead of actual hydrologic variables.[55] However, in a recent study,[56] rainfall and runoff time series were transformed into three sub-series for the purposes of one-step-ahead streamflow forecasting using wavelet decomposition: short, intermediate, and long wavelet periods.

It soon became apparent, from the increasing number of reported studies, that NN could provide accurate estimates of hydrologic variables, such as river level, stage, or discharge — even in the very simple case of having inputs that are restricted to some combination of antecedent rainfall depths and/or past river flow ordinates — leading to the subsequent application of NN to other catchment datasets around the world. In most early studies, the NN was applied as an independent model and its performance compared with other conventional models. The initial findings equated to proof-of-concept testing. In rainfall-runoff modeling, for example, rainfall information alone was found to be not sufficient to compute the runoff from a catchment, since antecedent conditions related to the state (soil moisture) of the catchment are important in determining runoff behavior.[37,57] Thus, in subsequent studies, discharge or water level measurement at previous time steps was, and continues to be, adopted as a surrogate input that is used to represent the soil moisture state of each catchment. It was also apparent that since network inputs included past observed records, from previous time steps, one could be considered to be modeling the change in flows rather than their absolute values, such that the process involved is one of updating.[37] This was a particular problem for one-step-ahead forecasting in that a near-linear relationship often holds. Shamseldin[23] and Rajurkar et al.[58] sought to address such issues in the latter case by means of a linear model coupled to the NN. No observed runoff and/or water level record for previous time steps is used as a direct input to the NN. Instead, the linear response function that relates rainfall to runoff is first derived from a set of corresponding rainfall and runoff datasets. Thereafter, runoff estimated from that linear model is used as a surrogate for catchment conditions, and operationalized alongside current and antecedent rainfall as inputs to the NN.

The conclusions from the initial studies were particularly important in establishing the credibility of the field in the 1990s, but papers of this type are now becoming less acceptable as more rigorous and alternative contributions to research are sought. The early studies involved developing connections, between mechanisms that represent the inner workings of the human brain and the field of hydrologic modeling, but the subsequent selling of a novel and untested product to a fresh audience also presented substantial challenges that did not go without a hitch! Gaining proper hydrologic accreditation was indeed a difficult, and sometimes painful, process, up and until the end of the millennium. The paper of Wilby et al.,[14] for example, that attempted to provide some hydrologic understanding of the internal components of various NN models was — after two attempts — rejected in 2002 from a top journal in the field on the grounds that it "does not make a tangible contribution to the fields of hydrology and water resources" (source withheld). The ISI Web of Knowledge, nevertheless, returned a list that contained 45 international citations of that paper at the end of 2008. The institutional situation has since improved somewhat, but convincing others that NN hydrologic modeling is part of water resources science and engineering and that it offers something interesting and worthwhile to think about or to understand is always going to be a major challenge for both *Innovators* and *Early Adopters*!

6.4. Evaluating the Field (2000–2001)

Three major reviews of published articles appeared at the end of the pioneering period. The influx of connectionist concepts into the hydrologic sciences and initial progress at this pivotal point is described in: American Society of Civil Engineers;[59,60] Maier and Dandy;[61] Dawson and Wilby[25] — with the last authors at that stage discussing around 50 papers that had been published on the subject of rainfall-runoff modeling and flood forecasting. Two illustrative, but less critical, compilations have also been published,[62,63] that were not formal reviews and did not purport to deliver a set of authoritative statements about the field, but instead sought only to convey a sense of how things stood.[62,63]

The 'millennium reviews' that are contained in such texts attempted to summarize and contextualize this emergent field at the turn of the century: (i) concluding, in broad statements and recommendations, on the need for neurohydrologists to address outstanding issues; and (ii) highlighting concerns that would need to be resolved in order for NN to be accepted by the hydrologic community. It is — for the record — perhaps worth noting that although such summaries have often been cited in subsequent publications on NN hydrologic modeling, only limited attention has been paid to the numerous criticisms and challenges that were reported at that point. It is clear that several members of the pioneering group have tried in subsequent endeavors to address some of the important issues that were raised, while in other cases the required activities were not pursued. It is difficult to second guess the reasons behind this: perhaps neurohydrologists are too interested in their software products; perhaps the substantial challenges that were posed are too difficult or too taxing for fresh entrants; perhaps the need for lateral thinking and moving 'outside the box' was not recognized. Instead, the 'scientific rationale' for numerous papers continues to be in keeping with a course of action that is perhaps best described in terms of the historical precedent of "let's try it on another dataset." The main issues that were raised in the millennium review papers are, nevertheless, worth revisiting, since a set of clear messages was presented, and such items can provide a reference point for understanding and appreciating the scenario that unfolded.

The eleven-member 'Task Committee of the American Society of Civil Engineers'[59,60] provided a mixed and cautious message: (i) in reporting that neural solutions had made a significant impact on the field; and (ii) by confirming that such tools could perform as well as existing models. For example, it did not state that neural solutions could produce superior results: something that could perhaps be interpreted as a challenge for some neurohydrologists to demonstrate that their tools can do a lot better! It was also argued that NN cannot be considered as a universal cure-all for hydrologic problems, since such tools are trained on paired input-output records in "the hope that [such tools] are able to mimic the underlying hydrologic process" (p. 134). The physics in the resultant model is locked up in a set of internal weights and threshold values, so that no fresh knowledge or insight is to be gained. It was also argued that such tools are not replacements for other modeling techniques, since there is no demand for such replacements. Existing mathematical models (i.e. equation-based solutions) are able to represent our limited understanding of the physics involved and, in problematic cases, the portion that is not well understood is either supplemented with empirical knowledge or is lumped under the various assumptions that may go into the development of a mathematical model. It was also pointed out that several studies[64-67] had warned about the pitfalls of NN and cautioned against their indiscriminate use. The committee suggested that before embarking on an ambitious use of neurocomputing methods it would be prudent to consider the issues that were raised in such papers. Neural networks, in the final sections of the paper were, nevertheless, perceived to be alternative modeling tools that were "worthy of further exploration" (p. 134) and, to that end, the committee posed a set of five fundamental questions that still needed to be addressed. Each of their questions is revisited and appraised next.

Q1. Can NN be made to reveal any physics?

The final product and its development process are not transparent. This was identified as a major detrimental issue for researchers and practitioners. If neural solutions are to gain wider acceptance and respect, then some sort of explanation — or partial explanation — must be provided as an integral part of the model. Two existing directions of enquiry were identified: knowledge-based networks and rule-extraction techniques. It was also pointed out that neither had been used in hydrologic applications at that point. Hsu *et al.*[68] was instead cited as having provided some 'heuristic functional relationships between input and output variables;' and knowledge-based networks were presented as a means for incorporating some theoretical knowledge in the model construction process that could be further refined using training examples.[69-71] Figure 6.4 depicts the two, largely independent, algorithms that would be involved in developing such knowledge-based solutions. This hybrid learning method maps problem-specific 'domain theories,' represented in propositional logic, into a NN and thereafter refines this reformulated symbolic knowledge using traditional FFNN procedures and BP. It seeks to use the information provided by one source to offset information missing from the other. It is also possible to extract symbolic rules from the final product: empirical testing has shown that such extracted rules: (i) closely reproduce the accuracy of the network from which they are extracted; (ii) are superior to the rules produced by methods that directly refine symbolic rules; (iii) are superior to those produced by previous techniques for extracting rules from trained NN; and (iv) are 'human comprehensible.'

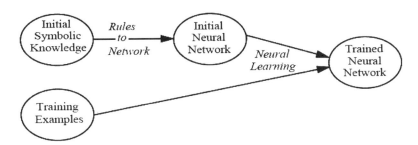

Fig. 6.4. Flow chart of theory-refinement by a knowledge-based neural network showing (1) rules-to-network translator; and (2) refiner, i.e. neural learning algorithm. Reprinted from *Artificial Intelligence*, 70(1-2), Towell, G. G. and Shavlik, J. W., Knowledge-Based Artificial Neural Networks, 119-165, Copyright 1994, with permission from Elsevier.

Sadly, however, the answer to this question remains something of an open issue, since neither of the two suggested options has been followed up. Thus, no physical rules or knowledge has so far emerged. Instead, a handful of researchers have attempted to 'open the box' (such activities are described in Sec. 6.6). It is clear, in retrospect, that a lot more effort should have been devoted to this particular topic, since it was identified as the biggest barrier to overall progression and wider acceptance of NN. It was, nevertheless, a tall order, since the products in question are designed to be used as black-box mechanisms, and neurocomputing enthusiasts might not be that interested in the underlying physics. It is also interesting to ponder about what such activities might be expected to deliver: are we seeking to find established physics (replication of equations) or different and/or improved physics (knowledge discovery)?

Q2. Can an optimal training set be identified?

There are no fixed rules regarding the selection or division of datasets for model development purposes. This is a classic challenge that still continues to plague traditional hydrologic modeling — for example, in the calibration of conceptual models — but to what extent is it a problem for NN? Most neurohydrologists had, at that point, responded to such challenges by adopting a simple 'trial and error' approach, of their own choosing, while at the same time attempting to keep to the old maxim that 'more is better.' The committee thus concluded, on the basis of such material, that NN are data intensive solutions: more records lead to slower training; having an insufficient number of records leads to poor generalization; and acquiring more observations could be costly or difficult to collect. This sequence of actions and outcomes, nevertheless, mixes together a number of inter-related points that required separate evaluations. The main argument is deceptive, since *more* does not necessarily equate to *better* – it is the information content that matters — a question of *quality* not *quantity*. The committee considered an optimal dataset to be one that resolved two competing properties: (i) it must possess sufficient records to fully represent the modeling domain; and (ii) it must contain the minimum number of data pairs needed for training. Learning about the potential advantages of assembling different datasets is, of course, a useful scientific endeavor that is to be encouraged. Neural network solutions and their modeling trade-offs, however, are about the provision of 'acceptable,' as opposed to 'optimal,' solutions, and the final product must reflect practical and operational issues in terms of the level of generalization that is delivered and the amount of time and effort that can be devoted to model development — both 'fitness-for-purpose' factors related to end user stakeholder requirements.

The impact of scarce information and poorer quality datasets was also raised by the committee in explaining that it was important to know under what circumstances a particular NN might fail. It was suggested that in physics-based approaches, the physics can be used to fill the gaps, in situations where observed measurements are not available. Neural network learning, however, is an iterative process, where a random model or partly-trained model is progressively improved through the process of calibration. It cannot fail in a strict sense, since it delivers different levels of generalization as it learns different aspects of the solution space. Supplemental material generated from physics-based knowledge could, of course, be used to fill in gaps in the material on which it is developed; better still, one could train a NN on a physics-related dataset, for example using input-output patterns derived from a 'first principles relationship,' and thereafter perform modifications or updates using real datasets. Identification of optimal datasets is related to the more general problem of measuring information content — a branch of applied mathematics and electrical engineering involving the quantification of information that is termed 'Information Theory.' This recognizes that the most important quantities involved are entropy (i.e. the amount of information that is contained in a random variable) and mutual information (i.e. the amount of information in common between two random variables). If we could first measure the content, it would be thereafter be possible to examine the impact of different datasets on NN.

It is, nevertheless, possible that the unearthing of optimum training and testing datasets could lead to substantial cost and resource savings, arising from the development of more efficient and effective models. For example, in one set of experiments,[72] dividing datasets into optimal subsets using a Genetic Algorithm (GA) and a Self-Organizing Map (SOM) indicated that poorer performance was primarily related to the data themselves, as opposed to model parameters or architectures, such that the method for constructing optimum sub-divisions remains a critical factor. In contrast, a GA has also been used to determine an optimal training dataset for an existing NN hydraulic model. Having first built the model, a GA was subsequently used to select an optimal, more balanced training set using subsets selected from the initial training set. Those observations that contributed the most to providing an improved solution resulted in a final training dataset that was half of the original size and a more accurate model.[73] For the most part, however, optimal division is considered to be more of a practical hindrance and not an operational priority, since good models are generally produced. It is also difficult to appreciate what the impact of resolving such matters might have in terms of the wider hydrologic acceptance of NN, since, as mentioned earlier, this problem is not specific to NN.

Q3. Can NN improve on time series analysis?

The committee accepted that NN could be used to predict future streamflows or river stages based on past measurements, but also commented that it offered, in that particular sense, nothing over traditional time series models. Indeed, apart from providing comparative statements about NN delivering superior output accuracies, it was recognized that no real insights into the modeling process had been provided — for example, through investigations of 'memory structures.' The main means of encompassing time dependencies involved having past inputs incorporated in a 'moving window,' the length of the window or lag being determined by correlation. It was suggested that such models were perhaps too simple to be worth the effort, while, on the other hand, more inputs would lead to a more complicated and less parsimonious model. Indeed, publications, at that time, suggested that neurocomputing catered for nothing other than the provision of 'time-homogeneous' solutions that were unable to account for the impact of structures or dependencies or permit situational changes to be incorporated (e.g. changes in land use, irrigation patterns, crop rotations). Neural networks are, however, based on different concepts to time series models — but important differences are not exploited such that, from a brief consideration of reported applications, one might assume that the two are rather similar. Indeed, there are no fixed rules regarding the selection of inputs and outputs, although some effort has been directed towards finding a more intelligent method for identifying an optimal set of predictors for a model — termed 'input selection.'[74]

Neural network modelers must typically work with a number of diverse hydrologic datasets that contain the required information in different formats. The modeler must process these datasets into a format that can be used by the NN and, at the same time, decide which predictors will produce the 'strongest' NN model. Most models use a number of standard input predictors (e.g. antecedent rainfall, runoff, evapotranspiration, temperature) to deliver the required output response (usually river stage or discharge). The modeler may try different combinations of such predictors, different lead (lag) times for each of the predictors (e.g. $T-1$, $T-2$ for a rainfall forecasting application, see Luk *et al.*[75]) and different functions applied to each predictor (e.g. moving averages computed over different temporal periods). The modeler might also investigate the inclusion of other rational predictors to improve model performance, such as the addition of a periodic counter (e.g. using sine and cosine of annual hour count to provide the model with some indication of season[76]).

Several attempts have been made to automate the selection process, such as the use of partial mutual information[77-79] and impulse response weights.[80] There are, however, still no clear rules on the selection of an appropriate set of predictors for modeling, and most NN modelers still use trial-and-error. It is also common for modelers to examine the cross-correlation, between a number of predictors and the predictand, and to select a set of predictors from such results. Unfortunately, by using cross-correlation as a measure of the relationship, the modeler invariably selects predictors that are related to the predictand in a linear manner. This can lead to overly parameterized models that, when compared with more simple linear regression models, appear to underperform. The identification of optimal inputs for runoff modeling is an interesting topic that requires further investigation and is an active field of ongoing research. The real challenge in such cases is to perhaps discover unknown dependencies and relationships, amongst pertinent variables, that will bring out important features of the input dataset that are not revealed in the use of other techniques and to find innovative methods of extracting this information from the mechanisms and procedures involved.

Making improvements to existing working practices presented an important topic that was doable; strangely, however, the number of neurohydrologists that rose to the challenge is limited to a handful of reported applications that sought to incorporate different surrogates for catchment conditions and/or meaningful 'hints.' For example, inclusion of low-frequency inputs to NN rainfall-runoff models that had already been optimized on fast response components delivered short-term forecasting improvements, i.e. using streamflow and rainfall as inputs.[81] Ten low-frequency candidate inputs were thereafter considered for inclusion: potential evapotranspiration, antecedent precipitation index computed at five time steps, and four soil moisture index time series calculated using the soil moisture accounting reservoir of a lumped conceptual rainfall-runoff model. This could be regarded as a hybridization of traditional hydrologic modeling and NN approaches.

Q4. Can NN training be made more adaptive?

Neural networks were seen as fixed entities — similar to compiled code — that once trained could not be updated with new information. It was argued that prolonged and time consuming network retraining and retesting operations would be required, if hydrologic conditions had changed, or if the model needed to be adapted to suit other catchments, regions or seasonal patterns. It was also suggested — returning to the physical argument — that if weights and thresholds could be assigned some physical interpretation, then it is likely that (re)training could be handled in an 'adaptive fashion.' The main argument in this case would appear to be that the internal representation of physical processes — assuming that such things exist — could, in some manner or other, be strengthened and that the resultant outcome would be a superior model.

The essential requirement to develop more adaptive solutions, that can be applied under different conditions, is clear, but how adaptive are their hydrologic counterparts, such as stochastic models derived from complex mathematical analysis, or conceptual models that might need to be reformulated, as opposed to something that is simply recalibrated? The argument for total retraining is also something of a 'red herring,' as is the case for attempting to include more 'adaptive physics,' since: (i) NN are perhaps the most adaptive models around; and (ii) it is, as explained later, possible to develop modular solutions that can be adapted or tweaked on a piece-meal basis. It is also clear that conceptual issues might have clouded the bigger picture, since most published models at that point offered complete solutions — meaning that specific inputs were used to create specific outputs at 'system level.' The committee — as stated earlier — once again failed to recognize the fact that NN solutions are developed through the iterative procedure of 'learning by construction,' such that subsequent learning and adjustment of the connection weights is permitted, and the solutions can be adapted. Moreover, in the quest to discover an optimal model, limited effort had (or has since) been directed towards: (i) investigating the operational merits of mixing and matching models developed to support different levels of generalization; and (ii) exploring the potential aspects of 'further training.'

The development of NN sub-systems or components was also not highlighted; it was seldom explored at that point but the power to develop NN components for inclusion as integrated components either inside, in parallel, or in serial with more traditional models is also permitted. For post-review examples, see Ochoa-Rivera et al.[82] and Chen and Adams.[83,84] Moreover, modular combinations, in which a discrete set of neural models can be programmed to act as a combined model, are possible. The committee recognized the use of modular NN for hierarchical decision-making processes but not the potential use of such mechanisms in a more general sense as providing internal sub-components for statistical, conceptual or distributed physical process models (p. 134).

The effort involved in (re)training was also somewhat overstated. Modern computers and superior optimization algorithms offer faster and more efficient modeling procedures in contrast to what was available at that point, e.g. Conjugate Gradient (CG)[85] and Levenberg-Marquet (LM).[86,87] The calibration time can indeed be reduced to a matter of minutes: so, perhaps the total operational effort that is involved should, henceforth, be compared to the effort and resources that would be required to develop and calibrate a mathematical model starting from scratch! Training is, nevertheless, an important part of the learning process, and the amount of effort needed is compounded by the fact that no fixed rules exist regarding inputs or architectures or assessment procedures, requiring a protracted trial-and-error process to deliver an 'optimal model.' However, it must be remembered that the true purpose of optimization is to calibrate a particular model, and not to search for an optimum model structure and, in this respect, most modest NN models will deliver an acceptable result.

The committee stated that the need for retraining would be especially true in cases where a process is dynamic, since a network trained on previous samples might not be suitable for some current hydrologic condition. However, such argument fails to consider the case for 'progressive construction' and, more to the point, if the power to incorporate fresh information is related to different dynamics or changing processes or conditions, then the real problem is not about calibration or training issues but one of poor initial model specification in which the developer has failed to appreciate the important states or factors involved. This problem is common to most types of model and would require all types of model — traditional or otherwise — to be rebuilt using an appropriate set of components and drivers, e.g. perhaps using surrogates or binaries to deliver different states and conditions in the case of NN. Examples include implicit coding of alternative land use scenarios in the manner of fuzzy set memberships;[28] and implicit coding of snow accumulation and snowmelt processes in the manner of logical values.[88] The power of using different inputs to differentiate between different states could also, for instance, be tested using before and after studies relating to dam construction projects where discharge and sediment characteristics are altered in the post-dam period.

Exploring and testing adaptive power and potential is another improvement to existing working practices that was clearly feasible; strangely, however, the number of neurohydrologists that rose to meet such challenges is scant. It is difficult to speculate on the reasons for this: perhaps it was thought that there was no case to answer; perhaps it was too simple a challenge; perhaps it required a difficult and complicated shift from digital hydrologic modeling mindsets to analog signal processing mentalities; perhaps testing the intricacies of minor positive or negative gains in the adaptive processing capabilities of different neural solutions was not of interest to the pioneers. The most plausible explanation is that the 'shakers and movers,' being few in number, were forced to choose between several competing alternatives and that the rewards and returns associated with extensive and repetitive experimentation were not high up on their list of priorities.

Q5. Are NN good extrapolators?
Neural networks are good interpolators and poor extrapolators: meaning, in this case, the extension of a relationship between two or more variables, beyond the range covered by our knowledge of that situation, or the calculation of a value outside the range of measured records. Models based on first principles — if available — will always be the

safest option for extrapolation purposes. The committee indeed reported that, although several published studies had confirmed the ability of neurocomputing solutions to perform well when faced with problems that fell within the domain of their training inputs, their performance deteriorated rapidly when the input vectors were far from the space of the inputs used for training purposes. From an analysis of published material, the report concluded that such tools are not very capable of extrapolation. This news, however, came as no big surprise to neurohydrologists: it remains a well-established dictum for the field that extrapolation beyond the range of the training data is not to be recommended. Indeed "interpolation can often be done reliably, but extrapolation is notoriously unreliable."[89]

This topic overlaps with some of the earlier issues, and it is important to have sufficient training pairs to avoid the need for extrapolation: but cases inside large 'holes' in the training dataset may also effectively require extrapolation, and such challenges are related to an earlier point about needing to understand when such tools will fail. For the record, NN will tend to fit a linear solution into unknown regions, as opposed to failing or crashing. There is, nevertheless, a certain irony in this final question: the committee are assuming that the power to perform extrapolation is always good! Extrapolation, however, is dangerous with any model, especially an empirical model, and some types of NN will tend to make particularly poor predictions in such circumstances. However, whether this is an advantage or disadvantage depends largely on the application, since the power to stay within known bounds precludes the production of wild and erroneous outputs — a strength! Indeed, uncritical extrapolation should be regarded as a bad point, and extrapolation far from the training data is usually dangerous and unjustified in all cases. This matter of making improvements to existing working practices once again presented an important topic that could be followed up; strangely, however, the number of neurohydrologists that rose to the challenge remains limited to a handful of reported applications that sought to experiment with different methods of extrapolation.

Most neurohydrologists rescale their input and output data streams to be consistent with the transfer function that is used inside the NN;

typically, such functions are constrained to a fixed range [0–1]. Extreme values in modeled datasets are thus scaled to fit that interval, but frequent users will also, as a matter of habit, incorporate some degree of latitude in the permitted input-output response by varying the range of their standardization procedure to, for example, 0.1–0.9 or 0.2–0.8, although the ability of the model to extrapolate remains somewhat limited.[37,90] The use of a linear transfer function in the output layer will offer a degree of extrapolation beyond fixed bounds, but this could lead to a different type of model, and, in practice, the amount of extrapolation is then limited by the nodes in the hidden layer, e.g. implementation of a linear[19,22] and clipped linear[91] transfer function in the output unit. It has also been asserted that, in order to assist the NN to extrapolate, a "cubic polynomial guidance system" should be added to the output layer.[92] However, the reported method for determining an appropriate form for the guidance system involved the use of the testing dataset, thereby disrupting the established protocol that requires the use of independent training and out-of-sample testing procedures, thus introducing even greater dependence on the available input and output records. Later, more elaborate solutions were suggested, including scaling the activation function so as to leave 'room for extrapolation,' but at the expense of introducing an additional, initially arbitrary, parameter.[93,94] The need for extrapolation has also been investigated based on an understanding that predictive abilities would be greatly enhanced if some method could be found for incorporating domain knowledge about peak events into the model development procedure — in addition to using whatever observed information is available for modeling purposes in the measured datasets.[95] One possible form of additional domain knowledge is the Estimated Maximum Flood (EMF), a notional event with a small but non-negligible probability of exceedence. This notional estimated value was included, along with recorded flood events, in the training of NN models for six catchments in the south west of England. The results demonstrated that, with prior transformation of the runoff dataset into logarithms of flows, the inclusion of domain knowledge in the form of such extreme synthetic events improved the generalization capabilities of the model and did not disrupt its training process. Thus, where guidelines are available for EMF estimation, the application of this approach is recommended as a potential means for overcoming the inherent extrapolation problem in multi-layered FFNN.

This leads to an interesting outcome for the question of extrapolation, since the answer was perhaps never in doubt. Neural networks are poor extrapolators, but the bigger question might be to what extent is extrapolation a real issue for such tools and, indeed, has the problem perhaps to some extent been solved in the case of streamflow estimation procedures? It could be that the main advantage of a particular model, from an operational standpoint, is that it can be used to extrapolate well, from a small amount of data in the usual range, to the entire region of coexistence — in both directions — up to some critical potential maximum and down to some critical potential minimum. However, in order to assess such things, one would need to identify the relevant critical points. No real attempt has so far been made to repeat the use of EMF; thus, in terms of a successful answer, it has not achieved the required acknowledgment. Finally, it should be recognized that the challenges associated with extrapolation are not limited to NN. This is a universal modeling problem. Extreme events are by definition rare and, therefore, provide limited opportunities for relevant parameter calibration and verification in all hydrologic modeling schemes.

The other two reviews were more practical in spirit. Maier and Dandy[61] reviewed the use of NN models for the prediction of water resources. Their review focused on the modeling processes that had been adopted. Most papers reported findings related to FFNN trained using the default procedures of BP (hereinafter referred to as BPNN), important issues related to the optimal division of datasets, the pre-processing of datasets, and the selection of appropriate modeling inputs were seldom considered. It was also reported that the selection of appropriate stopping criteria, and that the optimization of network geometry and internal network parameters, were generally poorly described or carried out in an inadequate manner. It was suggested that such activities might result in the production of non-optimal solutions, poorer model performance, and an inability for researchers to draw meaningful comparisons between different models. Their conclusions were even more damning, stating that, in a number of papers, the modeling process had been carried out incorrectly. Of particular concern was the inappropriate utilization of specific datasets during the training process; using a fixed number of iterations as a stopping condition; arbitrary selection of inputs, architectures, and internal model parameters; scaling of the input datasets to the extreme ranges of the transfer function; and the overall effect that such items could have on the modeling outputs. The need to consider the interdependencies that exist between different steps involved in developing a model was also highlighted. The review concluded that there is little doubt about the potential of NN to be a useful tool for the prediction and forecasting of water-related variables. However, in order to achieve significant advances, it was stressed that there needs to be a change in the mindset of users, from the simple, repeated application of models to an ever-increasing number of purposeful case studies that are designed to support the development and implementation of operational guidelines for neurohydrologists. It was also suggested that NN, at that point, tended to be applied to existing problems for which other methods had been found to be unsuccessful, more often than not adopting a set of similar strategies to their earlier counterparts. Neural networks should, however, be viewed as alternatives to more traditional approaches and not a replacement to an existing tool kit that works well in some situations and perhaps not so well in others. Likewise, in addition to seeking out topics where such tools can excel, it was suggested that research should be directed towards the identification of situations in which NN do not perform well — to help define the boundaries of their applicability — an opinion that others are starting to articulate.[16]

Dawson and Wilby,[25] in the third and final review, examined published literature on rainfall-runoff modeling and flood forecasting. Their survey considered different modeling practices and noted a lack of contributions in which alternative architectural configurations were compared and contrasted against one another. Moreover, of more critical importance, such models were seldom assessed against more conventional statistical approaches. For example, the results of two different NN rainfall-runoff models (FFNN and RBNN) were compared with a step-wise linear regression model and zero-order forecasts in experiments for the River Mole in England.[24] From their analysis, it was clear that no rigorous framework existed for the application of NN to hydrologic modeling, and so a protocol was provided comprising a flow chart of required steps, stages, and feedback loops for iterative model development. The paper concluded by commenting on the fact that little guidance had been provided regarding choice of error measures, but conceding that the same criticism might also be leveled at the wider discipline of hydrologic modeling. Their review called for more inter-model comparisons and more rigorous assessment of NN solutions against traditional hydrologic methods. The authors considered subsequent advances in the field to be contingent upon: (i) further refinement of objective guidelines for model construction; and (ii) development/use of standard measures of model skill, as provided in HydroTest (www.hydrotest.org.uk).[96,97]

It is interesting to speculate on the impact of such publications. The main item of interest in the initial peer-reviewed report was about physical factors or lack thereof: something that is perhaps always going to be a major stumbling block for most data-driven modeling solutions. The two practical reviews, in their closing sections also to some extent, provided some additional support and encouragement for undertaking the physics-based investigations that were recommended by the earlier eleven-member committee: Maier and Dandy[61] suggested that "the potential of rule extraction algorithms should be investigated" (p. 120); Dawson and Wilby[25] commented that "there is considerable scope for the extraction of hydrological rules from the connection weights using sensitivity analysis or rule extraction algorithms" (p. 102).

The second point to note is that the first of the reviews[59,60] listed a set of seemingly difficult or impossible to resolve problems, some being of a more general scientific nature and not specific to NN, while other questions presented a rather blinkered concept of NN. The second two reports were, in contrast, more upbeat and instructive, providing working solutions to identified problems and issues. Setting grand challenges is all well and good, but it must be remembered that NN hydrologic modeling was, and still is, in its infancy; it is supported by a handful of academic researchers, with limited access to funds and resources, and expecting fast results in terms of physical process identification equates to setting and demanding a rapid and conclusive answer to the neurohydrologic equivalent of a 'Turing Test.'[98]

Indeed, rather than encouraging interested parties to rise to such challenges, the act of raising a large number of difficult questions in such a manner could have produced a strong detrimental impact on existing players and potential entrants to that field — by reducing the scientific status of NN hydrologic modeling and curtailing related access to the vital sources of funding that are needed to stimulate further NN research. It is also interesting to draw an analogy between the reported situation in hydrologic modeling, following the millennium reviews, and that of the historical predicament of such tools following their failure to model the XOR (eXlusive OR) function.[5] The dismal record of distrust inflicted by past encounters and criticisms suggests that the impact of receiving anything other than a strong upbeat or optimistic report might perhaps have resulted in some sort of 'catastrophic fall-out.'

6.5. Enlarging the Field (2001–Present)

Figure 6.3 would suggest that the impact of the 'millennium reviews' produced nothing other than a minor blip in the relentless and near-linear increase that followed in the number of relevant papers that were published. The accumulated increased output is, in fact, slightly greater than linear but, sadly, the response curve did not achieve anything approaching the higher levels of growth that would be indicative of 'wider acceptance' and 'take-off' or reflect a large number of *Early Adopters* in the RTALC. Meanwhile, in conjunction with what might best be described as a dampened response, mainstream opposition had also begun to develop. Pleas for a halt to: (i) isolated curve-fitting case studies; and (ii) near-endless comparisons performed on certain topics were starting to appear in the academic press: stronger science and greater synthesis of results was instead requested. Aksoy *et al.*,[99] for example, pass critical comment on a series of published papers related to NN modeling of ET_0,[100-102] questioning the reported need for further studies to reinforce and strengthen the associated conclusions. Two more powerful attacks[23,103] concluded that: (i) for short-term forecasting purposes, neural solutions offered no real advantages over traditional linear transfer functions; (ii) the demands and complexities involved in the development of neural solutions made them difficult to use and therefore "uncompetitive;" (iii) there is still much to be done to improve our understanding about the uncertain nature and hydrologic characteristics of neural forecasters "before [such mechanisms] could be used as a practical tool in real-time operations;" and (iv) the potential merit of putting further resources into the development of black-box computational intelligence methodologies, such as FFNN, remains questionable since "the quest for a universal model requiring no hydrological expertise might well be hopeless." The question of needing to publish NN models in a transparent format, perhaps using explicit mathematical formulations, is something that has also been raised and attempted; in this respect, NN equations have been provided for estimating sediment yield[104] and daily CIMIS-ET_0[105] — albeit that the correctness of the latter equation has been placed in doubt.[106]

Figure 6.3 should, thus, be viewed as the initial segment of Fig. 6.1, comprising a gradual rise in *Innovators*. It would be reading too much into the dataset to suggest that one or other of the slight ups and downs that are observed in the annual volume of published papers could equate to anything that resembled a potential crossing of the anticipated gap that should occur between *Innovators* and *Early Adoptors*. That gap has still to be encountered and crossed. The length of time that a product spends in this particular position is another important aspect of the overall situation that has not been discussed. Time is against NN: *Innovators* do not stay around for long — they are fast moving individuals who are always in a hurry to explore and exploit the most recent computer science offerings. *Innovators* are indeed always on the look out for the latest gadget or gizmo and will rapidly become both disinterested, and fewer in number, as subsequent inventions and discoveries appear! Thus, slowdowns and delays will have a detrimental impact on the capabilities of a product to progress and, in consequence, on its eventual success. The changing pattern of international concentrations and spatio-temporal distributions in terms of interested parties both within and between different continents and application domains will also be a controlling factor.

Neural networks, to their credit, have nevertheless prevailed in a competitive environment, such that the trialing and testing of neural solutions for hydrologic modeling purposes has managed to maintain its momentum in a number of different directions, including: (i) application to traditional regression-type problems; (ii) merging of different artificial intelligence and soft computing technologies; (iii) combination of models in modular and ensemble modeling approaches; and (iv) development of hybrid models — including the use of different neuroemulation strategies for delivering more efficient tools.

6.5.1. *Traditional Regression-type Applications*

Traditional applications abound. Most papers included in our trawl of the literature discussed modeling different aspects of river discharge in one form or another (e.g. runoff, stage, volume), but a number of the publications report applications in which neural models are applied to other important hydrologic processes. Examples of other hydrologic variables modeled include reservoir inflows,[107,108] groundwater levels,[109-111] flood or storm events,[112-114] river basin classification,[115] water table depth,[116] and droughts.[117,118] Some applications have introduced established procedures into different aspects of hydrologic science — equating to a broadening of the application domain. Important topics in this expanding field of science would include issues related to modeling and predicting the impact of 'climate change,' such as the combined use of sea surface temperature records as an expression of global climate anomalies, spatio-temporal rainfall distributions from the Next Generation Radar, meteorologic data obtained from local weather stations, and historical stream data from gaging stations to forecast discharge for a semi-arid watershed in southern Texas;[119] methods for the spatial assessment of different climatic scenarios on the bioclimatic envelope of plant species in Great Britain;[120] and the downscaling of climatic variables.[45-47] Others, as reported earlier, have attempted to build on initial findings: equating to a deepening of the application domain. The main issues that have been addressed are: inclusion of the last observed record as an input;[58] extrapolation beyond the range of the calibration dataset;[92-95] and single-step-ahead or multiple-step-ahead forecasting using recurrent NN.[43,121-125] Numerous scientists have also sought to perform comparative studies that evaluate and contrast the modeling skill of one particular type of NN, both with other types of NN, and/or against more traditional or modern approaches — in what might appear to be an extended series of simple curve-fitting experiments. This, however, could also, in part, be interpreted as a direct response to matters raised in the millennium reviews about NN models having similar levels of skill to existing mathematical procedures — equating to a continuation of past activities in the hope of finding a superior model, e.g. using different algorithms, such as BP, LM, and CG, or using different types of NN. The FFNN, GRNN, and RBNN models have, for instance, in this respect been compared and contrasted in several related fields including sediment modeling,[126] evapotranspiration modeling,[127] and river flow modeling.[128-131] Perhaps more worrying is the fact that a sizeable number of outmoded studies on the potential application of a 'standard' NN model to different rivers and stations are still being submitted for publication in international outlets — in most cases comprising a software default BPNN. This type of 'repeat experiment' is run on a particular catchment and established inputs are used to model established outputs such that no fresh insights are gained. The NN are often compared with more established techniques (such as physical, conceptual, and/ or statistical models), and it is usual to conclude that NN will outperform such techniques by some small margin of error that is calculated on a limited number of global statistics (although the opposite situation has also been reported[128]). The conclusions from such studies were particularly important in establishing the credibility of the field in the 1990s, but such papers are now becoming less acceptable as more rigorous and alternative contributions from research are sought.

Most papers in this sub-set examined the application of NN to flow forecasting in one form or another, but some focused more on the development of improved training algorithms and structures of the NN models themselves. Papers that fall into this group are two-fold: (i) papers that have amalgamated different computational concepts and technologies (Sec. 6.5.2); and (ii) papers that have attempted to develop improvements or to promote alternative thinking to the standard method through the adoption of simpler strategies and mindsets. For example, an improved algorithm for selecting candidate models based on a consideration of forecasting extreme events has been proposed,[116] and the impact of stopped-training, regularization, stacking, bagging, and boosting on NN generalization capabilities related to streamflow modeling for six different types of catchment has been investigated.[132] In this context, stopped-training and regularization are two methods that can be used to improve generalization and prevent overfitting;[133] stacking is the use of an additional model to combine the results of member models to arrive at a final decision; bagging equates to bootstrap aggregating;[134] and boosting is a method for enhancing the accuracy of weak performers by training subsequent members of an ensemble of observations that were poorly predicted by previous members.[135] The use of a NN nonlinear perturbation model to deliver improved rainfall-runoff forecasting for eight catchments, that spanned a range of different climatic conditions and different sized watersheds, has also been reported.[136] Further examples continue to be published, but such conceptual testing is seldom followed up.

6.5.2. *The Merging of Technologies*

Neural networks were not the only technological wizardries to appear in the 1990s. Two other related fields of interest were also starting to emerge in the hydrologic literature: (1) Fuzzy modeling comprising Fuzzy Logic (FL) and Fuzzy Set Theory (FST); and (2) Evolutionary modeling that encompasses Evolutionary Algorithm (EA), Genetic Algorithm (GA), and Genetic Programming (GP) applications.

The novel concepts involved were, in each case, quickly adopted by enthusiastic pioneers in a number of different fields of science with innovative applications being devised and tested. The potential advantages of such techniques were soon recognized and incorporated within NN modeling. Each technique, in isolation, provides an effective tool, but when used in combination, the individual strengths of such mechanisms can be exploited in a synergistic manner for the construction of powerful intelligent systems. The main algorithmic benefit that was on offer to neurocomputing in both cases appeared to be superior optimization, of parameters and/or structures, leading to a fresh round of proof-of-concept papers, in which different options or software products could be evaluated.

6.5.2.1. Neuro-fuzzy Products

Fuzzy Logic and Fuzzy Set Theory are used to support decision-making processes by means of a set of logic rules and membership functions.[137] This is an extension to the classical notion of a set that will permit the gradual assignment of membership to a set, such assignments being described with the aid of a membership function valued in the real unit interval [0, 1]. Fuzzy Logic is a form of multi-valued logic derived from FST to deal with reasoning that is approximate rather than precise. In FL, the degree of truth of a statement can range between 0 and 1 and is not constrained to the two truth values {true, false} of classic predicate logic. If linguistic variables or linguistic hedges are used, these items can also be managed by specific functions. The use of FL and FST approaches for the estimation of water resources dates back to the start of the 1990s. The main motivation in such cases is to offer possibilistic, instead of probabilistic or deterministic, solutions; for example, use of fuzzy rule-based models to simulate the infiltration process[138] or movement of groundwater in the unsaturated zone.[139]

Maier and Dandy[61] suggested, in the concluding paragraph of their millennium review, that future research effort should be directed towards the use of neuro-fuzzy systems (NFS). The rationale is several-fold. Neuro-fuzzy systems can capture the potential benefits of both fields within a single 'unified framework.' Such tools eliminate the basic problem in fuzzy system design (obtaining a set of fuzzy IF-THEN rules) by effectively using the learning capability of a NN for automatic fuzzy IF-THEN rule generation and parameter optimization. From another perspective, such systems can utilize the linguistic information of a human expert as well as quantitative data for modeling purposes. The other principal advantage of a combined system that is often cited (but seldom put to the test) is that such solutions are transparent and so could be used to help address the major criticism that is leveled against NN: this being that such tools do not consider or explain the underlying hydrologic processes in a watershed. Hydrologic information in the form of fuzzy IF-THEN rules can, however, be extracted from a NFS; so, one could interpret the rules of the model to infer the system dynamics that are represented in it.

Testing has so far focused on one main type of NFS: the Adaptive Neuro-Fuzzy Inference System (ANFIS).[140] More esoteric applications have investigated the Hybrid Intelligent System, in which integration is achieved through representing fuzzy system computations inside a generic NN[141] and application of the Counter Propagation Neural-Fuzzy Network (CFNN).[142-145] The Adaptive Neuro-Fuzzy Inference System is available in the MATLAB 'Fuzzy Logic Toolbox' — something that could perhaps explain its: (i) growing popularity and reported testing as a hydrologic modeling tool; and (ii) regular associated comparison against other types of traditional or neurocomputing models that can be programmed in this particular software package. Example applications include: flood forecasting,[30,31,146-155] suspended sediment estimation,[156] precipitation forecasting,[157] regional flood frequency analysis,[158] and reservoir operation.[159,160] Most studies suggest that this method offers considerable promise, although it was not always found to be the most attractive option.[161] Moreover, despite a surge of interest, most reported applications equate to a set of limited curve-fitting optimization procedures being tested against one another. There is no meaningful reporting, discussion or comparison of the (different) rules that have so far been extracted from one or more datasets. The one notable exception to this is Chang et al.,[145] who presented a set of tabulated operational pumping station procedures. Thus, irrespective of its transparent nature, no fresh knowledge has so far been provided to help counter the 'physical criticism;' simple partitioning into different magnitudes is sometimes observed and the various rules are reported but there is little else to go on. For example, Firat and Güngör[154] provide an interesting, but unexplained, set of screenshots for the different membership functions and rules contained in one of their two models; in other cases, the important internal mechanisms were not even listed. The neuro-fuzzy system models are thus, in most cases, being used primarily as black-box tools, and important questions remain unanswered. In particular: (i) what relationship exists between the rules that are generated on the one hand and different models or architectures or catchment datasets on the other? and (ii) to what extent are superior findings the result of a simple rule-based partitioning divide-and-conquer method, applied in what equates to a more complex and less parsimonious solution, than that of a traditional NN? More effort is needed in such cases to present the transparent components and to thereafter position the multifaceted findings that are produced in their appropriate hydrologic context.

6.5.2.2. Neuroevolution Products

Neuroevolution is the application of evolutionary computation to NN development. In this particular instance: (i) genetic mechanisms and procedures are used to breed new individuals; and (ii) evolutionary mechanisms and procedures are used to support NN development through the optimum determination of external inputs, number of hidden nodes, and parameter values. Each such algorithm incorporates aspects of natural selection or survival of the fittest. Each maintains a population of structures — in most cases commencing with a randomly-generated set of initial members — that evolves according to rules of selection, recombination, mutation, and survival, referred to as genetic operators. The fittest individuals are more likely to be selected for reproduction (retention or duplication), while recombination and mutation modify those individuals, yielding potentially superior ones. Evolutionary and genetic algorithms have become popular hydrologic modeling mechanisms, and numerous operational implementations have been reported for different types of application including: the calibration of conceptual rainfall-runoff models,[162-170] the optimization of groundwater recharge,[171,172] decision support for watershed management,[173] and precipitation prediction.[174]

Neuroevolution has been used to support hydrologic applications in several ways. The most common approach has been to optimize an entire NN at once through evolution of its parameter weights. For example, a GA can be applied to determine the starting weights of a NN prior to further training using standard BP;[175] using a combined GA and conjugate gradient descent procedure to train a NN updating model that outperformed a linear autoregressive exogenous updating method;[176] using a real-coded GA to train a NN rainfall-runoff model for the Kentucky River Basin in the USA that outperformed a conventional NN, especially on low flows;[177] and using a GA combined with resilient propagation to breed a set of alternative rainfall-runoff models for the

Upper River Wye in the UK.[178] The use of a GA to select an optimal set of inputs[78] and to determine an optimal training dataset[72,73] for a NN was covered in Sec. 6.4.

More recent experimental studies have investigated the potential optimization merits of: (i) cooperative coevolution, as implemented within the Symbiotic Adaptive NeuroEvolution (SANE) algorithm;[179,180] and (ii) swarm intelligence, as implemented within the Particle Swarm Optimization (PSO) algorithm.[181,182]

The SANE algorithm supports a different approach in that a population of neurons is evolved, as opposed to a population of NN. To develop a complete solution, the individual neurons are grouped together so as to optimize one part of the solution space, while cooperating with other partial solutions. This produces increased specialization in the final model and provides a more efficient search of the solution space. It also maintains greater diversity throughout the population because convergence is not towards a single 'type' of individual. This approach has been applied to rainfall-runoff modeling of the River Ouse in England.[20,183-185] The results showed that better-performing models can be developed if different objective functions are used other than the conventional sum of errors squared. It was also discovered that targeting the objective function to different parts of the hydrologic record produced better results, especially for low-magnitude events, and that this method produced superior models for longer forecasting horizons in more complex and more challenging situations.

Particle Swarm Optimization is a stochastic, population-based, computer algorithm that has also been applied to hydrologic problem-solving and optimization of NN. This swarm intelligence method is based on socio-psychologic principles and can provide insights into social behavior, as well as contributing to engineering applications. It has roots in two main component methodologies: artificial life (e.g. bird flocking, fish schooling, and swarming) and evolutionary computation. The method is focused on collective behaviors that result from the local interactions of the individuals both with each other and with their environment. Particle Swarm Optimization shares many similarities with evolutionary computation techniques, such as the GA. The system is initialized with a population of random solutions and searches for optimum solutions by updating generations. However, unlike GA, PSO has no evolution operators, such as crossover and mutation. Instead, in PSO, simple software agents, called *particles*, move in the search space of an optimization problem such that the position of a particle represents a candidate solution to the problem at hand. Each particle searches for better positions in the search space, by following the current optimum solution particles and changing its velocity, according to rules originally inspired by behavioral models of bird flocking. This algorithm has been used to train NN to predict real-time downstream daily water levels at different lead times for Fo Tan on the Shing Mun River of Hong Kong using measured records related to an upstream gaging station at the point of forecast.[186,187]

Neuroevolution can also be used to optimize solutions on multiple objective functions, and this approach presents considerable promise for the development of better NN hydrologic models. Three reported examples[20,188,189] are mentioned here. The first set of experiments[20] applied a timing error correction procedure to the NN. The model was used to provide river level forecasts for the River Ouse in England. The objective function was modified to include a penalty term that was applied when timing error was detected. The results showed that timing errors could be corrected but at the expense of overall goodness of fit. The second set of experiments[188] performed a multi-objective comparison between a NN and the conceptual HBV rainfall-runoff model. Both models were calibrated using the popular Non-dominated Sorting GA II (NSGA-II) and adopted objective functions related to low flows, high flows, and 'hydrograph shape.' The third set of experiments[189] compared NN river discharge forecasting capabilities on two mesoscale catchments in different climatic regions trained using the Multi-Objective Shuffled Complex Evolution Metropolis–University of Arizona (MOSCEM–UA) and NSGA–II. The results were compared to models developed using single-objective LM and GA training methods; numerous objective functions were trialed. The study showed that multi-objective algorithms give competitive results compared with single-objective solutions. Furthermore, posterior weight distributions inside each model suggested that multi-objective algorithms are more consistent/reliable in finding optimal solutions. It was also shown that:

(i) network performance is sensitive to the choice of objective function(s); and (ii) including more than one objective function proved to be helpful in constraining the NN training.

6.5.3. *Modular and Ensemble Modeling*

Neural networks can be used in isolation — providing a single solution to a stated problem. However, as with equations, such tools can also be used in the manner of collectives to produce a combined model output. The concept of having one model that does everything is indeed perhaps outdated, since no single model has the power to provide a superior level of skill, across all catchments or different types of system behavior, particularly when the solution space is complex. If one is instead going to use a divide-and-conquer approach, in which the overall problem is partitioned into smaller segments that are easier to model, then the matter of selecting an appropriate means for dividing and recombining arises. Neural networks can be used to perform both operations as well as providing the independent mechanisms for modeling each individual part. The general principle is one of distilling complex relationships into simpler units or sub-sets that can be resolved in an efficient manner.

Neural networks are capable of approximating a continuous well-behaved relationship between their predictors and predictands, but might not be suitable for representing a fragmented or discontinuous relationship that exhibits significant internal variation over the solution space, since accommodating voids or barriers or cliffs or overhangs in the input-output mapping operation is difficult. The inability to map fragmented datasets is overcome by decomposing the complex mapping space into simpler sub-domains that can be learned with relative ease by the individual NN models. In this situation, both problem and dataset can be split into a number of different sub-sets, such that separate models can thereafter be built for each component. The individual models in such an arrangement are called local or expert models. The overall type of modular model that is constructed is sometimes called a 'committee machine' (CM).[190,191] Neural network committee machines are used to combine the responses of multiple expert NN into a single response and the resultant combined response is generally superior to that of its individual constituent expert NN. However, in terms of disadvantages, using more models will, of course, mean that there is a much greater danger of overfitting, given the much larger number of NN and NN–CM model parameters involved. The two key decisions that are required to implement a divide-and-conquer approach, as mentioned earlier, must be made when building a CM. The first is how to split the problem and its dataset. The second is how to combine the individual model outputs to produce a final output. Two approaches are possible at each end of the process: 'hard' decisions in which a separate expert model is used under specific conditions; and 'soft' decisions in which inputs and outputs to the expert models are mixed and merged in some meaningful or weighted manner to produce a final output.

The simplest method of divide-and-conquer is to use hard input and hard output local expert models. Each individual local model is trained in isolation on selected sub-sets of relevant instances contained in specific regions of the solution space and, thereafter, the output of only one specialized expert is taken into consideration, as and when required. The divisions can be implemented manually, by experts on the basis of domain knowledge, using sub-sets constructed from classification procedures that are applied prior to modeling that sub-set with a NN. Three reported examples[192-194] are mentioned herein. The first set of experiments[192] used a SOM to divide their input dataset into five clusters, where each cluster represented a different part of the hydrograph. Individual NN were then trained on each cluster. The second set of experiments[193] performed a similar exercise in that the hydrograph was decomposed into different components and each segment was modeled with a separate NN. The third set of experiments[194] likewise modeled two sub-processes (baseflow and excess flow) in a comparison of three different partitioning schemes: (i) automatic classification based on clustering; (ii) temporal segmentation of the hydrograph using an adapted baseflow separation technique; and (iii) application of an optimized baseflow separation filter. It is also possible to develop mechanical means of performing non-complex combinations by, for example, switching between different single modeling solutions at each time step adopting a crisp probabilistic method.[195] This would permit NN to be used as one or other of the numerous individual solutions that could be opted for. The alternative might be to switch between different single modeling solutions at different levels of discharge, adopting a range-dependent method.[196] In this approach, a threshold autoregressive model was first used to determine a set of specified ranges for individual training datasets, with individual NN thereafter being trained on each sub-set. The power to use different models to capture different aspects of a problem is also demonstrated in the reported use of a NN to model eight different output parameters for the purpose of constructing a runoff hydrograph:[197] (1) peak flow; (2) time of peak flow; (3, 4) baseflow at and (5, 6) timing of rising and recession limbs of the hydrograph; and (7, 8) width of the hydrograph at 75 and 50% of peak flow. Each of these studies demonstrated superior performance compared to a single global model.

In contrast, it is also possible to train an automated modular NN to cope with the discontinuities that arise in a rainfall-runoff mapping operation for a watershed that has significant variations in its solution space due to the different types of functional relationship that occur at low, medium, and high magnitudes of streamflow. For example, one such reported modular solution consisted of a gating network and a set of expert NN.[198] Each expert model was used to map the relationship pertaining to a sub-set of the overall solution space. The gating network was used to identify the best expert model(s) for a given input vector, and its output was the probability of a particular input vector being associated with each individual local expert. The output from that modular NN was thereafter calculated by multiplying each local individual expert response by the corresponding weight (probability) of the gating network. This complex modular solution, as expected, was found to be superior to an individual NN.

Later modeling of simple and complex hydrologic processes involved a spiking modular neural network (SMNN), comprising an input layer, a spiking layer, and an associator neural network layer (Fig. 6.5).[199] This type of model incorporates the concepts of self-organizing networks and modular networks and the modular nature supports the finding of domain-dependent relationships. The performance of the tool was tested using two diverse modeling case studies of: (i) streamflow (a single-input

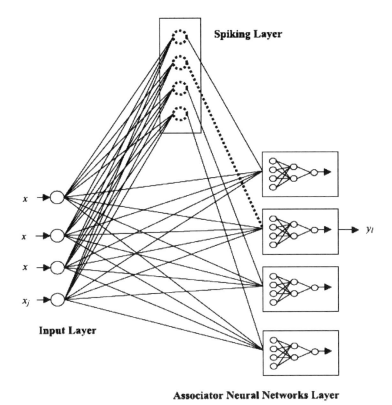

Fig. 6.5. Spiking Modular Neural Network. Reprinted from Parasuraman, K., Elshorbagy, A. and Carey, S. K., Spiking modular neural networks: A neural network modeling approach for hydrological processes, *Water Resources Research*, 42(5), W05412, 9 May 2006. Copyright 2006 American Geophysical Union. Reproduced by permission of American Geophysical Union.

single-output process); and (ii) actual evaporation measured via eddy covariance (a multiple-input single-output process). Two variants were tested in the spiking layer: competitive learning (CL) and a SOM. The modular method performed slightly better than a regular FFNN in streamflow modeling, but was much better than its competitor at modeling evaporation flux. For streamflow, it was, in fact, better at modeling high flows, but no better at modeling low flows. The results of this investigation confirmed that such mechanisms can be successful at breaking down a complex solution space into a relatively larger number of simpler mapping options, which can be modeled with relative ease. It thus supported earlier reported findings.[198] Of the two options, competitive learning was also, on both occasions, found to deliver slightly more accurate results. Thus, since SMNN (competitive) is designed to learn the distribution of input space in terms of a functional response, while SMNN (SOM) learns both the distribution and the topology of the input space, it can be concluded that modeling the topological relationship does not improve the performance of a SMNN model. Moreover, regrettably, since individual NN models are constructed for each pertinent cluster, the topological information that is learned during the classification process does not have a direct input into the performances of the associator NN. The Self-Organizing Linear Output (SOLO)[200] map is another SOM related mechanism that uses classification and mapping layers to transform inputs into outputs via a matrix of multivariate linear regression equations.

Ensemble modeling adopts a different approach to 'divide-and-conquer.' The problem is not reduced into different individual components. Instead, different models are developed to address the total problem, with numerous holistic solutions thereafter being combined to provide improved accuracies, precisions, and/or certainties. The objective of ensemble learning is to integrate a number of individual learning models, situated in an ensemble, so that the generalization performance of that ensemble is better than any of the individual component models. If the individual component models are created using the same learning algorithm, the ensemble learning schema is *homogeneous*, otherwise it is *heterogeneous*. Clearly, in the former case, it is essential that the individual learning models are not identical and, theoretically, it has been shown that, for such an ensemble to be effective, it is important that the individual component models be sufficiently accurate and *diverse* in their predictions. Having generated an appropriate set of models, a method of integration is required to create a successful ensemble, and the most popular method of integration in such cases is to combine the models using a learning model — a process that was described earlier as 'stacking.' The end product in this arrangement constitutes a meta-model created from a meta-dataset.

Two important questions arise in such matters: which models should be included and what method should be used to combine their predictions? No single model is perfect and, for example, in the case of rainfall-runoff modeling, each model will at best describe one or more particular stages, phases or mechanisms of the overall process that is involved. So, the prudent amalgamation of several discharge estimates obtained from different models could be expected to provide a more accurate representation of catchment response, in comparison to each individual model that is used to build a particular combination. The main objective, in performing this exercise, is to more heavily weigh the best solutions. Shamseldin et al.[201] were probably the first to explore the use of multi-model methods for hydrologic predictions, arguing that the existence of operational dualities provided theoretical and empirical justification for a multi-model approach. Three different methods of model mixing were used to combine simulated outputs, obtained from five single individual daily discharge rainfall-runoff models. Identical procedures were applied to eleven international catchments, of various sizes and climatic conditions, spanning different geographic locations around the world. The combined models were more accurate than the best performing original model; removal of the worst input model produced differing results — sometimes better, sometimes worse. It should also be stressed that ensemble modeling is not restricted to NN. For example, multi-model ensembles were used to analyze the simulation discharge outputs produced from numerous models that participated in the distributed model intercomparison project (DMIP),[202,203] and to combine the discharge outputs obtained from three conceptual hydrologic models: the Sacramento Soil Moisture Accounting (SAC–SMA) model, the Simple Water Balance (SWB) model, and the HYMOD model.[204]

Neural network ensemble modeling can involve the combination of numerous independent NN models developed on strategic permutations of fixed and varied components, including, for instance, solutions related to the implementation of different architectural parameters, training parameters, learning methods, starting conditions, or sub-sets of the main training dataset. For example, a NN has been used to combine 30 different NN models developed using bagging procedures for streamflow prediction purposes,[205] i.e. each network was trained on a randomly selected sub-set of the full training dataset. Several publications[195,201,206,207] have also reported proof-of-concept assessment and comparison of NN solutions against different mechanisms of model output combination, including soft computing methods, means, medians, and weighted-averages. The potential benefits of combining discharge forecasts obtained from several different types of individual models have also been demonstrated,[208] comprising NN, FL, 'conceptual model,' and 'persistence forecast' output. The combinatorial approaches that were trialed and tested in the last example included arithmetic-averaging, a probabilistic method in which the best model from the last time step is used to generate the current forecast, two different soft computing methodologies, and, last but not least, two different types of NN solution. The combined solutions, again, performed better than their individual counterparts with the NN and probabilistic methods performing best overall. Testing has also involved a comparison of six approaches for creating NN ensembles in pooled flood frequency analysis of the index flood and the 10-year flood quantile.[209] This investigation concluded that if the same method is used for creating ensemble members, then, combining member networks using stacking is generally better than using simple averaging. It was also suggested that an ensemble size of at least ten NN was needed to achieve sufficient generalization. Ensemble modeling will, doubtless, raise concerns about over-parameterization and the development of redundant components in a parallel modeling environment. The multi-model predictions that such methods have produced have, nevertheless, been shown to consistently perform better than single-model predictions when evaluated using predictive skill and reliability scores. The statistical techniques of bagging and boosting have, for example, been used to compare different approaches on six diverse catchments.[132] The computational techniques that were investigated in their reported experiments provided improved performance compared with standard NN. Moreover, superior results were achieved from the use of bagging and boosting procedures, over and above that of the other generalization techniques that were tested, comprising stacking, stop-training, and Bayesian regularization, albeit that the pattern of results across their catchments was not consistent.

6.5.4. *Neuro-hybrid Modeling*

The relationship between traditional modeling solutions and NN is not straightforward.[210] Neural networks offer more promise than a set of tools for the simple replacement of traditional modeling mechanisms. Sadly, however, such devices are often used to perform identical time-honored tasks! Neural networks can be used to include additional variables acquired from alternative or perhaps orthogonal data sources, irrespective of our current understanding about the relative hydrologic roles of such items or recordings. Neural network solutions can be run in serial or parallel configurations with existing models to provide either pre- or post-processing operations — perhaps acting as noise filters, error correctors, or complex linking mechanisms. Neural network code can be compiled to run as embedded functions inside another model. Neural networks can be used to clone existing models, in which case simple parallel processing operations would offer strong practical benefits, such as faster computational speed and increased robustness for operational applications. Trained products can also be developed for model reduction purposes, resulting in more efficient modeling operations, requiring fewer sources of costly or difficult-to-acquire input drivers, and perhaps producing more generalized results that could, nevertheless, be of some real practical benefit to interested parties, such as practitioners and resource managers in developing countries.

The development of hybrid solutions that mix and match different types of models is still in its infancy, but several topics of interest have emerged with respect to the manner in which a neuro-hybrid solution is developed, from the coupling of two or more different types of models. In most cases, a conceptual model is coupled with a standard NN model for the purpose of achieving a particular objective or goal — typically to obtain improved accuracies. For example, to produce improved streamflow forecasts, soil moisture accounting inputs computed in a conceptual model have been used to augment more established inputs to a NN.[81] In a similar manner, snow accumulation and soil moisture estimates, computed in a conceptual model, have been incorporated to provide superior forecasts for two rivers in a cold environment.[211] The

NN model in such cases is a simple serial recipient of internal variables extracted from the computational sub-routines of a traditional model.

It is also possible to develop a different type of serial recipient solution, in which the NN model receives external variables delivered from a distributed set of traditional (sub-)models.[83,84] In this case, the neural solution acts as a replacement for the more traditional 'integrator mechanism.' Figure 6.6 illustrates the process involved in integrating discharge outputs delivered from a set of semi-distributed sub-catchment conceptual models. The sub-catchment outputs provided the input to a NN that transformed its distributed inputs into an estimate of total runoff for the entire catchment, in a nonlinear manner, rather than the standard process of 'linear superposition.' The reported NN inputs might be arriving in parallel, but the overall process of serial coupling is, nevertheless, quite straightforward and, while of interest, does not represent one of the major research directions in neuro-hybrid modeling. Most reported neuro-hybrid modeling activities, instead, tend to focus on one or other of the following, more practical, issues.

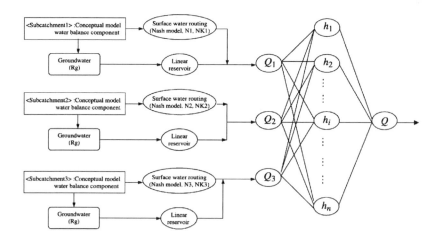

Fig. 6.6. Neuro-Hybrid Model. Reprinted from *Journal of Hydrology*, 318(1-4), Chen, J. and Adams, B. J., Integration of artificial neural networks with conceptual models in rainfall-runoff modeling, 232-249, Copyright 2006, with permission from Elsevier.

6.5.4.1. *Error Correction/Forecast Updating*

Neural networks can be used to correct an existing model. Operational real-time river flow forecasting models generally consist of a substantive rainfall-runoff model combined with an updating procedure. The principal purpose of updating is to provide an efficient mechanism for enhancement of the real-time estimated discharge of the rainfall-runoff model. These updating procedures are designed to determine the required adjustments that are needed to the internal variables inside the substantive model, based on an analysis of the most recent errors that arise between observed and simulated discharge hydrographs. The real-time river flow updating procedures that are used can be classified according to the different types of model variables that are to be adjusted. Pertinent adjustments could be made to: model input(s), the state of different internal storage elements, model parameters, and/or model output(s). There are four possible corresponding types of updating procedure: (i) input updating procedures; (ii) state updating procedures; (iii) parameter updating procedures; and (iv) output updating procedures. Each updating procedure can be viewed as a data-driven model that attempts to establish an empirical relationship between computed output errors and the model variables. The data-driven nature of the procedures also means that such mechanisms are ideal candidates for NN applications.

The first three types of updating procedure are used to adjust the relevant variables in the corresponding model, such that when the substantive rainfall-runoff model is re-run, using its modified set of variables, the estimated discharge closely matches the observed discharge. This is essentially an inverse modeling problem, in which the model variables are expressed as a function of model outputs. Solving the inverse problem requires a model that will reverse a complex forward relation. Neural networks can, in theory, be effectively used to approximate the inverse function — but no more than a handful of recent hydrologic publications deal with this type of application.[212,213]

The operation of the fourth updating procedure type (i.e. the output updating procedure) requires external modification of estimated discharges from the rainfall-runoff model, without re-running the substantive model, to obtain the updated discharge forecasts. Forecast updating in this procedure is, thus, seen to be a separate modeling exercise. Model output updating procedures are the most widely used updating procedures, as they are considered to be simpler to apply than the other updating procedures. However, despite their apparent suitability, only a handful of NN publications exist on this topic.

Model output updating procedures can be classified according to their method of providing an improved final output. The *direct* method involves using recent observed discharges and substantive model estimated discharges to provide updated discharges, but without an explicit need to calculate the errors of the substantive model. The alternative method requires the updated discharge to be obtained *indirectly* as the sum of (i) error forecasts and (ii) non-updated discharge forecasts.

In the indirect method, an add-on time series model is, in most cases, used to estimate the errors in the simulated/non-updated discharge forecasts of the substantive rainfall-runoff model. Thus, in terms of input-output modeling, the external inputs to the time series model are essentially the errors of the substantive model up to the time of issuing the forecasts. In this procedure, the updated discharge is obtained as the sum of the error forecasts and the non-updated discharge forecasts. Traditionally, univariate time series models, such as autoregressive (AR) and autoregressive moving average (ARMA) models, are used for forecasting the errors of the substantive rainfall-runoff model. Neural network updating equivalents have also been developed to provide a more general nonlinear form of these time series models for use in error forecasting and correcting operations.[214-217] The results of such experimental procedures were, nevertheless, rather mixed. For example, in one reported application,[214] the NN error updating procedure considerably enhanced the forecasting performance for short and long lead times, but in another,[215] the NN error updating procedure did not produce a better result than a traditional AR error updating procedure.

The direct output updating procedure is, in most cases, based on the input-output structure of linear and nonlinear autoregressive exogenous-input (ARX) time series models. These models provide forecasts of the future values of a time series, in which past values of that series, together with values of one or more exogenous time series, are used as external inputs to the model. In the context of model output updating, the external inputs to ARX models are the non-updated discharge forecasts of the substantive model as well as the current and the most recently observed discharge, just prior to the time of issuing the forecasts. Shamseldin and O'Connor[176] pioneered the use of NN to deliver a direct model output updating procedure and tested its performance using the daily discharge forecasts of a conceptual rainfall-runoff model for five catchments with different climatic conditions. In terms of real-time forecasting and operational requirements, the NN output updating procedure outperformed the traditional ARX output updating procedure, the poorer method being based on a linear structure and with significant improvements resulting for semi-arid catchments.

6.5.4.2. Embedded Products

Neural network models can be loosely- or tightly-coupled to an existing model, operating as an external mechanism, or neural solutions can be embedded inside an existing model as a finished product that is implemented in the manner of an internal component. The number of reported NN hydrologic modeling solutions that exist and can be run as standalone applications, remote to, and outside of their original software development environment, nevertheless, remains limited to a handful of specialist situations. Most such models exist as computational solutions that are housed inside or connected to a much larger software program, such as a decision support system (DSS). This situation could, in part, be related to practical and personal trust issues involved in switching existing procedures to run on a NN; for example, the need for effective communication between model designers and end-users, in addition to actual upgrade and installation matters, was identified as an important factor in implementing an operational model for a water treatment plant.[218]

There is a pressing need for decision support software that can incorporate expert knowledge and will permit decision makers to act upon a set of model results in a 'real-time environment,' or to use the program offline, for the purposes of performing 'what if' analyses of alternative scenarios. Several NN related applications have, as a result, been developed and reported. For example, a DSS was developed to assist operators and managers of a water supply system for drought characterization and management purposes.[219] The system had three components: a 'water demand forecasting' component for the City of Lexington in Kentucky; a streamflow forecasting component for the Kentucky River; and an integrated expert system component that permitted decision makers to explore a wide range of drought management scenarios. Five-day forecasts are first fed into the expert system; it then determines the drought potential and recommends a drought management policy. Several 'water demand forecast' modeling approaches were tested and the preferred solution that was incorporated into their DSS was a NN. The river flow forecasting component of that combined system, nevertheless, remained a conceptual rainfall-runoff model that was not replaced with an equivalent NN. Four other examples are also mentioned here. The first one is the reported development and testing of DEcision Support for Management Of Floods (DESMOF).[220] This decision-making toolbox was applied to the Red River basin in Manitoba, Canada, and showed successful results in both prediction and mitigation of the effects of flooding for the City of Winnipeg. Different types of models are incorporated inside DESMOF: the flood forecasting module comprised two five-input–four-output NN. This module provides the main driver that feeds other processes. It uses five hydrometerologic inputs to produce estimates of eight runoff-hydrograph characteristics. Sitting alongside the NN in DESMOF is the 'operation of flood control structures' module and the 'flood damage reduction options' module, functioning as both receivers and users of the neural outputs and being coupled to the NN in a serial arrangement. The power of hybrid solutions in control schemes that use machine-learning methods, in combination with physically-based hydrologic and hydrodynamic models to provide a faster nonlinear 'optimal control mechanism,' is likewise demonstrated in the Aquarius DSS, a physically-based generic modeling tool that can be used to build models of regional water systems and support their control.[221] The size and complexities of optimization problems in traditional dynamic system modeling can be very large, such that the adoption of a traditional optimization procedure for pumping stations would be far too slow for real-time optimal control of water systems, under fast-changing situations, or in the case of urbanized polders as in the Netherlands. Four NN were developed to provide a series of neural classifiers that predicted the required status of pumping stations, offering separate solutions for urban polders, glasshouse polders, rural polders, and the storage basin. Each NN was connected to the Aquarius DSS using a dynamic link library, leaving Aquarius to perform the simulation, while the trained NN decided an appropriate control action, with good results. The development of a spatial decision support system (SDSS) that will permit geographic information to be integrated with meteorologic and hydrologic monitoring stations, for integrated, real-time river basin flood control in a multi-purpose multi-reservoir system, has also been reported.[222] The flood forecasting module comprised a set of NN. The models were trained to forecast inflows into upstream reservoirs, as well as providing local inflow forecasts for the downstream reservoirs and at the flood control points for each forecast hour required. Although physical models could be applied, NN were preferred, since physical models would be limited in terms of forecast lead times and NN produced superior results to standard time series analysis methods. The forecasts provided real-time flood forecasting capabilities and distributed maps of discharge that are updated in real-time. The forecasts are fed into a dynamic programming module that determines optimal strategies of operation for the flood gates. The system was applied to the Han River Basin in Korea and vindicated on a large flood event for 1995: the results included a reduction in the impact of downstream floods, while at the same time maintaining sufficient supplies of water for other purposes. The final example in this list is the RAISON (Regional Analysis by Intelligent Systems ON microcomputers) DSS, which was developed by the National Water Research Institute, Environment Canada, and applied to the problem of toxic chemicals in the Great Lakes.[223] The main use of NN in this particular application is to estimate missing input records, something that is particularly useful in the case of meteorologic values required for the air trajectory model, although such tools are also used to perform various classification procedures. Neural network estimation of missing records for assorted purposes are described in other papers.[224-227]

6.5.4.3. *Neuroemulation Products*

Neural networks can be used to emulate an existing model. The overall goal of an emulator is to imitate the behavior of a model, and this can be done to varying degrees. However, unlike a simulator, an emulator does not attempt to precisely replicate the internal state or states of the device that is being emulated; rather, it only attempts to reproduce its external behavior. Full emulation would mean that the emulator used identical inputs and outputs; partial emulators can also be developed that operate on sub-sets of the original inputs. The fundamental principle is that the neuroemulator must reproduce some aspect of the original hydrologic model. Neuroemulators can be used in experimental proof-of-concept testing, or to provide plug-ins, that offer more efficient or effective computational processing. Emulation can be applied to an existing model, for model reduction purposes, or for rapid prototyping, sensitivity analysis, and bootstrapping operations. Hydrologic modeling can be categorized according to five different types of neuroemulator:

(i) *For proof-of-concept testing*: Neural network emulators, in these studies, are used to illustrate the practical application of a particular technique. This field of investigation started in the pioneering period. For example, modeling synthetic spatial and temporal rainfall output patterns produced using a mathematical simulation that conceptualized rainfall as a set of clustered cells,[33] or modeling predicted hydrograph responses for a series of synthetic storm events produced using a well-established conceptual model.[37] The latter case involved getting emulators to reproduce outputs that had been obtained from mathematical modeling procedures, but used different input drivers to the original model. The intention in such experiments is to explore the potential predictive capabilities of the tool under test in a controlled and simplified modeling environment.

(ii) *For structural exploration of models*: Neural network emulators can be used to compare and contrast internal structures or mechanisms, by analogy, through the process of attempting to match their inner setup against the internal components of the original model. For example, their internal components could be matched against those of their mathematical counterparts, perhaps leading to a better overall understanding of the internal dynamics inside both individual models. Studies in this area have concentrated on discovering internal structures that could be explained in terms of physical processes. For example, modeling predicted discharge from a conceptual model of the Test River Basin in Hampshire, England, and then examining the behavior of the hidden neurons in relation to the physical model inputs.[14] The testing on that occasion showed that different hidden neurons appeared to be associated with physical processes, i.e. baseflow, quickflow, and seasonal variations in the soil moisture deficit. Further modeling and testing explorations have also been conducted on predicted discharge for different rainfall events and catchment conditions produced using the Xinanjiang Rainfall-Runoff Model.[228-231] Full and partial emulations were developed using identical inputs and two different outputs. Use of different input combinations enabled the competencies of neural solutions developed on a reduced number of parameters to be assessed.

(iii) *To support more efficient and effective model development*: Neural network emulators can be used in support of model development, improved parameter estimation, and model testing procedures. These emulators can be orders of magnitude faster to run and offer a greater degree of independence from structures and methods related to the original model. Neural network emulators, for example, were used to model predicted runoff volumes for recorded single-burst storm events in Singapore produced using a dynamic rainfall-runoff simulation.[36] The motivation behind such a study was to develop a response surface that related calibration parameters to model output. Using neuroemulation, the simulation period for response surface development was reduced from seven hours to one minute or less. The emulator also produced lower output errors on test datasets. Neural network emulators have also been used to model estimated suspended sediment concentrations for recorded typhoon events in Taiwan.[232] In this case, the less complicated surrogate demanded fewer, and easier to acquire, input datasets than the original hydrologic model.

(iv) *To support more efficient and effective model coupling*: Model coupling can, in practice, be difficult for a number of reasons, e.g. formats, compatibilities, and the ability to modify source code. Neural network emulators can, instead, be developed to replace an existing set of models that feed into one another, in an organized sequence of events, for the sole purpose of enabling much simpler model coupling. This process is illustrated in the reported development of individual NN replacements for four separate physical models: a rainfall-runoff model, a hydraulic model, a salt intrusion model, and a model to calculate visibility.[233] The outputs predicted by the NN rainfall-runoff model were fed into the NN hydraulic model and, so on, cascading through each of the four models in a serial manner, providing what amounts to a coupled 'modular solution.' Further discussion on such matters has already been made in earlier sections. Three of the four emulator models performed well, illustrating that both neuroemulation and neuro-coupling can deliver practical combined solutions, although one model required improvements before it could be used in this context.

(v) *To deliver faster operational processing*: Neural network emulation of existing models can offer major improvements in processing speed. The need for faster processing is a critical factor in operational situations, but something that is seldom recognized as being of great importance in the hydrologic literature. For example, in reservoir optimization procedures, a neural plug-in has been used as a direct replacement for the slower-running MIKE-11 model.[234] Other studies provide more recent illustrations of the power of such tools to emulate a complex hydraulic simulation model (EPANET).[235-238] Neural network emulators were used in such experiments to capture the complex domain knowledge base of EPANET under different trial conditions. Speed-ups of 10 to 25 times were achieved depending on the test concerned.

Neuroemulation, to conclude, shows great promise, but the number of scoping studies remains limited. More investigations are required and stronger attention must be paid to post-model testing for robustness and sensitivities. The irregular spread of reported testing across the different groupings is also a matter of potential concern that will, at some point, need to be addressed — so as to avoid pigeonholing such opportunities into a single topic of research and development — since that might hamper the production of more novel applications and diminish the intuitive appeal of neuroemulation as a rewarding set of activities.

6.6. Taking Positive Steps to Deliver the Goods

It is clear that insufficient attention has been paid to resolving the problems and issues that were raised in the major reviews of 2000–01. The field is now well established, and it is generally accepted that NN can be applied effectively to modeling river discharge and its related parameters; however, researchers should not be complacent. There are a number of difficult questions that still need to be answered — but who will answer them? Is it the job of *Innovators* — technical enthusiasts who are few in number? Is it a job for the *Early Adopters* — pragmatic visionaries who have not yet arrived on the scene in large numbers? Maier and Dandy[61] commented that this was a very active research area, with fresh developments in network architectures and optimization algorithms being proposed on an ongoing basis. That was eight years ago, and the field of data-driven modeling has continued to expand, providing more novel and exciting tools, that offer untested software platforms for interested researchers to conduct their experiments on, for example, support vector machines[239-249] and gene expression programming.[250-254] Fresh opportunities will have impacts in terms of push and pull factors. New-fangled tools might offer more powerful or more informative solutions and will be a major attraction for *Innovators*. Technological enthusiasts will always be pulled towards the latest gadgets and gizmos that might enable them to pursue fresh fields of enquiry, and pushed away from unaccepted NN — the latter perhaps being viewed as being something of an 'outdated concept.' Technologists, likewise, would not be motivated to answer specific domain level criticisms that are levied at what, in their minds, are legacy products from the last decade. That job would be left to the *Early Adopters*: however, as mentioned earlier, the outstanding issue of needing to pursue physical interpretation is still a major barrier that is handicapping the technological transition of neurocomputing solutions from *Innovators* to *Early Adopters*. The fast-changing nature of technological innovation has, thus, created a dilemma, in which *Innovators* are moving on, but the technology is struggling to be accepted, since, having encountered substantial scientific opposition, the number of potential champions has been reduced to a small group of enthusiasts as opposed to a much larger group of *Early Adopters*. Moreover, in recent times, it has generally become the accepted norm that contributions from this field must do more than merely apply a standard BPNN to yet another catchment dataset. It is well-recognized that NN can provide accurate forecasts of river flow, provided sufficient data are available for training. Contributions to the field must now go much further: to ask new questions, or address important questions that remain unanswered. The quest for hard science has begun in earnest, but is still handicapped in terms of available resources and interested staff.

Expressed in terms of military scenarios, three existing fronts can be identified in an ongoing campaign to achieve a successful crossing of the divide that exists between *Innovators* and *Early Adopters*. To help move things forward, in the short term, it is suggested that neurohydrologists must operate in a more coordinated manner and develop a shared 'research agenda.' Medium-term goals should be to develop methods for discovering superior solutions, or focus on interpreting the resultant models, for the purpose of extracting hydrologic rules or gaining hydrologic insights, i.e. 'opening the box.' The other issue that was identified in the millennium reviews, but not acknowledged as a major factor, is the need for some measure of statistical confidence or uncertainty to be supplied, similar to what one might expect to be provided with statistical regression analysis. This shortcoming is often raised during audience questions at conferences. It is, as a result, highlighted in this chapter as being of substantial interest to 'rank and file' hydrologists. It presents a difficult challenge at this point and one that will not deliver a major shift in forward momentum; it is, as such, viewed as a long(er)-term objective. Each front is discussed next, in greater detail.

6.6.1. *Building a Collective Intelligence*

Koutsoyiannis[16] has raised a number of issues about the use and abuse of connectionist models in hydrologic modeling, including the need to report positive and negative findings. The main obstacle to wider acceptance and an improved standing for NN solutions across different hydrologic modeling communities could be related to a perceived notion of 'stagnation,' reflecting what appears in numerous publications to be a continual reworking of past reported applications and, in general, a substantial lack of overall progress. For example, in most papers, traditional inputs are used to produce traditional outputs, and no substantial innovations are attempted. There is, as a result, scant evidence and little or no rigorous testing of the purported advantages and acclaimed benefits of neural solutions in the hydrologic sciences: how well, for instance, do such tools handle noisy and/or error-prone datasets? Instead, numerous papers have focused on reporting simplistic curve-fitting exercises; different models developed on different algorithms, or different architectural configurations, are fitted to various datasets, and an assortment of metrics is used to select some sort of 'winner.' The scientific meaning or hydrologic merit of such differences is seldom discussed and little methodological progress is achieved.

Typical investigations constitute 'local case studies' based on one or perhaps two catchments, which will seldom, if ever, produce universal insight(s). This makes it difficult to compare and contrast reported findings from different investigations: the ability to establish the extent to which the published result is due to the method that was applied or to the dataset that was used is likewise lost. Most published investigations are, instead, designed to unveil the potential merit(s) of some specific tool or method, and so no common set of rules or guidelines is produced. For a general discussion on the ever-increasing trends of: (i) more and more complex hydrologic models; and (ii) comparison of different methods, see Sivakumar.[255] Further, most investigations are not repeated, no confirmation studies are performed on identical datasets, and the reported methodologies are seldom revisited or applied to a different set of catchments or datasets in a purposeful and constructive manner. For an exception to this rule, the interested reader is referred to the discussions of Aksoy et al.[256] and Kisi.[257]

To address substantive questions, and to help move matters forward, connectionist enthusiasts must return to 'first principles.' Reporting on positive and negative aspects will require hydrologic standards to be established using benchmark studies and datasets, e.g. four standard datasets are provided for the purpose of error metric evaluation on HydroTest (www.hydrotest.org.uk).[96,97] Effective comparison studies should be established and confirmation tests performed. The basic expectation is one of transparency: comprehensive documentation must be provided, together with detailed methodologies and datasets, for further scrutiny and experimental replication, i.e. best scientific practice!

The need for rigorous inter-comparison has been debated at meetings of the European Geosciences Union (formerly, European Geophysical Union). The scientific motivations that support such activities can be found in a series of past operational hydrologic forecasting inter-comparison projects conducted by the World Meteorological Organization that ended in 1974, 1984, and 1990.[258-260] It should also be noted that, while the timing of the reported international exercises coincided with the arrival of powerful desktop computers, it nevertheless pre-dated our first identified application of neurohydrologic modeling. It is important to stress that the object of such exercises is not to find the single best model that performs well under all circumstances. It is, instead, intended to give potential users a set of hard facts against which to compare the performances of different methodologies under various conditions. It also provides model developers with the opportunity to compare their latest offerings with earlier approaches. One option might be to establish a 'community hub' for sharing datasets, tools, and information in much the same way that the paleo-environmental research community now disseminate data from local reconstructions. The National Climate Data Center (NCDC) for Paleoclimatology, for example, maintains a useful archive of such material at: http://www.ncdc.noaa.gov/paleo/data.html.

Resolving the requirement for positive and negative appraisals will require an initial series of substantial benchmarking exercises, as well as case studies that can be extended and revisited. It is argued that, to support such activities, neurohydrologists will require several different items that do not at present exist. To provide a controlled environment necessitates a set of standardized and agreed: (i) problems; (ii) open-access datasets; (iii) pre-processing procedures; (iv) testing and reporting procedures; and (v) repositories of model outputs that will permit subsequent comparisons to be performed. To achieve this, consensus and commitment is required to enable the neurohydrologic community to work together. This initiative might be launched through specific meetings or dedicated sessions on this subject at core international conferences; or perhaps via targeted funding to support scientific networks or information hubs, of the type that was described in the preceding paragraph(s).

To date, four NN cooperative hydrologic modeling studies have been undertaken by enthusiasts in order to explore the range of different methods and procedures that are used and, perhaps more importantly, to help draw the international research community together. The first reported study was conducted in 2002.[261] Fourteen participants from around the world participated: a variety of different NN models were used to predict daily flow (measured as stage in cm) for a small, unidentified UK catchment of 66.2 ha at $T + 1$ and $T + 3$ days ahead. The software that was used to develop the models included off-the-shelf packages and custom-built, in-house tools that had been programmed by the participants. This study revealed limited overall differences in the performance of different models, perhaps because of the relative simplicity of the dataset that was used in the inter-comparison. The second study was reported in 2005.[262] Twelve participants were asked to model river stage in a much larger unidentified UK catchment of 331,500 ha using 15-min datasets. Each participant produced two NN models that predicted stage at a lead time of 24 hrs; one model was allowed to use upstream records as input, the other was not. The results were more varied; some models produced high Nash-Sutcliffe Index scores of well over 90%, but with others scoring below 80%.[263]

Two subsequent, but unpublished, cooperative studies were presented at the *European Geosciences Union Annual General Assembly* in 2007 and 2008. The first unpublished study involved six participants modeling 6-hr datasets for the Bird Creek catchment (USA). Models were developed for $T + 6$ and $T + 24$ hrs ahead. In this inter-comparison — due to marked linearities in the dataset — a simple multiple linear regression model outperformed the more sophisticated NN.[264,265] The second unpublished study involved five participants modeling six-hr datasets for the much larger River Ouse catchment (UK). Models were developed for $T + 1$ step ahead (6 hrs) and $T + 4$ steps ahead (24 hrs). In this case, the best model for predicting $T + 1$ step ahead records was found to be an MLP; but, surprisingly, multiple linear regression was best at predicting $T + 4$ step ahead records.[266,267]

6.6.2. *Deciphering of Internal Components*

Neural networks offer potential advantages in terms of resolving difficult computational problems, but the hydrologic modeling on offer is often overshadowed by negative impressions and demonization attributed to the tools in question being perceived as the 'ultimate black-box model.' The millennium reviews were united on the need to offer some sort of 'physical interpretation:' the eleven-member Task Committee of the American Society of Civil Engineers[59,60] considered this to be a major issue; Maier and Dandy[61] suggested, in their concluding paragraph, that future effort should be directed towards the extraction of knowledge; Dawson and Wilby[25] acknowledged the scope for investigating the meaning of internal components. This issue continues to hinder the wider appreciation and acceptance of NN as a meaningful alternative method for use in the numerous different sub-disciplines of hydrologic modeling. It is possible that such barriers might be conquered (or at least, to some extent, mitigated) if neurohydrologists could put more effort into interrogating the mechanisms that are produced with regard to their internal components and/or captured physical processes. Early work in different subject domains has focused on rule extraction as a method for understanding the behavior of a NN,[268-270] but sound hydrologic modeling examples can also be found.[271-273] Others have sought to obtain an improved understanding of NN hydrologic models through the application of investigative procedures, such as sensitivity analysis[274,275] and saliency analysis[276] or, more recently, through the explicit introduction of physical quantities and the use of a constrained network architecture to understand a model that predicted evapotranspiration.[277]

Several studies have tried to inspect the numerical behavior of the hidden processing units as opposed to the overall model or its individual connection weights, recognizing the fact that micro-, meso- and macro-level mechanisms might possess different properties of interest. For example, analytical procedures have been applied to NN models that were built to simulate the daily discharge outputs from a conceptual model calibrated for the Test River Basin in Hampshire, UK.[14] Hidden unit contributions to overall model external output were plotted against the components of the conceptual model. The use of antecedent precipitation and evaporation inputs resulted in two of the hidden units appearing to capture baseflow and quickflow components. The third hidden unit appeared to map important seasonal variations in soil moisture deficit, and exhibited clear threshold behavior.

The direct output of internal hidden units has also been examined on several occasions. For example, a NN was used to model daily flow in the Kentucky River watershed.[278] Model inputs comprised past flow and past and present precipitation records. The model contained four hidden units, which were then correlated against (i) the model inputs; and (ii) five simulated outputs from a conceptual rainfall-runoff model. These included total computed flow, baseflow, surface flow, soil moisture content, and infiltration. Two of the hidden units were found to be strongly correlated with past river discharge, baseflow, and soil moisture, and it was suggested that these units were clearly capturing a baseflow component. Findings for the other hidden units indicated that such mechanisms might be capturing quickflow and rainfall infiltration processes. Likewise, an autoregressive NN model was developed for the Narmada River watershed in India.[279] It modeled hourly discharge at the Mandala gaging station. In this investigation, the hidden unit outputs were ranked by discharge and plotted. The results showed units organized by discharge: one for higher magnitudes; one for lower magnitudes; and one in the central range, suggesting that the NN might be dividing the output into sub-sets of the predicted discharge. Further studies have been performed on a NN model that was developed for the River Ouse in North Yorkshire, UK.[280] The results from different methods of analysis applied to both internal and external outputs depicted a comparable internal organization of hidden units into baseflow, surface flow, and quickflow components.

These few studies mark the start of exciting opportunities that might help to expand the present boundaries of hydrologic representation. It has the potential to open up the black-box to practising hydrologists and could, ultimately, lead to the development of superior NN models.

6.6.3. *Estimating Confidence and Uncertainty*

Maier and Dandy[61] suggested, in their concluding paragraph, that future effort should be directed towards dealing with model and parameter uncertainties. The inherent black-box nature of NN models means that confidence in model performance is compromised, since such tools cannot explain their reasoning. In order to overcome this issue, a number of studies have examined the uncertainties involved. Two potential remedies have been proposed. The first suggestion is that instead of discarding the training dataset after a model had been calibrated, the records involved should be retained as an integral part of the model.[103] Thus, during its prediction phase, the uncertainty for each output prediction could be judged in real time by measuring the Euclidian distance between the input vector at prediction and the training dataset. The second suggestion is that Monte Carlo simulation should be used to investigate uncertainty with regard to the performance of NN models.[281]

In general, the specification of predictive uncertainty by developing confidence intervals and probability distributions for the model outputs provides vital information for risk-based decisions. For example, it enables "an authority to set risk-based criteria for flood watches, flood warnings, and emergency response; and enable the forecaster to issue watches and warnings with explicitly stated detection probabilities."[282] In spite of the practical importance of uncertainly analysis, the assessment of the predictive uncertainty associated with a hydrologic NN model remains largely an unexplored topic of research. The limited uptake of investigations in this area is not unexpected, however, since: (i) researchers still need more time in which to investigate the suitability of NN hydrologic models; and (ii) the complexity of the product, together with the closed source structure of most commercial software packages, does not support the user in modifying the code to perform uncertainty analysis.

Similar to traditional models, NN predictive uncertainty can be attributed to a number of different factors, such as model structure, poor estimation of model parameters, and errors in the input dataset. The analysis of NN parametric uncertainty would nevertheless be of great interest to hydrologic modelers, since it is often commented that neural solutions might suffer from 'over-parameterization' — potential failings in this respect being regarded by some as a major obstacle in the use of NN in hydrology.[22] The number of parameters in such models can indeed range from two in the case of NN sediment modeling[156] to the often ill-cited example of a Time Delay Neural Network (TDNN) that used 1105 parameters to fit a dataset that contained only 1000 training points — sufficient as the author pointed out to probably give a statistician a heart attack![283] The reported rationalization for having such a huge number of parameters in the latter case, however, is three-fold: first, the network was forced to learn a statistical outlier; second, degrees of freedom in that system did not correspond to the number of free parameters in a direct manner; and last, but not least, it was developed in a competitive environment – the sole purpose of that model being to deliver the highest possible accuracies in the *Santa Fe Institute Time Series Prediction Competition*. Modelers are also permitted to implement pruning algorithms as part of the training process that will deliver a more parsimonious final product.[178,284]

Assessment of NN uncertainties can be conducted using probabilistic approaches, such as mean-value first order analysis, Bayesian methods, and Monte Carlo simulation. For example, a Bayesian training method was used in the analysis of parametric uncertainty for a NN model of salinity forecasting, and it was concluded that accounting for NN uncertainty is very important.[285] Similarly, a NN trained using the Bayesian method was found to slightly outperform a standard NN, but with the added advantage of being able to provide confidence limits.[286]

Monte Carlo simulation is another mechanism that has been used to determine the uncertainties of a NN river flow forecasting model, allowing parameters to vary between three different increasing ranges.[281] The simulations were found to be very sensitive to the ranges involved and a large number of unrealistic simulations were produced. It was concluded that the use of more advanced versions of the Monte Carlo method, such as the Metropolis Algorithm, might overcome the problem of having a large number of unrealistic simulations. It is also possible to adopt a probabilistic bootstrap technique for the purpose of quantifying predictive uncertainty in a NN river flow forecasting model.[287] The application of possibilistic fuzzy-based approaches could also prove beneficial in providing meaningful uncertainty evaluations of NN hydrologic models.[288] This approach is more general and more flexible, than probabilistic approaches, since it can handle the uncertainties arising from imprecise and/or (especially) incomplete knowledge about a phenomenon of interest.

6.7. Final Thoughts: Searching for a 'Killer App'

This chapter has provided a synopsis of neural network hydrologic modeling. It has described the path that has been followed, evolving as it did from an initial set of tentative research explorations in the early 1990s to date. Several ongoing directions of research have also been discussed. Enthusiasts within the field should, nevertheless, perhaps adopt some bigger ideas for the strategic development of neurohydrology and aim to implement their preferred products as operational devices within the next few years.

Neural networks, despite the depth and breadth of research in this field, have still to be integrated into real-time hydrologic forecasting systems. It is suspected that such models have yet to be incorporated within live runoff and flood prediction systems, because uncertainties surrounding their predictions are not quantified, and due to the general lack of support for 'physical interpretation.' These are both issues of trust: in the tools that are built and in the scientists or organizations that promote or distribute such mechanisms! To put matters right is an ongoing challenge: such topics represent exciting opportunities in terms of fresh developments for the field, and a handful of researchers have started to explore some of the issues involved, but is that sufficient to propel matters forward in a timely manner?

Moore[13] explains what is needed to drive the start-up process and thereafter bridge the gap that exists between *Innovators/Technology Enthusiasts* and *Early Adopters/Visionaries*: a "killer app" — the emergence of a unique or compelling application that would drive the acceptance of one particular product over other, more established alternative solutions. It is important to stress that *Visionaries* in the adoption model are not looking for an 'improvement' — they are looking for a 'scientific breakthrough.' Failure to discover, or at least failing to articulate, a compelling application that will provide a step-change in benefits is the route to disappointment. It will also tend to drive them on to other things. Interested parties will purchase the software to test it out, but that product will never get incorporated into mainstream activities, since the rewards on offer will never quite measure up to the amount of effort that is required. The solution in such cases is to get very practical: to step down from the lofty theoretical pedestal that has been established and move away from the commercial hype that surrounds it. Neurohydrologists, to be successful, must not be sidetracked into re-working or re-visiting mature problems, such as struggling to sort out unanswerable hydrologic questions that have been around for several decades, or 'bogged down' in the minutiae of settling complex computational issues, that to hydrologists might seem unimportant, such as the number of internal processing units that are required to deliver a marginal increase in predictive power for some particular modeling scenario and/or dataset. The aim must be to focus on one particular compelling application and remove every obstacle to getting that application developed and accepted. Selecting a killer app is, however, very difficult, but the fundamental point is that such applications must be designed to exploit and deliver the stated advantages and rewards of neurocomputing. The maxim to be adopted in such cases is perhaps best expressed in the popular quote of Benjamin Franklin (1706-1790): "Hide not your talents. They for use were made. What's a sundial in the shade" (http://www.dictionary-quotes.com). The application selection process, in this case, is thus driven by an overriding desire to discover a meaningful hydrologic challenge that demands a nonlinear, supervised learning, pattern recognition, adaptive response, fault tolerant, parallel processing modeling environment. We conclude by outlining two potential lines of inquiry related to global warming and climate change in which improved modeling will have clear implications for hydrologic scientists:

Exploitation of computational efficiency: Neural networks could be used to model massively complex nonlinear systems. One candidate challenge might be to use a NN to emulate the behavior of a Regional Climate Model (RCM). The inputs to the NN would be a common suite of coarse-resolution outputs obtained from a General Circulation Model (GCM) — information on atmospheric conditions at the boundaries of the RCM domain, i.e. zonal and meridional wind speeds, moisture, etc. The NN outputs would be sets of RCM output computed across thousands of nodes and multiple levels in the atmosphere. The resulting emulator would enable regional downscaling in a fraction of the time, paving the way for more thorough explorations of relevant boundary forcing uncertainty. Alternatively, the emulator could be wrapped within a user-friendly interface, enabling more widespread use of the downscaled scenarios by decision-makers. The approach is already being trialed using linear statistical methods in the forthcoming UK Climate Projections (UKCP09).

Exploitation of pattern-recognition capability: Neural networks could be used for environmental change detection and attribution, to identify variability and trends in the climate system, and to ascribe these changes to specific causative factors, whether natural or man-induced. Concerns about long-term hydrologic change (for example, driven by land use or climate change) prompt questions, such as *when* and *where* do/might significant trends first emerge? However, as with conventional approaches, the problem is partly due to a lack of information on driving variables and partly due to the power of techniques to discern changes (at a given level of confidence) that lie outside the 'noise' of natural variability. The input data might be spatial (as in land cover variations) or temporal (as in heavy rainfall driving flood risk change at a critical location). In either case, subtle changes in NN weights or structures could help to disentangle the (time- or spatially-varying) significance of different drivers in situations where their interactions or transfer functions are poorly resolved by conventional methods. This potential application might represent a further use for the emerging Bayesian frameworks discussed in Sec. 6.6.2.

References

1. W. James, *Principles of Psychology* (Henry Holt, New York, 1890).
2. W. S. McCulloch and W. Pitts, *Bull. Math. Biol.*, 115 (1943).
3. A. M. Turing, *Intelligent Machinery* (unpublished report for National Physical Laboratory, 1948), subsequently published in *Machine Intelligence 5*, Eds. B. Meltzer and D. Michie (Edinburgh University Press, Edinburgh, 1969), p. 3.
4. F. Rosenblatt, *The Perceptron: A Perceiving and Recognizing Automaton* (Cornell Aeronautical Laboratory, 1957), Technical Report 85-460-1.
5. M. Minsky and S. Papert, *Perceptrons, An Introduction to Computational Geometry* (MIT Press, Cambridge, MA, 1969).
6. D. E. Rumelhart, G. E. Hinton and R. J. Williams, in *Parallel Distributed Processing: Explorations in the Microstructures of Cognition Vol. 1*, Eds. D. E. Rumelhart and J. L. McClelland (MIT Press, Cambridge, MA, 1986), p. 318.
7. S-I. Amari, *IEEE Trans. Electron. Comput.*, 299 (1967).
8. S-I. Amari, *Neurocomput.*, 185 (1993).
9. P. J. Werbos, *Beyond Regression: New Tools for Prediction and Analysis in the Behavioral Sciences* (Ph.D. Thesis, Harvard University, Cambridge, MA, 1974).
10. D. Parker, *Learning-logic* (Office of Technology Licensing, Stanford University, Stanford, CA, 1982).
11. S. Grossberg, *Letter to the Editor* (1994):
    http://www.cns.bu.edu/Profiles/Grossberg/Gro1994KohonenLetter.pdf
12. R. J. Abrahart, in *Proceedings of the First International Conference on GeoComputation* (University of Leeds, Leeds, 1996), p. 1.
13. G. A. Moore, *Crossing the Chasm: Marketing and Selling High-Tech Products to Mainstream Customers* (HarperCollins Publishers, New York, 1991).
14. R. L Wilby, R. J. Abrahart and C. W. Dawson, *Hydrol. Sci. J.*, 163 (2003).
15. F. Pappenberger and K. J. Beven, *Water Resour. Res.*, W05302 (2006).
16. D. Koutsoyiannis, *Hydrol. Sci. J.*, 832 (2007).
17. D. Koutsoyiannis, H. Yao and A. Georgakakos, *Hydrol. Sci. J.*, 142 (2008).
18. V. Klemeš, *Hydrol. Sci. J.*, 13 (1986).
19. N. J. De Vos and T. H. M. Rientjes, *Hydrol. Earth Syst. Sci.*, 111 (2005).
20. R. J. Abrahart, A. J. Heppenstall and L. M. See, *Hydrol. Sci. J.*, 414 (2007).
21. T. Wagener and N. McIntyre, *Hydrol. Sci. J.*, 735 (2005).
22. E. Gaume and R. Gosset, *Hydrol. Earth Syst. Sci.*, 693 (2003).

23. A. Y. Shamseldin, *J. Hydrol.*, 272 (1997).
24. C. W. Dawson and R. L. Wilby, *Hydrol. Sci. J.*, 47 (1998).
25. C. W. Dawson and R. L. Wilby, *Prog. Phys. Geog.*, 80 (2001).
26. Z. Jingyia and M. J. Hall, *J. Hydrol.*, 98 (2004).
27. D. P. Solomatine and Y. Xue, *ASCE J. Hydrol. Eng.*, 491 (2004).
28. R. J. Abrahart and S. M. White, *Phys. Chem. Earth*, **B**, 19 (2001).
29. A. Y. Shamseldin, *J. Hydroinform.*, 22 (2010).
30. M. Firat and M. Güngör, *Math. Comput. Simulat.*, 87 (2007).
31. M. Firat, *Hydrol. Earth Syst. Sci.*, 123 (2008).
32. T. M. Daniell, in *Proceedings of the International Hydrology and Water Resources Symposium* (Institution of Engineers - Australia, Barton, ACT, 1991), p. 797.
33. M. N. French, W. F. Krajewski and R. R. Cuykendall, *J. Hydrol.*, 1 (1992).
34. A. H. Halff, H. M. Halff, and M. Azmoodeh, in *Engineering Hydrology*, Ed. C. Y. Kuo (American Society of Civil Engineers, New York, 1993), p. 760.
35. K. W. Kang, C. Y. Park and J. H. Kim, *Korean J. Hydrosci.*, 1 (1993).
36. S-Y. Liong and W. T. Chan, in *Proceedings of the Third International Conference on the Application of Artificial Intelligence to Civil and Structural Engineering*, Eds. B. H. V. Topping and A. I. Khan (Civil-Comp Press, Edinburgh, 1993), p. 67.
37. A. W. Minns and M. J. Hall, *Hydrol. Sci. J.*, 399 (1996).
38. K-L. Hsu, H. V. Gupta and S. Sorooshian, *Water Resour. Res.*, 2517 (1995).
39. H. R. Maier and G. C. Dandy, *Water Resour. Res.*, 1013 (1996).
40. A. F. Atiya, S. M. El-Shoura, S. I. Shaheen and M. S. El-Sherif, *IEEE Trans. Neural Networks*, 402 (1999).
41. Ö. Kisi, *Turk. J. Environ. Sci.*, 9 (2005).
42. Y-M. Chiang, L-C. Chang and F-J. Chang, *J. Hydrol.*, 297 (2004).
43. E. C. Carcano, P. Bartolini, M. Muselli and L. Piroddi, *J. Hydrol.*, 291 (2008).
44. J. F. Adamowski, *J. Hydrol.*, 247 (2008).
45. P. Coulibaly, Y. B. Dibike and F. Anctil, *J. Hydrometeor.*, 483 (2005).
46. Y. B. Dibike and P. Coulibaly, *Neural Networks*, 135 (2006).
47. M. S. Khan, P. Coulibaly and Y. Dibike, *Hydrol. Process.*, 3085 (2006).
48. S. Lingireddy and G. M. Brion, Eds. *Artificial Neural Networks in Water Supply Engineering* (American Society of Civil Engineers, Reston, VA, 2005).
49. G. B. Sahoo, C. Ray and E. H. De Carlo, *J. Hydrol.*, 525 (2006).
50. H. Elhatip and M. A. Komur, *Environ. Geol.*, 1157 (2008).
51. B. Lin, M. Syed and R. A. Falconer, *Environ. Model. Softw.*, 729 (2008).
52. N. Hamidi and N. Kayaalp, *Clean*, 380 (2008).
53. J. Amorocho and W. E. Hart, *Trans. Am. Geophys. Union*, 307 (1964).
54. M. J. Hall and A. W. Minns, in *Proceedings of the Fourth National Hydrology Symposium* (British Hydrological Society, London, 1993), p. 5.51.
55. J. Smith and R. N. Eli, *J. Water Res. Pl. Manag.*, 499 (1995).
56. F. Anctil and D. G. Tape, *J. Environ. Eng. Sci.*, S121 (2004).
57. M. Campolo, P. Andreussi and A. Soldati, *Water Resour. Res.*, 1191 (1999).

58. M. P. Rajurkar, U. C. Kothyari and U. C. Chaube, *J. Hydrol.*, 96 (2004).
59. American Society of Civil Engineers *J. Hydrol. Eng.*, 115 (2000).
60. American Society of Civil Engineers *J. Hydrol. Eng.*, 124 (2000).
61. H. R. Maier and G. C. Dandy, *Environ. Model. Softw.*, 101 (2000).
62. R. S. Govindaraju and A. Ramachandra Rao, Eds. *Artificial Neural Networks in Hydrology* (Kluwer Academic Publishers, Dordrecht, 2000).
63. R. J. Abrahart, P. E. Kneale and L. M. See, Eds. *Neural Networks for Hydrological Modelling* (A. A. Balkema Publishers, Rotterdam, 2004).
64. C. Chatfield, *Int. J. Forecasting*, 1 (1993).
65. W. C. Carpenter and J. F. Barthelemy, *J. Comput. Civil. Eng.*, 345 (1994).
66. T. Hill, L. Marquez, M. O'Connor and W. Remus, *Int. J. Forecasting*, 5 (1994).
67. A. K. Jain and J. Mao, *IEEE Trans. Neural Networks*, 1 (1997).
68. K-L. Hsu, X. Gao, S. Sorooshian and H. V. Gupta, *J. Appl. Meteorol.*, 1176 (1997).
69. G. G. Towell, *Symbolic Knowledge and Neural Networks: Insertion, Refinement, and Extraction* (Ph.D. Thesis, University of Wisconsin, Madison, WI, 1991).
70. G. G. Towell and J. W. Shavlik, *Mach. Learn.*, 71 (1993).
71. G. G. Towell and J. W. Shavlik, *Artif. Intell.*, 119 (1994).
72. G. J. Bowden, H. R. Maier and G. C. Dandy, *Water Resour. Res.*, 1010 (2002).
73. R. G. Kamp and H. H .G. Savenije, *Hydrol. Earth Syst. Sci.*, 603 (2006).
74. P. Leahy, G. Kiely and G. Cocoran, *J. Hydrol.*, 192 (2008).
75. K. C. Luk, J. E. Ball and A. Sharma, *J. Hydrol.*, 56 (2000).
76. R. J. Abrahart and L. M. See, *Hydrol. Process.*, 2157 (2000).
77. G. J. Bowden, G. C. Dandy and H. R. Maier, *J. Hydrol.*, 75 (2005).
78. G. J. Bowden, H. R. Maier and G. C. Dandy, *J. Hydrol.*, 93 (2005).
79. R. J. May, G. C. Dandy, H. R. Maier and J. B. Nixon, *Environ. Model. Softw.*, 1289 (2008).
80. G. F. Lin and G. R. Chen, *Hydrol. Process.* 2524 (2008).
81. F. Anctil, C. Michel, C. Perrin and V. Andreassian, *J. Hydrol.* 155 (2004).
82. J. C. Ochoa-Rivera, R. García-Bartual and J. Andreu, *Hydrol. Earth Syst. Sci.* 641 (2002).
83. J. Chen and B. J. Adams, *J. Hydrol.*, 232 (2006).
84. J. Chen and B. J. Adams, *J. Hydrol. Eng.*, 408 (2006).
85. M. R. Hestenes and E. L. Stiefel, *J. Res. Natl. Bur. Stand. Sect.*, B, 409 (1952).
86. K. Levenberg, *Q. Appl. Math.*, 164 (1944).
87. D. Marquardt, *SIAM J. Appl. Math.*, 431 (1963).
88. Y. B. Dibike and P. Coulibaly, *J. Hydroinform.*, 289 (2008).
89. W. S. Sarle, Ed. *Neural Network FAQ - Part 3: Generalisation* (2002) ftp://ftp.sas.com/pub/neural/FAQ.html
90. A. W. Minns, in *Hydroinformatics'96: Proceedings of the Second International Conference on Hydroinformatics'96,* Ed. A. Müller (A. A. Balkema, Rotterdam, 1996), p. 207.

91. N. Karunanithi, W. J. Grenney, D. Whitley and K. Bovee, *J. Comput. Civil. Eng.*, 201 (1994).
92. C. E. Imrie, S. Durucan and A. Korre, *J. Hydrol.*, 138 (2000).
93. P. Varoonchotikul, M. J. Hall and A. W. Minns, in *Proceedings of the Second International Symposium on Flood Defence*, Eds. B. Wu, Z-Y. Wang, G. Wang, G. G. H Huang, H. Fang and J. Huang (Science Press New York Ltd., New York, 2002) p. 908.
94. P. Varoonchotikul, *Flood Forecasting Using Artificial Neural Networks* (A. A. Balkema, Rotterdam, 2003).
95. P. Hettiarachchi, M. J. Hall and A. W. Minns, *J. Hydroinform.*, 291 (2005).
96. C. W. Dawson, R. J. Abrahart and L. M. See, *Environ. Model. Softw.*, 1034 (2007).
97. C. W. Dawson, R. J. Abrahart and L. M. See, *Environ. Model. Softw.*, in press (2010).
98. A. M. Turing, *Mind*, 433 (1950).
99. H. Aksoy, A. Güven, A. Aytek, M. I. Yuce and N. E. Unal, *Hydrol. Process.*, 2715 (2008).
100. Ö. Kisi, *Hydrol. Sci. J.*, 1092 (2006).
101. Ö. Kisi, *Nord. Hydrol.*, 247 (2006).
102. Ö. Kisi, *Hydrol. Process.*, 1925 (2007).
103. D. Han, T. Kwong and S. Li, *Hydrol. Process.*, 223 (2007).
104. H-Y. Lee, Y-T. Lin and Y-J. Chiu, *J. Hydrol. Eng.*, 362 (2006).
105. A. Aytek, A. Guven, M. I. Yuce and H. Aksoy, *Hydrol. Sci. J.*, 893 (2008).
106. R. J. Abrahart, N. Ab Ghani and J. Swan, *Hydrol. Sci. J.*, 382 (2009).
107. P. Coulibaly, F. Anctil and B. Bobée, *J. Hydrol.*, 244 (2000).
108. Z. X. Xu and J. Y. Li, *Hydrol. Process.*, 2423 (2002).
109. I. N. Daliakopoulos, P. Coulibaly and I. K. Tsanis, *J. Hydrol.*, 229 (2005).
110. P. C. Nayak, Y. R. S. Rao and K. P. Sudheer, *Water Resour. Manag.*, 77 (2006).
111. B. Krishna, Y. R. Satyaji Rao and T. Vijaya, *Hydrol. Process.*, 1180 (2008).
112. D. Cameron, P. Kneale and L. See, *Hydrol. Process.*, 1033 (2002).
113. C. W. Dawson, R. J. Abrahart, A. Y. Shamseldin and R. L. Wilby, *J. Hydrol.*, 391 (2006).
114. K. T. Lee, W. C. Hung and C. C. Meng, *Water Resour. Manag.*, 67 (2008).
115. B. S. Thandaveswara and N. J. Sajikumar, *J. Hydrol. Eng.*, 290 (2000).
116. P. Coulibaly, F. Anctil, R. Aravena and B. Bobée, *Water Resour. Res.*, 885 (2001).
117. A. K. Mishra, V. R. Desai and V. P. Singh, *J. Hydrol. Eng.*, 626 (2007).
118. J. C. Ochoa-Rivera, *J. Hydrol.*, 174 (2008).
119. A. Makkeasorn, N. B. Chang and X. Zhou, *J. Hydrol.*, 336 (2008).
120. R. G. Pearson, T. P. Dawson, P. M. Berry and P. A. Harrison, *Ecol. Model.*, 289 (2002).
121. F-J. Chang, L-C. Chang and H-L. Huang, *Hydrol. Process.*, 2577 (2002).
122. L-C. Chang, F-J. Chang and Y-M. Chiang, *Hydrol. Process.*, 81 (2004).
123. D. N. Kumar, K. S. Raju and T. Sathish, *Water Resour. Manag.*, 143 (2004).

124. T. Pan and R. Wang, *Hydrol. Process.*, 3603 (2005).
125. E. C. Carcano, P. Bartolini and M. Muselli, in *Device Applications of Nonlinear Dynamics*, Eds. S. Baglio and A. Bulsara (Springer-Verlag, Berlin, 2006), p. 191.
126. Ö. Kisi, *Hydrol. Sci. J.*, 1025 (2004).
127. Ö. Kisi, *Hydrol. Process.*, 2449 (2008).
128. B. Sivakumar, A. W. Jayawardena and T. M. G. H. Fernando, *J. Hydrol.*, 225 (2002).
129. B. Sivakumar, M. Persson, R. Berndtsson and C. B. Uvo, *Water Resour. Res.*, 1011, 101.1029/2001WR000333 (2002).
130. Ö. Kisi and H. K. Cigizoglu, *Civ. Eng. Environ. Syst.*, 211 (2007).
131. Ö. Kisi, *Hydrol. Res.*, 27 (2008).
132. F. Anctil and N. Lauzon, *Hydrol. Earth Syst. Sci.*, 940 (2004).
133. O. Giustolisi and D. Laucelli, *Hydrol. Sci. J.*, 439 (2005).
134. L. Breiman, *Mach. Learn.*, 49 (1996).
135. Y. Freund and R. Schapire, *J. Comput. Sys. Sci.*, 119 (1997).
136. B. Pang, S. Guo, L. Xiong and C. Li, *J. Hydrol.*, 504 (2007).
137. L. Zadeh, *Inf. Control*, 338 (1965).
138. A. Bárdossy and M. Disse, *Water Resour. Res.*, 373 (1993).
139. A. Bárdossy, A. Bronstert and B. Merz, *Adv. Water Res.*, 237 (1995).
140. J-S. R. Jang, *IEEE Trans. Syst. Man. Cybern.*, 665 (1993).
141. P. C. Nayak, K. P. Sudheer and S. K. Jain, *Water Resour. Res.*, W07415 (2007).
142. J. Nie and D. A. Linkens, *Int. J. Control.*, 369 (1994).
143. F-J. Chang and Y-C. Chen, *J. Hydrol.*, 153 (2001).
144. F-J. Chang, H-F. Hu and Y-C. Chen, *Hydrol. Process.*, 219 (2001).
145. F-J. Chang, K-Y. Chang and L-C. Chang, *J. Hydrol.*, 24 (2008).
146. D. K. Gautam and K. P. Holz, *J. Hydroinform.*, 3 (2001).
147. P. C. Nayak, K. P. Sudheer, D. M. Rangan and K. S. Ramasastri, *J. Hydrol.*, 52 (2004).
148. P. C. Nayak, K. P. Sudheer, D. M. Rangan and K. S. Ramasastri, *Water Resour. Res.*, W04004 (2005).
149. K. W. Chau, C. L. Wu and Y. S. Li, *J. Hydrol. Eng.*, 485 (2005).
150. M. E. Keskin, D. Taylan and O. Terzi, *Hydrol. Sci. J.*, 588 (2006).
151. S-H. Chen, Y-H. Lin, L-C. Chang and F-J. Chang, *Hydrol. Process.*, 1525 (2006).
152. M. Aqil, I. Kita, A. Yano and S. Nishiyama, *J. Hydrol.*, 22 (2007).
153. A. El-Shafie, M. R. Taha and A. Noureldin, *Water Resour. Manag.*, 533 (2007).
154. M. Firat and M. Güngör, *Hydrol. Process.*, 2122 (2008).
155. M. Zounemat-Kermani and M. Teshnehlab, *Appl. Soft. Comput.*, 928 (2008).
156. Ö. Kisi, *Hydrol. Sci. J.*, 683 (2005).
157. T. Partal and O. Kisi, *J. Hydrol.*, 199 (2007).
158. C. Shu and T. B. M. J. Ouarda, *J. Hydrol.*, 31 (2008).
159. S. J. Mousavi, K. Ponnambalam and F. Karray, *Fuzzy Set Syst.*, 1064 (2007).
160. F-J. Chang and Y-T. Chang, *Adv. Water Res.*, 1 (2006).

161. P. Singh and M. C. Deo, *Appl. Soft. Comput.*, 968 (2007).
162. Q. J. Wang, *Water Resour. Res.*, 2467 (1991).
163. M. Franchini, *Hydrol. Sci. J.*, 21 (1996).
164. D. P. Solomatine, in *Hydroinformatics'98: Proceedings of the Third International Conference on Hydroinformatics*, Eds. V. Babovic and L. C. Larsen (A. A. Balkema, Rotterdam, 1998), p. 1021.
165. D. P. Solomatine, *J. Glob. Optim.*, 55 (1999).
166. M. Thyer, G. Kuczera and B. C. Bates, *Water Resour. Res.*, 767 (1999).
167. H. Madsen, *J. Hydrol.*, 276 (2000).
168. H. Madsen, *Adv. Water Res.*, 205 (2003).
169. J. G. Sardinas and M. D. J. Pedreira, *Ing. Hidraul. Mex.*, 55 (2003).
170. S. T. Khu and H. Madsen, *Water Resour. Res.*, W03004 (2005).
171. K. E. Harrouni, D. Ouazar, G. A. Waiters and A. H-D. Cheng, *Eng. Anal. Bound. Elem.*, 287 (1996).
172. M. M. Eusuff and K. E. Lansey, *Water Resour. Manag.*, 379 (2004).
173. M. K. Muleta and J. W. Nicklow, *J. Water Res. Pl. Manag.*, 35 (2005).
174. Z. Sen and A. Oztopal, *Hydrol. Sci. J.*, 255 (2001).
175. D. Whitley, T. Starkweather and C. Bogart, *Parallel Comput.*, 347 (1990).
176. A. Y. Shamseldin and K. M. O'Connor, *Hydrol. Earth Syst. Sci.*, 577 (2001).
177. A. Jain and S. Srinivasulu, *Water Resour. Res.*, W04302 (2004).
178. R. J. Abrahart, L. M. See and P. E. Kneale, *J. Hydroinform.*, 103 (1999).
179. J. Horn, D. E. Goldberg and K. Deb, *Evol. Comput.*, 27 (1994).
180. M. A. Potter, *The design and analysis of a computational model of cooperative coevolution* (Ph.D. Thesis, George Mason University, Fairfax, VA, 1997).
181. J. Kennedy and R. C. Eberhart, in *Proceedings of the IEEE International Conference on Neural Networks* (IEEE Service Center, Piscataway, NJ, 1995), p. 1942.
182. J. Kennedy, R. C. Eberhart and Y. Shi, *Swarm Intelligence* (Morgan Kaufmann Publishers, San Francisco, CA, 2001).
183. C. W. Dawson, L. M. See, R. J. Abrahart and A. J. Heppenstall, *Neural Networks*, 236 (2006).
184. R. J. Abrahart, L. M. See and A. J. Heppenstall, *J. Intell. Syst.*, 373 (2007).
185. A. J. Heppenstall, L. M. See, R. J. Abrahart, and C. W. Dawson, in *Practical Hydroinformatics*, Eds. R. J. Abrahart, L. M. See and D. P. Solomatine (Springer-Verlag, Berlin, 2008), p. 319.
186. K. W. Chau, *J. Hydrol.*, 363 (2006).
187. K. W. Chau, *J. Hydrol.*, 131 (2007).
188. N. J. De Vos and T. H. M. Rientjes, *Hydrol. Sci. J.*, 397 (2007).
189. N. J. De Vos and T. H. M. Rientjes, *Water Resour. Res.*, W08434 (2008).
190. S. Haykin, *Neural Networks: A Comprehensive Foundation* (Prentice Hall, Englewood Cliffs, NJ, 1991).

191. A. J. C. Sharkey, Ed. *Combining Artificial Neural Networks: Ensemble and Modular Multi-Net Systems* (Springer-Verlag, London, 1999).
192. L. M. See and S. Openshaw, *Hydrol. Sci. J.*, 763 (1999).
193. A. Jain and S. Srinivasulu, *J. Hydrol.*, 291 (2006).
194. G. Corzo and D. Solomatine, *Hydrol. Sci. J.*, 491 (2007).
195. L. M. See and S. Openshaw, *Hydrol. Sci. J.*, 523 (2000).
196. T. S. Hu, K. C. Lam and S. T. Ng, *Hydrol. Sci. J.*, 729 (2001).
197. S. Ahmad and S. P. Simonovic, *J. Hydrol.*, 236 (2005).
198. B. Zhang and R. S. Govindaraju, *Water Resour. Res.*, 753 (2000).
199. K. Parasuraman, A. Elshorbagy and S. K. Carey, *Water Resour. Res.*, W05412 (2006).
200. K.-L. Hsu, H. V. Gupta, X. Gao, S. Sorooshian and B. Imam, *Water Resour. Res.*, 1302 (2002).
201. A. Y. Shamseldin, K. M. O'Connor and G. C. Liang, *J. Hydrol.*, 203 (1997).
202. K. P. Georgakakos, D-J. Seo, H. Gupta, J. Schake and M. B. Butts, *J. Hydrol.*, 222 (2004).
203. N. K. Ajami, Q. Duan and S. Sorooshian, *J. Hydromet.*, 755 (2006).
204. Q. Duan, N. K. Ajami, X. Gao and S. Sorooshian, *Adv. Water Res.*, 1371 (2007).
205. D-I. Jeong and Y-O. Kim, *Hydrol. Process.*, 3819 (2005).
206. L. M. See and R. J. Abrahart, *Comput. Geosci.*, 987 (2001).
207. L. Xiong, A. Y. Shamseldin and K. M. O'Connor, *J. Hydrol.*, 196 (2001).
208. R. J. Abrahart and L. M. See, *Hydrol. Earth Syst. Sci.*, 655 (2002).
209. C. Shu and D. H. Burn, *Water Resour. Res.*, W09301 (2004).
210. H. F. P. Van den Boogaard and A. C. H. Kruisbrink, in *Hydroinformatics'96: Proceedings of the Second International Conference on Hydroinformatics*, Ed A. Müller (A. A. Balkema, Rotterdam, 1996), p. 471.
211. P. Nilsson, C. B. Uvo and R. Berndtsson, *J. Hydrol.*, 344 (2006).
212. L. A. Garcia and A. Shigidi, *J. Hydrol.*, 215 (2006).
213. R. Zou, W-S. Lung and J. Wu, *Water Resour. Res.*, W08427 (2007).
214. V. Babovic, R. Canizares, H. R. Jensen and A. Klinting, *J. Hydrol. Eng.*, 181 (2001).
215. L. Xiong and K. M. O'Connor, *Hydrol. Sci. J.*, 621 (2002).
216. L. Xiong, K. M. O'Connor and S. Guo, *Hydrol. Earth Syst. Sci.*, 247 (2004).
217. M. Goswami, K. M., O'Connor, K. P. Bhattarai and A. Y. Shamseldin, *Hydrol. Earth Syst. Sci.*, 394 (2005).
218. Q. J. Zhang, A. A. Cudrak, R. Shariff and S. J. Stanley, *J. Environ. Eng. Sci.*, S15 (2004).
219. A. Jain and L. E. Ormsbee, *Civ. Eng. Environ. Syst.*, 105 (2001).
220. S. Ahmad and S. P. Simonovic, *Water Resour. Manag.*, 391 (2006).
221. B. Bhattacharya, A. H. Lobbrecht and D. P. Solomatine, *J. Water Res. Pl. Manag.*, 458 (2003).
222. K. C. Shim, D. Fontane and J. W. Labadie, *J. Water Res. Pl. Manag.*, 190 (2002).

223. W. G. Booty, D. C. L. Lam, I. W. S. Wong and P. Siconolfi, *Environ. Model. Softw.*, 453 (2001).
224. M. Khalil, U. S. Panu and W. C. Lennox, *J. Hydrol.*, 153 (2001).
225. A. Elshorbagy, S. P. Simonovic and U. S. Panu, *J. Hydrol.*, 123 (2002).
226. M. J. Diamantopoulou, V. Z. Antonopoulos and D. M. Papamichail, *Water Resour. Manag.*, 649 (2007).
227. P. Coulibaly and N. D. Evora, *J. Hydrol.*, 27 (2007).
228. R. J. Abrahart and L. M. See, *Hydrol. Earth Syst. Sci. Discuss.*, 287 (2007).
229. R. J. Abrahart and L. M. See, *Hydrol. Earth Syst. Sci.*, 1563 (2007).
230. R-J. Zhao, Y-L. Zhang, L-R. Fang, X-R. Liu and Q-S. Zhang, in *Hydrological Forecasting: Proceedings of the Symposium on the Application of Recent Developments in Hydrological Forecasting to the Operation of Water Resource Systems* (International Association of Hydrological Sciences Press, Wallingford, 1980), p. 351.
231. R-J. Zhao, *J. Hydrol.*, 371 (1992).
232. C. Y. Hsu, H. Y. Lee and Y. L. Lin, in *River Basin Management II - Wessex Institute of Technology Transactions on Ecology and the Environment Vol. 60*, Ed. C. A. Brebbia (Wessex Institute of Technology Press, Southampton, 2003), p. 461.
233. R. G. Kamp and H. H. G. Savenije, *Hydrol. Earth Syst. Sci.*, 1869 (2007).
234. D. P. Solomatine and L. A. Avila Torres, in *Hydroinformatics'96: Proceedings of the Second International Conference on Hydroinformatics*, Ed. A. Müller (A. A. Balkema Publishers, Rotterdam, 1996), p. 201.
235. D. G. Jamieson, U. Shamir, F. Martinez and M. Franchini, *J. Hydroinform.*, 3 (2007).
236. Z. Rao and F. Alvarruiz, *J. Hydroinform.*, 15 (2007).
237. E. Salomons, A. Goryashko, U. Shamir, Z. Rao and S. Alvisi, *J. Hydroinform.*, 51 (2007).
238. F. Martínez, V. Hernández, J. M. Alonso, Z. Rao and S. Alvisi, *J. Hydroinform.*, 65 (2007).
239. Y. B. Dibike, S. Velickov, D. Solomatine and M. B. Abbott, *J. Comput. Civil. Eng.*, 208 (2001).
240. C. Sivapragasam, S-Y. Liong and M. F. K. Pasha, *J. Hydroinform.*, 141 (2001).
241. S-Y. Liong and C. Sivapragasam, *J. Am. Water Resour. Assoc.*, 173 (2002).
242. K. Y. Choy and C. W. Chan, *Int. J. Syst. Sci.*, 763 (2003).
243. M. Bray and D. Han, *J. Hydroinform.*, 265 (2004).
244. X. Yu, S-Y. Liong and V. Babovic, *J. Hydroinform.*, 209 (2004).
245. P-S. Yu, S-T. Chen and I-F. Chang, *J. Hydrol.*, 704 (2006).
246. C. Sivapragasam and S-Y. Liong, *Nord. Hydrol.*, 37 (2005).
247. T. Asefa, M. Kemblowski, M. McKee and A. Khalil, *J. Hydrol.*, 7 (2006).
248. J-Y. Lin, C-T. Cheng and K-W. Chau, *Hydrol. Sci. J.*, 599 (2006).
249. M. Behzad, K. Asghari, M. Eazi and M. Palhang, *Expert. Syst. Appl.*, 7624 (2009).
250. Ö. Terzi and M. E. Keskin, *J. Appl. Sci.*, 508 (2005).

251. A. Güven, A. Aytek, M. I. Yuce and H. Aksoy, *Clean*, 905 (2008).
252. A. Aytek and Ö. Kisi, *J. Hydrol.*, 288 (2008).
253. A. Aytek and M. Alp, *J. Earth Syst. Sci.*, 145 (2008).
254. A. Guven and A. Aytek, *ASCE J. Hydrol. Eng.*, 812 (2009).
255. B. Sivakumar, *Hydrol. Process.*, 4333 (2008).
256. H. Aksoy, A. Guven, A. Aytek, M. I. Yuce and N. E. Unal, *Hydrol. Sci. J.*, 825 (2007).
257. Ö. Kisi, *Hydrol. Sci. J.*, 829 (2007).
258. World Meteorological Organization, *Intercomparison of Conceptual Models used in Operational Hydrological Forecasting* (WMO, Geneva, 1975).
259. World Meteorological Organization, *Intercomparison of Models of Snowmelt Runoff* (WMO, Geneva, 1986).
260. World Meteorological Organization, *Simulated Real-time Intercomparison of Hydrological Models* (WMO, Geneva, 1992).
261. Artificial Neural Network Experiment Group, *Proceedings of the Eighth National Hydrology Symposium* (University of Birmingham, Birmingham, 2002), p. 75.
262. C. W. Dawson, L. M. See, R. J. Abrahart, R. L. Wilby, A. Y. Shamseldin, F. Anctil, A. N. Belbachir, G. Bowden, G. C. Dandy, N. Lauzon, H. Maier and G. Mason, in *Proceedings of the International Joint Conference on Neural Networks* (Institute of Electrical and Electronics Engineers, Piscataway, NJ, 2005), p. 2666.
263. J. E. Nash and J. V. Sutcliffe, *J. Hydrol.*, 282 (1970).
264. R. J. Abrahart, C. W. Dawson, D. Han, P. Coulibaly, A. Jain and A. Y. Shamseldin, *Geophys. Res. Abstr.*, 07353 (2007).
265. C. W. Dawson and R. J. Abrahart, *Geophys. Res. Abstr.*, 07301 (2007).
266. R. J. Abrahart and C. W. Dawson, *Geophys. Res. Abstr.*, 08394 (2008).
267. C. W. Dawson and R. J. Abrahart, *Geophys. Res. Abstr.*, 06602 (2008).
268. R. Andrews, J. Diederich and A. B. Tickle, *Knowl. -Based Syst.*, 373 (1995).
269. A. Lozowski, T. J. Cholewo and J. M. Zurada, in *Proceedings of the IEEE International Conference on Neural Networks* (Institute of Electrical and Electronics Engineers, Piscataway, New Jersey, 1996), Plenary, Panel and Special Sessions, p. 94.
270. J. M. Benitez, J. L. Castro and I. Requena, *IEEE Trans. Neural Networks*, 1156 (1997).
271. A. W. Minns, in *Hydroinformatics'98: Proceedings of the Third International Conference on Hydroinformatics,* Eds. V. Babovic and L. C. Larsen (A. A. Balkema, Rotterdam, 1998), p. 805.
272. A. W. Minns, *J. Hydroinform.*, 3 (2000).
273. G. B. Kingston, H. R. Maier and M. F. Lambert, in *MODSIM 2003: Proceedings of the International Congress on Modelling and Simulation,* Ed. D. A. Post (Modelling and Simulation Society of Australia and New Zealand, Canberra, 2003), p. 825.
274. H. R. Maier, G. C. Dandy and M. D. Burch, *Ecol. Model.*, 257 (1998).

275. K. P. Sudheer, *J. Hydrol. Eng.*, 264 (2005).
276. R. J. Abrahart, L. M. See and P. E. Kneale, *Comput. Geosci.*, 921 (2001).
277. A. Johannet, B. Vayssade and D. Bertin, *Proc. World Acad. Sci. Eng. Technol.*, 162 (2007).
278. A. Jain, K. P. Sudheer and S. Srinivasulu, *Hydrol. Process.*, 571 (2004).
279. K. P. Sudheer and A. Jain, *Hydrol. Process.*, 833 (2004).
280. L. M. See, A. Jain, C. W. Dawson and R. J. Abrahart, in *Practical Hydroinformatics: Computational Intelligence and Technological Developments in Water Applications*, Eds. R. J. Abrahart, L. M. See and D. P. Solomatine (Springer-Verlag, Berlin, 2008), p. 87.
281. A. Y. Shamseldin, C. W. Dawson and R. J. Abrahart, in *Hydroinformatics 2006: Proceedings of the Seventh International Conference on Hydroinformatics*, Eds. P. Gourbesville, J. Cunge, V. Guinot and S-Y. Liong (Research Publishing Services, Chennai, 2006), p. 741.
282. R. Krzysztofowicz, *J. Hydrol.*, 2 (2001).
283. E. A. Wan, in *Time Series Prediction: Forecasting the Future and Understanding the Past*, Eds. A. Weigend and N. Gershenfield (Addison-Wesley, Reading, MA, 1994), p. 195.
284. G. Corani and G. Guariso, *Neural Comput. Appl.*, 66 (2005).
285. G. B. Kingston, H. R. Maier and M. F. Lambert, *Math. Comput. Model.*, 499 (2006).
286. M. S. Khan and P. Coulibaly, *Water Resour. Res.*, W07409 (2006).
287. R. K. Srivastav, K. P. Sudheer and I. Chaubey, *Water Resour. Res.*, W10407 (2007).
288. A. P. Jacquin and A. Y. Shamseldin, *Water Resour. Res.*, W04425 (2007).

CHAPTER 7

EVOLUTIONARY COMPUTING IN HYDROLOGY

Vladan Babovic

Singapore-Delft Water Alliance, Department of Civil Engineering,
National University of Singapore, Singapore 117576
E-mail: cvebv@nus.edu.sg

Raghuraj Rao

Singapore-Delft Water Alliance, Department of Civil Engineering,
National University of Singapore, Singapore 117576
E-mail: cverr@nus.edu.sg

Many hydrologic processes are believed to be highly complex, nonlinear, time-varying, and spatially distributed. Hence, the governing mechanisms are not easily described by simple models. With unprecedented growth in instrumentation technology, recent investigations in hydrology are supported with immense quantities of data. In order to take full advantage of the information contained in such data, scientists are increasingly relying on a suite of data-driven techniques to understand the complex hydrologic processes. Evolutionary computing (EC) techniques, with a host of optimization and modeling tools, can contribute significantly to achieve the objectives of this knowledge-discovery exercise in hydrology. This chapter discusses the utility of these EC techniques in attempting data analysis and modeling problems associated with hydrologic systems. It introduces the concept and working principle of EC techniques in general and reviews their applications to different domains of hydrology. The study also illustrates different case studies of utilizing 'genetic programming' (GP) technique as a modeling, data assimilation, and model emulation tool.

7.1. Introduction

Systems and processes of scientific and engineering interest are often governed by complex behaviors (nonlinear, time-varying, multi-temporal/spatial scales) and experience chaotic disturbances and randomness. Observing these systems, understanding the processes influencing them, and establishing models that can closely resemble, represent, and reproduce the system behavior have been some of the basic objectives of scientific investigations since ages. It is essential to establish such a knowledge in the form of representative model for any design, troubleshooting, and operational activity involving these complex systems. Recent interest in solving problems related to health, safety, and environment and increasing need for designing efficient systems leading to better human comfort have further expanded the scope, and, in turn, the complexity, of research problems. It is becoming increasingly difficult to resolve issues and fully establish the underlying mechanisms of multi-dimensional, multiphysics, multiscale complex systems using the traditional theory-driven hypotheses or modeling techniques, which derive the mathematical representation of the system based on the physical/chemical/biological laws and theories in order to fit the observations made on the system. Alternately, data-driven techniques, which start from a dataset search for a set of rules/relations between the observed system variables, are becoming highly useful knowledge-discovery tools. Computer-assisted automatic modeling of systems with less-known/completely unknown mechanisms is gaining popularity, especially with the advent of information revolution in recent "peta byte age" that has enabled efficient large-scale data collection, storage, retrieval, and analysis activities. In this endeavor, the challenging question is: how to utilize and make sense out of this data overflow? Surely, such a "data to knowledge" exercise calls for a technique that can process the raw data from scratch and automatically build and improve the relationships between the variables based on the objective of the investigation.

Evolutionary principles, established in biology, provide intuitive basis for designing a suitable stable structure evolving from the data over many random variations and objective-based selection. In the 1960s, harnessing this idea of automated knowledge evolution, Lawrence Fogel[1] proposed evolutionary programming and John Holland[2] proposed genetic algorithm, which have gained increasing attention. Later, in 1992, John Koza[3] proposed another stream of similar approach called "genetic programming." They all independently initiated and explored the scope of using evolutionary principles into different modes of data-mining techniques, which are now collectively referred to as "evolutionary computing" (EC) techniques. This chapter reviews and revisits the application of data-driven evolutionary computing techniques in understanding and analyzing complex hydrologic processes.

The rest of the chapter is organized as follows. Section 7.2 introduces the basics of evolutionary computing. General evolution principle and its scope, different evolutionary computing techniques, and their general scheme of implementation are discussed. Section 7.3 provides an overview of applications of EC techniques to resolve issues relevant to different areas of hydrology; an overview of existing literature on the topic is also outlined. Section 7.4 illustrates three different applications of genetic programming (GP) as a modeling, data assimilation, and model emulator tool. Section 7.5 discusses some of the important issues pertaining to EC techniques and also highlights the precautions to be taken while applying the same to hydrologic processes. This is followed by conclusions in Sec. 7.6.

7.2. Evolutionary Computing

7.2.1. *Evolution Principle*

The principle of evolution is the primary unifying concept of biology, linking every organism together through a phylogenetic process. Every new creature in this chain of life is the product of a series of accidental "genetic events" that have been sorted out thoroughly under selective pressure from the environment. While evolution itself is merely an effect of physical laws acting on and within the populations and species, yet it is capable of "engineering" solutions to the problem of survival that are unique to each individual's circumstances. The essence of this time-tested design procedure is basically a two-step iterative sequence:

random variation (genetic mutations/distribution) followed by selection (based on the 'fittest survives' criterion). Over many generations, these repeated sequences of random variations and natural selections have shaped the behavior of individuals and species to fit the demands of their surroundings. Hence, this evolution process is the Nature's unique way of optimally designing complex, but stable, entities in the presence of internal and environmental constraints/stresses.

7.2.2. *Evolutionary Computing Techniques*

Evolutionary Computation (EC) is the general term for several computational techniques that are based, to some degree, on the principles of natural evolution. Evolutionary algorithms (genetic algorithms, genetic programming, evolutionary strategies, evolutionary programming techniques), swarm intelligence (ant colony, particle swarm methods), self-organization (self-organizing maps, growing neural gas), and differential evolution are just few example techniques that follow evolutionary computing principle. These techniques are being increasingly widely applied to a variety of problems, ranging from practical applications in industry and commerce to leading-edge scientific research.[1,4,5]

Figure 7.1 summarizes the scheme of information flow in a typical EC algorithm. Just as natural evolution started from an initial population of creatures some two billion years ago, the evolution in artificial media — 'in silico' — begins by creating an initial set (generation) of contending solutions (population in the form of mathematical equations, set of rules or sequences of numbers/patterns) for the given problem of interest, such as model design, parameter estimation, pattern recognition, and optimization. The set may be generated by randomly creating a population of initial solutions or by utilizing any available knowledge about the problem. The 'parent' solutions in a given generation then generate 'offsprings' by means of 'reproduction.' The 'crossover' operation (similar to sexual reproduction) produces a new population by exchanging 'parts' (chromosomes) from any two existing 'parent' solutions. The 'mutation' operation (asexual reproduction) builds a member for the new generation by randomly replacing a part of the individual 'parent' solution with a randomly generated new structure. The 'permutation' operation randomly switches two 'components' (genes) within the individual 'parent.' The resulting new generation 'offspring' solutions are evaluated for their effectiveness in solving the problem using a 'fitness criterion' which tests the ability to reproduce the

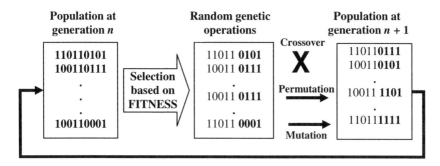

Fig. 7.1. Schematic representation of Evolutionary Computing algorithm.

known behavior (measured in terms of prediction error, pattern recognition accuracy, optimality, etc.). The population in the new generation undergoes selection based on the 'survival of the fittest' criterion and the best set of solutions that satisfy the criterion is chosen as parents for the next round of improvement (through subsequent reproduction). Typically, the evolutionary computation is run for either a pre-specified number of generations or until a desired performance is attained by the best solution.

7.3. Evolutionary Computing in Hydrology: An Overview

Different modes and stages of hydrologic systems are believed to be highly nonlinear, time-varying, and spatially distributed. Processes, such as evapo/convective transports, precipitation, drainage, and runoff, are complex in their mechanisms and governed by multiple factors. Hence, these processes are often difficult to describe fully by hydrodynamic models. The challenges of holistic analysis of these systems and limitations of traditional modeling approaches have attracted alternate techniques. With improved data acquisition and processing capabilities in the field, data-driven EC techniques have found applications for solving many different problems in hydrologic sciences.[6-8] These applications range from basic understanding of the kinetics and dynamics of the processes (for water quantity and quality) to real-time applications, such as forecasting, water collection/distribution system design-optimization-operation-control, and water resources management. Recent interest has also been on harnessing the power of EC techniques to bring the basic physical/numerical hydrodynamic models closer to reality. They have been extensively used to optimally tune the parameters of the local model (model calibration to fit the observed water quantity/quality data) and also to learn/predict the nature of model-observation mismatch (model error correction).

Table 7.1 provides an overview of these hydrologic applications and different EC techniques used to resolve the associated problems, along with relevant literature for further reference. Since the applications are numerous and the techniques are many, detailed discussion on each of these applications and techniques is near-impossible to provide. In the following sections, we elaborate on one important and widely used EC technique, called genetic programming (GP), and highlight example applications, resolving issues relevant to hydrology.

7.4. Genetic Programming and its Scope in Hydrology

Genetic programming (GP)[3] is a data-driven evolutionary computing method for automatically generating input-output relations. It differs from other black-box-type data-driven modeling techniques, such as Artificial Neural Networks (ANNs), Fuzzy Logic, and Regression Trees, in the sense that it provides mathematically meaningful structures (with optimum parameters) relating input-output variables of the system.[19,51] The basic GP algorithm follows the scheme presented in Fig. 7.1. The components and procedures used to build/operate/upgrade the model structures are detailed in Fig. 7.2. Given a set of observed data (training data) on input-output variables, GP generates a population of models with random structure without needing any prior knowledge of the mechanisms governing the process. These models are then evaluated using part of the initial data not used during training (validation data) using a suitable fitness measure. The probability of a given model surviving during the model evolution process is proportional to how well it predicts the output of the validation data. Components of successful models are continuously recombined with those of others to form new models. In each generation, GP optimizes the model structure, with a lower level nonlinear least-squares algorithm harnessed to estimate the associated model parameters.

Table 7.1. Overview of application of EC techniques in hydrologic sciences.

Area	Application	EC tools
Hydrologic process modeling	Transport processes	GP[9]
	Rainfall prediction	GA[10,11] GA/GP[12]
	Runoff prediction	GA[13] Chaos theory(+GA)[14-18] GP[19-25]
	Water quality modeling	GP[26-28] Chaos theory[29]
Hydrologic model calibration	Rainfall-runoff model parameter tuning	GA[30-35]
	Water quality model calibration	GA[36-38]
Water management	Reservoir management	GA[39-41]
	Water distribution systems	GA/GP[42]
	Water quality management	GA[43,44]
Data assimilation	Model error correction	GP[22,23,45-48]
	Data interpolation	GA[49] GP[50]

The main components of a GP tool are (as depicted schematically in Fig. 7.2): (i) the terminal set, which is a list of relevant input/output variables and constants (genes); (ii) the functional set (language table), which is a list of mathematical operators (e.g. +, - , /,*, ^, sqrt, sin, exp,

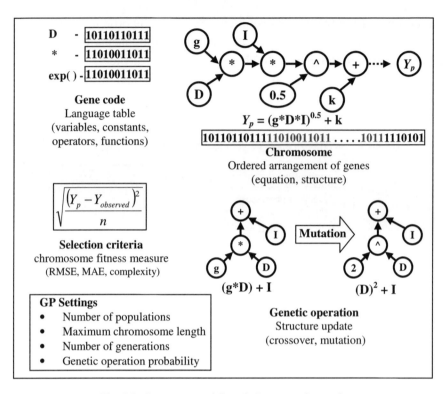

Fig. 7.2. Components of Genetic Programming tool.

log); (iii) the search space, which is a set of all possible models (chromosomes) that can be constructed using the functional and terminal sets; and (iv) the fitness function, which is a measure for evaluating the suitability of a candidate model in terms of its fitting/prediction capability. Measures, such as the root mean squared error (RMSE), are augmented by a term that penalizes complex (long or clumsy nonlinear terms) structures to form the fitness function. Relatively successful individuals are selected from the current population (in proportion to their fitness) to serve as parents for the next generation. This selection policy is called 'fitness proportional selection.' The genetic operators (crossover, mutation, and permutation) are used to obtain the new set of model structures (offsprings) that will undergo testing and selection in the next generation. Typical tunable parameters in a GP include: population size; number of generations (iterations); probabilities for crossover, mutation, and permutation; functional weightings (the probability of selecting specific operators); the fitness measure; and the selection policy.

In this chapter, we highlight the utility of this unique modeling tool to address three different classes of data mining problems in hydrology.

- GP as a modeling tool — directly learn the hydrologic observations;
- GP as a data assimilation tool — correct the numerical model error; and
- GP as a model emulator — representative numerical model

7.4.1. Modeling the Observations: GP as a Modeling Tool

As presented in Table 7.1, GP has been extensively applied to many data-driven modeling exercises in hydrology, specifically for rainfall-runoff (R-R) prediction.[24] The runoff formation process is believed to be highly nonlinear, time-varying, and spatially distributed, and not easily described by simple models. Considerable time and effort has been directed to model this process, and many hydrologic models have been built specifically for this purpose.[52] All of them, however, require significant amounts of data for their respective calibration and validation. Using physical models raises issues of collecting the appropriate data with sufficient accuracy. In most cases, it is difficult to collect all the data necessary for such a model. By using data-driven models, such as genetic programming (GP), one can attempt to model runoff on the basis of available local hydrometeorologic observations. This section demonstrates the use of GP for creating rainfall-runoff models on the basis of data alone.

Any rainfall-runoff (R-R) forecasting system is based on the past/current states of hydrometeorologic and catchment conditions as inputs (rainfall R, hydrodynamic variables, such as catchment geometry/area, bathymetry, and surface roughness) and catchment response variables (future values for runoff: discharge Q, water quality, etc.) as outputs. If we choose R and Q as the two important variables representing the local runoff process, then the 1-hr discharge forecast model, for example, can be mathematically represented as:

$$Q_f(t+1) = F[Q_{obs}(t), Q_{obs}(t-1),..., Q_{obs}(t-n), R_{obs}(t),$$
$$R_{obs}(t-1),..., R_{obs}(t-n)] \quad (7.1)$$

Here, F is the forecasting function that relates the variables Q and R, the subscripts 'f' and 'obs' refer to the forecast (future) and observed (past and current) values respectively; n (temporal memory of the variables) is generally the catchment's concentration time (in hours assuming 1-hr sampling interval for Q_{obs} and R_{obs}). There are two strategies for forecasting (e.g. k hrs) in future:

(i) Iterative forecasting, which uses 1-hr ahead forecaster k times by re-using the predicted output as the input in subsequent iterations to arrive at $Q_f(t+k)$:

$$Q_f(t+k) = F[Q_{obs}(t), Q_{obs}(t-1),..., Q_{obs}(t-n), R_{obs}(t), R_{obs}(t-1),...,$$
$$R_{obs}(t-n), Q_f(t+k-1), Q_f(t+k-2),..., Q_f(t+k-n)] \quad (7.2)$$

(ii) Direct forecasting, which directly forecasts $Q_f(t+k)$ using the present and past values of the variables:

$$Q_f(t+k) = F[Q_{obs}(t), Q_{obs}(t-1),..., Q_{obs}(t-n), R_{obs}(t),$$
$$R_{obs}(t-1),..., R_{obs}(t-n)] \quad (7.3)$$

Given the data for Q and R, the objective of any data-driven modeling exercise would be to establish the regression function F. For efficient forecasting, it is essential for F to capture the inherent process nonlinearity (representing the rainfall-runoff process' temporal/spatial dynamics). In the absence of any prior knowledge about the complexity of the system dynamics, modeling techniques, such as GP, which rely on natural evolution of structures, provide a better solution to the R-R forecaster (F) design problem.

A typical GP modeling procedure for designing F involves extraction of input-output data for selected forecast horizon k, setting the GP parameters and initial population, a training session (which runs for a prescribed number of generations) to obtain a sample F, a validation session to test (using the data that was not used during training) the forecasting efficiency of F, and re-running the training session (with different initial population and/or different GP settings) till F attains an acceptable prediction efficiency. The model F is then evaluated on a completely different test data and the performance is established using different prediction error measures, such as the root mean squared error (RMSE), mean absolute error (MAE), and correlation coefficient (CC) between the predicted and the actual Q series.[20-22,25] Since the observed Q time series shows strong autocorrelations, the GP might capture it as a linear relation between $Q_{obs}(t + k)$ and $Q_{obs}(t)$. To avoid this local optimum solution and the associated phase errors for long-term forecasting, Babovic and Keijzer[23] suggest a difference-variable approach that relates dQ with other variables. They also use forecasted future values for R as part of the input variable set.

$$dQ_f(t + k) = F\,[Q_{obs}(t),...,Q_{obs}(t - n), R_f(t + k - 1),..., R_f(t + 1),$$
$$R(t), ..., R_{obs}(t - n)] \quad (7.4)$$

Here, the model F predicts the temporal propagation of difference in discharge, $dQ_f(t + k) = Q_f(t + k) - Q_f(t + k - 1)$, rather than actual discharge quantity. Babovic and Keijzer[21] show that GP designs F with strong nonlinear function relating the present value of $dQ = Q_{obs}(t) - Q_{obs}(t - 1)$ and a square root function involving past values of Q and R.

A case study (as described in WMO[53] and used by Babovic and Keijzer[23]) is reconsidered here to illustrate the effect of few of the GP schemes described above. In brief, the system represents the R-R process on a catchment with 104 km^2 of mainly rural/forested area and drained majorly by Orgeval River. Storm events between 1972 and 1974 have been recorded and corresponding Q-R data are used as the training set. Data on six different storm events during 1979–80 are used as validation set. The problem is set up using a simple GP package[54] and the analysis is done for comparing different modeling/forecast schemes. The results are summarized in Table 7.2 and are compared with other modeling/forecasting techniques. Some of the R-R models designed by GP (referred as GPMs in the table) are shown below from Eqs. (7.5) to (7.7).

GPM1: GP model learned using only the past and present Q data.

$$Q_f(t + 1) = 0.002123 + Q_{obs}(t - 2) - 3.465 * (Q_{obs}(t - 1)$$
$$- 0.9985 * Q_{obs}(t)) \quad (7.5)$$

GPM2: GP model learned using the past and present Q and R inputs.

$$Q_f(t+1) = (-1.456 * Q_{obs}(t-1)) + (2.442 * Q_{obs}(t) \\ + 0.1015 * (R_{obs}(t-5) * R_{obs}(t-1))) \quad (7.6)$$

GPM3: GP model learned using dQ as output and Q and R as inputs.

$$dQ_f(t+1) = (Q_{obs}(t) + (-0.03 * ((36.99 * Q_{obs}(t-1) \\ + (-5.745 * Q_{obs}(t-2))) - R_{obs}(t)))) \quad (7.7)$$

Table 7.2. Performance comparison (RMSE in m³/s) of R-R forecast models. Iterative forecasting for prediction horizon $k > 1$.

Forecast horizon (k)	GPM1	GPM2	GPM3	Naïve forecast	Numerical simulation (NAM)
1 hr	0.056	0.021	0.026	0.051	0.431
2 hrs	0.079	0.034	0.025	0.135	0.430
6 hrs	0.127	0.073	0.054	0.294	0.429
12 hrs	0.521	0.118	0.092	0.538	0.430

From this simple analysis, we can observe that all the different GP models successfully capture the nonlinear dynamics of the system. Though allowed to choose upto 12 hrs of past values, GPM2 captures the term $R_{obs}(t-5)$, indicating the exact concentration time of 5 hrs. The GP models also provide better forecasts compared to the naïve (assuming a static error for the whole of forecast horizon) and numerical simulation models (NAM).[55] The worst forecast error magnitude ($k = 12$ hrs) provided by the best model GPM3 (RMSE = 0.092) is just 7% of the mean discharge recorded during the validation storm events. This is much better than the result obtained with NAM (30% relative error) and for 12-hr naïve forecast (37% relative error). These observations highlight that GP models can very well capture the dynamics of the R-R process and can thus be used as alternate modeling tools for real-time forecasting.

7.4.2. *Modeling the Model Error*: *GP as a Data Assimilation Tool*

The numerical simulation models are based on fundamental hydrodynamic principles and, hence, are the most realistic way of representing the hydrologic processes. Unlike data-driven models, which are designed and tuned for local predictions, the numerical models have the ability to provide spatial and temporal information about the process over a large set of grid points in the domain. They have an additional advantage of providing simultaneous and physically meaningful solution for water quantity and quality parameters. Nevertheless, they are also susceptible to errors, arising from incorrect parameter settings, incomplete local geometry/bathymetry, and wrong initial states. Hence, they often show deviations in their predictions from the actual observations. Data assimilation procedures[45,46] are employed to overcome this model-system mismatch and to update the model in real-time set-up. Such procedures either directly update the input variables (model forcing), model states, and model parameters or correct the error in the predicted output variables. Output error-correction procedures have been applied in hydrology for updating catchment runoff in conceptual rainfall-runoff models[22,45] and for updating discharge flows at specific locations.[16] In these cases, an error-correction forecast model is built using the observed model residual errors ($Q_{obs} - Q_{sim}$) and then superimposed on the simulation model output. Figure 7.3 gives a schematic representation of the error-correction strategy for real-time forecast systems. Here, we also demonstrate a small example of utilizing GP as a numerical model error-learning tool and its utility in updating the R-R forecasts given by the NAM model.

Error forecast model (GPE1) obtained using present/past error data:

$$e(t + \underline{1}) = (3 * e(t) + (-2.623 * e(t-1)) + 0.16008 * e(t-2)) \quad (7.8)$$

Error model (GPE2) obtained using present/past error and Q_{obs} data:

$$e(t + 1) = ((-1.602 * e(t-1)) + 2.255 * e(t)) \\ + (-0.1677 * ((e(t) + (-1.037 * e(t))) * Q_{obs}(t)) \quad (7.9)$$

The updated forecasts are obtained by correcting the simulation model (NAM) forecasts $Q_{sim}(t + 1)$ with the error forecast.

$$Q_f(t + 1) = Q_{sim}(t + 1) + e(t + 1) \quad (7.10)$$

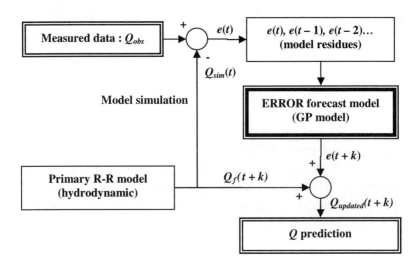

Fig. 7.3. Implementation scheme and data flow for data assimilation strategy using genetic programming based error forecast models.

Figure 7.4 provides a comparison of performances for the above analysis using GP models as data assimilation tools. It is observed, from the comparison of Q forecast for 800 hrs of simulation, that both the error forecast models can improve the prediction performance of numerical model. The GPE2 model, with terms involving both past errors and Q_{obs}, can provide an improvement of upto 80% in average error of the numerical model. The models GPE1 and GPE2 (as in Eqs. (7.8) and (7.9)) are designed using default settings of the GP tool and, hence, have further potential to improve the performance by tuning GP settings and also by incorporating past and forecasted R terms into the model structure, as suggested by Babovic and Keijzer.[23] These observations provide further evidence on the utility of GP technique as an effective data assimilation tool.

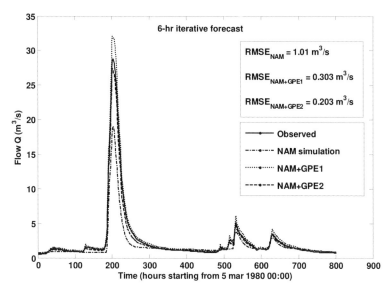

Fig. 7.4. Data assimilation using GP error forecast models. Comparison of GPE1 and GPE2 error model performance.

7.4.3. *Modeling the Model*: *GP as a Model Emulator Tool*

Many real-time water resources operational management systems rely on optimization routines which require very fast and efficient internal models to forecast water quantity and quality. This objective is difficult to achieve, as most of the real-time forecast models (especially for water quality) involve large-scale numerical simulations which take longer computing time. Hence, there is a need to develop alternate tools that are faster, and yet can repeat the performance of original models with sufficient accuracy. Model order reduction and model emulation are two of the options used to achieve this speed-up activity. The latter approach has attracted considerable attention in hydrology, mainly using data-driven techniques.[56-58] Though model emulation is not really a new topic in hydrology, applications so far have focused on relatively simpler processes, such as rainfall-runoff with a low-dimensional input and output. Emulators are generally regression-based simple transfer functions (in time or space) applied for emulating the state-of-the-art physics-based large-scale numerical models. These models are designed to analyze multivariate dynamic processes (3D flows and transport), with complex interaction of the sub-processes, several different spatial and temporal scales, nonstationary conditions, and multiple controls. Due to all these complex and multi-dimensional factors, the presently desired emulations of 3D processes (e.g. water quality) are by no means routine, and have the ingredients of a difficult, but very challenging, new and unique topic, with high practical significance. Evolutionary computing techniques can provide effective solutions to some of these issues. In this section, we envisage few strategies for implementing EC techniques for model order reduction and as model emulators. We also provide an illustrative example of one such possibility.

7.4.3.1. Model Order Reduction using GP Models as Spatial Interpolators

One possible approach in enhancing the computational swiftness of the numerical model during the real-time forecasting exercise (involving repeated data assimilation steps of prediction and correction) is to optimally reduce the number of computational points (downscaling of the numerical grid). Figure 7.5 suggests one such scheme of computational order reduction using data-driven GP models as coarse- to fine-grid predictors. The main idea is to identify and run the numerical simulation for only a fewer set of representative locations in space (lumped grid points with acceptable model accuracy) that can provide collective behavior of the region and, at the same time, are fast enough to complete the simulation within the timeframe, acceptable for real-time implementation. The desired spatial resolution of these predictions is then obtained using an ensemble of pre-trained transfer function models. Data-driven GP modeling tools can be used to design such transfer function models which can directly predict water quantity/quality at multiple fine-grid points using data from spatially nearby coarse-grid points as inputs. These models can be designed and tuned in 'offline' mode using the training data from the 'hindcast' simulation of both the coarse-grid and the fine-grid resolution models. In real-time setup, these models serve as spatial interpolators during the 'prediction' step and, in turn, significantly reduce the order of the numerical computation.

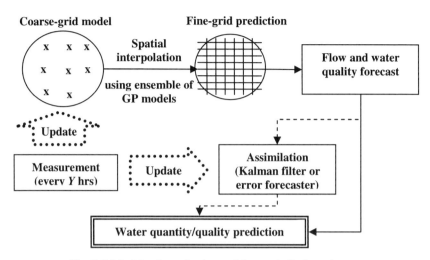

Fig. 7.5. Model order reduction and data assimilation scheme.

7.4.3.2. *GP as a Representative Model for Model Emulation*

Another approach that can enhance the computational efficiency of real-time forecasting set-up is to use model emulators which can reproduce the performance of numerical models (with acceptable accuracy) and yet take much less computational time. Offline simulation data can be used to design GP models that can closely emulate the time propagation of water quantity and quality parameters at desired locations (stations of interest) with finer prediction intervals (minutes to hourly scale) for the period during which original model updates are still running. Here, the data-driven models act as temporal interpolators and can be used to forecast entities with finer-time resolution using the numerical simulation output data available at coarser-time samples (e.g. twice a day) as initial conditions. Figure 7.6 depicts the scheme for implementing such a strategy for real-time forecasting setup. A sample model emulation example is demonstrated in Fig. 7.7.

GP is trained using the NAM simulation data (hourly time series Q_{sim}) as used in Sec. 7.4.1. One-hr Q_{sim} forecast model is built using $Q_{sim}(t)$ as input and $Q_{sim}(t + 1)$ as output variable (based on Q_{sim} data during 1972–74). The GP model emulator, with $Q_{sim}(t)$ as input, is then

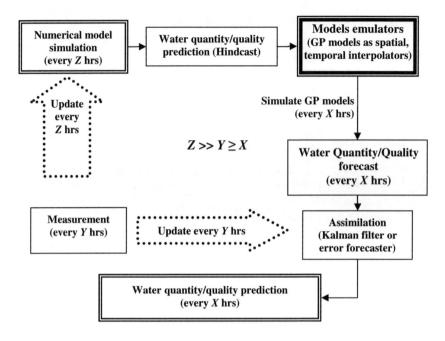

Fig. 7.6. Model emulation and data assimilation scheme.

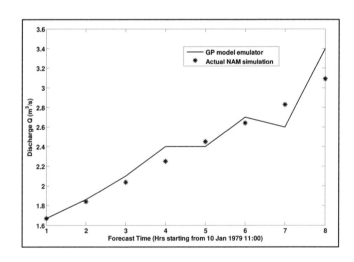

Fig 7.7. Iterative forecast using NAM model emulator.

used iteratively to forecast $Q_{sim,\ k}$ steps ahead in time (based on validation data during 1978–79). A sample forecast is presented in Fig. 7.7 at a selected time point as initial condition (10 Jan 1979 11:00). It is clearly seen that the GP model emulator can predict the actual NAM simulation quite well upto 6 hrs ahead in time. During these 6 hrs, the numerical model can be re-updated or calibrated using observed data (assimilation). At the end of 6 hrs, the new initial conditions can be taken into GP model and the next 6 hrs of forecast can be obtained. The model emulator shows similar performance for different modes of discharge profiles (increasing, decreasing, and flat). This shows the efficiency of GP model emulators in reproducing the short-term propagation of Q_{sim}. It also establishes the utility of using GP-trained model emulators as temporal interpolators for numerical models.

This type of model emulation exercises (using them as spatial and temporal interpolators) can significantly contribute to the computational swiftness and efficiency of hydrologic forecasting systems. This, in turn, bridges the gap between the computationally intense numerical models and real-time decision support systems used for optimized operation and control of water resources.

7.5. Important Issues Pertaining to Genetic Computing

In the context of hydrologic systems, the developments in EC algorithms, in particular those within the branch of genetic programming, have been shown to have good potential. It is also important, at the same time, to highlight some of the issues that might affect the success of EC techniques. Here, we briefly discuss some of the important issues.

Selection of datasets: Due to the inherent randomized nature, the EC search techniques are relatively less prone to data overfitting. Nevertheless, in order to obtain reliable and generalizable solutions using EC techniques, it is necessary to design training and validation datasets that provide and test different aspects of the hydrologic system.

EC settings: As shown in Figs. 7.1 and 7.2, the EC technique involves many parameters that govern the evolution process. The convergence of the EC algorithm to the best solution (global optimum) depends, to a great extent, on these factors. However, precaution needs to be taken while setting these EC parameters. The optimum settings vary from application to application. Hence, for a given problem, it is often necessary to attempt different re-runs using different settings and compare the solutions. Therefore, the settings that best suit the problem under consideration should be optimized using a separate cross-validation dataset during training.

Model structure: The mathematical expressions obtained using GP may approximate a broad scope of nonlinearities in the system. It can generate lengthy, complex nonlinear terms, which may not have any direct physical significance. One can introduce penalties in fitness function for such structures during the evolution or can restrict GP to generate dimensionally meaningful model structures.[51] Also, for a given problem, there might be many different model structures with the same degree of fitness, which call for domain expertise for further shortlisting.

Effect of noise: Presence of noise in the data needs special attention while applying data-driven methods, and EC techniques are no exception in this regard. Depending on the extent of noise, the GP technique needs to be tuned so as not to learn the noise as the main signal. One needs to take into account the variance and bias in the input data while modeling.[59]

7.6. Conclusions

Complete understanding of complex hydrologic processes and their governing mechanisms calls for alternate knowledge-discovery approaches. Evolutionary computing techniques provide promising solutions to many of such applications. Data-driven genetic computing techniques (GA and GP) have been applied successfully to a wide range of problems starting from modeling processes to water resources management. Genetic Programming concept has found special attention as a modeling tool. Illustrations have been shown for its utility in directly modeling the observations, as a model correction tool or as a possible replacement for numerical models. Though the EC methods are attractive, they warrant domain expertise and certain precautions before harnessing their full potential.

Acknowledgments

The authors gratefully acknowledge the support and contributions of the Singapore-Delft Water Alliance (SDWA).

References

1. D. B. Fogel, Evolutionary Computation. Toward a New Philosophy of MachineIntelligence (IEEE Press, Piscataway, NJ, 1995).
2. J. H. Holland, *Adaptation in Natural and Artificial Systems* (University of Michigan Press, Ann Arbor, 1975).
3. J. R. Koza, *Genetic Programming: On the Programming of Computers by Means of Natural Selection* (MIT Press, Cambridge, MA, 1992).
4. V. Babovic, *Emergence, Evolution, Intelligence: Hydroinformatics* (Balkema Publishers, Rotterdam, 1996).
5. A. E. Eiben and J. E. Smith, *Introduction to Evolutionary Computing* (Springer, Natural Computing Series, 1st edition, 2003).
6. D. Savic and S. T. Khu, in *Encyclopedia of Hydrological Sciences*, Part 2. Hydroinformatics (John Wiley & Sons, 2005).
7. M. B. Butts, J. Madsen and J. C. Refsgaard, *Encyclopedia of Physical Science and Technology* (2004).
8. S. H. Chen, A. J. Jakeman, J. P. Norton, *Math. Comp. Sim.*, 78 (2008).
9. K. Parasuraman, A. Elshorbagy and S.K. Carey, *Hydrol. Sci. J.*, 52 (2007).
10. F. Anctil, N. Lauzon, V. Andréassian, L. Oudin and C. Perrin, *J. Hydrol.*, 328 (2006).
11. M. Nasseri, K. Asghari, M. J. Abedini, *Expert Sys. App.*, 35 (2008).
12. X. Y. Yu, S-Y. Liong and V. Babovic, *J. Hydroinform.*, 6 (2004).
13. W. Rauch and P. Harremoës, *Urban Water*, 1 (1999).
14. V. Babovic and M. Keijzer, in *Proceedings of the Modelli complessi e metodi computazionali intensivi per la stimae la previsione Conference* (Italy, 1999).
15. V. Babovic and M. Keijzer, in *Coping with Floods: Lessons Learned from Recent Experiences*, Ed. J. Marsalek (Kluwer Academic, 1999).
16. V. Babovic, S-Y. Liong and X. Yu, *Proceedings of the XIII IAHR-APD Congress* (Singapore, (2002).
17. B. Sivakumar, *J. Hydrol.*, 227 (2000).

18. B. Sivakumar, A. W. Jayawardena and T. M. K. G. Fernando, *J. Hydrol.*, 265 (2002).
19. V. Babovic and M. B. Abbott, *J. Hyd. Res.*, 35 (1997).
20. D. A. Savic, G. A. Walters and J. W. Davidson, *Water Resour. Manage.*, 13 (1999).
21. P. A. Whigham and P. F. Crapper, *Math. Comp. Model.*, 33 (2001).
22. S. T. Khu, S-Y. Liong, V. Babovic, H. Madsen and N. Muttil, *J. Am. Water Resour. Assoc.*, 36 (2001).
23. V. Babovic, M. Keijzer, *Nord. Hydrol.*, 33 (2002).
24. S-Y. Liong, T. R. Gautam, S. T. Khu, V. Babovic, M. Keijzer and N. Muttil, *J. Am. Water Resour Assoc.*, 38 (2002).
25. A. Aytek and Ö. Kişi, *J. Hydrol.*, 351 (2008).
26. N. Muttil and J. H. W. Lee, *Ecol. Model.*, 189 (2005).
27. D. J. Hill, B. Minsker, A. J. Valocchi, V. Babovic and M. Keijzer, *J. Hydroinform.*, 9 (2007).
28. A. Aytek, M. Asce and M. Alp, *J. Earth Sys. Sc.*, 117 (2008).
29. B. Sivakumar, *J. Hydrol.*, 258 (2002).
30. V. Babovic, L. C. Larsen and Z. Wu, *Proceedings of the First International Conference on Hydroinformatics* (Delft, The Netherlands, 1994).
31. O. J. Wang, *Environ. Model. Softw.*, 12 (1997).
32. C. T. Cheng, C. P. Ou and K. W. Chau, *J. Hydrol.*, 268 (2002).
33. R. K. Agrawal and J. K. Singh, *Biosys. Engg.*, 86 (2003).
34. E. Agyei and K. Hatfield, *J. Hydrol.*, 316 (2006).
35. S. T. Khu, H. Madsen and F. Pierro, *Adv. Water Resour.*, 31 (2008).
36. G. J. Pelletier, S. C. Chapra and H. Tao, *Environ. Model. Softw.*, 21 (2006).
37. J. Shen, *Ecol. Model.*, 191 (2006).
38. R. Zou, W. S. Lung and J. Wu, *Ecol. Model.*, 220 (2009).
39. V. Babovic, *Computation Fluid Dynamics: Bunte Bilder in der Praxis?* (1996).
40. A. V. M. Ines, K. Honda, A. Das Gupta, P. Droogers and R. S. Clemente, *Agric. Water Manage.*, 83 (2006).
41. P. Chaves and T. Kojiri, *Adv. Water Resour.*, 30 (2007).
42. V. Babovic, *Scientific Presentations AIO Meeting '93* (The Netherlands, 1993).
43. J. T. Kuo, Y. Y. Wang and W. S. Lung, *Water Research*, 40 (2006).
44. R. Farmani, H. J. Henriksen and D. Savic, *Environ. Model. Softw.*, 24 (2009).
45. J. C. Refsgaard, *Nord. Hydrol.*, 28 (1997).
46. H. Madsen and C. Skotner, *J. Hydrol.*, 308 (2005).
47. K. Parasuraman and A. Elshorbagy, *Water Resour. Res.*, 44 (2008).
48. Y. K. Demissie, A. J. Valocchi, B. S. Minsker and B. A. Bailey, *J. Hydrol.*, 364 (2009).
49. C. L. Chang, S. L. Lo and S. L. Yu, *J. Hydrol.*, 314 (2005).
50. R. Kalra and M. C. Deo, *Appl. Ocean Res.*, 29 (2007).
51. M. Keijzer and V. Babovic, *Proceedings of the Genetic and Evolutionary Computation Conference* (Orlando, Florida, 1999).

52. X. Y. Yu and S-Y. Liong, *J. Hydrol.*, 332 (2007).
53. World Meteorological Organization (WMO), *WMO Operational Hydrology Report 38 – WMO No. 779* (World Meteorological Organization, Geneva, 1992).
54. S. Lakshminarayanan, K. Tun and H. Takada, *Chem. Engg. Comm.*, 195 (2008).
55. S. Nielsen and E. Hansen, *Nord. Hydrol.*, 4 (1973).
56. R. Grzeszczuk, D. Terzopoulos and G. Hinton, *Proceedings of SIGGRAPH '98* (Orlando, FL, 1998).
57. C. O. Proaño, A. Verwey, H. F. P. van den Boogaard and A. W. Minns, *Proceedings of Hydroinformatics '98* (Rotterdam, 1998).
58. H. Schiller, *Neural Networks*, 20 (2007).
59. M. Keijzer and V. Babovic, *J. Genetic Prog. Evol. Mach.*, 3 (2002).

CHAPTER 8

WAVELET ANALYSES IN HYDROLOGY

David Labat

*Laboratoire des Mécanismes et Transferts en Géologie
(CNRS-Univ. P. Sabatier-IIRD), 14 Avenue Edouard Belin,
31400 Toulouse, France
E-mail : labat@imtg.obs-mip.fr*

Most of the signals encountered in hydrology and earth sciences are characterized by intermittent or transient features that lead to nonstationary data series, such as rainfall, river flow, and seismic signals. Extracting information from such signals remains difficult with classical Fourier algorithms, since they do not include time localization. The main advantage of wavelet analysis is that the window size is not fixed but varies with the frequency, following the famous Heisenberg incertitude theorem. Then, the basic objective of the wavelet transform is to achieve a complete time-scale representation of localized and transient phenomena occurring at different timescales. This chapter presents a detailed review of one-dimensional wavelet analysis and its implementation in hydrology and earth sciences. After a presentation of the continuous wavelet transform and of the multi-resolution discrete wavelet transform, two recent advances in wavelet analyses are focused on: the wavelet entropy and the wavelet coherence analysis. The chapter also provides an extensive bibliography on wavelet use in hydrology and earth sciences, emphasizing recent advances in hydrologic modeling.

8.1. From Fourier to Wavelets

Analyzing a signal basically consists of looking for similarity between the signal and well-known mathematical functions usually underlined by a clear physical meaning. In mathematics, the degree of similarity between two square continuous-time integrable signals $x(t)$ and $y(t)$ can be indicated using the classical scalar product I_{XY} defined as:

$$I_{XY} = \int_{-\infty}^{+\infty} x(t)y(t)dt \qquad (8.1)$$

If $I(x,y)$ is equal to zero, the two functions are said to be orthogonal. Any signal $x(t)$ then can be projected and analyzed using a set of functions $y(t,\{p_i\}_{i=1,...,N})$, where $\{p_i\}_{i=1,...,N}$ is a set of:

$$I_{XY}(p_i) = \int_{-\infty}^{+\infty} x(t)y(t,p_i)dt \qquad (8.2)$$

In a reverse manner, $x(t)$ can be expanded in terms of a linear superposition of the set of functions $y(t,\{p_i\})$. If the functions $y(t,\{p_i\})$ are orthogonal to each other, the signal $x(t)$ is then characterized by the values of the scalar product between the signal $x(t)$ and the projection basis $y(t,\{p_i\})$. The Fourier transform can be viewed as a projection of the temporal domain into the frequency domain (ω) using the set of functions:

$$y(t,\omega) = \exp(i2\pi\omega t) \Rightarrow S(\omega) = \int_{-\infty}^{+\infty} x(t)y^*(t,\omega)dt \qquad (8.3)$$

where * stands for the complex conjugate.

The scalar product $S(\omega)$ is the spectral amplitude of the signal $x(t)$ at the frequency ω. It corresponds to the convolution of the signal over the trigonometric orthogonal basis defined by $\{\exp(i\omega_n t), n \in (R)\}$, where ω_n is a discrete frequency ($\omega_n = 2\pi n/T$). Comprehensive reviews of spectral analysis[1-4] and recent applications of spectral methods to discharge time series[5-7] are already available in the literature. A well-known "extension" of the spectral analysis consists of the multifractal theory that leads to a systematic analysis of all the statistical moments of the time series, including fractional moments.[8,9]

The first option proposed by the signal analysis community consists of the windowed Fourier transform or Gabor methods,[10] a time-frequency transform that allows a temporal discrimination of the frequency but with an arbitrary function of constant width in terms of both time and frequency. The windowed Fourier transform is then defined by:

$$y(t_1, t_2, \omega) = \exp(i2\pi\omega t) \cdot \Phi\left(\frac{t_1 - t_2}{\sigma}\right) \tag{8.4}$$

$$\Rightarrow SG(\omega, t_1 - t_2) = \int_{-\infty}^{+\infty} x(t) y^*(t_1, t_2, \omega) dt$$

where $\Phi\left(\dfrac{t_1 - t_2}{\sigma}\right)$ is the windowed Fourier function different from zero only around t_2 on a time domain $[-\omega/2 : \omega/2]$. A commonly used function is a Gaussian of width σ given by:

$$\Phi\left(\frac{t_1 - t_2}{\sigma}\right) = \exp\left(-i \cdot \sigma^2 \frac{(t_1 - t_2)^2}{2}\right) \tag{8.5}$$

8.2. The Continuous Wavelet Transform

The wavelet transform constitutes another type of projection that will become more developed since it constitutes the most valuable analysis when dealing with signals encountered in the earth sciences.[11]

The first step consists of changing the projection basis in order to provide a time-scale dependent window of projection. In continuous time, but on a finite interval $[0, T]$, the classical orthonormal complex Fourier basis is composed of the one-parameter functions $\{\exp(i\omega_n t), n \in R\}$, where ω_n is a discrete frequency ($\omega_n = 2\pi n/T$). This is replaced by a two-parameter basis $\{\psi_{a,\tau}(t), (a,\tau) \in (R^*_+ \times R)\}$, which allows a time-scale discrimination of processes. Therefore, the coefficients of the wavelet transform of a square-integrable continuous-time signal $x(t)$ are defined by the linear integral operator (corresponding to the classical L^2 inner product):

$$C_x(a, \tau) = \int_{-\infty}^{+\infty} x(t) \psi^*_{a,\tau}(t) dt \quad \text{with} \quad \psi_{a,\tau}(t) = \frac{1}{\sqrt{a}} \psi\left(\frac{t-\tau}{a}\right) \tag{8.6}$$

where * corresponds to the complex conjugate. The function $\psi(t)$, which can be real or complex, plays the role of a convolution-kernel, and is called a wavelet. The parameter a can be interpreted as a dilation ($a > 1$) or contraction ($a < 1$) factor of the wavelet function $\psi(t)$, corresponding to different scales of observation. The parameter τ can be interpreted as a temporal translation or shift of the function $\psi(t)$, which allows the study of the signal $x(t)$ locally around the time τ. A frequency interpretation of Eq. (8.1) is also of interest. Effectively, using the property that the temporal convolution of two functions is equal to the product of their Fourier transforms (the Parseval theorem), the wavelet coefficients of a continuous-time signal $x(t)$ can also be obtained via the relation:

$$C_x(a,\tau) = \frac{1}{2\pi}\int_{-\infty}^{+\infty}\hat{x}(\omega)\sqrt{a}\hat{\psi}(a\omega)e^{i\omega\tau}d\omega = \frac{1}{2\pi}\int_{-\infty}^{+\infty}\hat{x}(\omega)\Psi_{a,\tau}(\omega)d\omega$$

(8.7)

This formulation indicates that the wavelet coefficients correspond to the filtering of $\hat{x}$ (Fourier transform of $x(t)$) by a set of band-pass filters $\Psi_{a,\tau}$ related to the Fourier transform of the wavelet function $\psi(t)$ (denoted by $\hat{\psi}(\omega)$) via the relation:

$$\Psi_{a,\tau}(\omega) = \sqrt{a}\hat{\psi}(a\omega)e^{i\omega\tau}$$

(8.8)

From this point of view, the parameter a can be physically interpreted as a dilation or contraction of the filter (corresponding to different non-overlapped frequency intervals), and the parameter τ can be interpreted as a phase shift.

As a consequence, the main property of the wavelet transform is to make possible a time-scale discrimination of processes. This allows, for example, distinguishing between two signals, which have very similar Fourier spectra. An illustration is given in Fig. 8.1, where two quite different signals are analyzed. They both constitute two sinusoidal functions characterized by a single frequency. For the first signal, both frequencies are present during the whole interval of observation, and, for the second, they are temporally localized on the first and second halves of the time interval. The two Fourier spectra are very similar, whereas the wavelet coefficients exhibit clearly the temporal discrepancy of the two components. Indeed, concerning the first signal, high wavelet coefficients corresponding to the two scales (lighter color) are visible on the whole interval, whereas in the second case two temporally localized bands are clearly distinguishable corresponding to the two different sub-process scales.

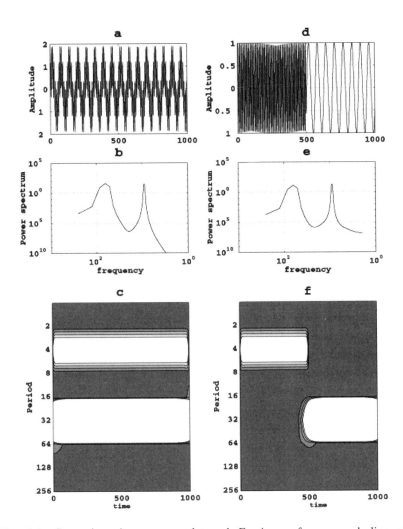

Fig. 8.1. Comparison between wavelet and Fourier performances dealing with nonstationary signals. Signals composed of two main frequencies but present at overlapped and non-overlapped time intervals ((a) and (d)); similar spectral densities of two signals ((b) and (e)) and continuous Morlet wavelet spectra ((c) and (f)), which display the time localization of the two different sub-processes.

The wavelet function must fulfil some strict mathematical conditions.[12,13] For example, the time-scale localization property requires that wavelet functions be characterized, in both time and frequency domains, by compact supports or, at least, by a sufficiently fast decay. Mathematically, these properties can be formalized by two conditions called "admissibility conditions." As mentioned above, one of these conditions is a rapid decrease of $\psi(t)$ around the origin of time, implying that:

$$\int_{-\infty}^{+\infty} \psi(t)dt = 0 \qquad (8.9)$$

This property can be transposed to the frequency domain (rapid decrease around the null frequency) as:

$$\int_{-\infty}^{+\infty} \frac{|\hat{\psi}(\omega)|^2}{|\omega|} d\omega = K_\psi < \infty \qquad (8.10)$$

In addition to a normalization of the wavelet function (the wavelet function must satisfy $\|\psi\| = 1$, where $\|\cdot\|$ is the classical L^2-norm), more restrictive conditions must be imposed on the wavelet function in order to obtain a reconstruction formula. More precisely, the continuous-time signal $x(t)$ can be retrieved from its wavelet decomposition coefficients $C_x(a,\tau)$ only if the wavelet function has N vanishing moments, that is:

$$\int_{-\infty}^{+\infty} t^k \psi(t) dt = 0, \quad k = 1, \ldots, N-1 \qquad (8.11)$$

The wavelet function, which fulfils Eq. (8.11), is then said to have regularity of order N. These conditions are called "regularity conditions." Conditioned by these restrictions, there exists a reconstruction formula that allows the synthesis of the continuous-time signal $x(t)$ through its wavelet coefficients $C_X(a,\tau)$:

$$x(t) = \frac{1}{K_\psi} \int_0^{+\infty} \int_{-\infty}^{+\infty} C_x(a,\tau) \psi_{a,\tau}(t) \frac{da \, d\tau}{a^2} \qquad (8.12)$$

Moreover, another property of the continuous wavelet transform is its ability to conserve energy between time domain and time-scale domain, that is:

$$\int_{-\infty}^{+\infty} x^2(t) \, dt = \frac{1}{K_\psi} \int_0^{+\infty} \int_{-\infty}^{+\infty} |C_x(a,\tau)|^2 \frac{da \, d\tau}{a^2} \qquad (8.13)$$

One disadvantage of these non-orthogonal wavelets is that the continuous wavelet transform of the signal is characterized by a redundancy of information among the wavelet coefficients. This redundancy implies correlation between coefficients, which is intrinsic to the wavelet-kernel and not to the analyzed signal. The wavelet coefficient interpretation is then strongly dependent on the chosen projection basis, which must be adapted to the specific problem at hand. For example, the Mexican Hat wavelet allows for a good temporal resolution but a poor frequency resolution, whereas the Morlet wavelet allows for a good frequency resolution although with a less precise temporal resolution.

One can, however, construct an orthonormal continuous time wavelet basis that eliminates this redundancy problem. Here, a discrete-time approach is adopted.

8.3. Discrete-time Wavelet Transform and Multi-resolution Analysis

For practical applications, the scientist or engineer does not have at his or her disposal a continuous-time signal process but rather only a discrete-time signal, which is denoted herein by $x(i)$. In order to apply the techniques introduced below, the time-scale domain needs to be discretized, and one can choose the general form ($\{a_0^{-j}, k.\tau_0.a_0^{-j}\}$, $(k, j) \in Z^2$), where a_0 ($a_0 > 1$) and τ_0 are constants.

8.3.1. Discrete Wavelet Transform

The discretized version of Eq. (8.1) defines the discrete wavelet transform coefficients, as follows:

$$C_x(j,k) = \int_{-\infty}^{+\infty} x(t)\psi^*_{j,k}(t)dt \quad \text{with} \quad \psi_{j,k}(t) = a_0^{j/2}\psi(a_0^j t - k\tau_0)$$

(8.14)

where * corresponds to the complex conjugate. The integer j is the scale factor (analogous to the parameter a) and the integer $k\tau_0$ is the translation factor (analogous to the parameter τ). As in the continuous-time wavelet transform, the reconstruction formula can be applied with appropriate choice of the coefficients a_0 and τ_0. This allows a representation of a discrete-time signal process in terms of series expansion and leads ultimately to the concept of multi-resolution analysis.

Once the time-scale domain has been discretized, some particular values for a_0 and τ_0, corresponding to an octave representation[14,15] of the scales, are chosen, namely: $a_0 = 2$ and $\tau_0 = 1$. The continuous grid is then replaced by a discrete diadic grid of the form: $(\{2^{-j}, k.2^j\}, (k,j) \in Z^2)$. The orthogonal discrete wavelet transform coefficients $C_{j,k}$ are then defined by the convolution product:

$$C_{j,k}^x = \int_{-\infty}^{+\infty} x(t)\psi^*_{j,k}(t)dt \quad \text{with} \quad \psi_{j,k}(t) = 2^{j/2}\psi(2^j t - k) \quad (8.15)$$

In such an orthonormal basis, all wavelets ψ are orthonormal to both their translates and their dilates, that is:

$$\int_{-\infty}^{+\infty} \psi_{m,n}(t)\psi_{m',n'}(t)dt = \delta_{m,m'}\delta_{n,n'} \quad (8.16)$$

where δ is the Kroneker delta symbol ($\delta_{i,j} = 1$ if $i = j$ and $\delta_{i,j} = 0$ if $i \neq j$). Therefore, it is possible to construct a complete basis under which all square integrable signals $x(i)$ can be expanded as a linear combination of translates and dilates of orthonormal wavelets:

$$x(i) = \sum_{j=0}^{+\infty} \sum_{k=-\infty}^{+\infty} C_{j,k}^x(i)\psi_{j,k}(i) \quad (8.17)$$

Equation (8.17) is the discrete-time formulation of signal synthesis; it can also be interpreted as successive approximations (in a mean square sense) of the analyzed discrete signal $x(i)$ by the sequence $\{x_n(i)\}_{n=1,...,N}$ defined by:

$$x_n(i) = \sum_{j=0}^{n-1} \sum_{k=-\infty}^{+\infty} C_{j,k}^x(i)\psi_{j,k}(i) \quad (8.18)$$

This constitutes the conceptual basis of multi-resolution analysis. Multi-resolution wavelet analysis[16] allows the decomposition of a function or signal in a progression of successive "approximations" and "details," corresponding to different scales j. The difference between the actual signal and its approximation of order n is called the "residual."

Intuitively, the approximation is relatively smooth and the detail, being composed of high-frequency components, is rougher. Note that the detail corresponds to the difference between two successive levels of approximation j and $j + 1$. One can then define more rigorously the concept of multi-resolution analysis, as follows.

A multi-resolution analysis is a sequence of included spaces $\{V_m\}_{m \in Z}$ and their respective orthogonal space $\{O_m\}_{m \in Z}$, satisfying two main mathematical properties of major relevance:

(P1): whatever the resolution m, the $\{O_m\}_{m \in Z}$ space is such that each sub-space contains all necessary, and only necessary, information to go from an approximation of resolution m (i.e. a space V_m) to the upper resolution $m + 1$ (i.e. a space V_{m+1}).

(P2): the successive approximations of a square integrable function converge as the resolution m increases, and the successive approximations contain less and less information as the resolution m decreases.

Therefore, it seems natural to construct orthonormal bases of wavelet-type on the successive spaces of functions: $\{V_m\}_{m \in Z}$ and $\{O_m\}_{m \in Z}$. Each successive sub-space $\{O_m\}_{m \in Z}$ is characterized by an orthonormal basis based on dilation and contraction of the function $\psi(t)$, called the wavelet function, and, likewise, each successive sub-space $\{V_m\}_{m \in Z}$ is characterized by an orthonormal basis obtained via dilation and contraction of the function $\varphi(t)$, called the "scale function." Finally, the following orthogonal properties hold for the wavelet functions ψ and the scale functions φ.

$$\begin{cases} \int_{-\infty}^{+\infty} \psi_{m,n}(t)\psi_{m',n'}(t)dt = \delta_{m,m'}\delta_{n,n'} \\ \int_{-\infty}^{+\infty} \varphi_{m,n}(t)\psi_{m',n'}(t)dt = 0 \\ \int_{-\infty}^{+\infty} \varphi_{m,n}(t)\varphi_{m',n'}(t)dt = \delta_{n,n'} \end{cases} \qquad (8.19)$$

Contrary to the wavelet functions ψ, which are orthogonal to their translates *and* dilates, the scale functions φ are orthogonal only to their translates. In what follows, algorithms for the construction of wavelet and scale functions are developed.

Using the discrete wavelet reconstruction formula, it is now possible to give the definition of successive "approximations" and "details" based on wavelet and scale functions making an arbitrary cut-off in the increasing scales.[16] Therefore, the approximation of a signal x at a resolution m (denoted by A^m) and the detail of the same function at a resolution m (denoted by D^m) are defined by:

$$A_x^m(i) = \sum_{k=-\infty}^{+\infty} \langle x, \varphi_{m,k} \rangle \cdot \varphi_{m,k}(i) \quad \text{and}$$

$$D_x^m(i) = \sum_{k=-\infty}^{+\infty} \langle x, \psi_{m,k} \rangle \cdot \psi_{m,k}(i) \tag{8.20}$$

Due to space mutual orthogonality and completeness, the sampled signal $x(i)$ can then be easily retrieved using the fundamental relation:

$$x(i) = A_x^m(i) + \sum_{i \geq m+1} D_x^m(i) = A_x^m(i) + R_x^m(i) \tag{8.21}$$

where R_x^m represents the residual at the level of approximation m.

To sum up, the multi-resolution analysis allows an orthogonal decomposition of a sampled hydrologic signal in terms of approximations and details of increasing order of resolution. Because of the orthogonality of the decomposition, it should be noted that the multi-resolution induces a more efficient and easier interpretation of energy distribution on the different scales of the decomposition. The multi-resolution analysis facilitates the temporal isolation of processes. This filtering and identification method is very easy to use, particularly for de-noising purposes. Figure 8.2(a) presents a simple implementation of a wavelet low-pass filtering, which corresponds to keeping only the lowest scales of the decomposition. For identification purposes, a wavelet band-pass filtering is implemented, which corresponds to keeping only some component (scales) of the decomposition. The quality of this method can be seen by inspecting the spectral density of the isolated process (e.g. in the pumping case).

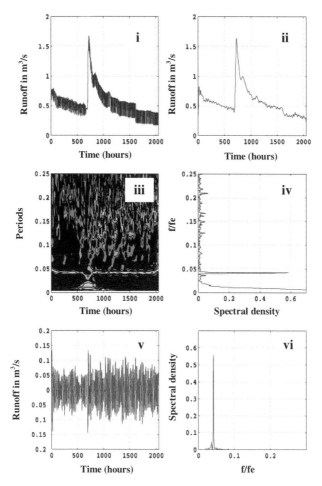

Fig. 8.2(a). Application of the multi-resolution analysis to a pumping signal superimposed to a flood at the outlet of a karstic spring. A periodic pumping, superimposed on a natural flood, observed at the outlet of the Esperelle system during the period from March 27, 1995 to June 20, 1995, which represents $2048 = 2^{11}$ data elements. Morlet wavelet analysis clearly identifies a periodic component with a constant frequency f_p, superimposed on a white noise process characterized by high-frequency structures. A low-frequency structure corresponding to the flood superimposed on the pumping process can also be observed. This first example allows the reader to see the correspondence between the continuous wavelet transform and Fourier analysis for a relatively simple (but measured) signal. In this case, since the pumping characteristic frequency is time-invariant, there is no problem of time isolation as an intermittent process (see Fig. 8.2(b)), where the characteristic frequency is time-variant. Nevertheless, the presence of high-frequency structures, which could be temporally localized by the wavelet transform and not by Fourier analysis, is visible.

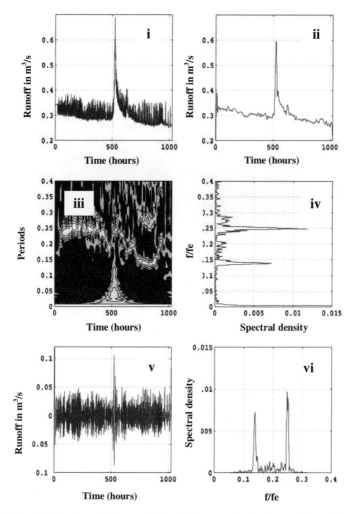

Fig. 8.2(b). Application of the multi-resolution analysis to an intermittent signal superimposed to a flood at the outlet of a karstic spring. An intermittent outflow, with short timescales, observed at the outlet of the Cernon spring, during the period from April 3, 1995 to May 16, 1995, which represents $1024 = 2^{10}$ data elements. The continuous wavelet transform clearly indicates the presence of two distinct periodic processes characterized by two frequencies f_1 and f_2. In particular, whereas the Fourier spectrum is not able to differentiate temporally between these two processes, the wavelet analysis clearly shows that the passage from the characteristic frequency f_1 to f_2 is consecutive to the flood. Moreover, for the phenomenon related to the frequency f_2, a secondary harmonic is also evident.

8.3.2. Orthogonal Wavelet Bases and Orthogonal Wavelet Expansions

The simplest wavelet function is the Haar function (Fig. 8.3), which is used for wavelet spectrum estimation in this chapter (presented later). It is given by:

$$\psi(x) = \begin{cases} 1 & for \quad 0 < x < 1/2 \\ -1 & for \quad 1/2 \leq x < 1 \\ 0 & elsewhere \end{cases} \quad (8.22)$$

Only N coefficients are needed to completely recover the signal, and this wavelet is often used because of its differencing properties. Moreover, Beyklin et al.[17] demonstrated a simple and direct relationship between details and approximations at different scales. Details and approximations can be computed directly from the original signal $A^0(i)$ of length 2^M using the recurrence relation:

$$D^{m+1}(i) = \frac{1}{\sqrt{2}} \left[A^m(2i-1) - A^m(2i) \right]$$
$$A^{m+1}(i) = \frac{1}{\sqrt{2}} \left[A^m(2i-1) + A^m(2i) \right] \quad (8.23)$$

for $i = 1, .., 2^{M-m-1}$; $m = 0, .., M-1$

Since ψ and φ belong to the function space V^1, it is possible to define two sets of coefficients $\{h_k\}$ and $\{g_k\}$, $k \in Z$, characterizing completely the wavelets and satisfying the fundamental dilation equations:[18]

$$\varphi(t) = \sqrt{2} \sum_k h_k \varphi(2t-k) \quad and \quad \psi(t) = \sqrt{2} \sum_k g_k \varphi(2t-k) \quad (8.24)$$

with the relation between $\{h_k\}$ and $\{g_k\}$ as:

$$g_k = (-1)^k h_{2N-k+1}, \forall k = 0, \ldots, 2N-1 \quad (8.25)$$

For example, for $N = 1$, taking $h_0 = h_1 = 1/2$ and $g_0 = 1/2$, $g_1 = -1/2$ and equating to zero all other coefficients, one can retrieve the Haar wavelet transform defined by Eq. (8.22), as presented below. In order to construct the basis functions φ and ψ that fulfil the dilation relation (Eq. (8.24)), several methods have been developed. In this respect, an important theoretical result should be emphasized, namely, the existence of

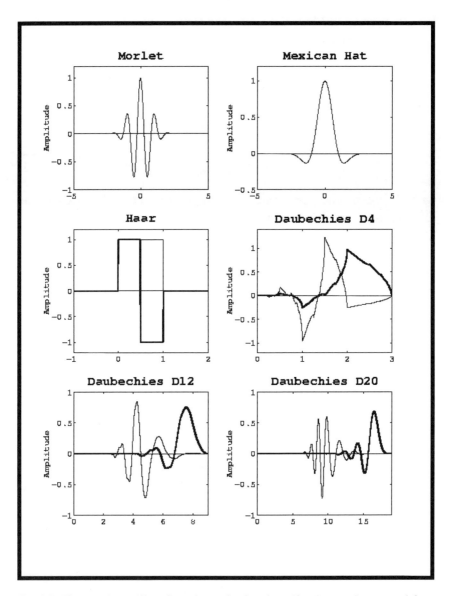

Fig. 8.3. Time representation of usual wavelet functions (from top to bottom and from left to right): Morlet wavelet; Mexican Hat wavelet; Haar wavelet (bold) and scale function; Daubechies 4 wavelet (bold) and scale function; Daubechies 12 (bold) and scale function; and Daubechies 20 (bold) and scale function.

compactly supported basis functions, whose iterative construction process was discovered by Daubechies.[15]

$$\begin{cases} \psi_K^n(i) = \sum_{m=1}^{K} \alpha_m \psi_K^{n-1}(2i-m) \\ \varphi_K^n(i) = \sum_{m=1}^{K} (-1)^m \alpha_m \psi_K^n(2i+m-K+1) \end{cases} \quad (8.26)$$

The Haar function constitutes the first term of the sequence $\{\psi_K^n\}_{n \in N}$ (namely ψ_K^0). Note that, as the number of coefficients (K) increases, the wavelet becomes more and more regular. One can see some examples of such Daubechies wavelets in Fig. 8.3. The successive approximations and details of order m of the signal $x(i)$ are then easily obtained via the following relations:[16]

$$\begin{cases} A_x^m(k) = \sum_{l=-\infty}^{l=+\infty} h(l-2k) A_x^{m+1}(l) \\ D_x^m(k) = \sum_{l=-\infty}^{l=+\infty} g(l-2k) \langle x, \varphi_{m+1}(l-2^{-j-1}l) \rangle \end{cases} \quad (8.27)$$

In hydrology, applications of this property can be found in Fig. 8.4(a) that presents a multi-resolution decomposition of the discharges in the Amazon River over the last century, and in Fig. 8.4(b) that presents the decomposition of the karstic watershed rainfall-runoff response into a rapid response (rainfall-runoff detail correlation) and a long-term response (rainfall-runoff approximation correlation).

8.4. Signal Energy Repartition in the Wavelet Frame

The repartition of energy in the signal allows for the determination of the scales which concentrate the essential dynamics of a signal. Therefore, a determination of the temporal variations of the distribution of energy across the scales is an important application of the wavelet transform.

Previously, the Fourier transform allowed for the determination of the distribution of energy across frequencies ω using the power spectrum density $P_{XX}(\omega)$ defined for a signal $x(t)$ by:

$$P_{XX}(\omega) = \left| \frac{1}{T} \int_0^T x(t)^2 \exp(-i 2\pi \omega t) dt \right|^2 \quad (8.28)$$

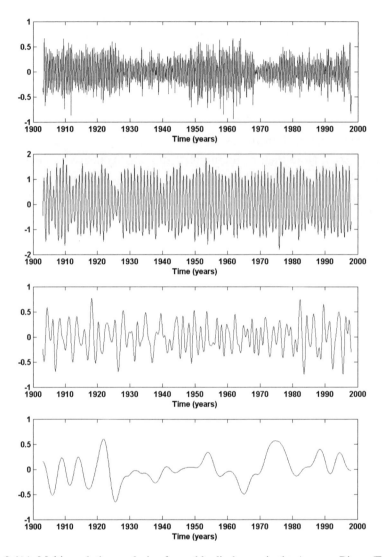

Fig. 8.4(a). Multi-resolution analysis of monthly discharges in the Amazon River. (From top to bottom) intra-annual, annual, 3- to 5-year multi-resolution details, and 5-year approximation.

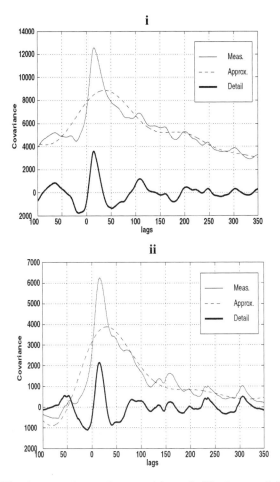

Fig. 8.4(b). Wavelet covariance decomposition of 30-minute rainfall-runoff data collected over Aliou watershed in France over two distinct periods. The covariance between the approximations and between the details is also calculated in order to extract two separable components in the global covariance function. A rapid response (rainfall-runoff detail covariance) corresponding to the mean peak of the global covariance function can be extracted. The rainfall-runoff approximation covariance corresponds to the tail of the global covariance function. The decomposition of the total covariance function in the details and approximations covariance leads to the identification of two components with very different behaviors. The covariance between the rainfall rates and the runoff details leads to the isolation of a rapid component corresponding to the one-day peak (**i**). The covariance between approximations indicates the presence of a slower component, corresponding to the tail of the global covariance function (**ii**). So, the scale decomposition leads also to a two-component interpretation of the global response in a specific event.

The companion concept of wavelet energy was defined first by Hudgins et al.[19] and then by Brunet and Collineau.[20] Other results concerning wavelet variance estimators and applications can also be found in the literature.[21-25] This concept is based on the conservation of energy between the time domain and the time-scale domain, that is:

$$\int_{-\infty}^{+\infty} x^2(t)\,dt = \frac{1}{K_\psi} \int_0^{+\infty} \int_{-\infty}^{+\infty} |C_x(a,\tau)|^2 \frac{da\,d\tau}{a^2} \qquad (8.29)$$

Therefore, the total signal variance can be distributed in the wavelet domain and the wavelet spectrum $W_X(a,\tau)$ of a continuous-time signal $x(t)$, which can be defined, by analogy with the Fourier analysis, as the modulus of its wavelet coefficients:[26]

$$W_X(a,\tau) = C_X(a,\tau) C^*x(a,\tau) = |C_X(a,\tau)|^2 \qquad (8.30)$$

This wavelet spectrum can also be averaged in time or in scale, but at the cost of a loss of information.[27] On one hand, time-averaging gives the variance signal distribution between the different scales, referred to as the global averaged wavelet power spectrum. On the other hand, scale-averaging allows the temporal identification of a particular component of the signal.

The determination of the characteristic periods of oscillation is achieved using the global wavelet spectrum. Statistical confidence limits of the peaks identified in the wavelet power spectrum have already been discussed.[27-30] Two statistical limits are classically considered in order to restrict analysis to statistically significant-only fluctuations (Fig. 8.5).

Only periodicity different from red noise is taken into consideration and 95% and 90% confidence intervals are generally depicted in the figures, following Torrence and Campo calculations.[27]

Due to the edge effects leading to the presence of a cone of influence, if T is the total length of the series, only periods smaller than $T/(2\sqrt{2})$ are considered as significant.

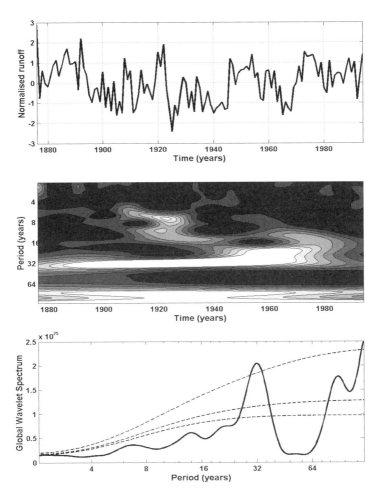

Fig. 8.5. Morlet wavelet analysis of the global annual freshwater discharge to oceans' fluctuations from 1875 to 1994.[31] From top to bottom: temporal fluctuations of continental discharge, two-dimensional Morlet wavelet spectrum, and global wavelet spectrum and global morlet wavelet spectrum. In the third plot, the dashed lines correspond to the 95%, 90%, and 80% statistical confidence interval from red noise.[32]

According to Meneveau,[33] the concept of wavelet variance can also be defined by using the orthogonal multi-resolution approach. The total energy (variance) $T_X(m)$ captured by the detail D_m at scale m of a signal $x(i)$ of length 2^M can be estimated as:

$$T_x(m) = \frac{1}{2^{M-m}} \sum_{j=1}^{2^{M-m}} \left(D_{m,j}^x\right)^2 \tag{8.31}$$

Note that these concepts were first implemented to study intermittent turbulent signal above the forest canopy.[34-36] One of the main advantages of the discrete orthogonal wavelet transform is its ability to concentrate the energy of the signal in a limited number of coefficients. A simple way to quantify this property is to represent the "Lorentz curve" $L(k)$ of the different signal representations (time, Fourier, and wavelet). The construction of the "Lorentz curve" $L(k)$, which ranges from 0 to 1, can be summarized as follows. First, the vector of length 2^N (time signal or Fourier transform or wavelet transform) is sorted in increasing order of the modulus. The remainder of the procedure consists in setting to zero all coefficients greater than the kth value of the sorted vector, taking the inverse transform of the new vector (no transform needed if the vector is the signal itself), and calculating its energy variance by cumulating the squared coefficients. Finally, the value of $L(k)$ is given by the ratio of this energy (at level k) by the total energy of the signal (at $k = k_{max} = 2^N$).

Labat et al.[37] show, for example, that the Haar wavelet transform allows the concentration of most of the variance of karstic spring runoff time series on only 100 to 500 coefficients, out of 8192 coefficients. Haar wavelet transform energy concentration for half-hourly runoffs is also quite powerful (especially when compared to the Fourier transform), since only 100 coefficients out of 2^{14} represent more than 96% of the total variance of the signal. This score is the best one can have, by comparison with the Fourier transform and data Lorentz curves.

Similarly, the signal compression technique developed by Donoho et al.[38] or Johnstone and Silverman[39] can be used here in order to expose the time-variability of the signals, i.e. time intervals of high activity or intermittence. The method involves three steps:

(i) take the discrete orthogonal wavelet transform of the signal;
(ii) apply a threshold γ on the wavelet coefficients (i.e. set to zero all the coefficients of modulus lower than the threshold); and
(iii) take the inverse discrete wavelet transform of the new vector.

The most common threshold value is the so-called "hard threshold," defined as:

$$\gamma = \sigma \cdot \sqrt{2\log(N)} \quad (8.32)$$

where σ is the estimated signal variance and N is the length of the signal. Note that other threshold values are also given in the literature, such as the "smooth threshold."[38] Among others, a didactic application wavelet de-noising can be found in Carniel *et al.*[40]

8.5. Statistical Entropy in the Wavelet Frame

The second important thermodynamic concept consists of entropy. Entropy is a measure of the degree of uncertainty, i.e. of the disorder or of the lack of information in a system characterized by a time series.[41] Note that the entropy concept has already been used in hydrology, especially in basin geomorphology and water resources management.[42]

If complete information is available on a system, its entropy will be equal to zero. If some uncertainty remains, it will be greater than zero. In considering a set composed of n events, uncertainty exists in the determination of which event will occur. This uncertainty can be quantified using two factors: the number of possible events and the probability distribution of the events.

The first definition of entropy was proposed by Shannon[43] as:

$$S = H_1 = -\sum_{i=1}^{N} p_i \cdot \ln(p_i) \quad (8.33)$$

with the convention $0*\ln 0 = 0$ by definition. This equation makes it possible to quantify the general disorder of a system, including all the values present in the system (in this instance, the hydrologic or climatic system). Shannon[43] introduced the entropy S, defined in Eq. (8.33), in order to interpret the thermodynamic concept of entropy in terms of information content. Maximum Shannon entropy is equal to $ln(N)$, where

N is the number of classes of the probability distribution. Shannon entropy S can also be defined in a continuous framework as:

$$S = \int_0^\infty - f(x) \cdot \ln(f(x)) dx \tag{8.34}$$

where $f(x)$ is the continuous probability density function of the variable x.

Given a frequency distribution of a dataset denoted $\{p_i\}$, Rényi[44] also defined a generalized entropy denoted as H_α:

$$H_\alpha = \frac{1}{1-\alpha} \ln\left(\sum_{i=1}^N p_i^\alpha\right) \tag{8.35}$$

where α is a parameter. Note that the special case of $\alpha = 1$ corresponds to Shannon entropy.

However, probability density is not the only kind of distribution which can give useful information about a signal. This definition has also been extended to other types of distributions. This chapter focuses on the extension of Shannon entropy to energy distribution.

Using a Fourier energy distribution, a new frequency distribution (FD) entropy denoted as S_{FD}

$$S_{FD}(\omega) = \int_0^\infty - P_{XX}(\omega) \cdot \ln(P_{XX}(\omega)) dx \tag{8.36}$$

can be defined, where $P_{XX}(\omega)$ is the Fourier spectrum of the variable x. This Fourier-based entropy has already been applied in hydrology by Sonuga[45] for flood frequency analysis and rainfall-runoff relationships. In a similar way, this generalization can also be achieved considering time-frequency distributions. Considering, for example, a time-frequency energy representation from Cohen's class $C(t,\omega)$ defined as

$$C(t,\omega) = \frac{1}{4\pi^2} \iiint \phi(\theta,\tau) x(u+\frac{\tau}{2}) x(u-\frac{\tau}{2}) \exp(i(\theta u - \theta t - \omega\tau)) du\, d\theta\, d\tau \tag{8.37}$$

the time-frequency distribution (TFD) entropy S_{TFD} can be defined as:

$$S_{TFD}(t,\omega) = \int_0^\infty - |C_x(t,\omega)|^2 \cdot \ln(|C_x(t,\omega)|^2) dt\, d\omega \tag{8.38}$$

This formulation is interesting, as it allows for a temporal evolution of the disorder of a system.

Since Shannon entropy constitutes a useful criterion for analyzing system's order or disorder using any energy probability distribution, a time-scale distribution called wavelet entropy can be introduced. A large majority of these concepts have been recently and successfully developed and applied in neurosciences.[46-52] The continuous wavelet entropy, denoted as S_W, is defined as:[53]

$$S_W(t) = \int_0^\infty -P_W(a,t) \cdot \ln(P_W(a,t)) da \quad (8.39)$$

where

$$P_W(a,t) = \frac{|W_\psi^x(a,\tau)|^2}{\int |W_\psi^x(a,\tau)|^2 da}$$ corresponds to the wavelet energy probability

distribution for each scale a at time t.

Wavelet entropy is minimum when the signal is characterized by an ordered activity, and wavelet entropy is maximum when the signal consists of a superimposition of a large number of processes (or "of a large number of processes"). When considering an ordered mono-frequency signal, a wavelet multi-resolution or continuous wavelet representation of such a signal is resolved around one unique scale or level. This particular level then concentrates around 100% of the total energy of the signal. The wavelet entropy then will be close to zero. When considering a signal generated by a white noise, for example, all levels will carry a certain amount of energy. These levels can also be expected to carry the same amount of energy and, consequently, the wavelet entropy will be maximal. Through this multiscale approach, relevant scale levels can automatically be detected, which include the main complex processes of the system.

A multi-resolution wavelet entropy can also be defined, allowing for a scale representation of the disorder of a system.[54] Once the multi-resolution coefficients $C_{XX}(j,k)$ have been calculated using Eq. (8.15), the energy at time k and for scale j, $E(j,k)$, can be approximated by $E(j,k) = |C_{XX}(j,k)|^2$. As previously seen, summing this energy for all discrete time k leads to an approximation of the energy content at scale j, denoted as $E(j)$:

$$E(j) = \frac{1}{N}\sum_k E(j,k) \qquad (8.40)$$

In order to follow the Shannon entropy framework, a probability density function must be defined as the ratio between the energy of each level and total energy:

$$p_E(j) = \frac{E(j)}{\sum_j E(j)} \qquad (8.41)$$

This corresponds exactly to the probability density distribution of energy across the scales. It must also be noted that, because of the orthogonal representation, the following relation holds:

$$\sum_j p_E(j) = 1 \qquad (8.42)$$

Finally, the multi-resolution wavelet $SMR(j)$ entropy at the scale j is defined by analogy with the continuous wavelet entropy:

$$SMR(j) = -p_E(j) \cdot \ln(p_E(j)) \qquad (8.43)$$

The temporal evolution of multi-resolution wavelet entropy is achieved by decomposing the signal and calculating on these intervals the entropy, as defined above.

8.6. Wavelet Analysis of the Time-scale Relationship Between Two Signals

In the previous sections, the importance of the introduction of wavelet energy and wavelet entropy to characterize the degree of complexity of a signal was highlighted. In this section, attention is given to the introduction of wavelet indicators to investigate the relationship between two given signals, e.g. rainfall rates/discharges and discharge/climate circulation indices.

In the same manner as in Fourier analyses, wavelet cross-correlation should be preferred to classical cross-correlation. This method effectively provides new insights into the scale-dependant degree of correlation between two given signals.[24,55] Classical cross-correlation makes it possible to quantify the degree of similarity between two signals, denoted as $x(t)$ and $y(t)$. The cross-correlation R_{XY} between two stochastic processes is defined as:

$$R_{XY} = \frac{C_{XY}}{\sqrt{C_{XX} \cdot C_{YY}}} \tag{8.44}$$

with $C_{XY} = E[x(t)y(t)]$, $C_{XX} = E[x(t)^2]$, $C_{YY} = E[y(t)^2]$. It must be recalled that under stationary and ergodic assumptions

$$E[x(t)y(t)] = \lim_{T \to \infty} \frac{1}{T} \int x(u)y(u)du \tag{8.45}$$

This notion has been generalized in order to introduce a time delay between two signals

$$R_{XY}(t_1 - t_2) = \frac{C_{XY}(t_1 - t_2)}{\sqrt{C_{XX}(0) \cdot C_{YY}(0)}} \quad \text{with} \quad C_{XY}(t_1 - t_2) = E[x(t - t_1)y(t - t_2)] \tag{8.46}$$

that indicates that cross-correlation analysis provides a quantitative measure of the relatedness of two signals, shifting in time with respect to each other. It allows for the identification of common components occurring at the same delay. However, if the signals include nonstationary components, the cross-correlation becomes invalid. The same conclusion arises if both signals are characterized by highly variable processes occurring over a wide range of scales. The continuous wavelet cross-correlation is introduced, which overcomes the limitations of classical correlations.

Considering two correlated processes $x(t)$ and $y(t)$, the wavelet cross-correlation function, denoted as $WC_{XY}(a,u)$, for a given scale a and a given delay time u can be defined as:[23]

$$WC_{XY}(a,u) = E[W_{XX}(a,\tau) \cdot W_{YY}(a,\tau+u)] \tag{8.47}$$

At this point, some authors recommend distinguishing the real part denoted as $RWC_{XY}(a,\tau)$ from the imaginary part $IWC_{XY}(a,\tau)$ of the wavelet cross-correlation, arguing that the real part alone is sufficient to quantify the strength of correlation between the two signals at a given scale a. The wavelet cross-correlation is then defined as:

$$WR_{XY}(a,u) = \frac{RWC_{XY}(a,u)}{\sqrt{RWC_{XX}(a,0) \cdot RWC_{YY}(a,0)}} \qquad (8.48)$$

Sello and Bellazzini[56] propose to consider only the real part of the wavelet transform and define a wavelet local correlation coefficient $WLCC(a,u)$ based on the wavelet cross-spectrum, $W_{XY}(a,u) = W_{XX}^{*}(a,u) \cdot W_{YY}(a,u)$, as:

$$WLCC(a,u) = \frac{Real(W_{XY}(a,u))}{|W_{XX}(a,u)| \cdot |W_{YY}(a,u)|}. \qquad (8.49)$$

This wavelet local correlation coefficient $WLCC(a,u)$ definition is based on the following relationship between the classical cross-correlation and the real part of the cross-wavelet-spectrum:

$$\int_{-\infty}^{+\infty} x(t)y(t)dt = \frac{1}{c_\psi} \int_{0}^{+\infty} \int_{-\infty}^{+\infty} Real(W_{XY}(a,\tau)) da\, d\tau \qquad (8.50)$$

However, real-part or imaginary-part wavelet analysis does not provide a clearly separate information. As in the Fourier analysis, only modulus and arguments are considered here as interpretative. This conclusion is confirmed by Mizuno-Matsumoto et al.,[57,58] who observed, in a preliminary analysis, no significant differences between the definitions. Wavelet cross-correlation is thus defined as:

$$WR_{XY}(a,\tau) = \frac{\sqrt{|RWC_{XY}(a,\tau)|^2 + |IWC_{XY}(a,\tau)|^2}}{\sqrt{|WC_{XX}(a,0)| \cdot |WC_{YY}(a,0)|}} \qquad (8.51)$$

where RWC_{XY} and IWC_{XY} are, respectively, the real and the imaginary part of the cross-wavelet correlation function. An illustration of the wavelet cross-correlation on multiscale synthetic signals is depicted in Fig. 8.6.

In conclusion, the introduction of wavelet cross-correlation allows a rapid identification of the degree of correlation between processes at a given scale and the determination of the delay between these processes.

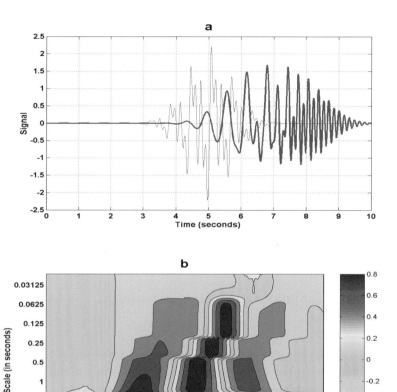

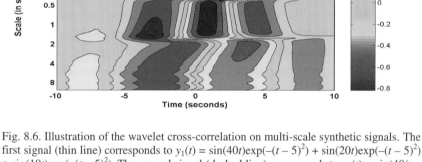

Fig. 8.6. Illustration of the wavelet cross-correlation on multi-scale synthetic signals. The first signal (thin line) corresponds to $y_1(t) = \sin(40t)\exp(-(t-5)^2) + \sin(20t)\exp(-(t-5)^2) + \sin(10t)\exp(-(t-5)^2)$. The second signal (dashed line) corresponds to $y_2(t) = \sin(40(t-3))\exp(-(t-8)^2) + \sin(20(t-2))\exp(-(t-7)^2) + \sin(10(t-1))\exp(-(t-6)^2)$. Then, the three scale processes (with periods, respectively, equal to 0.6, 0.3, and 0.08 second) are, respectively, delayed by 3, 2, and 1 seconds (a). The wavelet cross correlation (b) allows a rapid and efficient time-scale identification of these delays.

8.7. Wavelet Cross-spectrum and Coherence

By analogy with the Fourier cross-spectrum, Liu[26] defines the wavelet cross-spectrum $W_{XY}(a, \tau)$ between two signals $x(t)$ and $y(t)$ as:

$$W_{XY}(a,\tau) = C_X(a,\tau)C^*_Y(a,\tau) \qquad (8.52)$$

where $C_X(a,\tau)$ and $C^*_Y(a,\tau)$ are, respectively, the wavelet coefficient of the continuous-time signal $x(t)$ and the conjugate of the wavelet coefficient of $y(t)$. This formalism leads to a clear analogy with the Fourier analysis, whereby, for example, the "Wiener-Kintchine" wavelet representation theorem is applicable. Indeed, with these notations, the wavelet cross-spectrum is the multi-dimensional wavelet transform of the covariance function:

$$\begin{aligned}E\left[w_{XY}(a,\tau)\right] &= E\left[\int_{-\infty}^{+\infty} x(u)\psi^*_{a,\tau}(u)du \cdot \int_{-\infty}^{+\infty} y(v)\psi_{a,\tau}(v)dv\right] \\ &= \int_{-\infty}^{+\infty}\int_{-\infty}^{+\infty} E\left[x(u)\,y(v)\right]\psi^*_{a,\tau}(u)\,\psi_{a,\tau}(v)dudv \qquad (8.53) \\ &= \int_{-\infty}^{+\infty}\int_{-\infty}^{+\infty} Cov_{XY}(u-v)\,\psi^*_{a,\tau}(u)\,\psi_{a,\tau}(v)dudv\end{aligned}$$

The averaging techniques applied to the wavelet spectrum are also used here for expressing the cross-covariance of the signals x and y and its distribution through scales. However, Maruan and Kurths,[59] among others, demonstrated that wavelet cross-spectrum appears to be non-suitable to interpret interrelation between two processes (ENSO and NAO fluctuations in their example) and recommended the introduction and use of wavelet coherence analysis.

The notion of coherence in signal processing consists, from a general point of view, of a measure of the correlation between two signals or between two representations of these signals. In examining two continuous finite-energy signals $x(t)$ and $y(t)$, the classical cross-correlation, defined in the previous section, can be considered as a temporal coherence between two signals. Given the Fourier transform of these signals, the spectral coherence can be defined as:[60]

$$\rho(f) = \frac{|S_{XY}(f)|}{\sqrt{|S_{XX}(f) \cdot S_{YY}(f)|}} \qquad (8.54)$$

where $S_{XY}(f)$ is the cross-spectral density between the two processes $x(t)$ and $y(t)$. It may be recalled that, under stationary and ergodic assumptions, the cross-Fourier spectrum is defined as:

$$S_{XY}(f) = \int_{-\infty}^{+\infty} R_{XY}(\tau) \cdot \exp(-i2\pi f \tau) d\tau \tag{8.55}$$

Because of the Schwartz inequality, it may be noted that $\rho(f)$ takes a value between 0 and 1. If $\rho(f)$ is close to 1 for all frequencies, then $y(t)$ can be approximated as a linear time-invariant transformation of $x(t)$. In these definitions, the stationary assumption appears as being highly restrictive. However, when dealing with nonstationary processes, as encountered in hydrology, the use of a time-frequency representation of the signals is suggested. Such coherence highlights the temporal variations of the correlation between two signals and allows for the detection of transient high covariance. To overcome the problems inherent to nonstationary signals, it has been proposed to introduce the wavelet coherence.[61,62] The main issue is that an application of a similar formula to Eq. (8.52) leads to a coherence always equal to 1, as pointed out by Labat[32] and Labat et al.[63] Attempts that have been made to avoid this problem essentially differ on two points: the fundamental definition of wavelet coherence and the estimation of the wavelet cross-spectrum.

On one hand, Sello and Bellazzini[56] introduce the cross wavelet coherence, defined as:

$$CWCF(a,\tau) = \frac{2 \cdot |W_{XY}(a,\tau)|^2}{|W_{XX}(a,\tau)|^4 + |W_{YY}(a,\tau)|^4} \tag{8.56}$$

which measures the intensity coherence of the signals. However, this definition is too distant from the original Fourier formulation. On the other hand, Torrence and Webster[61] propose to determine wavelet coherence using a smooth estimate of the wavelet spectrum and to define a smooth wavelet spectrum and cross spectrum, denoted as $SW_{XX}(a,\tau)$ and $SW_{XY}(a,\tau)$, respectively:

$$SW_{XX}(a,\tau) = \int_{t-\delta/2}^{t+\delta/2} W_{XX}^*(a,\tau) W_{XX}(a,\tau) da d\tau \tag{8.57}$$

$$SW_{XY}(a,\tau) = \int_{t-\delta/2}^{t+\delta/2} W_{XX}^*(a,\tau) W_{YY}(a,\tau) da d\tau \tag{8.58}$$

The scalar δ represents the size of the two-dimensional filter.[62] It constitutes an important parameter of the wavelet coherence and must be adequately determined for an acceptable estimation of the wavelet coherence. The wavelet coherence can then be defined by analogy with Fourier coherence as:

$$WC(a,\tau) = \frac{\left|SW_{XY}(a,\tau)\right|}{\sqrt{\left[\left|SW_{XX}(a,\tau)\right| \cdot \left|SW_{YY}(a,\tau)\right|\right]}} \quad (8.59)$$

Schwarz inequality still ensures that WC takes a value between 0 and 1. Statistical confidence limits of wavelet coherency based on red-noise background are discussed in the literature.[30,59,64]

An extension of the wavelet-based estimation of coherence is the wavelet-based bi-coherence analysis that allows identifying transient second-order relationships. Bi-coherence is a well-known technique in Fourier analysis that allows identifying phase-coupling between two signals in order to extract nonlinear information. Given two signals $x(t)$ and $y(t)$, the wavelet cross-bispectrum is given by:

$$B_{xxy}^{W}(a_1, a_2, t) = \int W_x(a_1, t) W_x(a_2, t) W_y(a, t) d\tau \quad (8.60)$$

where $\dfrac{1}{a} = \dfrac{1}{a_1} + \dfrac{1}{a_2}$

and the wavelet bi-coherence is calculated as follows:

$$b_{xxy}^{W}(a_1, a_2, t) = \sqrt{\frac{\left|B_{xxy}^{W}(a_1, a_2, t)\right|^2}{\int \left|W_x(a_1, \tau) W_x(a_2, \tau)\right| d\tau \cdot \int \left|W_y(a, \tau)\right|^2 d\tau}} \quad (8.61)$$

An application of wavelet bi-coherence can be found in Savin et al.,[65] which studies the properties of the turbulent boundary layer over polar caps. An application of wavelet coherence in global hydrology is depicted in Fig. 8.7.

8.8. Applications of Wavelet Transforms in Hydrology and Earth Sciences

A review of wavelets in geosciences can be found in Kumar and Foufoula-Georgiou.[66] Wavelet analyses are a common tool in many

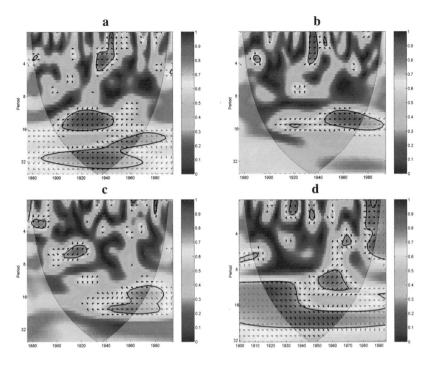

Fig. 8.7. Squared wavelet coherence showing the relations between annual global continental discharge fluctuations and several climatic indices: (a) NAO; (b) AO; (c) SOI; (d) PDO; (e) NINO12; (f) NINO3; (g) NINO34; and (h) NINO4. The relative phase relationship is shown with arrows (with in-phase pointing right, anti-phase pointing left). The thick black contour designates the 95% confidence level against AR(1) red noise, and the cone of influence, where edge effects are not negligible, is shown as a lighter shade. At the global scale, coherence with NAO is put in evidence over the 7- to 10-year band from 1905 to 1940. Then, at larger scales, the band of coherence between global continental discharge and NAO increase from 25- to 33-year band from 1890 to 1970 to a 20- to 25-year band from 1950 to 1990. A relationship between global discharge variability and AO is shown over the 10- to 16-year band considering the 1920–1990 interval. The SOI and global discharge exhibit also some high coherent fluctuations during 1875–1910 in the 2- to 8-year band but also over the 15- to 20-year band from 1940 to 1990. The highest coherent values are obtained with PDO over the 10- to 25-year band over the entire interval. At multi-decadal scales, NINO4 appears as the Pacific SST index that controls the 8- to 32-year variability from 1925 to the current period.

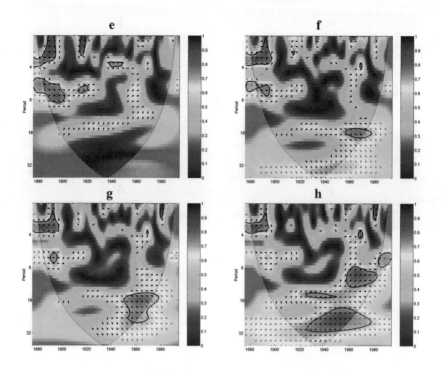

Fig. 8.7 (Continued)

fields of physics, especially in fluid mechanics with isolation of coherent structures in turbulent flows.[34,67-72]

Wavelet analysis has become more and more popular in the earth sciences, especially in order to identify coherent structures in time series. In fact, there is a good chance that wavelets will be used by scientists from all different fields dealing with time series that clearly exhibit transient processes. This popularity is possible largely because of the existence of free wavelet programs (especially in MATLAB) made available through the Web [see, for example, www.amara.com/current/wavelet.html, atoc.colorado.edu/research/wavelets, or Uvi wave toolbox[73] for univariate analyses, and www.pol.ac.uk/home/research/waveletcoherency/ for wavelet bivariate analyses). Table 8.1 provides some important references of wavelet applications in several earth sciences fields. In what follows, a brief review of wavelet applications in hydrology is made.

Table 8.1. Studies on the applications of wavelets in the earth sciences.

Application domain	References
Oceanography	22, 74–80
Mineralogy and Petrography	81, 82
Paleoclimate transitions and Paleo proxies	83–93
Climatology	77, 80, 94–112
Solar cycles and Astrophysics	53, 90, 113–115
Solid Earth	116–120
Precipitation fields and Micrometeorology	35, 36, 121–128
Hydrology	30–32, 37, 99, 129–139
Hydrologic modeling	140–152

In hydrology, wavelets can, at the fundamental level, be used to detect transient or periodic processes in relevant time series. For example, Whitfield and Dohan[131] analyze temperature and conductance time series over two small streams in Canada in order to identify rapid transient processes. A multiscale moving t-test procedure is described by Fraedrich et al.[130] in order to identify rupture in River Nile flood levels. Sujono et al.[144] also propose a discrete wavelet analysis of recession curves in order to discriminate baseflow from direct flow component.

Continuous wavelet transform is well-suited for identification of interannual variability in hydrologic and meteorologic time series. For example, Lafrenière and Sharp[136] apply this technique to characterize the multi-annual variability of two catchments in Canada and relate this variability to El Niño episodes introducing cross-wavelet spectra. The same method is also applied by Gaucherel[135] to flow data from 10 French Guyana rivers relating the multi-annual variability to Atlantic circulation indices, by Labat et al.[37,133] to three French karstic watersheds, by Anctil and Coulibaly,[107] Coulibaly and Burn,[108] and Coulibaly[110] to several watersheds in Québec including both discharge and precipitation time series analyses, and also by Partal and Küçük[111] to identify trends in annual precipitation time series in Turkey. Kang and Lin[147] present an early application of wavelets to study water quality fluctuations in an agricultural watershed, with a case study in Pennsylvania, USA.

Some applications of wavelets in hydrology also concern much longer periods, ranging from a century to several million years. For example, Labat[138] examines the continental runoff variability over the last century, and Labat[139] proposes an extensive wavelet analysis of annual runoff variability of the largest rivers of the world. Saco and Kumar[101] also perform a similar study based on multi-resolution analysis, but restricted to the United States. Smith et al.[132] use both the continuous "Mexican Hat" wavelet analysis and the multi-resolution analysis (more precisely, the repartition of energy across scales) to study streamflow variability over 91 rivers in the US and on tropical islands, towards distinguishing, for example, western snowmelt from "rainstorm" island. Schaefli et al.[30] present a comprehensive review of the applications of all continuous wavelet methods (wavelet spectrum, cross-wavelet spectrum, and wavelet coherence) to daily discharge, temperature, and precipitation over the Vispa river in Switzerland.

Multi-resolution concepts can also be used as a stochastic tool to generate synthetic time series, as pointed out by Aksoy et al.,[142] who propose a statistical model based on Haar multi-resolution analysis that allows producing synthetic suspended sediment discharge over a US watershed. The original procedure of this can be found in Bayazit and Aksoy[153] and Bayazit et al.[140] dealing with streamflow simulation together with comparison to the Fourier transformation. Comparisons of this procedure with some well-known stochastic generation techniques are also presented in the literature.[109,143,154] Recent developments in wavelet hydrologic modeling are also based on the introduction of artificial neural networks (ANNs) coupled to wavelet models, the so-called hybrid wavelet-ANN models.[145,151,155,156]

Further, several studies use the multi-resolution analyses also as a modeling tool to reconstruct or to predict river discharge fluctuations. For example, Labat et al.[31] use cross-correlation coupled with multi-resolution analyses to estimate the annual continental runoff over the last century. Cannas et al.[145] propose a wavelet-based pre-processing method in a neural network application for flow forecasting in the Tisro River basin in Italy. Partal and Cigizoglu[152] present a similar analysis for suspended sediment data simulation. Still other wavelet-based models are presented for estimation of daily discharges in Taiwan catchments,[141,146] for snowmelt flood forecasting in Canadian watersheds,[149] for flow forecasting in the Skrwa Prawa River in Poland,[150] and for evaluation of rainfall-runoff models.[148]

8.9. Challenges and Future Directions

In conclusion, wavelet transforms appear as a powerful tool to identify transient processes in hydrologic (and other geophysical) time series, whereas multi-resolution analysis, for now, appears as restricted to isolate a given component corresponding to a given scale. Wavelet entropy and wavelet coherence are also now being extensively used to provide more detailed information about the timescale. In these ways, wavelet analysis provides a better investigating tool than classical Fourier analysis, and thus deserves far more attention.

However, continuous wavelet transforms are normally used mainly to identify transient periodicity, whereas the global wavelet spectrum allows a precise identification of the corresponding oscillation scale. Classical cross-correlation between isolated components allows identification of correspondence between time-scale processes. However, it must be kept in mind that wavelets cannot be considered as the "Grail," since the interpretation of the results from the wavelet analysis must rely on physical processes.

In my opinion, wavelet research in hydrology should be dedicated to the development of wavelet-based models taking into account the intrinsic multiscale nature of many physical relationships, for example, between rainfall and runoff or sediment fluxes in hydrology. Several models that couple wavelets with other techniques (e.g. artificial neural networks) already exist (see Table 8.1), but appropriate improvments must be made for real-time forecasting. In this way, wavelets will become a more complete and better simulation tool for the study of multiscale properties of hydrologic (and other geophysical) processes.

Acknowledgments

Amazon discharge data are provided by the HYBAM (Hydrology and Geodynamics of the Amazonian Basin) program database. The HYBAM program is a shared partnership between the ANA (National Agency for Water) and the UNB (University of Brasilia) in Brazil and the IRD (Research Institute for the Development) in France. Discharge data for the other various karstic watersheds are from Parc Naturel Regional des Grands Causses and Station d'ecologie Expérimentale de Moulis (CNRS – a. Mangin).

References

1. J. S. Walker, *Notices of the American Mathematical Society*, 658 (1997).
2. A. Papoulis, *Probability, Random Variables and Stochastic Processes* (McGraw Hill, 1964).
3. M. B. Priestley, *Spectral Analysis and Time Series* (Academic Press, 1981).
4. M. Ghil, M. R. Allen, M. D. Dettinger, K. Ide, D. Kondrashov, M. E. Mann, A. W. Robertson, A. Saunders, Y. Tian, F. Varadi and P. Yiou, *Rev. Geophys.*, 1 (2002).
5. J. D. Pelletier and D. L. Turcotte, *J. Hydrol.*, 198 (1997).
6. Y. Tessier, S. Lovejoy, P. Hubert, D. Schertzer and S. Pecknold, *J. Geophys. Res.* **D21**, 26427 (1996).
7. P. Pekarova, P. Miklanek and J. Pekar, *J. Hydrol.*, 62 (2003).
8. A. Davis, A. Marshak, W. Wiscombe and R. Cahalan, *J. Geophys. Res.* **D4**, 8055 (1994).
9. G. Pandey, S. Lovejoy and D. Schertzer, *J. Hydrol.*, 62 (1998).
10. D. Gabor, *J. Inst. Elect. Eng.*, 429 (1946).
11. A. Grosmann and J. Morlet, *SIAM J. Math. Anal.*, 723 (1984).
12. C. E. Heil and D. F. Walnut, *SIAM Rev.*, 628 (1989).
13. B. Jawerth and W. Sweldens, *SIAM Rev.*, 377 (1994).
14. I. Daubechies, *IEEE Trans. Information Theory*, 961 (1990).
15. I. Daubechies, *Ten Lectures on Wavelets. CSBM-NSF Series Applied Mathematics* (SIAM Publications, 1992).
16. S. Mallat, *IEEE Trans. Pattern Analysis and Machine Intelligency*, 674 (1989).
17. G. Beyklin, R. Coifman and V. Rokhin, *Comm. Pure Applied Mathematics*, 141 (1991).
18. G. Strang, *SIAM Rev.*, 614 (1989).
19. L. Hudgins, C. A. Friehe and M. E. Mayer, *Phys. Rev. Lett.*, 3279 (1993).
20. Y. Brunet and S. Collineau, in *Wavelets in Geophysics*, Eds. E. Foufoula-Georgiou and P. Kumar (Academic Press, New York, 1995), p. 129.
21. D. Percival, *Biometrika*, 619 (1995).
22. D. Percival and H. O. Mofjeld, *J. Am. Stat. Assoc.*, 868 (1997).
23. H. Li and T. Nozaki, *JSME Int. J.* **B**, 58 (1997).
24. M. Onorato, M. V. Salvetti, G. Buresti and P. Petagna, *European J. Mech.* **B**, 575 (1997).
25. J. Poggie and A. J. Smits, *AIAA J.*, 1597 (1997).
26. P. C. Liu, in *Wavelets in Geophysics*, Eds. E. Foufoula-Georgiou and P. Kumar (Academic Press, New York, 1995), p. 151.
27. C. Torrence and G. P. Campo, *Bull. Am. Meterol. Soc.*, 61 (1998).
28. Z. Ge, *Annales Geophysicae*, 2259 (2007).
29. Y. Liu, X. S. Liang and R. H. Weisberg, *J. Atmos. Oceanic Tech.*, 2093 (2007).
30. B. Schaefli, D. Maruan and M. Holschneider, *Adv. Water Resour.*, 2511 (2007).

31. D. Labat, J. Ronchail, J. Callède, J. L. Guyot, E. de Oliveira and W. Guimarães, *Geophys. Res. Lett.*, L02501 (2004).
32. D. Labat, *J. Hydrol.*, 275 (2005).
33. C. Meneveau, *J. Fluid Mech.*, 469 (1991).
34. G. G. Katul, M. B. Parlange and C. R. Chu, *Phys. Fluids*, 2480 (1994).
35. B. J. Turner, M. Y. Leclerc, M. Gauthier, K. E. Moore and D. R. Fitzjarrald, *J. Geophys. Res.* **D1**, 1919 (1994).
36. G. G. Katul and M. B. Parlange, *Water Resour. Res.*, 2743 (1995).
37. D. Labat, R. Ababou and A. Mangin, *J. Hydrol.*, 149 (2000).
38. D. L. Donoho, I. M. Johnstone, G. Kerkyacharan and D. Picard, *J. R. Stat. Soc.* **B**, 301 (1995).
39. I. M. Johnstone and B. W. Silverman, *J. R. Stat. Soc.* **B**, 319 (1997).
40. R. Carniel, P. Malisan, F. Barazza and S. Grimaz, *Soil Dynamics and Earthquake Engineering*, 321 (2008).
41. J. E. Bates and H. K. Shepard, *Phys. Lett.* **A**, 416 (1993).
42. V. P. Singh, *J. Water Soc. America*, 1 (2000).
43. C. E. Shannon, *Bell System Tech. J.*, 379 & 623 (1948).
44. A. Rényi, in *Selected Papers of Alfred Rényi*, 526 (1976).
45. J. O. Sonuga, *J. Hydrol.*, 81 (1976).
46. D. Lemire, C. Pharand, J.-C. R. Rajaonah, B. Dubé and A.-R. LeBlanc, *IEEE Trans. Biomedical Engineering*, 1 (2000).
47. N. Wessel, A. Schimann, A. Schirdewan, A. Voss and J. Kurths, in *Lecture Notes in Computer Sciences*, 78 (1933).
48. R. Q. Quiroga, O. A. Rosso, E. Basar and M. Schurmann, *Biological Cybernetics*, 291 (2001).
49. O. A. Rosso, S. Blanco, J. Yordanova, V. Kolev, A. Figliola, M. Cchürmann and E. Basar, *J. Neuroscience Methods*, 65 (2001).
50. O. A. Rosso, M. T. Martin and A. Plastino, *Physica A*, 497 (2003).
51. O. A. Rosso and M. L. Mairal, *Physica A*, 469 (2002).
52. J. Yordanova, V. Kolev, O. A. Rosso, M. Schurmann, O. W. Sakowitz and E. Ozgoren, *J. Neuroscience Methods*, 99 (2002).
53. S. Sello, *New Astronomy*, 105 (2002).
54. S. Blanco, A. Figliola, R. Q. Quiroga, O. A. Rosso and E. Serrano, *Phys. Rev. E*, 932 (1998).
55. D. Labat, R. Ababou and A. Mangin, *Comptes Rendus de l'Académie des Sciences de Paris, Géosciences*, 551 (2002).
56. S. Sello and J. Bellazzini, in *8th European Turbulence Conference* (Barcelona, 2000).
57. Y. Mizuno-Matsumoto, T. Yoshimine, Y. Nii, A. Kato, M. Taniguchi, J. K. Lee, T. S. Ko, S. Date, S. Tamura, S. Shimojo, K. Shinosaki, T. Inouye and M. Takeda, *Epilepsy and Behaviour*, 288 (2001).

58. Y. Mizuno-Matsumoto, G. K. Motamedi, W. R. S. Webber and R. P. Lesser, *Clinical Neurophysiology*, 33 (2002).
59. D. Maruan and J. Kurths, *Nonlinear Proc. Geophys.*, 505 (2004).
60. W. A. Gardner, *Signal Processing*, 113 (1992).
61. C. Torrence and P. J. Webster, *J. Climate*, 2679 (1999).
62. J. P. Lachaux, A. Lutz, D. Rudrauf, D. Cosmelli, M. Le Van Quyen, J. Martinerie and F. Varela, *Neurophysiology Clinic*, 157 (2002).
63. D. Labat, R. Ababou and A. Mangin. *Comptes Rendus de l'Académie des Sciences de Paris* **Serie IIa**, 873 (1999).
64. A. Grinsted, J. C. Lmoore and S. Jevrejeva, *Nonlinear Proc. Geophys.*, 561 (2004).
65. S. Savin, J. Büchner, G. Consolini, B. Nikutowski, L. Zelenyi, E. Amata, H. U. Auster, J. Blecki, E. Dubinin, K. H. Fornacon, H. Kawano, S. Klimov, F. Marcucci, Z. Nemecek, A. Pedersen, J. L. Rauch, S. Romanov, J. Safrankova, J. A. Sauvaud, A. Skalsky, P. Song and Y. Yermolaev, *Nonlinear Proc. Geophys.*, 443 (2002).
66. P. Kumar and E. Foufoula-Georgiou, *Rev. Geophys.*, 385 (1997).
67. F. Argoul, G. Arnéodo, G. Grasseau, Y. Gagne, E. J. Hopfinger and U. Frish, *Nature*, 51 (1989).
68. J. Liandrat and F. Moret-Bailly, *European J. Mech.* **B**, 1 (1990).
69. M. Farge, *Ann. Rev. Fluid Dyn.*, 395 (1992).
70. S. R. Long, R. J. Lai, N. E. Huang and G. R. Spedding, *Dynamics Atmos. Oceans*, 79 (1993).
71. H. Higuchi, J. Lewalle and P. Crane, *Phys. Fluids*, 297 (1994).
72. M. Farge and G. Rabreau, *Comptes Rendus Acad. Sc. Paris* **II**, 1479 (1998).
73. S. S. Gonzales, P. N. Gonzales and G. S. Garcia. *Uvi-Wave Wavelet Toolbox for use with Matlab* (University of Vigo, Department of Communication Technology, 1996), pp. 196.
74. E. P. Flinchem and D. A. Jay, *Estuarine, Coastal and Shelf Science*, 177 (2000).
75. S. D. Meyers, B. G. Kelly and J. J. O'Brien, *Mon. Wea. Rev.*, 2858 (1993).
76. S. D. Meyers and J. J. O'Brien, *Geophys. Res. Lett.*, 777 (1994).
77. K. Higuchi, J. Huang and S. Shabbar, *Int. J. Climatol.*, 1119 (1999).
78. S. Minobe and N. Mantua, *Prog. Ocean.*, 163 (1999).
79. O. A. Lucero and N. C. Rodriguez, *Atmos. Res.*, 219 (2001).
80. L. C. Breaker, P. C. Liu and C. Torrence, *Continental Shelf Research*, 727 (2001).
81. A. Fowler, A. Prokoph, R. Stern and C. Dupuis, *Geochimica and Cosmochimica Acta*, 311 (2002).
82. G. S. Wallace and G. W. Bergantz, *Earth Planet. Sci. Lett.*, 133 (2002).
83. E. W. Bolton, K. A. Maasch and J. M. Lilly, *Geophys. Res. Lett.*, 2753 (1995).
84. J. Jiang, D. Zhang and K. Fraedrich, *Int. J. Climatol.*, 969 (1997).
85. A. Prokoph and J. Veizer, *Chemical Geology*, 225 (1999).
86. Y. Guyodo, P. Gaillot and J. E. T. Channell, *Earth Planet. Sci. Lett.*, 109 (2000).
87. Y. Yokoyama and T. Yamazaki, *Earth Planet. Sci. Lett.*, 7 (2000).

88. D. Heslop, M. J. Dekkers and C. G. Langereis, *Paleogeography Paleoclimatology Paleoecology*, 1 (2002).
89. H. Lu, F. Zhang and X. Liu, *Global Planetary Change*, 67 (2002).
90. N. R. Rigozo, D. J. R. Nordeman, E. Echer, A. Zanandrea and W. D. Gonzales, *Adv. Space Res.*, 1985 (2002).
91. N. R. Rigozo, D. J. R. Nordeman, E. Echer, H. E. da Silva, M. P. Souza Echer and A. Prestes, *J. Atmos. Solar-Terrestrial Phys.*, 449 (2007).
92. N. R. Rigozo, D. J. R. Nordeman, M. P. Souza Echer, E. Echer, H. E. da Silva, A. Prestes and F. L. Guarnieri, *J. Atmos. Solar-Terrestrial Phys.*, 1049 (2007).
93. O. M. Raspopov, V. A. Dergachev, J. Esper, O. V. Kozyreva, D. Frank, M. Ogurtsov, T. Kolström and X. Shao. *Paleogeography Paleoclimatology Paleoecology*, 6 (2008).
94. S. Baliunas, P. Frick, D. Sokoloff and W. Soon, *Geophys. Res. Lett.*, 1351 (1997).
95. T. Benner, *Int. J. Climatol.*, 391 (1999).
96. O. A. Lucero and N. C. Rodriguez, *Atmos. Res.*, 177 (1999).
97. R. H. Compagnucci, S. A. Blanco, M. A. Figliola and P. M. Jacovkis, *Environmetrics*, 251 (2000).
98. S. Janicot and B. Sultan, *Geophys. Res. Lett.*, 523 (2000).
99. M. R. Jury and J. L. Melice, *Theoretical Appl. Climatology*, 161 (2000).
100. J. R. Kulkarni, *Int. J. Climatol.*, 89 (2000).
101. P. Saco and P. Kumar, *Water Resour. Res.*, 1049 (2000).
102. H.-S. Jung, Y. Choi and G.-H. Lim, *Int. J. Climatol.*, 1327 (2002).
103. O. A. Lucero and D. Rozas, *Atmos. Res.*, 35 (2002).
104. P. D. Tyson, G. R. J. Cooper and T. S. McCarthy, *Int. J. Climatol.*, 1105 (2002).
105. S.-K. Min, W.-T. Kwon, E.-H. Park and Y. Choi, *Int. J. Climatol.*, 223 (2003).
106. D. Zheng, X. Ding, Y. Zhou and Y. Chen, *Global Planetary Change*, 89 (2003).
107. F. Anctil and P. Coulibaly, *J. Climate*, 163 (2004).
108. P. Coulibaly and D. H. Burn, *Water Resour. Res.*, W03105, doi:10.1029/2003WR002667 (2004).
109. N. E. Unal, H. Aksoy and T. Akar, *Stoch. Environ. Res. Risk Assess.*, 245 (2004).
110. P. Coulibaly, *Adv. Water Resour.*, 1846 (2006).
111. T. Partal and M. Küçük, *Phys. Chemistry Earth*, 1189 (2006).
112. B. Su, M. Gemmer and T. Jiang, *Quaternary International*, 22 (2008).
113. R. W. Komm, *Solar Physics*, 45 (1995).
114. R. C. Willson and A. V. Mordvinov, *Geophys. Res. Lett.*, 3613 (1999).
115. L. T. Liu, H. T. Hsu, B. X. Gao and B. Wu, *Geophys. Res. Lett.*, 3001 (2000).
116. B.-L. Li and C. Loehle, *Geophys. Res. Lett.*, 3123 (1995).
117. P. Kovacs, V. Carbone and Z. Vörös, *Planetary Space Sci.*, 1219 (2001).
118. L. Vecsey and C. Matyska, *Geophys. Res. Lett.*, 395 (2001).
119. A. T. Y. Lui, *J. Atmos. Solar-Terrestrial Phys.*, 125 (2002).
120. D. Cowan and G. Cooper, *Exploration Geophys.*, 63 (2003).
121. P. Kumar and E. Foufoula-Georgiou, *Water Resour. Res.*, 2515 (1993).

122. N. Takeuchi, K. Narita and Y. Goto, *J. Geophys. Res.* **D5**, 10751 (1994).
123. P. Kumar, *J. Geophys. Res.*, 393 (1996).
124. V. Venugopal and E. Foufoula-Georgiou, *J. Hydrol.*, 3 (1996).
125. G. G. Katul, J. Schieldge, C.-I. Hsieh and B. Vidakovic, *Water Resour. Res.*, 1265 (1998).
126. G. G. Katul, C.-T. Lai, K. Schäfer, B. Vidakovic, J. D. Albertson, D. Ellsworth and R. Oren, *Adv. Water Resour.*, 1119 (2001).
127. J. Szilagyi, M. B. Parlange, G. G. Katul and J. D. Albertson, *Adv. Water Resour.*, 561 (1999).
128. A. C. Furon, C. Wagner-Riddle, C. R. Smith and J. S. Warland, *Agri. Forest Meteorol.*, 1305 (2008).
129. K. Dohan and P. H. Whitfield, *Water Sci. Tech.*, 325 (1997).
130. K. Fraedrich, J. Jiang, F.-W. Gerstengarbe and P. C. Werner, *Int. J. Climatol.*, 1301 (1997).
131. P. H. Whitfield and K. Dohan, *Water Sci. Tech.*, 337 (1997).
132. L. C. Smith, D. L. Turcotte and B. L. Isacks, *Hydrol. Process.*, 233 (1998).
133. D. Labat, R. Ababou and A. Mangin, *Ground Water*, 605 (2001).
134. D. Labat, J. Ronchail and J.-L. Guyot, *J. Hydrol.*, 289 (2005).
135. C. Gaucherel, *J. Hydrol.*, 101 (2002).
136. M. Lafrenieres and M. Sharp, *Hydrol. Process.*, 1093 (2003).
137. R. M. Lark, S. R. Kaffka and D. L. Corwin, *J. Hydrol.*, 276 (2003).
138. D. Labat, *Earth Planet. Sci. Lett.*, 143 (2006).
139. D. Labat, *Adv. Water Resour.*, 109 (2008).
140. M. Bayazit, B. Önöz and H. Aksoy, *Hydrol. Sci. J.*, 623 (2001).
141. C.-M. Chou and R.-Y. Wang, *Hydrol. Sci. J.*, 721 (2002).
142. H. Aksoy, T. Akar and N. E. Unal, *Nord. Hydrol.*, 165 (2004).
143. H. Aksoy, Z. F. Toprak, A. Aytek and N. E. Unal, *Renewable Energy*, 2111 (2004).
144. J. Sujono, S. Shikasho and K. Hiramatsu, *Hydrol. Process.*, 403 (2004).
145. B. Cannas, A. Fanni, L. M. See and G. Sias, *Physics Chemistry Earth*, 1164 (2006).
146. C.-M. Chou, *J. Hydrol.*, 174 (2007).
147. S. Kang and H. Lin, *J. Hydrol.*, 1 (2007).
148. S. N. Lane, *Hydrol. Process.*, 586 (2007).
149. J. F. Adamowski, *J. Hydrol.*, 247 (2008).
150. J. F. Adamkowski, *Hydrol. Process.*, 4877 (2008).
151. Ö. Kisi, *Hydrol. Process.*, 4142 (2008).
152. T. Partal and H. Cigizoglu, *J. Hydrol.*, 317 (2008).
153. M. Bayazit and H. Aksoy, *J. Appl. Stat.*, 157 (2001).
154. H. Aksoy and N. E. Unal, *ASCE J. Hydrol. Eng.*, 699 (2007).
155. R. Wang and T. Lee, *Proceedings of Statistical Methods for Hydrologic Systems*, 7, 08651 (1998).
156. T. W. Kim and J. B. Valdes, *ASCE J. Hydrol. Eng.*, 319 (2003).

CHAPTER 9

NONLINEAR DYNAMICS AND CHAOS IN HYDROLOGY

Bellie Sivakumar

School of Civil and Environmental Engineering,
The University of New South Wales, Sydney, NSW 2052, Australia
E-mail: s.bellie@unsw.edu.au
and
Department of Land, Air and Water Resources,
University of California, Davis, CA 95616, USA
E-mail: sbellie@ucdavis.edu

Ronny Berndtsson

Department of Water Resources Engineering, Lund University,
Box 118, S-221 00, Lund, Sweden
E-mail: ronny.berndtsson@tvrl.lth.se

During the last two decades, applications of nonlinear dynamic and chaos theories to hydrologic systems and processes have been on the rise. Early applications of these theories focused mainly on the investigation and prediction of chaos in rainfall and river flow dynamics. Subsequent years witnessed their applications for other purposes (e.g. data disaggregation, missing data estimation, reconstruction of system equations) and to other processes (e.g. rainfall-runoff, sediment transport). More recently, additional inroads have been made through their applications to the problems of scaling, groundwater contamination, parameter estimation, and catchment classification. The outcomes of these studies are certainly encouraging, especially considering the exploratory stage of the concepts in hydrologic sciences. The objectives of this chapter are: (1) to provide a comprehensive review of the applications of nonlinear dynamic and chaos theories in hydrology; and (2) to discuss the hope and scope for the future and also the challenges that lie ahead. In regards to the challenges, particular emphasis is given to discussing the need to improve our understanding of these largely less-understood concepts and to find appropriate ways for integrating them with other concepts that are already in existence or emerging. With the clear recognition that none of the existing one-sided 'extreme-view' modeling approaches is capable of solving the hydrologic problems, an urgent call for a balanced 'middle-ground' approach that can integrate different methods is also made.

9.1. Introduction

The nonlinear nature of hydrologic phenomena has been known for many decades.[1-3] However, much of early applications of data-based (i.e. time series) methods in hydrology (particularly during 1960s–1980s), constrained by the lack of computational power and measurement technology, essentially resorted to linear stochastic approaches.[4-8] Although the linear approaches continue to be prevalent in hydrology, advances in computational power and measurement technology during the last two decades or so have facilitated formulation of nonlinear approaches as viable alternatives. The nonlinear approaches that are popular in hydrology include: nonlinear stochastic methods,[9] artificial neural networks,[10] data-based mechanistic models,[11] and deterministic chaos theory.[12] The outcomes of applications of nonlinear approaches for hydrologic modeling and forecasting are certainly encouraging, especially considering the fact that we are still in the 'exploratory stage' in regards to these approaches, as opposed to the much more established linear stochastic approaches.

Among the nonlinear approaches, chaos theory seems to be the simplest (at least in its view of 'complex' systems) yet also remains the most controversial (see Schertzer et al.[13] and Sivakumar et al.[14] for a discussion in hydrologic studies). In the nonlinear science literature, the term 'chaos' is normally used to refer to situations where complex and random-looking behaviors arise from simple nonlinear deterministic systems with sensitive dependence on initial conditions[15-18] (the converse also applies). The three fundamental properties inherent in this definition: (i) nonlinear interdependence; (ii) hidden determinism and order; and (iii) sensitivity to initial conditions are highly relevant in hydrologic systems and processes. For example: (i) components and mechanisms involved in the hydrologic cycle act in a nonlinear manner and are also interdependent; (ii) daily cycle in temperature and annual cycle in river flow possess determinism and order; and (iii) contaminant transport phenomena in surface and subsurface waters largely depend upon the time (i.e. rainy or dry season) at which the contaminants were released at the source, which themselves may not be known. The first property represents the 'general' nature of hydrologic phenomena, whereas the second and third represent their 'deterministic' and 'stochastic' natures, respectively. Further, despite their complexity and random-looking behavior, hydrologic phenomena may also be governed by a very few degrees of freedom (e.g. runoff in a well-developed urban catchment depends essentially on rainfall), another fundamental idea of chaos theory.

These properties, by their very nature, facilitate pattern recognition and accurate short-term predictions but, at the same time, limit long-term predictions. Therefore, one of the purposes behind the investigation of chaos in hydrologic time series is to try to obtain better short-term predictions using simpler models than those possible using other approaches, which are either probabilistic both in the short-term and in the long-term (e.g. stochastic approaches) or more complex. In view of these, chaos theory has been finding increasing applications in hydrology in recent times, amid the controversy.

Early applications of chaos theory in hydrology focused mainly on the investigation and prediction of chaos in rainfall, river flow, and lake volume time series in a purely single-variable data reconstruction sense.[19-27] During the subsequent years, chaos theory was applied for other purposes, such as scaling and data disaggregation, missing data estimation, and reconstruction of system equations,[28-32] and other processes, such as rainfall-runoff and sediment transport.[33-36] They also addressed some of the important issues that had been, and continue to be, perceived to significantly influence the outcomes of chaos methods when applied to real hydrologic data, including issues of minimum data size, data noise, presence of zeros, selection of optimal parameters, and multivariable data reconstruction.[28,37-45] Further, they investigated the 'superiority' of chaos theory, if any, over other theories, such as stochastic methods and artificial neural networks, for prediction purposes.[40,41,46-49] More recently, studies have applied the ideas of chaos theory to either advance the earlier studies (on scaling, for example) or to address yet other hydrologic processes and problems, including groundwater contamination, parameter estimation, and catchment classification,[50-61] while at the same time also continuing investigations into the potential problems with chaos identification methods.[62-66]

The realization and recognition, in the aftermath of the encouraging outcomes from most of the above studies, that chaos theory could provide new perspectives and alternative avenues towards modeling and forecasting hydrologic systems and processes have been important driving forces for its ever-increasing applications, despite the continuing skepticisms that are thrown away from some quarters of the hydrologic community citing possible 'blind' applications of these 'less-understood' concepts without recognizing their potential limitations for real hydrologic data (resulting in 'false claims'). Although this is indeed heartening, we must also not lose sight of the fact that the true potential of chaos theory in hydrology can only be realized when it is attempted to solve the more challenging problems we are faced with today (e.g. hydrologic scaling, model parameter estimation, generalization framework) and will be in the future (e.g. incorporating the effects of global climate change), rather than simply chaos detection and prediction (for historical data, to be more precise). Identification of these challenging problems and evaluation of how chaos theory (either independently or in combination with others in an integrated manner) can be helpful towards solving them are crucial for true progress in hydrology. These issues are the motivation for this chapter.

The rest of this chapter is organized as follows. First, some of the methods commonly used for chaos identification and prediction (with specific reference to hydrologic time series) are introduced. Next, a comprehensive review of chaos studies in hydrology is presented. This then leads to a discussion of the current status and the scope for further research, including potential directions and challenges that lie ahead. A strong case is finally made (both philosophically and scientifically) for the urgent need to formulate a 'middle-ground' approach towards a more balanced and realistic representation of all the relevant properties of hydrologic systems and processes, rather than sticking to the one-sided 'extreme views' that unfortunately prevail in our current research practice; the vital role of chaos theory in this formulation is also highlighted.

9.2. Chaos Identification and Prediction

9.2.1. *Autocorrelation Function and Power Spectrum*

In the analysis of time series for identification of system properties, it is customary to use two basic linear tools: autocorrelation function and power spectrum. The autocorrelation function is a normalized measure of the linear correlation among successive values in a time series, and its utility lies in its ability to determine the degree of dependence present in the values. For a purely random process, the autocorrelation function fluctuates randomly about zero, indicating that the process at any certain instance has no 'memory' of the past at all, and for a periodic process, the autocorrelation function is also periodic, indicating the strong relation between values that repeat over and over again; the autocorrelation function of signals from a chaotic process, however, is expected to decay exponentially with increasing lag, because the states of a chaotic process are neither completely dependent nor completely independent of each other (although this is not always the case). The power spectrum, on the other hand, is particularly useful for identifying the regularities/irregularities in a time series. For a purely random process, the power spectrum oscillates randomly about a constant value, indicating that no frequency explains any more of the variance of the sequence than any other frequency; for a periodic or quasi-periodic sequence, only peaks at certain frequencies exist, measurement noise adds a continuous floor to the spectrum and, thus, in the spectrum, signal and noise are readily distinguished; chaotic signals may also have sharp spectral lines but even in the absence of noise there will be continuous part (broadband) of the spectrum, which is an immediate consequence of the exponentially decaying autocorrelation function.

Although autocorrelation function and power spectrum provide compelling distinctions between random and periodic (or quasi-periodic) signals, they are not reliable for distinguishing between random and chaotic signals. This is demonstrated herein by employing them to two artificially generated time series (Fig. 9.1(a) and (b)) that look very much alike (both look 'complex' and 'random') but nevertheless are the outcomes of systems (equations) possessing significantly different dynamic characteristics. The first series (Fig. 9.1(a)) is the outcome of a pseudo random number generation function:

$$X_i = rand(\ ), \tag{9.1}$$

which yields independent and identically distributed numbers (between 0 and 1). The second (Fig. 9.1(b)), however, is the outcome of a fully deterministic simple two-dimensional map:[67]

$$X_{i+1} = a - X_i^2 + bY_i \qquad Y_{i+1} = X_i, \tag{9.2}$$

which yields irregular solutions for many choices of a and b, but for $a = 1.4$ and $b = 0.3$, a typical sequence of X_i is chaotic. The initial values of X and Y used for this data series are 0.13 and 0.50, respectively.

Figure 9.1(c) and (d) shows the autocorrelation functions for these two series, while Fig. 9.1(e) and (f) presents the power spectra. It is clear that both the tools fail to distinguish between the two series. The failure is not just in 'visual' or 'qualitative' terms, but also in 'quantitative' terms: for instance, for both series, the time lag at which the autocorrelation function first crosses the zero line is equal to 1 (especially no exponential decay for the chaotic series) and the power spectral exponent is equal to 0 (indicating pure randomness in the underlying dynamics of both).

9.2.2. *Phase Space Reconstruction*

Phase space is another useful tool for representing the evolution of a system in time. It is essentially a graph or a coordinate diagram, whose coordinates represent the variables necessary to completely describe the state of the system at any moment.[68] The trajectories of the phase space diagram describe the evolution of the system from some initial state, which is assumed to be known, and, hence, represent the history of the system. The 'region of attraction' of these trajectories in the phase space provides at least important qualitative information on the 'extent of complexity' of the system.

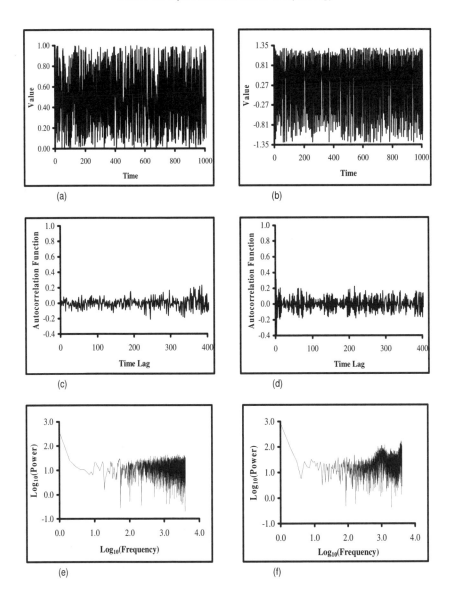

Fig. 9.1. Random versus chaotic data: (a) and (b) time series; (c) and (d) autocorrelation function; (e) and (f) power spectra.

For a dynamic system with known partial differential equations (PDEs), the system can be studied by discretizing the PDEs, and the set of variables at all grid points constitutes a phase space. One difficulty in constructing the phase space for such a system is that the (initial) values of many of the variables may be unknown. However, a time series of a single variable may be available, which may allow the attractor (a geometric object characterizing the long-term behavior of a system in the phase space) to be reconstructed. The idea behind such a reconstruction is that a nonlinear system is characterized by self-interaction, and a time series of a single variable can carry the information about the dynamics of the entire multivariable system. It must be noted that phase space may also be reconstructed using multiple variables when available.[69]

Many methods are available for phase space reconstruction from an available time series. Among these, the method of delays[70] is the most widely used one. According to this method, given a single variable series, X_i, where $i = 1, 2, ..., N$, a multi-dimensional phase space can be reconstructed as:

$$Y_j = (X_j, X_{j+\tau}, X_{j+2\tau}, ..., X_{j+(m-1)\tau}) \quad (9.3)$$

where $j = 1, 2, ..., N - (m - 1)\tau$, m is the dimension of the vector Y_j, called embedding dimension; and τ is an appropriate delay time (an integer multiple of sampling time). A correct phase space reconstruction in a dimension m generally allows interpretation of the system dynamics in the form of an m-dimensional map f_T, given by:

$$Y_{j+T} = f_T(Y_j) \quad (9.4)$$

where Y_j and Y_{j+T} are vectors of dimension m, describing the state of the system at times j (current state) and $j + T$ (future state), respectively. With Eq. (9.4), the task is basically to find an appropriate expression for f_T (e.g. F_T) to predict the future.

To demonstrate its utility for system identification, Fig. 9.2(a) and (b) presents the phase space plots for the above two series (Fig. 9.1(a) and (b)). These diagrams correspond to reconstruction in two dimensions ($m = 2$) with delay time $\tau = 1$, i.e. the projection of the attractor on the plane $\{X_i, X_{i+1}\}$. For the first set, the points (of trajectories) are scattered all over the phase space (i.e. absence of an attractor), a clear indication of a 'complex' and 'random' nature of the underlying dynamics and potentially of a high-dimensional system. On the other hand, the projection for the second set yields a very clear attractor (in a well-defined region), indicating a 'simple' and 'deterministic' (yet non-repeating) nature of the underlying dynamics and potentially of a low-dimensional system.

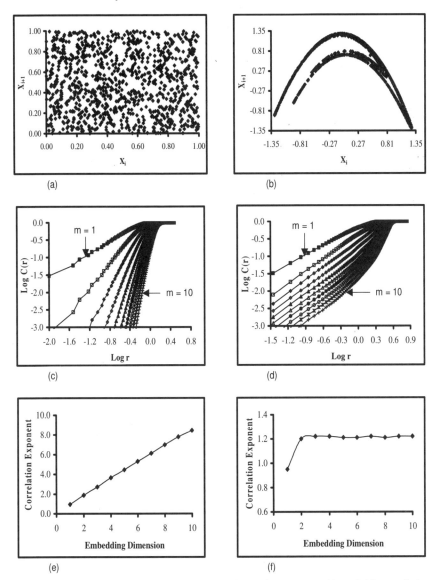

Fig. 9.2. Random versus chaotic data: (a) and (b) phase space; (c) and (d) correlation integral; (e) and (f) correlation exponent.

9.2.3. Correlation Dimension Method

The dimension of a time series is, in a way, a representation of the number of dominant variables present in the evolution of the corresponding dynamic system. Dimension analysis reveals the extent to which the variations in the time series are concentrated on a sub-set of the space of all possible variations. Correlation dimension is a measure of the extent to which the presence of a data point affects the position of the other points lying on the attractor in (a multi-dimensional) phase space. The correlation dimension method uses the correlation integral (or function) for determining the dimension of the attractor in the phase space and, hence, for distinguishing between low-dimensional and high-dimensional systems. The concept of the correlation integral is that a time series arising from deterministic dynamics will have a limited number of degrees of freedom equal to the smallest number of first-order differential equations that capture the most important features of the dynamics. Thus, when one constructs phase spaces of increasing dimension, a point will be reached where the dimension equals the number of degrees of freedom, beyond which increasing the phase space dimension will not have any significant effect on correlation dimension.

Many algorithms have been formulated for the estimation of the correlation dimension of a time series. Among these, the Grassberger-Procaccia algorithm[71,72] has been and continues to be the most widely used one. The algorithm uses the concept of phase space reconstruction for representing the dynamics of the system from an available single-variable time series, as presented in Eq. (9.3). For an m-dimensional phase space, the correlation function $C(r)$ is given by

$$C(r) = \lim_{N \to \infty} \frac{2}{N(N-1)} \sum_{\substack{i,j \\ (1 \leq i < j \leq N)}} H\left(r - \|Y_i - Y_j\|\right) \qquad (9.5)$$

where H is the Heaviside step function, with $H(u) = 1$ for $u > 0$, and $H(u) = 0$ for $u \leq 0$, where $u = r - \|Y_i - Y_j\|$, r is the vector norm (radius of sphere) centered on Y_i or Y_j. If the time series is characterized by an attractor, then $C(r)$ and r are related according to:

$$C(r) \underset{\substack{r \to 0 \\ N \to \infty}}{\approx} \alpha r^{\nu} \qquad (9.6)$$

where α is a constant and ν is the correlation exponent or the slope of the Log $C(r)$ versus Log r plot. The slope is generally estimated by a least square fit of a straight line over a certain range of r (scaling regime) or through estimation of local slopes between r values.

The distinction between (low-dimensional) determinism and (high-dimensional) stochasticity can be made using the ν versus m plot. If ν saturates after a certain m and the saturation value is low, then the system is generally considered to exhibit (low-dimensional) deterministic dynamics. The saturation value of ν is defined as the correlation dimension (d) of the attractor, and the nearest integer above this value is generally an indication of the number of variables dominantly governing the dynamics. On the other hand, if ν increases without bound with increase in m, the system under investigation is generally considered to exhibit (high-dimensional) stochastic behavior.

To demonstrate the utility of the dimension concept, Fig. 9.2(c) and (e) presents the correlation dimension results for the first set, whereas those for the second set are shown in Fig. 9.2(d) and (f). In each case, embedding dimensions from 1 to 10 are used for phase space reconstruction. It is clear that the first set is indeed the outcome of an infinite-dimensional system, i.e. absence of saturation in dimension (Fig. 9.2(e)), whereas the second set is the outcome of a low-dimensional system (with a correlation dimension value of 1.22, Fig. 9.2(f)).

9.2.4. *Nonlinear Prediction Method and Inverse Approach*

The nonlinear prediction method serves two different (yet related) purposes in the context of chaos analysis: (i) prediction of the future, normally using a local approximation technique; and (ii) identification of chaos from the prediction results, also termed as the 'inverse approach.' This inverse approach seems more reliable than the correlation dimension (or any other invariant) method for chaos identification, since it is essentially based on predictions, which is one of the primary purposes behind characterizing a system (stochastic or deterministic) for appropriate model selection, among others. This is particularly the case when prediction accuracy against lead time is used as an indicator,[73] since a rapid decrease in prediction accuracy with increase in lead time is a typical characteristic of chaotic systems. However, the nonlinear prediction method is used mostly as a prediction tool (rather than an identification tool) in chaos studies (especially in hydrology), and thus the emphasis herein is also the same.

In the local approximation prediction method, the f_T domain (Eq. (9.4)) is sub-divided into many sub-sets (neighborhoods), each of which identifies some approximations F_T, valid only in that sub-set. In this way, the underlying system dynamics are represented step by step locally in the phase space. The identification of the sets in which to sub-divide the domain is done by fixing a metric $\| \ \|$ and, given the starting point Y_j from which the forecast is initiated, identifying neighbors Y_j^p, $p = 1$, 2, ..., k, with $j^p < j$, nearest to Y_j, which constitute the set corresponding to Y_j. The local functions can then be built, which take each point in the neighborhood to the next neighborhood: Y_j^p to Y_{j+1}^p. The local map F_T, which does this, is determined by a least squares fit minimizing

$$\sum_{p=1}^{k} \| Y_{j+1}^p - F_T Y_j^p \|^2 \qquad (9.7)$$

The local maps can be learned in the form of local averaging[74] or local polynomials.[75] The local averaging technique has an important advantage over the local polynomial technique in the sense that it is computationally inexpensive, but the local polynomial technique seems to provide more accurate results.

The prediction accuracy can be evaluated using a variety of measures, such as correlation coefficient, root mean square error, and coefficient of efficiency. In addition, direct time series plots and scatter diagrams can also be used to choose the best prediction results among a large combination of results obtained with different embedding dimensions (m) and number of neighbors (k). The distinction between deterministic and stochastic systems can be assessed using the inverse approach, as follows:

(i) General performance: A high prediction accuracy may be an indication that the underlying dynamics are deterministic, whereas a low prediction accuracy is expected if the dynamics are stochastic; and

(ii) Prediction accuracy against the parameters involved:

(a) Embedding dimension (m): If the time series exhibits determinism, then the prediction accuracy would increase to its best with the increase in the embedding dimension up to a certain point (low value of m), called the optimal embedding dimension (m_{opt}), and would remain close to its best for embedding dimensions higher than m_{opt}. For stochastic time series, there would be no increase in the prediction accuracy with an increase in the embedding dimension and the accuracy would remain the same for any value of m;[76]

(b) Lead time (T): For a given embedding dimension, predictions in deterministic systems deteriorate considerably faster than in stochastic systems when the lead time is increased. This is due to the sensitivity of deterministic chaotic systems to initial conditions;[73] and

(c) Neighbors (k): Smaller number of neighbors would give the best predictions if the system dynamics are deterministic, whereas for stochastic systems the best predictions are achieved when the number of neighbors is large.[77]

9.2.5. Some Other Methods

Other methods that have been employed for identification and prediction of chaos in hydrologic time series include: the false nearest neighbor algorithm,[78] the Lyaponov exponent method,[79] the Kolmogorov entropy method,[80] and the surrogate data method[81] (to detect nonlinearity). A brief account of the concepts underlying these methods is presented next.

The false nearest neighbor algorithm provides information on the embedding dimension of the phase space (in other words, number of dominant variables) required for representing the system dynamics. It examines, in dimension m, the nearest neighbor Y_j^{NN} of every vector Y_j, as it behaves in dimension $m + 1$. If the vector Y_j^{NN} is a true neighbor of Y_j, then it comes to the neighborhood of Y_j through dynamic origins. On the other hand, if the vector Y_j^{NN} moves far away from vector Y_j as the dimension is increased, then it is declared a false nearest neighbor as it arrived in the neighborhood of Y_j in dimension m by projection from a distant part of the attractor. When the percentage of these false nearest neighbors drops to zero, the geometric structure of the attractor has been unfolded and the orbits of the system are now distinct and do not cross (or overlap). A key step in the false nearest neighbor algorithm is to determine how to decide upon increasing the embedding dimension that a nearest neighbor is false (see, for example, Kennel et al.[78] and Sangoyomi et al.[25] for additional details).

Lyapunov exponents are the average exponential rates of divergence or convergence of nearby orbits in the phase space. Since nearby orbits correspond to nearly identical states, exponential orbital divergence means that systems whose initial differences that may not be possible to resolve will soon behave quite differently, i.e. predictive ability is rapidly lost. Any system containing at least one positive Lyaponov exponent is defined to be chaotic, with the magnitude of the exponent reflecting the timescale on which system dynamics become unpredictable. Wolf et al.[79] presented an algorithm for computation of Lyapunov exponents from a time series, which is still widely used.

Kolmogorov entropy is the mean rate of information created by the system. It is important in characterizing the average predictability of a system of which it represents the sum of the positive Lyapunov exponents. The Kolmogorov entropy quantifies the average amount of new information on the system dynamics brought by the measurement of a new value of the time series. In this sense, it measures the rate of information produced by the system, being zero for periodic or quasi-periodic (i.e. completely predictable) time series, and infinite for white noise (i.e. unpredictable by definition), and between the two for chaotic system. The algorithm proposed by Grassberger and Procaccia[80] is widely used for Kolmogorov entropy estimation in hydrology.

The surrogate data method makes use of the substitute data generated in accordance to the probabilistic structure underlying the original data. This means that the surrogate data possess some of the properties (e.g. mean, standard deviation, cumulative distribution function, power spectrum), but are otherwise postulated as random, generated according to a specific null hypothesis. The null hypothesis here consists of a candidate linear process, and the goal is to reject the hypothesis that the original data have come from a linear stochastic process. The rejection of the null hypothesis can be made based on some discriminating statistics. In studies on chaos identification, it would be desirable to use any of the above 'invariants' (e.g. correlation dimension, prediction accuracy, Lyapunov exponent, Kolmogorov entropy). If the discriminating statistics obtained for the surrogate data are significantly different from those of the original data, then the null hypothesis can be rejected, and the original data be considered to have come from a nonlinear process. On the other hand, if the discriminating statistics obtained for the original data and surrogate data are not significantly different, then the null hypothesis cannot be rejected, and the original time series is considered to have come from a linear stochastic process.

9.2.6. *Remarks on Chaos Identification and Prediction Methods*

The reliability of the chaos identification and prediction methods for distinguishing between low-dimensional (chaotic) and high-dimensional (stochastic) systems has been under considerable debate, in view of their potential limitations when applied to real data, and much of the criticism has been directed at the Grassberger-Procaccia correlation dimension algorithm. Some of the relevant issues are: data size, data noise, presence of zeros, delay time selection, and even stochastic processes yielding low correlation dimensions.[13,24,43,66,82-89] Many studies have extensively responded to these concerns and discussed the reliability of the methods.[12,14,28,40,64,65,88,90] Addressing these issues is beyond the scope of this chapter. Suffice it to say, however, that the observations made earlier (Figs. 9.1 and 9.2) present testimony to the utility of these methods (and their superiority over some linear methods) for identification of nonlinear properties of time series and their predictions.

9.3. Chaos Theory in Hydrology: A Review

The ideas of chaos theory have thus far been applied to study numerous hydrologic systems, processes, and problems. Extensive reviews of these applications and the associated issues have already been made available in the literature.[12,91,92] In this section, some selected studies are highlighted, with reference to the hydrologic process/problem studied.

9.3.1. Rainfall

The first study, to our knowledge, on chaos in rainfall was conducted by Hense,[93] who applied the correlation dimension method to a series of 1008 values of monthly rainfall observed in Nauru Island and reported the presence of chaos on observing a low correlation dimension (between 2.5 and 4.5). Rodriguez-Iturbe et al.[19] analyzed a storm event recorded at 15-s intervals in Boston, USA and a weekly rainfall series from Genoa, Switzerland in their chaos investigation of rainfall. Employing the correlation dimension method, they reported the presence of chaos in the storm data (correlation dimension 3.78) and absence of chaos in the weekly rainfall data (no finite correlation dimension). They further supported their claim on the presence of chaos in the storm event through the observation of a positive Lyapunov exponent. The chaotic nature of storm events was also verified by Sharifi et al.,[94] who examined fine-increment data from three storms observed in Boston and reported low correlation dimensions of 3.35, 3.75, and 3.60, respectively. Islam et al.[95] reported the presence of chaos in rainfall by analyzing rainfall intensity data generated at 10-s time steps from a three-dimensional cloud model and observing a correlation dimension of 1.5. Tsonis et al.[88] studied data representing the time between successive raingage signals each corresponding to a collection of 0.01 mm of rain and reported the presence of chaos based on observation of a correlation dimension of 2.4.

Jayawardena and Lai[22] employed four different chaos identification methods (correlation dimension, Lyapunov exponent, Kolmogorov entropy, and local approximation prediction) to daily rainfall data from three stations in Hong Kong. The study provided convincing evidence regarding the presence of chaos in these rainfall data sets, and also the local approximation method was found to be superior to the linear autoregressive moving average (ARMA) method. Waelbroeck et al.[96] attempted prediction of daily tropical rainfall using local approximation method and observed that the prediction accuracy for daily rainfall dropped off quickly within a timescale of two days but the prediction was much better for the 10-day rainfall accumulations. Berndtsson et al.[21] studied, using the correlation dimension method, a 238-year monthly rainfall series recorded in Lund, Sweden, for the presence of chaos. They found no evidence of chaos in the raw rainfall series, but the noise-reduced series (first of its kind in hydrology in the context of chaos theory) was found to exhibit chaos with a dimension less than 4.

Analyzing data from 11 storm events in Iowa City, USA, Georgakakos et al.[97] reported the presence of chaos in all the events except one, with dimensions ranging from 2.8 to 7.9 in the high-intensity region and from 0.5 to 1.6 in the low-intensity region. Puente and Obregon[26] reported the presence of chaos in the Boston storm event[19] through application of correlation dimension, Kolmogorov entropy, false nearest neighbors, and Lyapunov exponent methods. Through presentation of a deterministic fractal-multifractal (FM) approach for modeling the storm event, they also hinted that a stochastic framework for rainfall modeling might not be necessary. Koutsoyiannis and Pachakis,[24] however, observed that a synthetic rainfall series (rainfall depths measured at 15 min) generated by a well-structured stochastic rainfall model might be practically indistinguishable from a historic rainfall series even if one used tools of chaotic dynamic theory for characterization.

Sivakumar et al.[98] reported preliminary evidence as to the presence of chaos in the daily rainfall series from Singapore, through observation of low correlation dimensions for data from six stations; they also addressed the issue of minimum data size requirement for correlation dimension estimation by analyzing rainfall records of different lengths. Sivakumar et al.[38] then extended the above investigation using the local approximation prediction method, including the deterministic versus stochastic (DVS) approach,[77] and presented further evidence to the presence of chaos. For the first time in hydrology, the surrogate data method was employed in this study to detect the absence of linearity in rainfall series. Studying the influence of noise in the rainfall data on correlation dimension estimate and prediction accuracy, Sivakumar et al.[39] proposed a systematic approach for noise reduction by coupling a noise level determination method[99] and a noise reduction method.[100]

As part of their search for chaos in rainfall-runoff, Sivakumar et al.[33,34] analyzed the monthly rainfall data observed over a period of 131 years in the Göta River basin in Sweden. Employing the correlation dimension method and the local approximation method, they reported presence of chaos in the monthly rainfall, with a correlation dimension value of 6.4. Studying rainfall data observed at four different temporal scales (i.e. daily, 2-day, 4-day, and 8-day) in the Leaf River basin in the USA, Sivakumar[28] suggested possible presence of chaos in rainfall process at each of these scales and, hence, in the scaling or fractal relationship between them. The study, however, also revealed a potential limitation of the correlation dimension method in the presence of a large number of zeros, and also that the issue of (minimum) data size requirement was not as severe as it was believed to be. Furthering this study, through application of both chaos and fractal methods to the above daily rainfall series from the Leaf River basin, Sivakumar[29] suggested the possibility of a chaotic multifractal approach for modeling rainfall process, rather than just a stochastic (i.e. random cascade) fractal approach that has been prevalent in hydrology. Encouraged by these results, Sivakumar et al.[30] investigated the utility of the concept of chaos for rainfall disaggregation purposes. Analyzing distributions of weights between rainfall process at six successively doubled temporal resolutions (6-hr, 12-hr, 24-hr, 48-hr, 92-hr, and 192-hr) from the Leaf River basin and observing evidence of chaotic behavior, they formulated a chaotic approach for rainfall disaggregation.

Tsonis and Georgakakos[101] addressed the problem of rainfall estimation from satellite imagery while investigating the possibility of deriving useful insights about the variability of the system from only a part of the complete state vector. Postulating a low-order observable vector and a system response as linear functions of portions of the state vector, they first conducted a numerical study on a toy model representing a low-dimensional dynamic system (Lorenz map) and then applied the approach to satellite images (spatial resolution: 4 km x 4 km; temporal resolution: 3 hrs) over the Des Moines River basin in Iowa, USA. They showed that, while reducing the number of observables reduces the correlation between actual and inferred rainfall amounts, good estimates for extremes are still recoverable. Gaume et al.[102] studied rainfall weights to investigate the presence of chaos, in a somewhat similar manner to the study by Sivakumar et al.[30] Employing the correlation dimension method to an 8-year rainfall weight series obtained by disaggregating a 10-min series to a 5-min series, they neither observed low-dimensional chaotic behavior nor that the data were composed of independent and identically distributed random variables. The results also suggested that the correlation dimension method could be an effective tool for exploring data also in other contexts in addition to chaos analysis. As part of the investigation on the potential problems in the application of chaos identification methods to real hydrologic series, Koutsoyiannis[66] analyzed the daily rainfall series from Vakari in western Greece (and also daily streamflow series from Pinios River in Greece) and reported absence of chaos (in both). It was also shown, through theoretical analyses, that specific peculiarities of hydrologic processes on fine timescales (e.g. asymmetric, J-shaped distribution functions, intermittency, and high autocorrelations) are synergistic factors that could lead to misleading conclusions regarding the presence of low-dimensional deterministic chaos. Sivakumar et al.[103] investigated the dynamic nature of rainfall in California's Sacramento Valley. They studied rainfall data observed at four different temporal scales between daily and monthly scales (i.e. daily, weekly, biweekly, and monthly). Employing the correlation dimension method, they reported that the rainfall dynamics were dominated by a large number of variables at all these scales but also that dynamics at coarser resolutions were more irregular than at finer resolutions. Comparison of all-year and winter rainfall, in an attempt to investigate the effects of zeros, showed that winter rainfall had a higher variability.

9.3.2. *River Flow*

To our knowledge, the first studies on chaos in river flow were conducted by Wilcox et al.,[20,104] who investigated daily snowmelt runoff measured from the Reynolds Mountain catchment in the Owyhee Mountains of southwestern Idaho, USA. As the application of the correlation dimension method to this series did not yield a low dimension, they concluded that the random-appearing behavior of snowmelt runoff was generated from the complex interactions of many factors, rather than low-dimensional chaotic dynamics. Jayawardena and Lai,[22] however, reported presence of chaos in river flow dynamics, based on their analysis of daily streamflow series from two stations in Hong Kong. The study was also the first ever to attempt predictions of river flow using chaos theory and to compare the results with those from stochastic (ARMA) method. Porporato and Ridolfi[105] reported further evidence on the presence of chaos in river flow through analysis of the flow data from Dora Baltea, a tributary of the River Po in Italy. Employing the correlation dimension and local approximation prediction (and also phase space and Poincare section) methods to the daily flow series, they observed clues to the existence of a strong deterministic component. The study also led to a more detailed analysis of the flow phenomenon,[27] including noise reduction, interpolation, and nonlinear prediction, which provided important confirmations regarding its nonlinear deterministic behavior.

Applying the local approximation prediction method to the daily streamflow series observed at 28 selected stations in the continental United States, Liu et al.[106] reported that streamflow signals spanned a wide dynamic range between deterministic chaos and periodic signal contaminated with additive noise. Analysis of the unregulated streamflow series of six rivers in the Canadian prairies provided possible signs of deterministic chaotic behavior, as Wang and Gan[37] reported correlation dimensions of about 3.0 for these series. However, observing a consistent underestimation of the correlation dimension for the randomly re-sampled data by an amount of 4 to 6, they interpreted that the actual dimensions of these streamflow series should be between 7 and 9, and that the streamflow process might be stochastic. Investigation of the Scandinavian river flow regimes for possible presence of chaos was attempted by Krasovskaia et al.,[107] who analyzed river flow series observed in a number of stations in the region. The results revealed a variety of fractal and intrinsic dimensions for the series that were well in agreement with the stability character of the investigated regime types, i.e. the less stable the regime, the higher the fractal and intrinsic dimensions and the number of variables required for its description. Stehlik[108] studied the runoff process at the experimental basin Uhlirska in the Jizera Mountains, Czech Republic, for identifying chaos. Analyzing data recorded at daily and 30-min timescales using the correlation dimension method, he reported presence of chaos in the 30-min series with a correlation dimension of 2.89 but absence of chaos in the daily series.

Pasternack,[109] studying daily flow series observed in Western Run in upper Baltimore County, Maryland in the Gunpowder River basin in USA, reported absence of chaos in river flow process. Citing the pitfalls in the implementation of the correlation dimension algorithm, he criticized the earlier studies that had reported presence of chaos in river flow process in particular, and hydrologic processes in general. Liaw et al.,[110] however, questioned the study by Pasternack,[109] commenting that the embedding parameters (mainly the delay time) for phase space reconstruction had not been properly selected and, therefore, that the attractor reconstructed in the correlation dimension analysis and surrogate data analysis could not actually represent the dynamic behavior of underlying dynamics of the system. Jayawardena and Gurung[41] made a systematic effort to address the issue of presence/absence of chaos in river flow (and other hydrologic) dynamics, by employing both nonlinear dynamic and linear stochastic approaches to flow series from the Chao Phraya River and the Mekong River in Thailand. The nonlinear dynamic analysis included correlation dimension estimation, noise reduction, and local approximation prediction, and the linear stochastic analysis included model identification, formulation, diagnostic tests, and prediction. Based on the predictions and determinism test results, they concluded that the flow phenomena were indeed nonlinear deterministic. Also at the same time, Lambrakis et al.,[46] through application of the correlation dimension and surrogate data methods, presented evidence to nonlinearity and chaotic nature of the discharge process of the spring of Almyros, Iraklion, Crete, Greece. They subsequently attempted short-term forecasting of this series using local approximation prediction method and artificial neural networks (ANNs), and reported satisfactory predictive ability of both, with neural networks performing slightly better.

In their study on the dynamic nature of rainfall-runoff process in the Göta River basin in Sweden, Sivakumar et al.[33,34] provided convincing evidence as to the presence of chaos in the monthly flow series, as they observed a low correlation dimension of 5.5 and also extremely good predictions. As part of their study on forecasting daily river flow process of the Adige River in Italy, Lisi and Villi[47] checked for evidence of chaotic behavior using correlation dimension and Lyapunov exponent as indicators. Observing a nonlinear deterministic chaotic type dynamics in the flow process, they employed the local approximation method for predictions. They also compared the results with the prediction results obtained using a stochastic (ARIMA) model, and reported superior performance of the former by a margin of 12 to 18% for up to 3-day ahead predictions. Sivakumar et al.[111] employed the local approximation method to the monthly flow series in Coaracy Nunes/Araguari River basin in northern Brazil: (i) to predict the runoff dynamics; and (ii) to detect possible presence of chaos in the runoff dynamics using the prediction results. The local approximation method yielded extremely good predictions for the flow series. Based on observations of a low optimal embedding dimension and a clear decrease in prediction accuracy with increasing lead time, they concluded that the flow process exhibited low-dimensional chaotic behavior. These results were subsequently verified and confirmed by Sivakumar et al.[40] through the very good predictions achieved using neural networks and low correlation dimension of 3.62 obtained for the flow series. These results also facilitated addressing the issue of data size requirement for a reliable estimation of correlation dimension, an issue that continues to be at the center of criticisms on chaos studies in hydrology.[13,66,112]

Porporato and Ridolfi[42] extended the phase space reconstruction and local approximation prediction concepts to a multivariate form to include information from other time series in addition to that of river flow. The effectiveness of their proposed multivariate prediction approach was tested on the daily river flow phenomenon at Dora Baltea in Italy, for which signs of low-dimensional determinism had been found earlier.[27,105] Using rainfall and temperature as additional variables in the multivariate approach, they reported that such an approach was much better than the univariate approach for river flow predictions, especially in the prediction of highest flood peaks. This nonlinear multivariate approach was also found to be superior than a stochastic ARMAX (Autoregressive

Moving Average with Extraneous inputs) model. The flexibility of the multivariate model to adapt to the different sources of information was also observed. Elshorbagy et al.[31,113] employed the principles of chaos theory for estimation of missing streamflow data. In the study by Elshorbagy et al.,[113] missing data in one streamflow series were estimated from another complete and cross-correlated flow series, whereas Elshorbagy et al.[31] used only a single streamflow series for estimating consecutive missing values in the series. Two general steps were followed: (i) identification of chaos in streamflow series; and (ii) estimation of missing data. Streamflow series observed from the Little River and the Reed Creek in Virginia, USA were considered in Elshorbagy et al.,[113] wherein the correlation dimension method was used for chaos identification and the missing data were estimated by fitting one global model and multiple local models; the superior performance of the local models was attributed to the chaotic behavior of the two series. In the study by Elshorbagy et al.,[31] flow series from the English River in Ontario, Canada, was studied; chaos identification was done through four different methods and missing data were estimated using artificial neural networks and K-nearest neighbors (K-nn). These studies, in a way, also led them to address the issue of noise reduction.[43]

Phoon et al.[45] proposed a practical inverse approach for optimal selection of the parameters involved in phase space reconstruction, chaos identification, and prediction (i.e. delay time, embedding dimension, and number of neighbors) of river flow series. Considering optimum prediction accuracy as a single definite criterion, they first demonstrated the effectiveness of their approach on a theoretical chaotic time series (the Mackey-Glass series) and then tested it on two river flow series observed in the Tryggevaelde catchment in Denmark and in the Altamaha river in Georgia, USA. The approach was found to perform better than the standard approach (wherein one or more parameters are kept constant) both for prediction and system characterization. In a similar context, Jayawardena et al.[44] presented a new criterion, based on the generalized degrees of freedom, for optimum neighborhood selection for local modeling and prediction of river flow (and other hydrologic) series. It was first demonstrated on three artificial chaotic series (Lorenz map, Henon map, and Logistic map) and then tested on three river flow series: Chao Phraya River in Thailand and Mekong River in Thailand and Laos; noise reduction was also performed on the flow series before employing the neighborhood selection scheme. The approach was found to be superior to the traditional one that arbitrarily fixes the number of neighbors.

An extensive analysis towards characterizing and predicting the river flow dynamics at the Lindenborg catchment in Denmark was performed by Islam and Sivakumar.[114] Through a host of both standard statistical techniques and specific nonlinear and chaotic dynamic techniques, they reported presence of chaos in this river flow dynamics with a low correlation dimension of 3.76. The study was the first one to employ the false nearest neighbor algorithm for chaos identification in river flow. Applying a local approximation prediction method, the study also reported near-accurate prediction results for the flow series and provided further support as to the utility and suitability of a chaotic approach for modeling and predicting river flow dynamics. Zhou et al.[32] investigated the presence of chaos in the last 500-year annual flood series in the Huaihe River Basin in eastern China. Employing power spectrum and correlation dimension techniques, they reported presence of chaotic behavior in the flood dynamics with an attractor dimension of 4.66. They also attempted, using the concepts of chaos theory and the inverted theorem of differential equations, reconstruction of the flood series.

As part of their investigation of chaos in the sediment transport phenomenon, Sivakumar and Jayawardena[36] analyzed the daily flow data observed at the St. Louis gaging station in the Mississippi River basin, USA. Observing a low correlation dimension value of 2.32, they suggested possible presence of chaos in the river flow dynamics. Sivakumar et al.[48] employed the nonlinear local approximation approach for forecasting the flow dynamics of the Chao Phraya River (Nakhon Sawan station) in Thailand. The analysis was performed in tandem with the forecasting of the flow series using artificial neural networks, in order to compare the performance of the two approaches. Attempting 1-day and 7-day ahead forecasts, they reported reasonably good prediction performance of both the approaches. However, the local approximation approach was found to be superior to neural networks; this result was attributed to the representation of the flow dynamics in the phase space step by step in local neighborhoods in the former approach, rather than a global approximation as done in the latter. In a similar vein, Laio et al.[49] employed local approximation method and artificial neural networks for forecasting flood (water stages) of River Tanaro in Alba, Italy, and reported slightly better performance of the former at short forecast times and the reverse situation for longer times. Sivakumar[115] studied the dynamic behavior of monthly streamflow in the western United States, through application of the local approximation prediction method to data observed at 79 stations across 11 states. The analysis was carried out by grouping the 79 stations under three categories on the basis of the magnitude of mean streamflow as: (i) low-flow stations; (ii) high-flow stations; and (iii) medium-flow stations. The study yielded good predictions of streamflow dynamics irrespective of the flow regime, an indication of the appropriateness of the local approximation approach. The results also revealed that predictions for the low-flow stations were relatively better than those for the medium-flow and high-flow stations.

Regonda et al.[52] investigated the type of scaling behavior (stochastic or chaotic) in the temporal dynamics of river flow, employing the correlation dimension method. Analyzing daily, 5-day, and 7-day flow series from each of three rivers in the United States, they reported the presence of chaotic scaling behavior in the flow dynamics of the Kentucky River (Kentucky) and the Merced River (California), and stochastic scaling behavior in the flow dynamics of the Stillaguamish River (Washington state). They also observed an increase in the dimensionality (or complexity) of the flow dynamics with the scale of aggregation. Sivakumar et al.[55] introduced a chaotic approach for streamflow disaggregation, through making appropriate modifications to the approach proposed earlier for rainfall.[30] The approach was employed for disaggregation of streamflow observed at the St. Louis gaging station in the Mississippi River basin, USA. Disggregation was attempted for four successively doubled resolutions: 2-day to daily, 4-day to 2-day, 8-day to 4-day, and 16-day to 8-day. The excellent agreement between the actual and the disaggregated values clearly revealed the suitability of the chaotic approach. The best results were achieved for low embedding dimension (2 or 3) and small number of neighbors (< 50), suggesting the presence of chaotic dynamics in the streamflow transformation process.

Salas et al.[59] investigated the effects of aggregation and sampling of hydrologic data on the identification of the dynamics underlying the system. They analyzed daily streamflows from two rivers in Florida, USA: the St. Marys River near Macclenny and the Ocklawaha River near Conner. Based on estimates of delay time, delay window, and correlation integral,[116] they reported that the Ocklawaha River, which has a stronger basin storage contribution, departs significantly from the behavior of a chaotic system, while the departure was less significant for the St. Marys River, which has a smaller basin storage contribution. The study also suggested, through analysis of Lorenz map, that increase in aggregation and sampling scales deteriorates chaotic behavior and eventually ceases to show evidence of low-dimensional determinism. Sivakumar et al.[56] used the phase space reconstruction concept to analyze a host of river flow series from different geographic locations and/or at different scales (including some analyzed in earlier studies) in their attempt towards formulating a catchment classification framework (more details on this are presented below).

9.3.3. Rainfall-Runoff

Despite the large number of applications to rainfall process and river flow process separately, chaos theory has surprisingly not found many applications to their combination, i.e. rainfall-runoff process. The only studies, to our knowledge, that have employed the concept of chaos theory to study the rainfall-runoff process were those conducted by Sivakumar et al.[33,34] and Dodov and Foufoula-Georgiou.[57]

Sivakumar et al.[33,34] investigated the presence of chaos in the rainfall-runoff process at the Göta River basin in Sweden, by analyzing the rainfall series and the runoff series first separately and then jointly (using runoff coefficient). The reason behind analyzing rainfall and runoff series separately was that their individual behaviors (input and output, respectively) could provide important information about the behavior of the joint rainfall-runoff process (input-output relationship), whereas the runoff coefficient (defined as the ratio of runoff to rainfall) was considered as a better representative of the rainfall-runoff process as a whole. Employing the correlation dimension method in Sivakumar *et al.*[34] and the nonlinear local approximation prediction method in Sivakumar *et al.*,[33] they suggested the presence of chaotic behavior in the rainfall-runoff process. However, in view of the assumptions made in the use of rainfall, runoff, and runoff coefficient series for characterizing the rainfall-runoff process, they also cautioned on the interpretations of the results and recommended further verifications and confirmations. In particular, the usefulness of runoff coefficient as a parameter connecting rainfall and runoff was questioned.

Dodov and Foufoula-Georgiou[57] studied the nonlinear dependencies of rainfall and runoff and the effects of spatio-temporal distribution of rainfall on the dynamics of streamflow at flood timescales in two basins in Central North America. They proposed a framework based on 'hydrologically-relevant' rainfall-runoff phase space reconstruction, but with specific acknowledgment that rainfall-runoff is a stochastic spatially extended system rather than a deterministic multivariate one. Three specific tasks were attempted: (i) quantification of the nonlinear dependencies between streamflow dynamics and the spatio-temporal dynamics of precipitation; (ii) study of how streamflow predictability is affected by the trade-offs between the level of detail necessary to explain the spatial variability of rainfall and the reduction of complexity due to the smoothing effect of the basin; and (iii) exploration of the possibility of incorporating process-specific information, in terms of catchment geomorphology and an *a priori* chosen uncertainty model, into nonlinear prediction. The results indicated the potential of using this framework for streamflow predictability and limits to prediction, as a function of the complexity of spatio-temporal forcing related to basin geomorphology, via nonlinear analysis of observations alone and without resorting to any particular rainfall-runoff model.

9.3.4. *Lake Volume*

Studies that have employed the concept of chaos theory for understanding and predicting the dynamic changes in lakes have been, to our knowledge, limited to the Great Salt Lake (GSL) in the United States, and especially to its volume time series at the biweekly scale. Sangoyomi *et al.*[25] reported presence of low-dimensional chaotic behavior in the GSL biweekly volume series, based on results from the application of the correlation dimension method, the nearest neighbor dimension method, and the false nearest neighbor dimension method. Observing a correlation dimension value of 3.4, they also suggested that the GSL biweekly volume dynamics might be described by four variables. Abarbanel and Lall[23] further verified the presence of chaos in the GSL volume series through application of the Lyapunov exponent method; they determined the average predictability of the series as a few hundred days. They also attempted forecasting of the GSL series using local approximation method and constructing local polynomial maps, and testing the forecast skill for a variety of GSL conditions, such as lake average volume, near the beginning of a drought, near the end of a drought, and prior to a period of rapid lake rise. The results indicated excellent short-term predictions for the GSL series, but also revealed degrading predictions for longer time horizons. Abarbanel et al.,[117] subsequently, extended the above predictability study by also attempting multivariate adaptive regression splines (MARS) and comparing the prediction results with those obtained using the local polynomials. Further details on the application of chaos theory to the GSL volume series can be found elsewhere.[75]

Regonda et al.[53] proposed a nonparametric approach based on local polynomial regression for ensemble forecast of time series, and tested its effectiveness on the GSL biweekly volume series, among others. They selected a suite of combinations of the four parameters involved in this approach (i.e. embedding dimension, delay time, number of neighbors, and polynomial order) based on an objective criterion, called the Generalized Cross Validation (GCV). This ensemble approach (also providing the forecast uncertainty) yielded improved performance over the traditional method of providing a single mean forecast, and its superior performance was particularly realized for short noisy data. In the specific context of 'optimum' parameter selection, their approach was somewhat similar to the one adopted by Phoon et al.[45] for river flow series, with an important extension with the addition of the order of polynomial.

9.3.5. Sediment Transport

Initial studies on the applications of chaos theory for sediment transport phenomenon were conducted by Sivakumar[35] and Sivakumar and Jayawardena.[36] Sivakumar[35] employed the local approximation method for one-step ahead predictions of daily suspended sediment concentration dynamics at the St. Louis gaging station in the Mississippi River basin, USA, and reported extremely good results (with correlation coefficient > 0.99 and coefficient of efficiency > 0.97). Observing that a small embedding dimension (i.e. 3) yielded the best prediction results, Sivakumar[35] also suggested that the suspended sediment concentration dynamics were dominantly influenced by only three variables. Despite these encouraging outcomes, it was also realized that studying only the sediment concentration series (which is just only one component of the river system) would not be sufficient for an accurate understanding of the entire sediment transport phenomenon, since establishing relationships between river water discharge, suspended sediment concentration, and suspended sediment load would be crucial for such an understanding. In light of this, Sivakumar and Jayawardena[36] analyzed time series of all these three components observed at the St. Louis gaging station, assuming that their individual behavior might give clues to the dynamic behavior of the overall sediment transport phenomenon. Application of the correlation dimension method to these series yielded low dimensions of 2.32, 2.55, and 2.41, respectively, suggesting the presence of chaotic behavior in the dynamics of each of these components and also possibly in the complete sediment transport phenomenon.

Following up on these studies, Sivakumar and Wallender[118] introduced a chaotic approach for disaggregation of suspended sediment load data. This approach was based on the chaotic disaggregation approach proposed by Sivakumar *et al.*[30] for rainfall data, and certain modifications were made to suit the suspended sediment load data. It was tested on the suspended sediment load data observed at the St. Louis gaging station, with disaggregations attempted for four successively doubled resolutions: 2-day to daily, 4-day to 2-day, 8-day to 4-day, and 16-day to 8-day. The study revealed the possible nonlinear deterministic nature of the sediment load transformation process at these scales, as the best results were achieved for low phase space dimension (< 4) and relatively small number of neighbors (< 100). Sivakumar and Wallender[119] conducted a more comprehensive study on the predictability of daily river flow and suspended sediment transport processes at the St. Louis gaging station. This study was an extension of the study by Sivakumar[35] in two specific ways: (i) predictions of river flow and suspended sediment load series were also made, in addition to the suspended sediment concentration series; and (ii) predictions were made not just for one-day ahead but up to ten days ahead. The local approximation method yielded extremely good one-day ahead predictions for all the series, but the accuracy decreased with increasing lead time, and more so for suspended sediment concentration and suspended sediment load. The results also indicated the number of variables dominantly governing the dynamics as three (optimal embedding dimension = 3) for each of the series, supporting the correlation dimension results reported by Sivakumar and Jayawardena[36] and the presence of chaos. Sivakumar and Chen[120] investigated, through the application of the correlation dimension method, the dynamic behavior of suspended sediment load transport at different temporal scales at the St. Louis gaging station. They analyzed data corresponding to five different temporal scales (daily, 2-day, 4-day, 8-day, and 16-day). The results indicated the presence of low-dimensional determinism in the suspended sediment transport series at each of these five scales, hinting at possible scale invariance in the suspended sediment load dynamics, which was explored earlier in different ways by Sivakumar and Wallender[118] and Sivakumar.[121] The number of variables dominantly governing the dynamics was identified to be in the order of 3 or 4, which provided further support to many of the earlier studies.

9.3.6. *Groundwater*

While surface water hydrology had witnessed a large number of chaos theory applications in the 1990s, subsurface hydrology had eluded the attention of chaos studies until earlier in this century. To our knowledge, Faybishenko[122] was the first to introduce the concept of chaos in subsurface hydrology through one of his studies on complex flow processes in heterogeneous fractured media. He analyzed the time series of pressure fluctuations from two water-air flow experiments in replicas of rough-walled rock fractures under controlled laboratory conditions,[123] using a host of methods, including correlation dimension, global and local embedding dimensions, and Lyapunov exponents. The results were then also compared with the chaotic analysis of laboratory dripping-water experiments in fracture models and field-infiltration experiments in fractured basalt. Based on this comparison, it was conjectured that: (i) the intrinsic fracture flow and dripping, as well as extrinsic water dripping (from a fracture) subjected to a capillary-barrier effect, are deterministic-chaotic processes with a certain random component; and (ii) the unsaturated fractured rock is a dynamic system that exhibits chaotic behavior because the flow processes are nonlinear, dissipative, and sensitive to initial conditions, with chaotic fluctuations generated by intrinsic properties of the system, not random external factors.

Sivakumar et al.[60] investigated the potential use of chaos theory to understand the dynamic nature of solute transport process in subsurface formations. They analyzed, using the correlation dimension method, time series of solute particle transport in a heterogeneous aquifer medium (which was simulated using an integrated transition probability/Markov chain model, groundwater flow model, and particle transport model, for varying hydrostratigraphic conditions), with the western San Joaquin Valley aquifer system in California as a reference system. The results generally indicated the nonlinear deterministic nature of solute transport dynamics (dominantly governed by only a very few variables, on the order of 3), even though more complex behavior was found to be possible under certain extreme hydrostratigraphic conditions. Later, Hossain and Sivakumar[61] studied the spatial patterns of arsenic contamination in the shallow wells (< 150 m) of Bangladesh, employing the correlation dimension method. Particular emphasis was given to the role of regional geology (Pleistocene vs. Holocene) on the spatial dynamics of arsenic contamination. The results, with correlation dimensions ranging between 8 and 11 depending on the region, suggested that the arsenic contamination dynamics in space is a medium- to high-dimensional problem. These results were further verified using logistic regression, with an attempt to explore possible (physical) connections between the correlation dimension values and the mathematical modeling of risk of arsenic contamination.[124] Eleven variables were considered as indicators of the aquifer's geochemical regime with potential influence on arsenic contamination, and a total of 2,048 possible combinations of these variables were included as candidate logistic regression models to delineate the impact of the number of variables on the prediction accuracy of the model.

9.3.7. *Other Hydrologic Problems*

9.3.7.1. *Scale and Scale Invariance*

Hydrologic processes result from interactions between climatic inputs and landscape characteristics that occur over a wide range of scales both in space and in time. The tremendous variability in climatic inputs (e.g. rainfall, temperature, wind velocity) and heterogeneities in landscape properties (e.g. basin area, soil type, land use, slope) may give rise to hydrologic processes with high variability at all scales and across scales. However, hydrologic systems and processes also exhibit scaling or scale invariance,[125-128] which serves as an important basis for transformation of data from one scale to another (e.g. disaggregation), among others. Many studies have employed the concept of chaos theory to study hydrologic scale invariance,[28,30,52,55,56,59,103,120] albeit in different contexts and employing different methodologies to different systems and processes, including rainfall, river flow, and suspended sediment load, as presented earlier. In general, these studies, while revealing the possibility of chaotic scaling behavior (and stochastic scaling in one or two cases), seem to suggest an increase in the dimensionality (or complexity) of the system dynamics with the scale of aggregation; in other words, dynamics changing from a less complex (more deterministic) behavior to a more complex (more stochastic) behavior with aggregation in time. However, there may indeed be exceptions to this situation with no trend possibly observed in the 'scale versus complexity' relationship,[30] since this relationship essentially depends on, for example, rainfall characteristics (e.g. intensity, duration) and catchment properties (e.g. size of basin, land use). While further investigation is obviously needed for a more reliable interpretation and conclusion on this relationship one way or another, the presence of chaotic dynamics in scaling behavior of rainfall, river flow, and suspended sediment load has important implications in hydrology, since it has been a common practice to employ stochastic (random) cascade approaches for scaling investigations and for data transformation.

9.3.7.2. *Parameter Estimation*

With the ever-increasing complexities of hydrologic models,[129] which require more details about processes and more parameters to be calibrated, parameter optimization has become an extremely challenging problem.[130] Constructive discussions and debates on this issue, especially on the uncertainties in hydrologic models and identification of the best optimization technique, have just started to come to the fore.[131-135] While these are certainly positive signs to the long-term health of hydrology, the basic problem lies essentially with our tendency (often driven by our technological and methodological advances) to develop more complex models than that may actually be necessary.[135,136] In an attempt towards model simplification, Sivakumar[54] proposed an approach that incorporates and integrates the chaos concept with expert advice and parameter optimization techniques. The simplification was brought out primarily through the determination of the 'number' of dominant variables governing the system (using the correlation dimension method), with the use of only a limited amount of data (often data corresponding to a single variable) representing the system. Hossain *et al.*,[50] in their study of Bayesian estimation of uncertainty in soil moisture simulation by a Land Surface Model (LSM), presented a simple and improved sampling scheme (within a Monte Carlo simulation framework) to the Generalized Likelihood Uncertainty Estimation (GLUE) by explicitly recognizing the nonlinear deterministic behavior between soil moisture and land surface parameters in the stochastic modeling of the parameters' response surface. They approximated the uncertainty in soil moisture simulation (i.e. model output) through a Hermite polynomial chaos expansion of normal random variables that represent the model's parameter (model input) uncertainty. The new scheme was able to reduce the computational burden of random Monte Carlo sampling for GLUE in the range of 10–70%, and it was also found to be about 10% more efficient than the nearest-neighborhood sampling method in predicting a sampled parameter set's degree of representativity.

9.3.7.3. Catchment Classification

The realization that hydrologic models are often developed for specific situations and that their extensions to other situations and generalizations are difficult has recently led some researchers to call for a catchment classification framework.[137-139] These researchers also suggest, largely motivated by the proposal of the dominant processes concept,[140] that identification of dominant processes may help in the formation of such a classification framework. With this idea, Sivakumar[54] introduced a classification framework, in which the extent of complexity or dimensionality (determined using nonlinear tools, such as the correlation dimension method) of a hydrologic 'system' is treated as a representation of the (number of) dominant processes. Following up on this, Sivakumar et al.[56] explored the utility of the phase space reconstruction, in which the 'region of attraction of trajectories' in the phase space was used to identify the data as exhibiting 'simple' or 'intermediate' or 'complex' behavior and, correspondingly, classify the system as potentially low-, medium- or high-dimensional. The utility of this reconstruction concept was first demonstrated on two artificial time series possessing significantly different characteristics and levels of complexity (random vs. chaotic series, discussed in Secs. 9.2.1 to 9.2.3), and then tested on a host of river-related data (flow, suspended sediment concentration, and suspended sediment load) representing different geographic regions, climatic conditions, basin sizes, processes, and scales. The ability of the phase space to reflect the river basin characteristics and the associated mechanisms, such as basin size, smoothing, and scaling, was also observed.

9.3.7.4. Sundry

There have also been several other studies that have, in one way or another, looked into the applications of nonlinear dynamic and chaos theories in hydrology. These include: applications to yet other hydrologic processes, proposals of new ways for hydrologic data analysis, and investigations on the reliability of chaos methods to hydrologic data. A very brief account of some selected studies is presented next, not in any specific order.

Rodriguez-Iturbe et al.[141] studied the soil moisture balance equation over large spatial territories at seasonal timescales, with explicit consideration of local recycling of precipitation and dynamic effects of soil moisture in the generation and modification of mesoscale circulation patterns for parameterization that also incorporates time delays. They showed that the dynamics of the equations are quite complex, being capable of exhibiting fixed point, limit cycle, and chaotic type behavior. Manzoni et al.[51] studied the soil carbon and nitrogen cycles from a dynamic system perspective, wherein the system nonlinearities and feedbacks were analyzed by considering the steady-state solution under deterministic hydroclimatic conditions. Laio et al.[49] employed the DVS method to daily river discharge from three Italian rivers in their investigation of nonlinearity in rainfall-runoff transformation. Khan et al.[63] and Sivakumar[64,65] investigated the reliability of the correlation dimension method in the detection of chaos in hydrologic time series. They addressed, among others, the effects of data size, and random and seasonal components, and zeros. Jin et al.[58] studied the nonlinear relationships between Southern Oscillation index (SOI) and local precipitation and temperature (in Fukuoka, Japan), by representing this joint hydroclimatic system using a nonlinear multivariate phase space reconstruction technique. Nordstrom et al.[142] proposed the construction of a dynamic area fraction model (DAFM) that contains coupled parameterizations for all the major components of the hydrologic cycle involving liquid, solid, and vapor phases. Using this model, which shares some of the characteristics of an Earth System Model of intermediate complexity, they investigated the nature of feedback processes in regulating earth's climate as a highly nonlinear coupled system. Still other studies of interest are those by Sivakumar,[143,144] Phillips and Walls,[145] She and Basketfield,[146] Phillips,[147] and Kim et al.,[148] among others.

9.4. Chaos Theory in Hydrology: Current Status and Challenges

As the foregoing indicates, we have made some sincere efforts during the last two decades to explore the potential of nonlinear dynamic and chaos concepts for modeling and prediction of hydrologic systems. The outcomes of these efforts are certainly encouraging, considering that we are still in the state of infancy in regards to these concepts when compared to the much more mature and prevalent linear stochastic concepts (this is not to say that we have achieved the 'full-fledged' status with the stochastic concepts). The inroads we have made, especially in recent years, in the areas of, for example, scaling, groundwater contamination, parameter estimation and optimization, and catchment classification, are significant, albeit their preliminary nature, since these are arguably some of the most important topics in hydrology at the current time.

With these positives, however, we must not forget the challenges that lie ahead on our way to further progress. Among these challenges, two are noteworthy: (i) improving our understanding of these largely less-understood chaos concepts for hydrologic applications; and (ii) finding ways to integrate these concepts with the others, either already in existence or emerging in the future. The former is important for avoiding 'blind' applications of the related methods (simply because the methods exist and are there to apply!) and 'false' claims (either in support of or against their utility); and the latter is important for taking advantage of the merits of the different approaches for their 'collective utility' to solve hydrologic problems rather than for their 'individual brilliance' as perceived. The rest of this section presents some examples to the potential limitations of the above studies and to the challenges ahead.

The outcomes of the studies by Sivakumar et al.[33,34] on the rainfall-runoff phenomenon (at the Göta River basin) are encouraging towards an overall watershed modeling problem, as they suggested the presence of chaotic behavior. However, the studies also possessed some important drawbacks. First, observing chaotic behavior in rainfall and runoff series independently does not necessarily mean that the joint rainfall-runoff phenomenon is chaotic. Second, the runoff coefficient was found to be inadequate to connect rainfall and runoff (as many of the runoff coefficient values were greater than 1.0), which, in turn, necessitated another parameter to represent the joint rainfall-runoff phenomenon; how to obtain such a parameter and how to utilize it within the context of chaos theory to understand the rainfall-runoff phenomenon are still unresolved questions. Similarly, questions may be raised on the studies on the Mississippi River basin sediment transport system[36,119] through analysis of the river discharge, suspended sediment concentration, and suspended sediment load series independently. For instance: how to derive a relationship between these components, based on ideas gained from chaos theory, in order to understand the sediment transport phenomenon as a whole?

Porporato and Ridolfi[42] employed a multivariate phase space reconstruction approach for predicting river flow dynamics (at Dora Baltea), using rainfall and temperature as additional variables (apart from river flow), and reported that such an approach was significantly better than the univariate one.[27,105] They also observed the flexibility of the multivariate model to adapt to the different sources of information. Some obvious questions, however, are: Would similar results be achieved for other phenomena as well, such as rainfall and temperature? Can any particular criterion be adopted to select the (number of) additional variables or one would still be forced to a (trial-and-error) sensitivity analysis?

The attempt by Elshorbagy et al.[113] to estimate missing data in one streamflow series from another complete cross-correlated series was based on the observation (and assumption) of chaotic behavior in both the series, which might have also been the reason for the better performance of the multiple local models compared to a single global model. The following questions are inevitable: Would chaotic behavior of two cross-correlated streamflow series be a necessary condition for missing data estimation in one of the series from the other? If only one series exhibits chaotic behavior, would it be better to estimate missing values in the non-chaotic series from the chaotic series or otherwise?

Would the multiple local models still perform better than a single global model?

In their study on the reconstruction of flood series (in the Huaihe River Basin) using the concepts of chaos theory and inverted theorem of differential equations, Zhou et al.[32] attempted the reconstruction in three dimensions and with linear and 2nd power nonlinear terms. As the correlation dimension of the flood series was estimated as 4.66 (i.e. requiring at least five-dimensional reconstruction), they stressed on the need for reconstruction in higher dimensions. However, since even a three-dimensional reconstruction (and with 2nd power terms) involves as large as 27 coefficients, it is important to ask how complex and complicated the equation(s) would be when higher-dimensional (say, five or more) reconstruction is attempted (and also with higher power nonlinear terms)? In a similar reconstruction study on sunspot numbers, Jinno et al.[149] used a modified form of the Rössler equation[150] as reference system equations, since the sunspot numbers' attractor, amplitude, and pseudoperiod were found to be similar to that of the Rössler equation. While the predictions achieved (up to 8 months ahead) are promising, the basis for choosing the Rössler equation to form the reference system of equations is questionable and has much wider implications. For example, how to reconstruct hydrologic phenomena that do not exhibit attractors and other properties similar to that of any artificial chaotic system?

Some studies[28,52,56,59,103] provide interesting insights into the problem of scaling and effects of data aggregation. They essentially suggest that complexity of the system dynamics increases (often from a more deterministic nature to a more stochastic nature) with aggregation in timescale. While this may indeed be the case in certain situations, its generalization is often difficult to make, since the system's dynamic complexity depends on the climatic inputs and the catchment characteristics. For example, the catchment area (and, hence, the time of concentration, not to mention the rainfall characteristics) plays a vital role in defining the relationship between scale and dynamic complexity. In fact, depending upon the catchment, the dynamic complexity may increase with aggregation in time up to a certain point (probably, somewhere close to the concentration time) and then decrease with further aggregation, possibly implying a 'scaling regime' (see Sivakumar et al.[30,55] in the context of rainfall and streamflow disaggregation).

The studies by Sivakumar et al.[60] and Hossain and Sivakumar,[61] searching for nonlinear deterministic dynamics in solute transport in a heterogeneous aquifer and arsenic contamination in shallow wells, are certainly interesting. However, they are, at best, crude one-dimensional approximations to the complex three-dimensional groundwater flow and transport processes. The studies only consider the time or the space (as the case may be), while what is actually needed is a spatio-temporal perspective. Further, although there is no 'mathematical' constraint, the 'philosophical' merit behind the use of phase space reconstruction concept in a spatial context (with its delay parameter defined in space), as is done in Hossain and Sivakumar,[61] is an issue to ponder.

The proposal by Sivakumar[54] for integration of different concepts/ methods to study hydrologic systems, and more specifically to simplify our modeling and parameter estimation practices, is a notable move forward, since different concepts possess different advantages and limitations. However, the utility and effectiveness of this proposal are yet to be seen through implementation. Moreover, recognition of the advantages and limitations of each of the existing concepts (speaking in a more general sense) in itself is a tremendously challenging task, as this requires adequate knowledge of all the concepts in the first place. This problem probably makes the idea of integration of different concepts less appealing, especially with our increasing emphasis on individual concepts in our research.[151]

The proposal on the use of the phase space reconstruction approach for 'system classification' and also its effective demonstration and testing on synthetic and river-related data[56] seem to provide strong clues to the potential of such an approach for formulation of a catchment classification framework. What remains to be studied, however, is the incorporation of the catchment characteristics into this classification framework and establishing connections between data (usually at the catchment scale) and the actual catchment physical mechanisms (at all scales) for this classification framework to be successful (see Wagener et al.[152] for further discussion on catchment classification framework, especially in the context of hydrologic similarity). It must also be noted that nonlinearity and chaos is not just about small changes leading to large effects and complex-looking outputs coming out of simple systems but it is also about large changes leading to small effects and simple-looking outputs coming out of complex systems. Whether or not the phase space reconstruction approach can also perform equally well for this latter situation remains to be seen.

As pointed out by Koutsoyiannis[66] (see also Tsonis *et al.*[153] and Sivakumar[29]), the presence of periods (or a large number) of zeros in a time series could result in an underestimation of the correlation dimension and thus (in the absence of any other analysis) potentially lead to the conclusion that chaos exists, when actually it does not. This can turn out to be a serious issue in chaos studies in hydrology, since zero values are a common occurrence in hydrologic time series (especially high-resolution rainfall). The fact that zero values are intrinsic to the system dynamics and thus must not be removed in any hydrologic analysis (possible exception may be in disaggregation analysis[30]) makes the problem even more complicated. This does not mean that the correlation dimension method must not be employed to hydrologic series, because dimension is simply a representation of the variability of the time series values (zeros included). What is required to realistically deal with this problem is an adequate definition of what constitutes 'periods' (or a large number) of zeros; in other words, what is the 'threshold' for the number (or percentage) of zeros in a time series that can still yield a reliable estimation of correlation dimension? This is a difficult question to answer, because determination of the 'sensitivity' of correlation dimension to the number of zeros is not straightforward, even for artificially generated time series (let alone real series). In fact, this question is not just limited to zeros but goes well beyond, since it is simply the question of 'repetition' of one or more values and that such repetitions may occur in many different ways depending upon the system (e.g. minimum streamflow, maximum/minimum water level in a reservoir, water release from a reservoir, daily suspended sediment load).

Global climate change will have, it is believed, threatening consequences for our water resources, and there are already noticeable indications with increases in abnormal events (e.g. floods, droughts) around the world. Since Global Climate Models (GCMs) provide climate data only at much coarser spatial (and temporal) scales than that are required for hydrologic predictions at regional and local levels, 'downscaling' of GCM outputs is essential. The existing statistical and dynamic downscaling techniques can indeed provide some success (see, for example, Fowler et al.[154] for a recent review of downscaling techniques), but the assumption of linearity inherent in almost all of these techniques is too simplistic and may greatly constrain their effectiveness, since the climate systems and the associated processes are essentially nonlinear, and possibly chaotic.[15,155] In view of this, there is increasing realization on the urgent need for formulation of nonlinear downscaling approaches (with explicit consideration to the system's chaotic properties), but unfortunately nothing seems to have been done yet. This downscaling problem is also a complex spatio-temporal problem (and probably much more complex than groundwater flow and contamination, and also the rainfall-runoff problem, discussed above), and thus may necessitate significant modifications to the single variable, and even multivariable, phase space reconstruction approach that has widely been adopted in hydrologic and climatic studies thus far. The nonlinear-dynamic based disaggregation approach proposed by Sivakumar et al.[30] for rainfall may provide some useful clues to deal with this problem, but even that too will only be of limited use and in a purely temporal sense. Although research efforts in this direction are currently underway, there is certainly a long way to go.

Finally, since the methods that have thus far been employed by studies on chaos in hydrologic systems are essentially data-based, their relevance to the actual physical mechanisms and system dynamics may be questioned. For example, the basic idea in phase space reconstruction is that a (nonlinear hydrologic) system is characterized by self-interaction and that a time series of a single variable can carry the information about the dynamics of the entire multivariable system. Although it is possible to provide convincing explanation of the relevance of this idea to some overall scenarios (e.g. input-output, rainfall-runoff, river discharge-sediment load), explanation as to the relevance of the parameters to specific system components is very difficult. This may be elucidated through an example here. There is sufficient information in the history of runoff data about the rainfall properties and the catchment characteristics over a period of time, and thus a single-variable phase space reconstruction (with runoff) should be able, at least theoretically, to reflect the system's dynamic changes. At the same time, how the delay time in the embedding procedure[70] is related to any of the system components and/or dynamics is hard to explain. The absence of consensus on the appropriate method for the selection of optimum delay time[82,83,85,116] makes this issue only more difficult to deal with. This issue, and similar ones, must be resolved for any hope towards establishing links between data and physics and thus to truly reflect the advances that can be made with nonlinear dynamic and chaos concepts. One must also recognize, however, that this problem is not just specific to chaos methods; it is much more widespread, and is applicable to literally all time series methods (see, for example, Kirchner[156] for some relevant details on models versus measurements); but that must not be a solace in any case.[136]

9.5. Closing Remarks: Chaos Theory as a Bridge Between Deterministic and Stochastic Theories

There have been two prevalent approaches in modeling and forecasting hydrologic systems and processes: deterministic and stochastic. On one hand, the use of deterministic approach may be supported on the basis of, for example, the 'permanent' nature of the earth, the ocean, and the atmosphere and the 'cyclical' nature of mechanisms or processes that happen within and between them. The stochastic approach, on the other hand, may be favored because of the 'highly irregular and complex nature' of hydrologic phenomena and our 'limited ability to observe' the detailed variations. With these two contrasting principles, the question of whether hydrologic phenomena are deterministic or stochastic (or more specifically, whether hydrologic phenomena are better modeled using a deterministic approach or a stochastic approach) is meaningless. Such a question is really a philosophical one that has no general answer, but it is better viewed as a pragmatic one, which has an answer only in terms of specific situations.[157] These specific situations are viewed in terms of the process of interest, the scale (time and/or space) of interest, and the purpose of interest, which collectively form the hydrologic system.[136] For some situations, both the deterministic approach and the stochastic approach may be equally appropriate; for some other situations, the deterministic approach may be more appropriate than the stochastic approach; and for still others, the stochastic approach may be more appropriate than the deterministic approach. And, there may still be some situations, where neither the deterministic approach nor the stochastic approach is appropriate when viewed independently but their combination may turn out to be appropriate.

It is reasonable to contend that, for most (if not all) hydrologic phenomena, both the deterministic approach and the stochastic approach are actually complementary to each other, despite (or because of) their contrasting principles. This may be supported by our observation of both 'deterministic' and 'random' nature of hydrologic phenomena at 'one or more scales' in time and/or space. For example, it is common to observe a significant deterministic nature in river flow in the form of seasonality and annual cycle, while the interactions of the various mechanisms involved in the river flow phenomenon and their various degrees of nonlinearity bring randomness. Our theoretical and practical experience with the use of deterministic and stochastic approaches for hydrologic phenomena provides additional support to this. For instance, the depth-averaged flow at a given site can be adequately described using a classical deterministic model, whereas the contaminant transport at the same site may require a stochastic model for adequate representation.[157] The underlying message from these examples is that use of a coupled deterministic-stochastic approach, incorporating both the deterministic and the stochastic components, will likely produce a higher 'probability of success' in our modeling efforts compared to either a deterministic approach or a stochastic approach when employed independently. An immediate question is: how to devise such a coupled deterministic-stochastic approach? Chaos theory, with its three fundamental principles, namely nonlinear interdependence, hidden determinism and order, and sensitivity to initial conditions, could provide an avenue for this. In fact, as Sivakumar[91] argued, chaos theory should not be viewed as a separate theory, but rather as a theory that connects the deterministic approach and the stochastic approach.

It is also relevant to note that (regardless of deterministic or stochastic approach) we are starting to place more emphasis on the applications of specific concepts/methods rather than creation of collective efforts to address the most challenging hydrologic problems affecting us all.[151] There is also increasing realization that none of the currently available tools (linear or nonlinear, deterministic or stochastic) by itself is adequate for satisfactory resolution of our hydrologic problems; for example, no model (how sophisticated it may be) has been shown to reliably capture the extreme hydrologic events at a given location, and extensions and generalizations of a model (developed for some situations) to other situations are often difficult. The following is only a small sample of the numerous questions that need to be asked about our existing research activities to tackle the challenges in hydrology: (i) how are we going to incorporate the nonlinear deterministic components that are inherent in hydrologic systems and processes into our linear, and nonlinear, stochastic methods? (ii) how are we going to address the property of sensitive dependence of hydrologic system dynamics on the initial conditions, when the initial conditions themselves cannot be known? (iii) how are we going to explain the 'random' and unpredictable system behavior using our nonlinear deterministic techniques? (iv) how are we going to estimate the uncertainty in the parameters that serve as important inputs to our complex models and, even worse, how are we going to define 'uncertainty' in the first place?[135] and (v) how are we going to establish the 'connections' between our 'data-based' techniques and the 'process-based' techniques, especially when there are 'disconnections' (intentional or unintentional) in our research?

These are difficult questions to answer, and the only way to do this is to find some 'common grounds' in our research ideas or concepts. This does not mean that one concept has to 'give up' making way for another, but this certainly requires some kind of 'compromise' and 'sacrifice' for the betterment of hydrologic research. To this end, bringing together different concepts could be a good starting point, as they could supplement and complement one another (and also could be used in an integrated manner), for reliably representing all the relevant properties of hydrologic systems and processes (linear or nonlinear, deterministic or stochastic). This, however, is an extremely challenging task, especially with our fascination for specific concepts for their 'individual brilliance' that reflects, more often than not, only our one-sided 'extreme' and 'biased' views. What is urgently needed, therefore, are sincere efforts to explore the 'collective utility' of different concepts for studying hydrologic systems. This warrants, as experience suggests, a change in our research paradigm and attitude;[158] that is, willingness to recognize the potential of different concepts and openness to accept the outcomes. Discussions and debates are obviously essential to this change, and constructive criticisms and useful inputs from all sides would certainly help towards defining and demarcating the travel path for this.

A general statement on the need for a 'middle-ground' approach is an easy thing to make, but the real challenge lies in identifying specific hydrologic problems and appropriate concepts that could, in combination, yield the desired effectiveness and efficiency. It is premature, at this stage, to pinpoint these areas, mainly because of our inadequate knowledge of all the concepts and, to some extent, of the hydrologic problems themselves. However, the situation is not that bleak, and there are already some encouraging signs. For example, the study by Sivakumar et al.[55] suggests that nonlinear deterministic techniques generally provide better results for streamflow disaggregation over finer resolutions than over coarser ones, while parametric and nonparametric stochastic approaches have been reported to provide very good results over coarser resolutions.[159,160] With these observations, it might be worthwhile to couple the nonlinear deterministic approach and the linear stochastic approach for streamflow disaggregation over a much wider range of scales (e.g. between daily and annual) than the range of scales studied thus far with these approaches independently (see Sivakumar et al.[55] for some details on the similarities and differences between these approaches). Such an approach could help towards a better view of the streamflow disaggregation problem, and also provide new avenues to the study of scale problems in hydrology at large. Similarly, other possible coupling or integration of concepts may be looked into study other challenging hydrologic problems.

Acknowledgments

This chapter significantly benefited from three earlier review papers by Bellie Sivakumar: *J. Hydrol.*, 227(1–4), 1–20; *Chaos Soliton. Fract.*, 19(2), 441–462; *Stoch. Environ. Res. Risk Assess.*, 23(7), 1027–1036.

References

1. C. F. Izzard, *J. Geophys. Res.*, 4811 (1966).
2. J. Amorocho, *Water Resour. Res.*, 861 (1967).
3. J. Amorocho and A. Brandstetter, *Water Resour. Res.*, 1087 (1971).
4. H. A. Thomas and M. B. Fiering, in *Design of Water Resource Systems*, Eds. Mass et al. (Harvard University Press, Cambridge, Massachusetts, 1962), p. 459.
5. A. A. Harms and T. H. Campbell, *Water Resour. Res.*, 653 (1967).
6. D. Valencia and J. C. Schaake, *Water Resour. Res.*, 580 (1973).
7. V. Klemeŝ, *Adv. Hydrosci.*, 285 (1978).
8. J. D. Salas and R. A. Smith, *Water Resour. Res.*, 428 (1981).
9. M. L. Kavvas, *ASCE J. Hydrol. Eng.*, 44 (2003).
10. American Society of Civil Engineers, *ASCE J. Hydrol. Eng.*, 124 (2000).
11. P. C. Young and K. J. Beven, *Environmetrics*, 335 (1994).
12. B. Sivakumar, *J. Hydrol.*, 1 (2000).
13. D. Schertzer, I. Tchiguirinskaia, S. Lovejoy, P. Hubert and H. Bendjoudi, *Hydrol. Sci. J.*, 139 (2002).
14. B. Sivakumar, R. Berndtsson, J. Olsson and K. Jinno, *Hydrol. Sci. J.*, 149 (2002).
15. E. N. Lorenz, *J. Atmos. Sci.*, 130 (1963).
16. H. Tong, *Non-linear Time Series: A Dynamical System Approach* (Oxford University Press, Oxford, 1990).
17. S. J. Goerner, *Chaos and the Evolving Ecological Universe* (Gordon and Breach Publishers, 1994).
18. H. Kantz and T. Schreiber, *Nonlinear Time Series Analysis* (Cambridge University Press, Cambridge, 1997).
19. I. Rodriguez-Iturbe, F. B. De Power, M. B. Sharifi and K. P. Georgakakos, *Water Resour. Res.*, 1667 (1989).
20. B. P. Wilcox, M. S. Seyfried and T. H. Matison, *Water Resour. Res.*, 1005 (1991).
21. R. Berndtsson, K. Jinno, A. Kawamura, J. Olsson and S. Xu, *Trends Hydrol.*, 291 (1994).
22. A. W. Jayawardena and F. Lai, *J. Hydrol.*, 23 (1994).
23. H. D. I. Abarbanel and U. Lall, *Climate Dyn.*, 287 (1996).
24. D. Koutsoyiannis and D. Pachakis, *J. Geophys. Res.*, 26441 (1996).
25. T. B. Sangoyomi, U. Lall and H. D. I. Abarbanel, *Water Resour. Res.*, 149 (1996).
26. C. E. Puente and N. Obregon, *Water Resour. Res.*, 2825 (1996).

27. A. Porporato and L. Ridolfi, *Water Resour. Res.*, 1353 (1997).
28. B. Sivakumar, *Hydrol. Earth Syst. Sci.*, 645 (2001).
29. B. Sivakumar, *Hydrol. Process.*, 943 (2001).
30. B. Sivakumar, S. Sorooshian, H. V. Gupta and X. Gao, *Water Resour. Res.*, 61 (2001).
31. A. Elshorbagy, S. P. Simonovic and U. S. Panu, *J. Hydrol.*, 123 (2002).
32. Y. Zhou, Z. Ma and L. Wang, *J. Hydrol.*, 100 (2002).
33. B. Sivakumar, R. Berndtsson, J. Olsson, K. Jinno and A. Kawamura, *Hydrol. Earth Syst. Sci.*, 407 (2000).
34. B. Sivakumar, R. Berndttson, J. Olsson and K. Jinno, *Hydrol. Sci. J.*, 131 (2001).
35. B. Sivakumar, *J. Hydrol.*, 149 (2002).
36. B. Sivakumar and A. W. Jayawardena, *Hydrol. Sci. J.*, 405 (2002).
37. Q. Wang and T. Y. Gan, *Water Resour. Res.*, 2329 (1998).
38. B. Sivakumar, S.-Y. Liong, C.-Y. Liaw and K.-K. Phoon, *ASCE J. Hydrol. Eng.*, 38 (1999).
39. B. Sivakumar, K. K. Phoon, S. Y. Liong and C. Y. Liaw, *J. Hydrol.*, 103 (1999).
40. B. Sivakumar, M. Persson, R. Berndtsson and C. B. Uvo, *Water Resour. Res.*, 1011, 10.1029/2001WR000333 (2002).
41. A. W. Jayawardena and A. B. Gurung, *J. Hydrol.*, 242 (2000).
42. A. Porporato and L. Ridolfi, *J. Hydrol.*, 109 (2001).
43. A. Elshorbagy, S. P. Simonovic and U. S. Panu, *J. Hydrol.*, 147 (2002).
44. A. W. Jayawardena, W. K. Li and P. Xu, *J. Hydrol.*, 40 (2002).
45. K. K. Phoon, M. N. Islam, C. Y. Liaw and S. Y. Liong, *ASCE J. Hydrol. Eng.*, 116 (2002).
46. N. Lambrakis, A. S. Andreou, P. Polydoropoulos, E. Georgopoulos and T. Bountis, *Water Resour. Res.*, 875 (2000).
47. F. Lisi and V. Villi, *J. Am. Water Resour. Assoc.*, 271 (2001).
48. B. Sivakumar, A. W. Jayawardena and T. M. G. H. Fernando, *J. Hydrol.*, 225 (2002).
49. F. Laio, A. Porporato, R. Revelli and L. Ridolfi, *Water Resour. Res.*, 10.1029/2002WR001551 (2003).
50. F. Hossain, E. N. Anagnostou and K. H. Lee, *Nonlinear Proc. Geophys.*, 427 (2004).
51. S. Manzoni, A. Porporato, P. D'Odorico, F. Laio and I. Rodriguez-Iturbe, *Nonlinear Proc. Geophys.*, 589 (2004).
52. S. K. Regonda, B. Sivakumar and A. Jain, *Hydrol. Sci. J.*, 373 (2004).
53. S. K. Regonda, B. Rajagopalan, U. Lall, M. Clark and Y. I. Moon, *Nonlinear Proc. Geophys.*, 397 (2005).
54. B. Sivakumar, *Hydrol. Process.*, 2349 (2004).
55. B. Sivakumar, W. W. Wallender, C. E. Puente and M. N. Islam, *Nonlinear Proc. Geophys.*, 383 (2004).
56. B. Sivakumar, A. W. Jayawardena and W. K. Li, *Hydrol. Process.*, 2713 (2007).

57. B. Dodov and E. Foufoula-Georgiou, *Adv. Water Resour.*, 711 (2005).
58. Y. H. Jin, A. Kawamura, K. Jinno and R. Berndtsson, *Nonlinear Proc. Geophys.*, 67 (2005).
59. J. D. Salas, H. S. Kim, R. Eykholt, P. Burlando and T. R. Green, *Nonlinear Proc. Geophys.*, 557 (2005).
60. B. Sivakumar, T. Harter and H. Zhang, *Nonlinear Proc. Geophys.*, 211 (2005).
61. F. Hossain and B. Sivakumar, *Stoch. Environ. Res. Risk Assess.*, 66 (2006).
62. F. Laio, A. Porporato, L. Ridolfi and S. Tamea, *Nonlinear Proc. Geophys.*, 463 (2004).
63. S. Khan, A. R. Ganguly and S. Saigal, *Nonlinear Proc. Geophys.*, 41 (2005).
64. B. Sivakumar, *Hydrol. Sci. J.*, 591 (2005).
65. B. Sivakumar, *J. Hydroinform.*, 175 (2005).
66. D. Koutsoyiannis, *Hydrol. Sci. J.*, 1065 (2006).
67. M. Henon, *Commun. Math. Phys.*, 69 (1976).
68. N. H. Packard, J. P. Crutchfield, J. D. Farmer and R. S. Shaw, *Phys. Rev. Lett.*, 712 (1980).
69. L. Cao, A. I. Mees and K. Judd, *Physica D*, 75 (1998).
70. F. Takens, in *Dynamical Systems and Turbulence*, Eds. D. A. Rand and L. S. Young (Springer-Verlag, Berlin, 1981), p. 366.
71. P. Grassberger and I. Procaccia, *Physica D*, 189 (1983).
72. P. Grassberger and I. Procaccia, *Phys. Rev. Lett.*, 346 (1983).
73. G. Sugihara and R. M. May, *Nature*, 734 (1990).
74. D. J. Farmer and J. J. Sidorowich, *Phys. Rev. Lett.*, 845 (1987).
75. H. D. I. Abarbanel, *Analysis of Observed Chaotic Data* (Springer-Verlag, New York, 1996).
76. M. Casdagli, *Physica D*, 335 (1989).
77. M. Casdagli, *J. Royal Stat. Soc. B*, 303 (1992).
78. M. B. Kennel, R. Brown and H. D. I. Abarbanel, *Phys. Rev. A*, 3403 (1992).
79. A. Wolf, J. B. Swift, H. L. Swinney and A. Vastano, *Physica D*, 285 (1985).
80. P. Grassberger and I. Procaccia, *Phys. Rev. A*, 2591 (1983).
81. J. Theiler, S. Eubank, A. Longtin, B. Galdrikian and J. D. Farmer, *Physica D*, 77 (1992).
82. A. M. Fraser and H. L. Swinney, *Phys. Rev. A*, 1134 (1986).
83. J. Holzfuss and G. Mayer-Kress, in *Dimensions and Entropies in Chaotic Systems*, Ed. G. Mayer-Kress (Springer, New York, 1986), p. 114.
84. J. W. Havstad, and C. L. Ehlers, *Phys. Rev. A*, 845 (1989).
85. W. Liebert and H. G. Schuster, *Phys. Lett.*, 386 (1989).
86. A. R. Osborne and A. Provenzale, *Physica D*, 357 (1989).
87. M. A. H. Nerenberg and C. Essex, *Phys. Rev. A*, 7065 (1990).
88. A. A. Tsonis, J. B. Elsner and K. P. Georgakakos, *J. Atmos. Sci.*, 2549 (1993).
89. T. Schreiber and H. Kantz, in *Predictability of Complex Dynamical Systems*, Eds. Yu. A. Kravtsov and J. B. Kadtke (Springer, Germany, 1996), p. 43.

90. E. N. Lorenz, *Nature*, 241 (1991).
91. B. Sivakumar, *Chaos Soliton. Fract.*, 441 (2004a).
92. B. Sivakumar, *Stoch. Environ. Res. Risk Assess.*, 1027 (2009).
93. A. Hense, *Beitr. Phys. Atmos.*, 34 (1987).
94. M. B. Sharifi, K. P. Georgakakos and I. Rodriguez-Iturbe, *J. Atmos. Sci.*, 888 (1990).
95. S. Islam, R. L. Bras and I. Rodriguez-Iturbe, *J. Appl. Meteor.*, 203 (1993).
96. H. Waelbroeck, R. Lopex-Pena, T. Morales and F. Zertuche, *J. Atmos. Sci.*, 3360 (1994).
97. K. P. Georgakakos, M. B. Sharifi and P. L. Sturdevant. in *New Uncertainty Concepts in Hydrology and Water Resources*, Ed. Z. W. Kundzewicz (Cambridge University Press, New York, 1995), p. 114.
98. B. Sivakumar, S. Y. Liong and C. Y. Liaw, *J. Am. Water Resour. Assoc.*, 301 (1998).
99. J. C. Schouten, F. Takens and C. M. van den Bleek, *Phys. Rev. E*, 1851 (1994).
100. T. Schreiber, *Phys. Rev. E*, 2401 (1993).
101. A. A. Tsonis and K. P. Georgakakos, *Nonlinear Proc. Geophys.*, 195 (2005).
102. E. Gaume, B. Sivakumar, M. Kolasinski and L. Hazoumé, *J. Hydrol.*, 56 (2006).
103. B. Sivakumar, W. W. Wallender, W. R. Horwath, J. P. Mitchell, S. E. Prentice and B. A. Joyce, *Hydrol. Process.*, 1723 (2006).
104. B. P. Wilcox, M. S. Seyfried, W. H. Blackburn and T. H. Matison, in *Symposium on Watershed Management* (American Society of Civil Engineering, 1990).
105. A. Porporato and L. Ridolfi, *Int. J. Mod. Phys.* **B**, 1821 (1996).
106. Q. Liu, S. Islam, I. Rodriguez-Iturbe and Y. Le, *Adv. Water Resour.*, 463 (1998).
107. I. Krasovskaia, L. Gottschalk and Z. W. Kundzewicz, *Hydrol. Sci. J.*, 705 (1999).
108. J. Stehlik, *J. Hydrol. Hydromech.*, 271 (1999).
109. G. B. Pasternack, *Adv. Water. Resour.*, 253 (1999).
110. C. Y. Liaw, M. N. Islam, K. K. Phoon and S. Y. Liong, *Adv. Water. Resour.*, 575 (2001).
111. B. Sivakumar, R. Berndtsson and M. Persson, *Hydrol. Sci. J.*, 377 (2001).
112. P. Ghilardi and R. Rosso, *Water Resour. Res.*, 1837 (1990).
113. A. Elshorbagy, U. S. Panu and S. P. Simonovic, *Hydrol. Sci. J.*, 781 (2001).
114. M. N. Islam and B. Sivakumar, *Adv. Water Resour.*, 179 (2002).
115. B. Sivakumar, *Environ. Model. Softw.*, 721 (2003).
116. H. S. Kim, R. Eykholt and J. D. Salas, *Physica D*, 48 (1999).
117. H. D. I. Abarbanel, U. Lall, Y. I. Moon, M. Mann and T. Sangoyomi, *Energy*, 655 (1996).
118. B. Sivakumar and W. W. Wallender, *Water Resour. Res.*, W05403, doi:10.1029/2004WR003152 (2004).
119. B. Sivakumar and W. W. Wallender, *Earth Surf. Process. Landf.*, 665 (2006).
120. B. Sivakumar and J. Chen, *Earth Surf. Process. Landf.*, 269 (2007).
121. B. Sivakumar, *Earth Surf. Process. Landf.*, 414 (2006).

122. B. Faybishenko, *Adv. Water Resour.*, 793 (2002).
123. P. Persoff and K. Pruess, *Water Resour. Res.*, 1175 (1995).
124. J. Hill, F. Hossain and B. Sivakumar, *Stoch. Environ. Res. Risk Assess.*, 47 (2008).
125. V. K. Gupta and E. C. Waymire, *J. Geophys. Res.*, 1999 (1990).
126. G. Blöschl and M. Sivapalan, *Hydrol Process.*, 251 (1995).
127. I. Rodriguez-Iturbe and A. Rinaldo, *Fractal River Basins: Chance and Self-Organization* (Cambridge University Press, Cambridge, 1997).
128. V. K. Gupta, *Chaos Solitons and Fract.*, 357 (2004).
129. B. Sivakumar, *Hydrol. Process.*, 4333 (2008).
130. K. J. Beven, in *Environmental Foresight and Models: A Manifesto*, Ed. M. B. Beck (Elsevier, The Netherlands, 2002), p. 227.
131. K. J. Beven and P. C. Young, *Water Resour. Res.*, 10.1029/2001WR001183 (2003).
132. H. Gupta, M. Thiemann, M. Trosset and S. Sorooshian, *Water Resour Res.*, 10.1029/2002WR001405 (2003).
133. K. J. Beven, *Hydrol. Process.*, 3141 (2006).
134. J. A. Vrugt, C. J. F. ter Braak, H. V. Gupta and B. A. Robinson, *Stoch. Environ. Res. Risk Assess.*, 1011 (2009).
135. B. Sivakumar, *Hydrol. Process.*, 893 (2008).
136. B. Sivakumar, *Stoch. Environ. Res. Risk Assess.*, 737 (2008).
137. R. A. Woods, *Hydrol Process.*, 1111 (2002).
138. M. Sivapalan, G. Blöschl, L. Zhang and R. Vertessy, *Hydrol. Process.*, 2101 (2003).
139. J. J. McDonnell and R. A. Woods, *J. Hydrol.*, 2 (2004).
140. R. B. Grayson and G. Blöschl, Eds. *Spatial Patterns in Catchment Hydrology: Observations and Modelling* (Cambridge University Press, Cambridge, 2000).
141. I. Rodriguez-Iturbe, D. Entekhabi, J. S. Lee and R. L. Bras, *Water Resour. Res.*, 1907 (1992).
142. K. M. Nordstrom, V. K. Gupta and T. N. Chase, *Nonlinear Proc. Geophys.*, 741 (2005).
143. B. Sivakumar, *Hydrol. Sci. J.*, 523 (2002).
144. B. Sivakumar, *Earth Surf. Process. Landf.*, 969 (2007).
145. J. D. Phillips and M. D. Walls, *Nonlinear Proc. Geophys.*, 371 (2004).
146. N. She and D. Basketfield, *Nonlinear Proc. Geophys.*, 461 (2005).
147. J. D. Phillips, *Geomorphology*, 109 (2006).
148. H. S. Kim, K. H. Lee, M. S. Kyoung, B. Sivakumar and E. T. Lee, *Stoch. Environ. Res. Risk Assess.*, 907 (2009).
149. K. Jinno, S. Xu, R. Berndtsson, A. Kawamura and M. Matsumoto, *J. Geophys. Res.*, 14773 (1995).
150. O. E. Rössler, *Phys. Lett. A*, 397 (1976).
151. B. Sivakumar, *Environ. Model. Softw.*, 515 (2005).
152. T. Wagener, M. Sivapalan, P. A. Troch and R. A. Woods, *Geography Compass*, doi:10.1111/j.1749-8198.2007.00039.x (2007).

153. A. A. Tsonis, G. N. Triantafyllou, J. B. Elsner JB, J. J. Holdzkom II and A. D. Kirwan Jr, *Bull. Amer. Meteor. Soc.*, 1623 (1994).
154. H. J. Fowler, S. Blenkinsop and C. Tebaldi, *Int. J. Climatol.*, 1547 (2007).
155. A. A. Tsonis and J. B. Elsner, *Nature*, 545 (1988).
156. J. W. Kirchner, *Water Resour. Res.*, W03S04, 10.1029/2005WR004362 (2006).
157. L. W. Gelhar, *Stochastic Subsurface Hydrology* (Prentice-Hall, New Jersey, 1993).
158. V. K. Gupta, C. Duffy, R. Grossman, W. F. Krajewski, U. Lall, M. McCaffrey, B. Milne, R. Pielke Sr., K. Reckow and R. Swanson. A Framework for Reassessment of Basic Research and Educational Priorities in Hydrologic Sciences, *A Report to the US National Science Foundation* (National Science Foundation, Washington, DC, 2000).
159. G. F. Lin, *J. Hydrol.*, 65 (1990).
160. D. G. Tarboton, A. Sharma and U. Lall, *Water Resour. Res.*, 107 (1998).

CHAPTER 10

SUMMARY AND FUTURE

Bellie Sivakumar

School of Civil and Environmental Engineering,
The University of New South Wales, Sydney, NSW 2052, Australia
E-mail: s.bellie@unsw.edu.au
and
Department of Land, Air and Water Resources,
University of California, Davis, CA 95616, USA
E-mail: sbellie@ucdavis.edu

Ronny Berndtsson

Department of Water Resources Engineering, Lund University,
Box 118, S-221 00, Lund, Sweden
E-mail: ronny.berndtsson@tvrl.lth.se

10.1. Summary of Findings

Advances in measurement technology and computer power during the past half a century or so have facilitated the proposal of many new scientific theories and the development of numerous mathematical techniques for modeling and forecasting natural and physical systems. These advances have also provided new impetus to some earlier scientific theories that could not be progressed further at the time of their discoveries due essentially to the lack of reliable data (in both quantity and quality) and appropriate mathematical techniques; chaos theory and artificial neural networks, for example, fall in this category of scientific theories. Almost all of these scientific theories and mathematical techniques have, in one form or another, found their applications in hydrologic modeling and forecasting, as deemed appropriate.

This book presented details of eight different data-based approaches in hydrology: stochastic methods, parameter estimation, scaling and fractals, remote sensing, artificial neural networks, evolutionary computation techniques, wavelets, and nonlinear dynamics and chaos. Indeed, these are only a small number of the very many data-based approaches that have found applications in hydrology. Nevertheless, they also form the majority of the most popular data-based approaches, with fuzzy logic and fuzzy set theory, support vector machines, and principal component analysis being some others.

The preceding eight 'state-of-the-art review' type chapters clearly reveal the progress we have made in the applications of these data-based approaches in hydrology. Thus far, these approaches have been applied to study different hydrologic systems, processes, and problems/purposes. Hydrologic systems in different geographic regions, climatic conditions, and possessing different catchment and land use characteristics have been studied. Among the hydrologic processes studied are rainfall, streamflow, rainfall-runoff, evapotranspiration, soil moisture, groundwater flow and contaminant transport, and sediment transport. The hydrologic problems studied include system identification, modeling, prediction, synthetic data generation, data disaggregation/downscaling, data measurement, noise reduction, missing data estimation, model parameter estimation and uncertainty, reconstruction of system equations, regionalization, and catchment classification, among others. These applications have aimed at capturing the various salient characteristics of hydrologic systems and processes, such as determinism, stochasticity, linearity, nonlinearity, complexity, scale, self-organized criticality, thresholds, and sensitivity to initial conditions.

Speaking in an overall perspective, it is a fair assessment to say that the outcomes of the applications of these data-based approaches in hydrology are encouraging. It is indeed true that we are still in the 'exploratory stage' as far as almost all of these data-based approaches are concerned and, thus, any conclusion on their ability, or lack thereof, to solve our hydrologic problems is perhaps premature. However, based on what we have learned thus far, and continue to do so, we can confidently say that they have great potential. Whether or not we will realize their true potential, however, will depend largely on two important, and often interconnected, aspects: (i) selection of a suitable approach for obtaining the most reliable and realistic modeling and forecasting outcomes possible, rather than application of just any approach; and (ii) study of the challenging broader-scale hydrologic issues for evaluating the general applicability of these approaches, rather than addressing a specific problem in an individual catchment that may often have only limited value. These are briefly discussed next.

As for a suitable approach for hydrologic modeling and forecasting, this book, ironically, offers many and yet does not (and cannot) offer any. It offers many because each of the eight approaches has been found to perform well for one situation or another. At the same time, it also does not offer any because not one single approach has been found to consistently perform the best for *all* situations. As some of the comparison studies reveal, different approaches (may) perform better for different hydrologic systems, processes, and problems, and also depending on the quantity and quality of available data. For instance, neural networks have been reported to perform better than chaos methods for some situations, while the opposite has been reported for some other situations.[1-3] Therefore, the suitability of a given data-based approach has been, and is, viewed largely from the perspective of a specific system/process/problem.

In view of these, we do not believe that it is possible to select the most suitable data-based approach for *all* hydrologic modeling and forecasting situations; such a selection is not the purpose of this book either. In fact, asking a general question about the most suitable, or the best, data-based approach is more of a philosophical interest than of practical relevance (such a question may, however, have some merits for specific individual situations). A more pertinent thing to do in utilizing the different data-based approaches in a general hydrologic modeling and forecasting perspective may be to look into the potential to possibly combine/integrate two or more data-based approaches (and/or with others), as appropriate, for better outcomes. Some studies have indeed proposed and/or performed such coupling/integration,[4-6] such as the combination of neural networks with conceptual models or fuzzy logic or wavelets, for various purposes. However, there is still a long way to go in this direction, especially when considering the broader hydrologic modeling and forecasting spectrum.

In order to reliably assess the true potential of data-based approaches in hydrology, it is absolutely essential to apply these approaches to challenging large-scale hydrologic issues, rather than just to specific local problems, a point also emphasized earlier in regards to the use of neural networks (in Chapter 6). Unfortunately, an examination of the hydrologic literature does not provide convincing evidence that this has happened so far; rather, data-based approaches have mainly found their applications at local scales and problems. Part of this unfortunate situation has to do with our inability to clearly identify and/or define the broader-scale hydrologic challenges; for instance: (i) what are the global-scale challenges in hydrology now and will be in the future? and (ii) what is the role of water researchers in the planning, development, and management of transboundary waters? Another part of this situation, however, has to do with our tendency to limit the application of approaches to local scales and/or our inability to study and find ways to extend them to address larger-scale issues; for instance: (i) can neural networks represent the land-atmospheric system? (ii) are stochastic methods useful for catchment classification? and (iii) how useful are data-based approaches for predictions in ungaged basins or in transboundary waters, when data either do not exist or are sometimes 'not (made) available' even when they do?

Whether the progress in data-based approaches in hydrology has been hindered by our inability to integrate them or to identify/define the major large-scale hydrologic challenges or a combination of both, one factor seems to significantly contribute to the situation: our tendency to 'specialize' in a sub-field of hydrology and/or a scientific theory/mathematical technique. By definition, 'specialization' means focusing our view on to something very specific; in other words, 'compartmentalization.' Although this is helpful to better identify and resolve specific issues, it also inevitably pushes the broader and general issues to the background. This is not the only danger, however. Specialization also leads to difficulties in communication of hydrologic theory, research, and practice, since different specialized disciplines (sub-field of hydrology and/or scientific theory) oftentimes adopt different terminologies or jargons, as they fit. As a result of this, it is also possible that different groups specialized in different disciplines may be studying the same hydrologic issue but without even realizing it, since they adopt different approaches and use different terminologies to go about it; an example for this is the 'threshold' concept.[7] There are also concerns that some terminologies that are very widely used in hydrologic modeling and forecasting still lack clear definitions and meanings and, thus, result in inconsistencies in our presentation, understanding, interpretations, and conclusions; some examples for these are model calibration, confirmation, verification, and validation.[8]

The above observations clearly emphasize the need to have a proper balance in the way we perform modeling and forecasting activities in hydrology: that is, a balance between our tendency for 'specialization' on one hand and the 'generalization' that is required in hydrology on the other. For obvious reasons, achieving such a balance is tremendously challenging; it requires compromises and sacrifices in our research expertise, funding opportunities, time, and other relevant resources/factors. Nevertheless, there is certainly great potential for this, as we can strive for 'generalization' *with the help of* 'specialization,' and vice versa. For instance, we may ask ourselves the following questions to extend our specialization for generalization purposes: (i) how to represent the land-atmospheric system or the river networks using neural networks, and what sort of modifications and/or extensions to the structure of neural networks would be required then? and (ii) are the nonlinear dynamics and chaos concepts suitable for formulation of a catchment classification framework or for predictions in ungaged basins, and can these concepts be used in conjunction with scaling theory, which would account for 'similarity' and 'sensitive dependence' properties together, to achieve better results?

It is clear, from these questions, that achieving generalization with the help of specialization, and vice versa, requires consideration of at least three important aspects: (i) establishing accurate links between hydrologic data and catchment physics; (ii) careful integration of different data-based (and other) approaches, as appropriate; and (iii) proper communication among the different specialized disciplines, which is crucial for both (i) and (ii). Unfortunately, however, the current state of growth in regards to these aspects is not that encouraging, if not outrightly dismal. In what follows, a brief discussion of these aspects is made, with some examples from the past and a look towards the future.

10.2. Data vs. Physics

One of the major criticisms on the applications of data-based approaches in hydrology is that they, by their very nature, do not adequately represent the physical mechanisms that actually happen in real catchments. This is indeed correct, and an important reason is that the scales (temporal and spatial) at which measurements are normally, and possibly, made (usually at the catchment scale) are often different from the scales at which the catchment physical mechanisms take place (at many different scales). Nevertheless, it must also be emphasized that almost all the data-based approaches are designed in such a way that their basic concepts (and many of the parameters involved in the individual techniques) take into account this factor. For instance, use of only a single variable streamflow time series for phase space reconstruction in the context of chaos theory (or in other contexts, such as in neural networks and in stochastic methods, for that matter) to represent the dynamics of the multivariable catchment system can be supported by the fact that the catchment is a nonlinear system characterized by self-interaction and that an appropriately chosen single variable often carries important information about the dynamics of that multivariable system. This concept is certainly intuitive, since there is sufficient information in the history of streamflow data about the rainfall properties and the catchment characteristics over a reasonably long period of time, and it also works well in practice. Granted that this concept may not allow a deeper understanding of the catchment physics, it nevertheless is useful for making predictions (through data training) and also for obtaining crucial information about the level of complexity of the catchment dynamics. This inevitably brings up another issue that needs to be discussed and debated when it comes to data-based approaches in hydrology: 'prediction vs. understanding.'

Quite simply, prediction is different and understanding is different, although they may be connected. Prediction is an engineering problem.

Understanding, in general, is a scientific problem.[9] Understanding may help in prediction, but it is neither necessary nor sufficient for prediction. The data-based approaches are mainly designed for modeling and prediction; understanding, if any, is only a secondary input or a by-product, depending upon the context. Therefore, any criticism on data-based approaches in hydrology, from the viewpoint of their inability to represent physics, must not come before asking, and answering to, a fundamental question: Is hydrology 'engineering' or 'science?' Although answering this question is beyond the scope of this book, a brief account here may be helpful.

Hydrology has been, and still is, largely a sub-division of civil engineering or water resources engineering, rather than a field of science. Consequently, a significant number (if not an absolute majority) of academicians, researchers, and practitioners of hydrology have been, and continue to be, taught and trained as engineers. Since engineering teaching and training essentially focus on finding 'engineering solutions' to hydrologic and water resources problems, rather than delving into basic scientific issues, it should not surprise anyone that hydrologic research continues to be more of an engineering endeavor than a scientific one. Therefore, needless to say, much of the interest in hydrology has been on 'predictions' in catchments to perform various 'engineering purposes' (e.g. design and construction of hydraulic and water resources structures), and not in 'scientific understanding' of the working of catchments. This is arguably one of the main reasons for the proliferation of data-based approaches and their applications in hydrology, with their developments facilitated by measurement and computer technologies. With the creation of 'Hydrology' departments or sub-divisions within the 'Faculty of Science' in many academic institutions around the world in recent times, the above situation may possibly change in the future, but it will likely take several decades, if not generations, realistically speaking. Engineering profession remains one of the top professions around the world and is also generally considered better than most scientific disciplines; this is especially the case in developing and under-developed countries, where a majority of the population lives. From an economic viewpoint alone, engineering profession remains far more attractive than many science disciplines and, therefore, there is not much incentive to change the status quo, unless something drastically changes in the near future. In view of all these, there is no convincing reason to think that the above situation will change anytime soon.

This is not to say that we have not cared about studying the 'science' and connecting the 'dots' between hydrologic data and catchment physics. We indeed have, and the attempts have been very many. The physical basis for stochastic time series methods in hydrology has been studied ever since their inception over forty years ago.[10-13] Many studies have attempted to explain the physics behind scaling theory and related concepts to understand catchment mechanisms.[9,14-20] The structure and working of artificial neural networks have been interpreted for identifying catchment processes and explaining their dynamics.[21-23] There have also been other studies that have addressed, in one form or another, the links between data-based concepts and catchment physics.[7,24] Nevertheless, it is fair to say that the progress we have made in this aspect is too little too slow, and there is a very long way to go to achieve a level of satisfaction. It is our hope that research in this direction will move at a much faster pace now on, and there are indeed some promising indications, especially in scaling theories and artificial neural networks.

10.3. Integration of Different Approaches

No single data-based (or any other type of) approach can resolve all the hydrologic modeling and forecasting issues, especially those at larger scales. By their very nature, different approaches are based on, and focus on, different aspects of hydrologic systems and processes, and thus no single approach can completely represent *all* the salient characteristics of catchments and processes. For instance, stochastic methods can adequately represent the stochastic characteristics of hydrologic processes, but they do not possess the ability to capture the determinism inherent in them; in fact, in many applications of stochastic methods, the deterministic components (e.g. trend, periodicity) are deliberately removed to make the system/data suit the stochastic assumption! Similarly, a data-based approach that has its roots in linear worldview cannot be expected to capture the nonlinear characteristics of hydrologic processes; the inability of linear stochastic approaches to represent the nonlinear, and chaotic, dynamic characteristics of hydrologic processes is an example for this situation. It is indeed true that some modifications can be made to an approach to possibly represent even other characteristics that are not part and parcel of its fundamental notion (e.g. deterministic methods with uncertainty estimation to represent stochasticity), but the fruit of such an effort is often difficult to realize; moreover, with such modifications (especially if significant), the approach may no longer be considered to belong to its original category!

Despite their different notions, and indeed because of them, combining or integrating two or more of the data-based approaches may help towards better representation of hydrologic systems and processes. The following are just a few of the numerous possibilities as to this integration: (i) coupling a deterministic approach and a stochastic approach could help to capture both the deterministic and stochastic components inherent in hydrologic processes; (ii) combining a linear approach and a nonlinear approach is more suitable for capturing both the linear and nonlinear aspects of hydrologic processes, when compared to either a linear approach or a nonlinear approach adopted independently; (iii) an evolutionary computing method can be incorporated in a stochastic approach to optimize the many parameters involved for making better hydrologic predictions; and (iv) singular spectrum analysis, artificial neural networks, and support vector machines may be integrated to improve hydrologic predictions. It is not known how many of these numerous integration possibilities have been investigated thus far, but there are already a significant number of examples in the literature.[4-6,25-29]

However, combining or integrating two or more data-based approaches is also a tremendously challenging task. An important reason for this is that adequate knowledge of all the different methods is a prerequisite for any effort towards integration. Unless knowledge of the basic ideas as well as the specific details of different approaches is obtained, the overall thought process to integrate the ideas and the specific ways and strategies to implement them are not at all possible. Our tendency to 'specialize' in a particular data-based approach or technique, rather than finding common grounds, makes this situation even far more difficult. At the same time, it is also unreasonable and unrealistic to expect everyone of us to obtain adequate knowledge of all the different approaches, which are already too many. In view of these, there is a need to find an alternative to deal with this situation. One possible way is to improve the existing opportunities and find new ones for discussions and debates among the different specialized groups to identify potential areas for integration. The existing journals, conferences, and other activities related to hydrology and water resources are certainly helpful to bring the different groups together, but they are not enough. What is needed further are activities that strive for a broader view of hydrology through strong interdisciplinary efforts (both among the different sub-groups in hydrology and among different science and engineering disciplines). To achieve this, the ability of the different groups to 'communicate' among themselves is key.

10.4. Communication

One of the unfortunate by-products of our tendency to specialize in a particular data-based approach or technique in hydrology (and in others) is the adoption of different terminologies and jargons, as appropriate to the respective approach. This should be fine as long as the different terminologies represent different aspects and, thus, have different definitions and meanings, since such would not cause any confusion at all. However, this is not exactly the case, as far as the different data-based approaches in hydrology are concerned. In this respect, two problems are glaring, among others: (i) different terminologies adopted under different data-based approaches may represent a single aspect and convey the same meaning; and (ii) inconsistencies and lack of clear definitions, even when the terminologies are common to all the data-based approaches. An example to each of these is briefly presented next.

In common parlance, the term 'threshold' refers to a point or level that separates areas having different characteristics or responses or similar traits, especially with a great degree of magnitude (e.g. flood level). Consistent with this general adoption, the term 'threshold' has indeed been explicitly used in hydrologic literature, with respect to rainfall, flows, and their effects.[30-32] However, this particular property has also been represented implicitly by many other terminologies in many different data-based approaches in hydrology; for instance, it has been represented by 'critical states' in scaling and self-organization studies,[33,34] by 'characteristic patterns' or 'characteristic modes' in studies on self-organizing maps and artificial neural networks,[35] and by 'regimes' in nonlinear deterministic and chaos studies.[36]

In regards to the inconsistencies and lack of clear definitions for commonly used terminologies in hydrology, there are several examples. Model, calibration, confirmation, validation, verification, training, and testing are some of the widely used terms, but they are also oftentimes very loosely defined and/or used. Even in the same data-based approach (as in the case of artificial neural networks or optimization procedures), the terms 'validation' and 'verification' are oftentimes used with different, and sometimes interchangeable, meanings in different studies. It is also common to see either two steps (i.e. training and testing) or three steps (i.e. training, validation, and testing) being used in the prediction/optimization procedures, not to mention that 'training' often refers to 'calibration' and 'testing' often refers to 'validation.' The confusion indeed arises from both semantic and philosophical considerations,[37] often originated elsewhere; however, unfortunately, we have not made any serious effort to rectify the situation and to bring clarity in hydrologic modeling and forecasting.

The difficulties associated with the above two situations, and perhaps with many others, and their effects on our larger hydrologic modeling and forecasting endeavors are clear. In the former case, not many of us are familiar with *all* of the relevant concepts (e.g. scaling and self-organization, self-organizing maps and neural networks, nonlinear dynamics and chaos) and, thus, it is difficult to know that all those different studies are addressing essentially one common aspect (threshold) of hydrologic systems and processes. In the latter situation, however, almost all of us have a rough idea about what the terms refer to, but still cannot consistently define them or clearly distinguish among themselves. Obviously, these difficulties have serious implications for the way we perform hydrologic teaching, research, and practice, and for the growth of hydrology.

There is also some positive news to report, however, as some efforts have already been made to address these difficulties. The important role of thresholds in hydrologic modeling and forecasting has been explained by Sivakumar[7] through an analogy between catchment behavior and human behavior. Such an analogy is simple and easily understandable by almost everyone, including those that do not have a strong mathematical background (which, unfortunately, cannot be said about the data-based approaches). The inconsistencies among terminologies used in hydrologic modeling has been extensively discussed by Refsgaard and Henriksen,[8] who have also proposed a framework for quality assurance guidelines, including a consistent terminology and a foundation for a methodology bridging the gap between scientific philosophy and pragmatic modeling. Among other things, clear definitions of calibration, validation, simulation, and performance criteria have been provided. These seem to be acceptable to all of us, regardless of the methods we use. These efforts should be augmented further to make improvements in these directions and also extended to still other areas that require proper attention.

10.5. Conclusion

Over the past half a century or so, data-based approaches have been playing an important role in hydrologic theory, research, and practice. They are already in the 'mainstream' hydrology in terms of the number of approaches, techniques, models, applications, and publications. However, in terms of physical explanation, understanding, integration, and communication, they are yet to gain the mainstream status. This situation needs to change in the near future if their true potential in hydrologic modeling and forecasting is to be realized. Sincere efforts and advances in these areas would certainly help in addressing many broader-scale hydrologic issues that are encountered today and will be in the future; the large-scale issues include formulation of a general (modeling) framework in hydrology, predictions in poorly-gaged and ungaged basins, planning and management of transboundary waters, and assessment of climate change impacts on water resources, among others.

There are already some positive indications that this is starting to happen,[19,38-48] but there is still a long way to go.

There is no doubt that data-based approaches in hydrology will continue to proliferate, especially with the almost daily growth in measurement technology, computer power, networking facilities, and other resources. Such new developments and moving forward are indeed welcome. However, to paraphrase Winston Churchill: "The further backward we look, the further forward we can see." Therefore, in tandem with our new developments, it is also important to take a good look back in time to make a more informed and balanced assessment of our progress in hydrology. To this end, we certainly agree with McDonnell[49] that while recent published material make "keeping up" easier, there is a body of important early work that is often ignored by researchers in hydrology (especially young researchers). We also believe that such "old" material would help hydrology immensely, especially when given a "new" and fresh look.

References

1. N. Lambrakis, A. S. Andreou, P. Polydoropoulos, E. Georgopoulos and T. Bountis, *Water Resour. Res.*, 875 (2000).
2. B. Sivakumar, A. W. Jayawardena and T. M. G. H. Fernando, *J. Hydrol.*, 225 (2002).
3. F. Laio, A. Porporato, R. Revelli and L. Ridolfi, *Water Resour. Res.*, 10.1029/2002WR001551 (2003).
4. F.-J. Chang, H.-F. Hu and Y.-C. Chan, *Hydrol. Process.*, 219 (2001).
5. F. Anctil and D. G. Tape, *J. Env. Engg. Sci.*, S121 (2004).
6. P. Nilsson, C. B. Uvo and R. Berndtsson, *J. Hydrol.*, 344 (2006).
7. B. Sivakumar, *Environ. Model. Softw.*, 515 (2005).
8. J. C. Refsgaard and H. J. Henriksen, *Adv. Water Resour.*, 71 (2004).
9. D. R. Dawdy, *ASCE J. Hydrol. Eng.*, 1 (2007).
10. V. M. Yevjevich, *Hydrology Paper 1* (Colorado State University, Fort Collins, Colorado, 1963), p. 1.
11. V. Klemeš, *Adv. Hydrosci.*, 285 (1978).
12. J. D. Salas and R. A. Smith. *Water Resour. Res.*, 428 (1981).
13. M. B. Parlange, G. G. Katul, R. H. Cuenca, M. L. Kavvas, D. R. Nielsen and M. Mata, *Water Resour. Res.* 2437 (1992).
14. V. K. Gupta, O. J. Mesa and D. R. Dawdy, *Water Resour. Res.*, 3405 (1994).

15. V. K. Gupta and D. R. Dawdy, *Hydrol. Process.*, 347 (1995).
16. J. S. Robinson and M. Sivapalan, *Water Resour. Res.*, 1045 (1997).
17. V. K. Gupta and E. C. Waymire, in *Scale Dependence and Scale Invariance in Hydrology* (Cambridge University Press, New York, 1998), p. 88.
18. V. K. Gupta, *Chaos Solitons Fractals*, 357 (2004).
19. M. Sivapalan, in *Encyclopedia of Hydrological Sciences* (Wiley, London, 2005), p. 193.
20. R. Mantilla, V. K. Gupta and O. J. Mesa, *J. Hydrol.*, 155 (2006).
21. R. L. Wilby, R. J. Abrahart and C. W. Dawson, *Hydrol. Sci. J.*, 163 (2003).
22. A. Jain, K. P. Sudheer and S. Srinivasulu, *Hydrol. Process.*, 571 (2004).
23. K. P. Sudheer and A. Jain, *Hydrol. Process.*, 833 (2004).
24. B. Schaefli, D. Maruan and M. Holschneider, *Adv. Water Resour.*, 2511 (2007).
25. T. W. Kim and J. B. Valdes, *ASCE J. Hydrol. Eng.*, 319 (2003).
26. B. Sivakumar, *Hydrol. Process.*, 2349 (2004).
27. J. Chen and B. J. Adams, *J. Hydrol.*, 232 (2006).
28. M. E. Keskin, D. Taylan and O. Terzi, *Hydrol. Sci. J.*, 588 (2006).
29. B. Sivakumar, *Stoch. Environ. Res. Risk Assess.*, 1027 (2009).
30. M. Crozier, *Landslides: Causes, Consequences and Environment* (Croom Helm, London, 1986).
31. N. Caine, *Geogr. Ann.* **A**, 23 (1990).
32. P. Reichenbach, M. Cardinali, P. De Vita and F. Guzzetti, *Environ. Geol.*, 146 (1998).
33. R. Rigon, A. Rinaldo and I. Rodriguez-Iturbe, *J. Geophys. Res.* **B6**, 11971 (1994).
34. I. Rodriguez-Iturbe and A. Rinaldo, *Fractal River Basins: Chance and Self-organization* (Cambridge University Press, Cambridge, 1997).
35. K.-L. Hsu, H. V. Gupta, X. Gao, S. Sorooshian and B. Imam, *Water Resour. Res.*, 1302 (2002).
36. B. Sivakumar, *Environ. Model. Softw.*, 721 (2003).
37. E. R. Rykiel, *Ecol. Model.*, 229 (1996).
38. C. Y. Xu, *Prog. Phys. Geog.*, 229 (1999).
39. R. B. Grayson and G. Blöschl, Eds., *Spatial Patterns in Catchment Hydrology: Observations and Modelling* (Cambridge University Press, Cambridge, 2000).
40. V. K. Gupta, C. Duffy, R. Grossman, W. K. Krajewski, U. Lall, M. McCaffrey, B. Milne, R. Pielke Sr, K. Reckow and R. Swanson, *A Report to the U S National Science Foundation* (National Science Foundation, Washington, DC, 2000), p. 1.
41. R. L. Wilby and T. M. L. Wigley, *Geophys. Res. Lett.*, 1135, doi:10.1029/2001GL013048 (2002).
42. M. Sivapalan, K. Takeuchi, S. W. Franks, V. K. Gupta, H. Karambiri, V. Lakshmi, X. Liang, J. J. McDonnell, E. M. Mendiondo, P. E. O'Connell, T. Oki, J. W. Pomeroy, D. Schertzer, S. Uhlenbrook and E. Zehe, *Hydrol. Sci. J.*, 857 (2003).
43. J. W. Kirchner, *Water Resour. Res.*, 10.1029/2005WR004362 (2006).
44. B. Sivakumar, A. W. Jayawardena and W. K. Li, *Hydrol. Process.*, 2713 (2007).

45. T. Wagener, M. Sivapalan, P. A. Troch and R. A. Woods, *Geography Compass*, 901 (2007).
46. B. Sivakumar, *Stoch. Environ. Res. Risk Assess.*, 737 (2008).
47. B. Sivakumar, *Hydrol. Process.*, 4333 (2008).
48. B. Sivakumar, *Stoch. Environ. Res. Risk Assess.*, in press (2010).
49. J. J. McDonnell, in *Streamflow Generation Processes* (IAHS, Oxfordshire, UK, 2006), p. v.

INDEX

Abnormal events 450
Accuracy 98, 161, 199, 246-247, 253, 260-263, 295, 302, 351, 362-363, 422-423, 425, 427, 432-433, 440, 442
Activation function 7, 294
Adaptive neuro-fuzzy inference system (ANFIS) 304
Adaptive response 336
Additive 63, 84, 93, 128, 174, 179, 181, 190, 192, 194-195, 233, 430
Agents 307
Aggregate 20-21, 23, 37, 41, 44, 53-54, 56, 60, 63, 70, 76, 96
Aggregation 20, 31, 37, 54-56, 87, 115, 249, 435-436, 442, 448-449
Agricultural decision support 260
Akaike information criteria 22
Algebraic
    behavior 156, 233
    exponent 206
    tail 154-156, 161
Algorithm
    ant colony 8
    backpropagation 269
    evolutionary 302, 305, 350
    false nearest neighbor 423-424, 434

Grassberger-Procaccia 420, 425
    genetic 5, 8, 67, 90, 98, 287, 302, 305, 349
    shuffled complex evolution 5, 67
    shuffled complex evolution Metropolis 89-90
Allometric hypothesis 214
Amazon River 385
Amplitude 116, 119, 123, 125, 164, 181, 195-196, 211, 220, 230-232, 372, 448
Analytical 65, 88, 164-165, 232, 331
Anisotropic 112, 122-123, 200, 205
Antecedent
    conditions 281
    precipitation/rainfall 281, 288-289, 331
Approximation 25, 27, 59, 69, 73-75, 86, 88, 161, 164, 228, 379-380, 385, 422, 426-428, 430-435, 437-440, 449
Aquifer 32, 146, 441-442, 449
Area 111-113, 121, 151, 202, 212-216, 218, 220, 223-226, 355, 442, 445, 448

Arrival time 31, 261
Arsenic contamination 441-442, 449
Artificial
    intelligence 268, 279, 300
    life 306
    media 350
    neural networks 2-4, 7, 251, 268, 352, 404-405, 412-413, 431, 433-435, 463-464, 470-471, 473
    time series 433, 444, 450
Asymptotic slope 184, 188
Atlantic multidecadal oscillation 46
Atmosphere 123, 197, 199, 202, 205, 246-251, 254-255, 263, 336, 452, 466-467
Attractor 113-114, 418, 420-421, 424, 431, 434, 448
Autocorrelation 19, 357, 415-416, 429
Autoregressive
    coefficient 35
    exogenous (ARX) 318
    exponent 35
    model 32, 310, 318
    moving average 5, 318, 426
    process 125, 131, 258
Averaging 98, 164, 186, 196-198, 201-202, 205, 221, 227, 314, 388, 398, 422

Backpropagation 269
Bagging 301-302, 313-314
Band pass filter 374, 380
Bandwidth 26-30, 40
Baseflow 62, 309-310, 323, 331-332, 403

Basin
    area 121, 216, 220, 225, 442
    of attraction 65, 67
    experimental 430
    gaged 57
    scale 56
    sub-basin 210-211, 217
    ungaged 57, 466-467, 474
Bathymetry 355, 359
Bayesian
    estimation 5, 46, 98, 443
    method/model 62, 98, 333, 337
Bernoulli process 35
Bias 28, 40, 80, 87, 183-191, 254
Bifurcation 215
Bimodality 25, 39, 44
Binary 45, 137, 154, 156, 186
Binomial 23, 91, 156-157
Biology 348-349
Bisquare 27
Black-box 5, 86, 285, 299, 304, 330, 332, 352
Boosting 302, 314
Bootstrap 5, 29-31, 39-41, 302, 322, 334
Boundary layer 400
Bounded 27, 113, 145-147, 176-177, 180, 207, 222, 232
Box and Jenkins 34
Brain activity 268
Branching 113, 177, 214-215
Brightness temperature 251, 256
Broken line model 35
Brownian motion 110, 114, 126, 131-132, 138-140, 142-143, 145-146, 149, 151, 217, 230
B-type machines 268

Calibration 6, 55, 57-58, 61, 65,
  251-252, 257, 286-287, 291,
  305, 355, 473
  automated 55, 61-63, 67,
    71-72, 76, 79, 81
  manual 61-62
  model 6, 54, 56, 58-60, 63, 67,
    352-353, 467
Canonical 136
Capillary-barrier effect 441
Cascade 38
  bare 137, 160
  binary 137, 154, 156
  bounded 145, 147, 176-177
  discrete 135-137, 152-153,
    158, 177
  dressed 137, 160
  lognormal 155, 164, 170, 174
  multiplicative 133, 135-137,
    142, 147, 159, 178
  random 428, 443
  stationary 147
  stochastic 443
Catchment 82, 164, 221, 281, 469
  area 224, 226, 448
  characteristics/properties 442,
    448, 449, 452, 468
  classification 6, 10, 414-415,
    436, 444, 446, 449, 464,
    466-467
  geometry 355
  modeling *see* Model
  scale 5, 82, 449, 468
  urban 413
Channel 81, 208, 210, 212,
  215-216, 220-221, 353,
  411

Chaos, Chaotic 2, 4, 9-10, 21, 108,
  114, 247
  deterministic 412, 429-430
  disturbance 348
  process 415, 441
  scaling 435, 442
  system 422-424, 436, 448
  theory 9-10, 108, 353,
    412-415, 426-427, 430,
    433-434, 436-442,
    446-448, 452-453, 463,
    468
Characteristic mode 473
Characteristic pattern 473
Chemical 2, 71, 321, 348, 442
Chi-square test 24
Chromosome 350, 354
Circulation
  indices 394, 403
  models 5, 175, 262, 336
Classification 214, 309, 312, 321,
  449
  catchment *see* Catchment
  river basin 300
Classifier 268, 321
Climate 19, 43, 46, 442
  anomalies 300
  change 5-6, 43, 45, 277,
    300, 414, 450
  prediction center morphing
    (CMORPH) 252
  Global model *see* Model
Cloud 122, 199
  cover 255
  model 262, 426
  reflectivity 197
  resolving model (CRM) 251

top temperature 251-252
type 254
Cluster 20-21, 23, 115, 176-177, 309, 312
Code 275, 354
Codimension 168
Coefficient
   autoregressive 35
   binomial 157
   correlation 39, 357
   of variation 202, 223-224, 422
   runoff 226, 436-437, 447
   wavelet 135, 182, 195-196, 374, 376-378, 388, 396, 398
Coherence
   bi 400
   spectral 398
   temporal 398
   wavelet 371, 398-400, 404
Committee machine 308
Communication 2, 319, 466-467, 472
Competitive evolution 68
Competitive learning 311
Complex, Complexity 4, 8-9, 69, 416, 435, 442, 448, 464
   behavior 348, 441, 444
   conjugate 372-373, 377
   evolution 5, 55, 68
   model 6, 22, 324
   shuffling 68
   system 9, 18, 38, 336, 412, 450
Compound Poisson process 19
Composite index *see* Index
Concentration time 221, 356, 358, 448

Conceptual 5, 32-33, 55-57, 65, 93, 276, 286, 290-291, 307
Conductance 403
Conductivity *see* Hydraulic
Cone of influence 388, 401
Confidence interval 272, 332, 388
Confidence limit 333, 388, 400
Conjugate gradient 291, 305
Connection weights 290, 297, 331
Connectionist concept/model 282, 327
Constraints 83, 220
Continuous process 19, 21, 158
Contraction 112-113, 119, 122, 127, 223
Convective 21, 204-206, 226, 247-248, 253-254
Convergence 67-68, 91, 424
Convolution 155, 167, 372-374, 378
Cooperative coevolution 306
Correlation 33
   auto *see* Autocorrelation
   coefficient 39, 125, 131, 231, 357, 396, 422, 436
   cross 36, 289, 394-396, 404-405
   exponent 419, 421
   function 258, 395-396, 415-416, 420
   integral 419-420, 436
   structure 33-35, 87, 258
   wavelet 135, 182, 195-196, 374, 376-377, 388, 391, 398
Covariance 34, 37, 94, 140, 149, 387, 398
Cox process 21

Cramer's theorem 159
Critical state 473
Crop 18, 42-43, 288
Crossover 91, 307, 350-351, 354
Cross validation *see* Validation
Curve fitting 298, 301, 304

Darcy flow 123
Darcy law 228
Data
    -based 3, 412, 454, 464-475
    -driven 3, 18, 28, 297, 317, 347-349, 351, 355, 361
    -mining 3, 349, 355
    assimilation 54-55, 82, 93, 175, 245-247, 255, 257, 347, 358-359, 362
    chaotic *see* Chaos
    disaggregation 4, 411, 413, 464
    hydrologic 4-5, 259, 413, 445, 467, 470
    infilling 7
    input 94, 293, 296, 309, 333, 336
    input-ouput 356
    interpolation 353
    length/size 45, 413, 425, 427, 432, 445
    measurement 4, 76, 464
    meta data 312
    meteorologic 246, 300
    missing data estimation 7, 411, 413, 433, 447
    noise/noisy 96, 192, 272, 413, 425, 438, 464
    output 293
    precipitation 24-25, 29, 37

proxy 46, 246
quantization 190, 192
radar 202-203, 248, 254
rainfall 20, 248, 254, 258, 261, 426-429
reconstruction 413
remotely sensed 7, 245-246, 258, 263
river flow/runoff 281, 294, 429-438
satellite 251, 254, 256, 258, 261
seasonal 37
spatial 18
spatio-temporal 57
streamflow 5, 66, 301, 403
surrogate data 423-425, 431
suspended sediment 7, 9, 323, 404, 439-440, 442-444, 447, 450
temporal 18, 404, 439-440
training *see* Training
transformation 39, 442
unbounded 27
Decay 33, 217, 228, 232, 376, 415-416
Decision support system (DSS) 260, 262, 319, 321, 365
Decomposition 83, 205, 218, 281, 308, 376, 378, 380, 385
    wavelet *see* Wavelet
Degrees of freedom 333, 413, 420, 433
Delay time/time delay 395, 418, 425, 431, 433, 436, 438, 445, 452
Delay window 436
Denoising 380

Density 88, 90
   average 137, 143, 147
   bare 133, 135, 137, 155, 158, 165
   drainage 113, 216
   dressed 137, 154-155, 161, 167
   kernel 5, 25-26, 39
   k-nn 30-31
   local 27
   measure 133, 135-137, 142-143, 147, 168-169, 172-174, 179, 182, 191
   multifractal 169
   nonparametric 31
   probability 21, 25-26, 30, 87, 94, 155, 167, 392, 394
   spectral 143, 148, 150, 152, 171, 200-201, 217, 229, 380, 385, 399
Dependence, Dependent 18, 23, 147
   domain 310
   geometric 215
   inter 9, 70, 280, 296, 412-413
   long-range 150
   Markov 174, 178
   range 310
   scale 255
   sensitive 9, 412, 454, 467
   serial 25
   time-scale 373
   weak 150
Deposition 208, 219
Design 37, 303, 350, 352, 356-357, 469
Detection/early detection 262
   chaos 414
   lightning 254
   rain 258
Determinism, Deterministic 4, 9, 108, 412, 427, 452
   approach 452-454
   chaos 412, 423, 429, 432, 473
   function 125, 135
   model 453
   process 441, 454
   set 113
   system 9, 412, 419, 421, 423
   theory 10
Diadic grid 378
Diagnosis, Diagnostic 91, 97
   test 431
Differential
   equation 420, 434, 448
   evolution 90, 98, 350
   evolution adaptive Metropolis 55, 90
   partial differential equation 57, 417
Diffusion 115
Dilation 112-114, 119-122, 128, 211, 223, 370, 373-374, 379, 383
Dimension
   attractor 434
   box 112, 168
   characteristic 221
   co 143, 168-169, 181, 229
   correlation 420-421, 425-436, 438, 440-443, 445, 448, 450
   embedding 418, 421-424, 433, 438-441

false nearest neighbor 438
fractal 110-111, 113, 118, 149, 151, 168, 202, 213-214, 280
  Hausdorff 110-112, 151
  intrinsic 430
  space/spatial 142, 154, 161, 164, 169, 229, 231
  topological 111
Dimensionality 70, 83-84, 92, 97-98, 246, 435, 442, 445
Dimensionality reduction 54-55
Direct search 65
Disaggregation *see* Data
Disaster 260, 262
Discharge 66, 78, 216-217, 225, 272, 274, 292, 310, 313-314, 317-319, 323, 355, 357, 359, 385, 389, 394, 401, 403-404, 431
  profile 365
  river 259, 281, 288, 300, 307, 325, 331, 404, 439, 445, 447, 451
  specific 228
  streamflow 64
  suspended sediment 9, 404
  time series 220-222, 372
  volume 81
Discontinuous 67, 308
Discretization 136, 217, 377, 417
Disorder 391, 393
Dispersion, Dispersivity 139, 197, 207, 216, 224, 228, 232
Dissipation, Dissipative 199, 441
Distance
  correlation 131, 174
  Euclidean 30, 40, 332
  flow 216, 218-219
  travel 228, 232
Distributed model 71, 78, 81, 95
Distributed model intercomparison project (DMIP) 82, 313
Distribution 6, 18
  α-stable 139
  binomial 23
  conditional 24, 114, 175, 178
  dressed 154
  energy 123
  exponential 225
  extreme value 45
  gamma 24, 34
  joint 137
  J-shaped 429
  Levy 142, 156, 197-198
  lognormal 24, 137, 156, 186
  log-Pearson 24
  marginal 24, 34, 44, 137, 154, 232
  normal 24, 39, 44
  palm 114
  parameter 84-85
  Poisson 19-20
  probability 22-25, 44, 87-88, 112, 181, 332
  rainfall 175, 300
  spatial 20, 81-82, 216
  stationary 89
  target 88, 91
  time-frequency 392
  uniform 68, 73, 91
Divergence 152-154, 160-161, 195, 226, 228, 231, 255, 424
Divide-and-conquer method 304, 308-309

Domain
    application 299-300
    frequency 372, 376
    model 286
    normal 25, 44
    theory 284
    time/temporal 372-373, 376-378, 388
Dominant processes concept 444
Dominant variables 420, 423, 443
Double trace moment (DTM) 181, 197, 201
Downscaling 5, 7, 37, 43, 108, 169, 175-177, 277, 300, 336, 362, 451, 464
Drainage 351
    area 251
    density 113, 209, 216
    network 112-113, 209, 212, 219-220
Drought 5, 9, 40, 262, 273, 277, 300, 320, 438, 450
Dry spell 23-24, 29
Duplication 305
Duration 20, 31, 191, 200-201, 225, 227, 442
Dynamic, Dynamics
    area fraction model (DAFM) 445
    identifiability analysis 55
    programming 32
    nonlinear *see* Nonlinear

Early adopters 268, 270-271, 274-275, 282, 335
Earth 263, 452
    observation 202, 263
    orbiting platform 245, 251

sciences 9, 262, 371, 373, 401-403
    system model 445
Ecohydrology 2
Ecology 2, 199
Eddy covariance 311
Efficiency 58, 68, 89, 94, 258-259, 336, 356, 361, 422
Eigenvalue 83, 122
El Niño southern oscillation (ENSO) 46
Elevation 206, 209-210, 219, 260
Embedding
    dimension 418, 421-423
    optimal dimension 423, 432, 440
    space 113
Empirical 24, 31, 37, 182, 186, 251, 284
Emulation, Emulator 276, 322-323, 347, 355, 361-363
    neuro 268, 300, 322
Energy 33, 56, 219, 390
    cycle 246, 249-250
    dissipation 199
    distribution 123, 380, 392-393
    potential 205
    repartition 385
    spectra 204
Engineering 1, 37, 277, 282, 287, 306, 348-349, 468-469, 472
Ensemble 42, 93, 108, 175, 232, 249, 302, 362
    forecast 256, 438
    Kalman filter *see* Kalman filter
    modeling 93, 268, 300, 308, 312-314
    moment 183

Entropy 287, 391
    generalized 392
    Kolmogorov 423-427
    Shannon 391-394
    wavelet 371, 393-394, 405
Environment 93, 95, 216, 247, 255, 279, 328, 336, 349-350
Eolian transport 208
Erosion 57, 107, 199, 208-211
Error
    additive 93
    approximation 27
    average mean integrated square 28
    correction 82, 307, 317, 352, 359
    estimate 258
    forecast 93, 257
    input 257
    magnitude 358
    mean absolute 357
    mean square 27
    measurement 59, 66, 87, 93, 176
    model/structural 87, 93-94, 176, 249-250, 257-259, 352, 355, 358
    multiplicative 194
    observational 257
    output 317, 323, 359
    prediction 250, 351, 357
    propagation 250, 262
    rainfall *see* Rainfall
    range 193
    relative 358
    residual 64, 359
    retrieval 246
    root mean square 64, 186, 354, 357, 422
    source 247
    static 358
    structure 249, 257-258
    update 318
    variance/covariance 66, 94, 179, 183, 257
Ergodic, Ergodicity 90, 134, 395, 399
Estimated maximum flood 294
Estimation, Estimator
    average 188
    Bayesian *see* Bayesian
    kernel density 5, 25-26, 28, 30
    local 28, 39, 82
    minimum variance 189
    missing data *see* Data
    nonparametric 25, 182-183, 185, 188, 190
    parameter/parametric *see* Parameter
    probability *see* Probability
    uncertainty *see* Uncertainty
    weighted moment 20
Euclidean distance *see* Distance
Evaporation 4, 277, 311, 331
Evapotranspiration 4, 7-8, 56, 59, 78, 246, 277, 288-289, 301, 331, 464
Evolution, Evolutionary
    algorithm 68, 73, 74, 302, 305, 350
    computation, computing 2, 4, 8, 305-306, 347, 349-352, 362, 464, 471

differential 55, 90, 98, 350
dynamic 81,
   mechanism 305
   modeling 302
   natural 8, 350, 356
   neuro 305-307
   operator 307
   population 69-70, 90, 98
   principle 8, 348-349
   programming 349-350
   process 350, 353, 365
   search 55, 70, 74
   shuffled complex 5, 55, 67, 74, 89, 307
   strategies 86, 350
   system 416, 420
   temporal/time 93-94, 393, 394, 416
   topography 209, 211, 216
Excess flow 309
Exogenous 305, 318
Expert knowledge/model 62, 96-97, 308-310, 319-320
Expert neural network *see* Neural network
Exponent 116
   correlation 419, 421
   Hack's 216-217
   Holder 118
   Lyapunov 423-427, 432, 438, 441
   moment scaling/scaling 124, 144, 202-205
   periodic autoregressive 35
   singularity 169
   spectral 200-201, 222, 416
   threshold 168
Exponential decay 33, 415-416

Exponential distribution *see* Distribution
Exponentiation 124, 138, 143, 145, 198
Extraction 71, 284, 330, 356
Extrapolation, Extrapolator 8, 27, 42, 181, 292-295, 300
Extremes/extreme value 5-6, 20, 23, 25, 45, 108-109, 136, 156, 159, 161, 163, 167-168, 200, 220, 227, 294-296, 301, 428, 441, 454
Extreme view 412, 415, 455

False alarm 258
False nearest neighbor 423-424, 427, 434, 438
Feedback 246, 249, 297, 445
Feed forward neural network *see* Neural network
Fidelity 246
Filter, Filtering 98, 138-139, 141, 143, 170-172, 192, 221, 309, 315, 400
   band pass 374, 380
   function 141, 170-171, 172
   Kalman *see* Kalman filter
   linear *see* Linear
First principles 199, 287, 292, 328
Fitness 351
   criterion 351
   function 354, 366
   measure 352, 354-355
Fixed point 445
Flood 5-7, 54, 62, 77, 107, 121, 164, 199, 220, 223, 277, 304, 450
   control 320-321

damage 320
forecast 7, 54, 245, 259-261, 274, 321
frequency 44, 223, 226, 276, 304, 314, 392
hydrograph 276
inundation 81
level 260-261, 472
management 260
maximum 224, 294
model 77
peak 223, 225, 432
plain 208
prediction 54, 261-262, 334
prone 259-261
scale 226
warning 259, 262, 332
wave 225, 261

Flow
Darcy 123
excess 309
flow-duration 80
forecast 7, 9, 34, 42, 62, 262, 281, 315, 320, 432, 434
high 62, 64, 77, 80, 307, 311, 435
low 64, 76, 80, 305, 307, 311, 435
medium 310, 435
peak 223, 276, 310
river *see* River
regime 430, 435

Fluid mechanics 402

Fluvial
drainage network 113
erosion/topography *see* Erosion
geomorphology 233

Flux 71, 96, 225-226, 248, 257, 311, 405
Forcing 46, 59, 93, 247-250, 256-259, 336, 359, 437
Forecast, Forecasting 2-3, 7-8, 17, 19, 22, 54, 169, 256, 290, 299, 301, 314, 328, 352, 357, 412, 414, 436, 465, 467, 473, 474
direct 356
discharge 272, 307, 314, 318
drought 9
ensemble 93, 256, 438
error *see* Error
flood *see* Flood
flow/streamflow *see* Flow/Streamflow
function 356
horizon 306, 356, 358
hydrologic *see* Hydrologic
iterative 356
lead time 174
model 334, 360
naïve 358
precipitation 7, 245, 247-248, 254-255, 304
rainfall *see* Rainfall
rainfall-runoff 302, 355
real-time 319, 358-359, 363
salinity 334
seasonal 42
system 261-262, 359
time series 277
uncertainty *see* Uncertainty
updating 317-318
water demand 320
water quality/quantity 361, 363, 364
weather 249, 255, 260

Fourier
   algorithm 371
   coherence 400
   energy distribution 392
   function 373
   method 9
   series 221, 281
   space 121, 143, 172, 228
   spectrum 374, 392, 398
   transform 9, 142, 170-172, 372, 374, 390
Fractal, Fractality 6, 107, 110, 113-114, 210, 214, 216, 428, 464
   deterministic fractal-multifractal 427
   dimension 110-111, 113, 118, 148, 151, 168, 202, 213-214, 430
   geometry 107
   mono 115
   method 4, 107-108, 168, 428
   multi see Multifractal
   set 109-110, 168
   sums of pulses 129
   theory 2, 199, 209
Fractional
   Brownian motion see Brownian motion
   differentiation 201
   Gaussian noise 35
   integration 124, 138, 141, 143, 201
   Levy motion 142
   moment 372
   stable motion 140-141

Frequency
   flood see Flood
   high 133, 174, 221, 229, 245, 247, 250, 252, 261, 379, 381
   low 40, 46, 132, 221, 256, 289, 381
Function
   activation 7, 294
   codimension 168
   correlation 258, 395-396, 415-416, 420
   covariance 141, 149-150, 170, 387, 398
   filter 141, 170-172
   Haar 135, 383, 385
   Heaviside 172, 420
   moment scaling 143, 160, 181, 192, 195-197, 227
   normalized 216
   scaling 116, 127
   spectral density 152, 172, 200
   transfer 32, 54, 150, 268, 293-294, 296, 298, 361-362
   wavelet 135, 373-374, 376, 379, 383
Fractured media 441
Freshwater 259, 389
Functional response 59, 312
Fuzzy
   inference 276
   logic 2, 10, 302-303, 352, 464-465
   membership 271, 292
   modeling 302
   set theory 2, 10, 302-303, 464
   system 303-304
   toolbox 276, 304

Gabor method 372
Gamma distribution *see*
  Distribution
Gating network 310
Gaussian
  kernel 27
  noise 35
  process 125, 138, 140, 213
Gene 351, 353-354
Gene expression programming 325
General circulation model *see*
  Model
General regression neural network 276
Generalized cross validation 438
Generalized likelihood uncertainty estimation (GLUE) 5, 79, 443-444
Generalized linear modeling 44
Generalized sensitivity analysis 78-79
Genetic
  algorithm 5, 8, 67, 74, 90, 98, 287, 302, 305, 349-350
  computing 365-366
  distribution 350
  event 349
  mechanisms 305
  mutation 350
  operation, operator 305, 351, 354
  programming 8, 302, 347, 349-350, 352, 355
Geochemical regime 442
Geocomputation 269
Geographic information system 2
Geology 57, 60, 82, 97, 208, 216, 441

Geometry 114, 359
  catchment *see* Catchment
  fractal 107, 109
  network 295
Geomorphology 208, 219, 233, 391, 437
Geophysics 175
Geostationary orbit 251
Geostationary satellite 6, 245-246, 250
Global
  climate model *see* Model
  precipitation measurement 252, 261
  search 64, 67, 76-77, 86
Goodness of fit 24, 63, 307
Gradient 64
  amplitude 181, 195-196, 220, 231
  descent 305
  hydraulic 122, 229, 230
Grassberger-Procaccia algorithm 420, 425
Great Salt Lake (GSL) 434, 438
Grid 142, 247-248, 362-363, 378
Groundwater 8, 277, 440
  contamination 4-6, 10, 411, 414, 446, 451, 464
  flow 5-6, 441, 449, 451, 464
  hydrology 1, 98
  level 300
  recharge 305
  storage 32
  table depth 7

Haar wavelet *see* Wavelet
Hack's exponent *see* Exponent
Hack's law 214

Hausdorff dimension *see* Dimension
HBV model 56, 77, 307
Heisenberg incertitude theorem 371
Henon map 416, 433
Hermite polynomial 443
Heterogeneity, Heterogeneous 23, 55-56, 96, 312, 441-442, 449
Heteroscedastic 66, 87
Heuristic functional relationship 40, 284
Hidden layer 294
Hidden node 305
Hindcast 362, 364
Histogram 26, 84
Holocene 441
Homogeneity, Homogeneous 22, 54, 64, 111-112, 206, 223, 228, 288, 312
Horton's ratio 214-215
Human nervous system 7
Humanities 277
Humidity 248, 255
Hybrid/Hybridization 77, 284, 290, 300, 304, 315-316, 320, 404
Hydraulic 43, 82
   conductivity 123, 146, 228-229, 232
   gradient 122-123, 229-231
   head 228, 232
   model 287, 324
   parameter 65
   shear stress 211
   structure 43, 469
Hydro, Hydrologic, Hydrology 1, 198, 245, 256, 351, 371, 403, 411
   behavior 54, 62

chemical 2
climate 2, 18, 35, 445
component 2
condition 59, 63, 87, 290, 445
cycle 2, 281, 413, 445
data 4-6, 259, 288, 413-414, 436, 445, 467
design 37
domain 347
dynamics 216, 320, 351-352, 358, 360
eco 2
event 5-6, 454
forecast 8, 328, 365, 463
graph *see* Hydrograph
geology 1
groundwater 1, 98
informatics/information 2, 260, 303
input 320
isotope 2
meteorology/meteor 2, 7, 251, 253, 355
model/modeling 5, 8, 10, 18, 53-56, 59, 74, 255, 257, 259, 263, 267-268, 272-274, 276-277, 286, 292, 313, 320, 322, 353, 403-404, 412, 443, 463
monitoring 321
neuro *see* Neuro, Neural
paleo 2
parameter 75, 260
phenomena 9, 412-413, 452-453
physical 2, 280
prediction 58, 313, 451, 471

process 5, 9-10, 18, 21, 33, 35-37, 54, 81, 250, 283, 311, 347, 349, 353, 442, 473
regime 77
relevant 79, 82, 97, 437
response 56, 76, 78
rule 297, 326
scaling 414
signal 9, 380
similar/similarity 58, 449
simulation 42
snow 2
space 2
state 246
stochastic 17-18, 25, 37
stratigraphy 441
subsurface 1, 98, 441
surface 1
system 4-5, 10, 54, 347, 365, 392, 411-412, 444, 473
theory 10, 57, 467
time series 17-18, 403, 413-414, 423, 429, 445, 450
unit 224-225
urban 2
vadose zone 2
variable 5, 7, 18, 63, 82, 250, 281, 300
watershed 6, 54
Hydrograph 32, 64, 76-77, 80, 307, 309-310, 317, 322
    flood 276
    recession/recession limb 32-33, 62, 310, 403
    rising limb 32, 310
    unit 216

HYMOD model 313
Hypothesis 213-214, 263, 273, 348, 424-425

Ice 46, 248, 251, 255
Identification
    chaos 414-415, 423, 425
    model 78-79, 431
    parameter 55, 62
    system 4, 78, 418, 464
Impulse response weight 289
Increment
    dependent 140
    generalized 196
    independent 139-140
    local 218
    multifractal 119, 121, 133, 140, 142, 145, 152, 181, 195, 233
    order 119-121, 195
    reduction 192
    self-similarity of the increments 117-119
    stationary 121, 125, 127-129, 131-132, 135, 145, 148, 181, 195
    topographic 209, 220
Index 38, 80, 116, 124, 130, 139-140, 148-149, 169, 197, 201, 206, 209, 220, 226, 251, 289, 314, 329, 401, 445
Infiltration 78, 226, 277, 303, 331, 441
Inflow 277, 300, 321
Information
    content 71, 78, 82, 287, 391
    mutual 287, 289
    partial mutual 289

spatial 42
temporal 359
theory 287
tree ring 42
Infrared sensor 245, 261
Initialization 60, 93, 247-249, 256
Innovation, Innovators 268-271, 274-275, 282, 289, 325, 335
Input
   data 94, 290, 293, 296, 309, 333, 336
   error 257
   layer 310
   model 57
   output 55-56, 61, 82, 268, 271, 283, 287, 294, 318, 352-253, 436
   selection 8, 288
   space 312
   system 54, 87
   variable 7, 284, 353, 359
Integrated response 82
Integration 250, 255-256, 304, 312, 449, 465, 470-471
   fractional *see* Fractional
Integrator mechanism 316
Intensity 172, 177, 399
   bare rainfall *see* Rain
   rainfall *see* Rain
Intensity-duration-frequency 163
Interaction
   land-atmosphere 246, 249
   land-vegetation-atmosphere 246
   parameter 58, 61-62
   radiation-hydrometeor 253
   self 4, 8, 451, 468
Interannual 19, 29, 40, 403

Interdependence, Interdependent *see* Dependence
Intermittence, Intermittent 32, 35, 176, 201, 206, 371, 381, 390, 429
Interpolation, Interpolator 252, 255, 292, 353, 362-365, 430
Invariance, Invariant 58, 93, 381, 399, 421, 425
   scale *see* Scale
Inverse
   approach 421-422, 433
   function 317
   modeling problem 83, 317
   transform 390-391
Irregular, Irregularity 66, 74, 110, 415
Isotope hydrology *see* Hydrology
Isotropic, Isotropy 108, 110, 112, 122, 152, 171, 200, 204-205, 232
Iteration, Iterative
   procedure/process 58, 61, 144, 154, 286, 290, 385
   forecasting 356, 358, 365
   search 59, 64

Jargon 467, 472

Kalman filter 54-55, 93-94, 98, 170, 173-174, 247, 257, 363-364
Karstic spring/watershed 221, 381-382, 385, 390, 403
Kernel density 5, 25-31, 41, 373-374
Kinetics 352
K-nearest neighbor (K-NN) 5, 30-31, 40, 42-43

Knowledge-based networks 284
Knowledge discovery 285, 347-348
Knowledge evolution 349
Kolmogorov entropy *see* Entropy

Lag, Index lag, Time lag 19, 24, 40-42, 80, 125, 130, 149, 288, 415
Lake, Lake volume 18, 45, 321, 413, 437-438
Land-atmosphere interaction 246, 249
Land-atmosphere system 466-467
Land cover 57, 60, 336
Land-data assimilation *see* Data
Landform 208
Landsat satellite 7
Landscape 208, 442
Landscape roughness 7
Landslide 208
Land surface model 74, 86, 246-250, 256, 258, 443
Land surface parameter *see* Parameter
Land surface temperature 7
Land use 45, 259, 288, 292, 336, 442, 464
Latent heat 255
Latent state 173-174
Latin hypercube sampling 74
Layer 268-269, 294-295, 310-312
Lead time 174, 260, 277, 288, 307, 318, 321, 329, 422, 432, 440
Learning 7, 83, 269, 274, 280, 284, 286, 290, 311-312, 320, 336, 359

Least squares 20, 63, 183, 188, 196, 353, 421-422
Legendre transform 159-161
Length
    data *see* Data
    period 23
    spell 23, 31
    stream 291
Levenberg-Marquet algorithm 291
Levy distribution *see* Distribution
Levy motion 140, 142
Levy noise 140
Levy process 142-143, 195
Levy variable 142-143
Lightning 245, 247-248, 254
Limit 111-112
    confidence *see* Confidence limit
    cycle 445
    distribution 89, 92
    multifractal 119
    renormalization 109, 126, 128
    rough 159, 168
    scaling 132
    self-similarity 140
    small-scale 159
Linear, Linearity 4, 454, 464, 470
    approximation 228
    classifier 268
    correlation 415
    drift 126, 132
    dynamics 216
    filter 138-140, 150
    fractional stable motion 140-141
    generalized 44
    integral operator 373

496　　　　　　　　　　　　　　*Index*

　　interpolation 252
　　method/model 39, 276, 318,
　　　　330, 336, 412, 415, 426,
　　　　431, 446, 455, 471
　　multiplicity 153, 160
　　perturbation 232
　　piecewise 114, 280-281
　　prediction/predictor 169-170,
　　　　176
　　process 425
　　regression 33, 182, 184, 289,
　　　　296, 312, 330
　　reservoir 32
　　resolution 112, 161, 163
　　size 111, 204, 221, 226
　　solution 293
　　superposition 316, 372
　　Taylor series 162
　　theory 232
　　transfer function 294, 298
　　transformation 60, 122, 142,
　　　　399
　　updating 94, 305
Linguistic hedge 303
Linguistic information 303
Linguistic variable 303
Liquid 228, 255, 455
Local 26, 57
　　approximation 27, 421-422,
　　　　426-428, 430-432, 435, 437
　　averaging 422
　　correlation coefficient 396
　　density 27
　　estimation 28, 39, 82
　　function 44, 422
　　localization 55, 371, 376
　　map 422
　　maxima 191

　　model 308-309, 352, 433, 447
　　neighborhood 111, 435
　　optima/optimization 64, 67,
　　　　95, 357
　　polynomial 42, 45, 422, 438
　　prediction 359, 426-427,
　　　　430-432
　　search 64-67, 77, 86
　　slope 115, 212, 218, 421
Lognormal 24, 38, 126, 136-137,
　　142, 154, 156, 164-165,
　　170-171, 174, 180, 202,
　　208, 229-230, 232
Log Pearson 24, 276
Logistic map 433
Logistic regression 442
Longitudinal 146, 232
Lorentz curve 390
Lorenz map 428, 433, 436
Low Earth orbit 251
Low flow *see* Flow
Low frequency *see* Frequency
Lumped model *see* Model
Lyapunov exponent *see* Exponent

Mackey-Glass series 433
Macrodispersivity 232
Magnitude 5, 31, 46, 66, 76, 258,
　　261, 304, 306, 310, 323, 331,
　　424, 435, 472
Management 54, 262, 272-273,
　　320, 353, 361
　　drought *see* Drought
　　flood *see* Flood
　　water *see* Water
Mapping 271, 280, 308, 310
Marginal probability distribution
　　see Distribution

Markov chain/model/process 5, 21-24, 29, 32, 42, 46, 54-55, 86, 88-90, 168, 170, 172-174, 177, 229, 441
Mathematical operator 353
MATLAB 84, 276, 304, 402
Matrix 22
    covariance 174
    parameter 38
    random 123
    rotation 41, 230
    transition probability 22
Maximization 164
Maximum likelihood 22, 66, 98
Mean 26, 33, 35, 46, 77, 84, 121, 123, 174, 219, 225, 230, 314, 333, 424-435, 438
Mean square error *see* Error
Measure 39, 60, 62-63, 71, 108
    α-stable measure 124, 138-139, 142
    bare *see* Density
    density *see* Density
    dressed *see* Density
    error 297
    fitness 353-355
    gradient amplitude 195
    multifractal *see* Multifractal
    normalized 415
    performance 70-71, 79, 96-97
    random 117-118, 131, 139, 215
    signature 79-80, 97
    stationary 121, 126, 142, 145, 153, 181, 195
    statistical 77, 327
    uniform 172

Measurement 2-3, 7, 54, 56
    data *see* Data
    error *see* Error
    uncertainty 58
Mechanism 2, 17-18, 20, 23, 57, 97, 114, 124, 128, 142, 205, 262, 269, 282, 285, 304-305, 308, 313, 316, 319-320, 331, 347, 413, 449, 451-452, 468, 470
Median 80, 314
Membership 271, 292, 303
    Membership function 303-304
Memory 415
    structure 288
    temporal 356
Merging 250, 257, 300, 302
Mesoscale 205, 307
    circulation pattern 445
    convective system 205-206, 226
    model 247-248
    precipitation 177
Meta data 312
Meta model 312
Meteorology, Meteorologic 21, 94, 247, 273
    data/time series 246, 256, 300, 403
    forcing 256
    hydro *see* Hydro
    micro 403
    model 247-248, 254
    parameter 248
    radar 202
    satellite 255
    station 321
    variable 247

Method of moments 20, 22, 33, 36-37
Metric 40, 107, 113, 214, 327-328, 422
Metropolis algorithm 55, 74, 78, 88-91, 307, 334
Microwave sensor 245, 251-253, 256
Mineralogy 403
Minimization 27-28, 60, 72, 189, 209, 422
Mismatch 59, 352, 359
Missing data estimation *see* Data
Mitigation 320
Model, Modeling
    alternating renewal 5, 23,
    autoregressive *see* Autoregressive
    autoregressive moving average *see* Autoregressive
    Bayesian *see* Bayesian
    behavioral 79, 307
    calibration *see* Calibration
    cluster 21
    complex *see* Complex
    continuous 19, 21, 158
    coupling 324
    crop 18
    dynamic area fraction 445
    earth system 445
    emulation/emulator *see* Emulation/emulator
    error *see* Error
    forcing *see* Forcing
    forecast 320, 355, 359-361, 363
    General circulation/Global climate 5, 262, 336, 450
    groundwater flow 441
    HBV 56, 77, 307
    hydraulic *see* Hydraulic
    hydrologic *see* Hydrologic
    integration 312, 449, 465, 467, 470-471
    input 57
    land surface *see* Land surface model
    lumped 55, 62, 96, 290
    Markov chain *see* Markov Chain
    mesoscale 247-248
    mixed 173
    modular 308
    nonparametric *see* Nonparametric
    normal 24, 208
    order 22, 32-33, 37, 119, 125, 191, 258, 361-362
    output 59-60, 70, 83, 87, 93, 273, 318
    parameter *see* Parameter
    particle transport 441
    physical 5, 56-57, 95, 114, 246, 320
    Poisson rectangular pulse 20
    Poisson white noise 19, 20
    precipitation *see* Precipitation
    process 18, 21
    radiative transfer 251, 255
    rainfall *see* Rainfall
    rainfall-runoff *see* Rainfall-runoff
    residue 360
    salt intrusion 324
    selection 422

simulation 39, 62, 77, 95, 247, 249, 324, 358-360, 364
stochastic *see* Stochastic
streamflow *see* Streamflow
structure 31-32, 38, 54, 56, 58, 61-62, 66, 78, 79, 291, 333, 352-354, 366
training *see* Training
uncertainty 58, 94, 249
validation *see* Validation
water balance 313
Module 320-321
Modulus 190-191, 388, 390-391, 396
Moment
    divergence 152-154, 160-161, 195
    double trace 181, 197-198, 201
    dressed 156-157
    empirical 182, 184, 193, 196
    ensemble 183
    fractional 372
    statistical 94, 203, 372
    weighted 20, 224
Moment divergence *see* Divergence
Moment scaling 118, 151, 159, 190, 200, 203, 205, 209, 223-224
    function *see* Function
Monsoon 260, 262
Monte Carlo 51, 54-55, 78-79, 94, 257, 332-334, 443-444
Monte Carlo Markov chain 5, 88
Morelet wavelet *see* Wavelet

Moving average 5, 36, 164, 289, 318, 426, 433
Moving window 288
Multi-affinity 115, 121, 135, 181, 195-196
Multialgorithm genetically adaptive 55, 70
Multifractal
    cascade 145, 147, 169, 225
    chaotic 428
    field 122, 171, 201, 232
    fractal-multifractal 427
    increment 119, 121, 133, 152
    limit 119
    measure 121, 135-136, 151, 153, 159, 169-170, 181-182, 190, 197
    model 176, 200-202, 206 227
    parameter 181, 201-202
    process 107, 118, 121, 131, 141, 151, 199
    scaling/scale-invariance 118, 120, 122, 146, 206, 212, 222, 227, 229
    spectrum 206
    theory 372
    universal 142-143, 169, 181, 195, 197, 201, 206,
Multifractality 115, 118-119, 121, 147, 152, 159, 168, 182, 197, 211, 220, 223
Multimodal 70, 90, 92
Multiobjective complex evolution 73
Multiplicative 34, 36, 84, 128, 133, 135-137, 142, 147, 159, 178, 192, 194, 197, 207, 258

Multiplier 76, 84, 173-174
Multi-resolution 179, 371, 377-380, 385, 390, 393-394, 404
Multiscale 9, 37, 107, 348, 393, 396, 403, 405
Multisensor 250
Multi step ahead 300
Multistep automatic calibration 55
Multivariate, Multivariable 24-25, 29-30, 36, 41, 86, 312, 362, 413, 418, 432-433, 445, 447, 451, 468
Multivariate adaptive regression splines (MARS) 438
Mutation 305, 307, 313, 350, 354-355
Mutual information *see* Information

Nash-Sutcliffe index 329
Nature, Natural 350
   evolution 8, 350, 356
   hazard 54
   sciences 279
   selection 69, 305, 350
   variability 336
Nauru island 426
Nearest neighbor 30-31, 40, 423-424, 438, 444,
   false *see* False nearest neighbor
   k-nearest neighbor (k-NN) *see* K-nearest neighbor
Near surface soil moisture 7
Neighborhood 26, 86, 422-424, 433-434
   local 111, 435
   nearest 444
Network architecture 325, 331

Network geometry 295
Neural, Neuro, Neuron 268, 278, 306, 350
   classifier 321
   computing 269, 284-285, 288, 293, 302-304, 326, 335
   emulation/emulator 268, 300, 322-325
   evolution 305-307
   fuzzy 303-304
   hidden 323
   hybrid 315-316
   hydrology 269, 334
   model 269-270, 291, 300, 313, 319, 328
   networks *see* Neural Networks
   science 393
   solution 268, 280, 283-284, 286, 292, 298-299, 315-316, 319, 323, 327, 333
Neural networks 2, 7, 31, 251-252, 267-268, 271, 280, 290, 292, 295, 308, 317, 336, 352, 404, 412-413, 431-432, 433-434, 435, 463-468, 471, 473
   approach/method 272
   committee machine 308
   feedforward (FFNN) 276, 284, 295-296, 299, 301, 311
   general regression (GRNN) 276, 301
   learning 286
   model *see* Neural
   modular 310
   radial basis (RBNN) 276, 296, 301
   simulator 275

solution *see* Neural
time delay (TDNN) 333
toolbox 276
Neyman-Scott
model 20
process 21
rectangular pulse 20
white noise 20
Nile River 18, 34, 36, 272, 403
Nitrogen cycle 445
Node 95, 294, 305, 336
Noise 336, 366
additive 174, 181, 190, 192, 430
data 96, 192, 272, 413, 425, 438, 464
dynamic 93
Gaussian 35, 140, 171
level 427
measurement 415
multiplicative 192
random 170, 257
red 388, 400
reduction 10, 427, 430-431
white 19-20, 124, 138, 140, 169-171, 174, 381, 393, 424
Non-behavioral 79, 81
Nonhomogeneous 46, 112
Nonlinear, Nonlinearity 4, 9, 39, 90, 92, 219, 247, 256, 263, 316, 318, 366, 412, 454, 464, 471
analysis 232, 437, 471
approach/method 412, 432, 454

characteristic/property 425, 470
dependence 152, 250, 412, 437, 453
detection 423
determinism 412, 430-431, 437, 441, 443, 449, 454, 473
disaggregation/downscaling 451
dynamics 2, 4, 9, 21, 280, 358, 411, 431, 446, 452, 464, 467, 473
effect 261
feature 39
function 221, 357
information 400
interaction 249
least squares 353
local prediction 421-422, 430, 434
mapping 280
model 93, 254-255, 302, 318
optimization 64
perturbation 302
process 347-348, 356, 425, 441, 451
propagation 249-250
reconstruction 445
relation 7, 32, 445
reservoir 276
science 412
system 336, 348, 351, 418, 445, 451, 468
transformation 60, 84-85, 122

Nonparametric 5, 18, 25-26, 29, 31, 39, 40-46, 182-183, 185, 188, 190, 438, 455
Nonstationary 43, 46, 64, 66, 87, 117, 126, 137, 150, 152, 178, 180, 362, 371, 395, 399
Normal
  distribution *see* Distribution
  domain *see* Domain
  model *see* Model
  transformation 24
Normalization 198, 376
  orthonormalization 41
  renormalization 109, 115, 123, 126, 128, 212-213, 220, 231-232
Nowcasting 7, 250, 262
Null hypothesis 424-425
Numerical
  analysis 209
  computation 362
  derivative 65
  discretization 217
  estimate 162, 212
  experiment 69
  hydrodynamic model 352
  model 219, 254, 355, 358-359, 360, 362-365
  procedure 167
  simulation 184, 358, 361-362
  weather prediction (NWP)/ forecasting 247, 260

Objective function 60, 62-64, 68, 77, 83, 86, 306-308
Ocean 94, 197, 199, 202, 257, 389, 403, 452
Octave representation 378

Offspring 68-69, 74-75, 350-351, 354
Operation, Operator 93, 122, 186, 255, 352, 354
  averaging 186, 198
  bootstrap 322
  crossover 350
  error correction 318
  evolution 307
  filtering 143, 192, 221
  forecast/warning 37, 62, 255, 261-262, 272, 317, 328
  genetic *see* Genetic
  integral 373
  mapping 308, 310
  mathematical 353
  measurement 93
  model 319
  observation 255
  processing 324
  real-time 299
  reservoir 304
  scaling/scale-change 112, 115, 122
Optimal control mechanism 320
Optimal embedding dimension 423, 432, 440
Optimization 55, 62, 71, 93, 303, 306, 324, 350, 443, 446
  constrained 85-86
  global 67, 89
  local 64, 83
  parameter *see* Parameter
Order 9, 215-217, 412-413, 453
  approximation 69, 378
  basin/stream 215, 217
  differential equation 420
  fractional integration 195

Index   503

model *see* Model
moment 37, 94, 119, 152, 161-162, 183
polynomial 218, 438
process 125, 131, 258
singularity 118, 168, 376
transition 184
Order reduction 361-362
Organism 349
Orographic 206
Orthogonal 315, 372, 377-380, 383, 390-391, 394
Orthonormal 41, 373, 377-379
Oscillation 46, 388, 405, 445
Outlier 25, 91, 333
Output 5, 54, 56-57, 82, 87, 268, 271
    layer 294
    model *see* Model
    variable 7, 64, 284
Overfitting 302, 309, 365
Overland 245, 250
Overland rain 245
Over-parameterization 8, 314, 333

Pacific decadal oscillation 46
Paleo
    climate 328, 403
    environment 328
    hydrology *see* Hydrology
    proxy 46
Parallel
    computing 54-55, 95, 98
    modeling 314
    processing 315, 336
Parameter, Parametric 20, 27, 54, 130
    catchment 276

codimension 143
delay 449
dimensionality reduction 54
distribution 85
embedding 431
estimation 4-6, 8, 10, 20, 33, 36, 54, 56, 63, 280, 350, 414, 443, 446, 464
field 82, 84
function 373
identification 62, 78
interaction 61-62
land surface 246, 260, 263
matrix 36-38
model 25, 33, 36, 54, 57, 59, 65, 79, 250, 287, 317, 333, 353, 444, 464
multifractal *see* Multifractal
optimization 303, 443
parameterization 8, 66, 83, 247, 248, 255, 314, 333, 445
periodic 34, 36
physical 114, 205
proxy 252
pulse size 129
regularization 54
scaling *see* Scale
selection 8, 438
set 17, 43, 56, 64, 72, 79
skewness 142
smoothing 31
space 63-65, 67-68, 73, 82
transformation 66
tuning 353
uncertainty 4, 6, 54-55, 59, 86-87, 90, 93, 98, 249, 333, 444, 454

Parseval theorem 374
Parent 74, 90, 136, 147, 179, 195, 207, 350-351
Pareto
   algorithm 74
   optimum 72-75, 77, 79
   rank 74
   region 73-74, 76-77
   solution 73-74, 77
   surface 72, 75
   tail 167
Partial differential equation *see* Differential equation
Partial mutual information *see* Information
Particle, Particles 110, 214, 251, 307
   swarm method 8, 306, 350
   transport 441
Passive microwave sensor 245-246, 250-253, 256-257
Pathway(s) 246
Pattern 82, 84, 97, 268, 276, 288, 322, 350
   branching 113
   characteristic 473
   circulation 445
   input-output *see* Input
   rainfall *see* Rainfall
   recognition 3, 271, 336, 350-351, 413
   search 65
   seasonal 290
   signature 97
   spatial 82, 84, 441
   weather 262
Peak flow *see* Flow
Perceptron 268-269

Percolation 66, 199
Performance
   criteria 72, 76, 79
   evaluation 62
   measure 61, 70, 79, 97
   model 62, 71-72, 80, 84, 96-97, 289, 332, 422
   trade-off 61, 77
Periodic 23, 415, 424, 430
   autoregressive 32-33, 35
   autoregressive moving average 33-35
   correlation 33-35
   function 135
   model 39
   periodicity 21, 23, 405, 470
   process 35, 382, 403, 415
   quasi-periodic 415, 424
Permutation 313, 351, 354-355
Persistence 23, 25, 29, 32, 314
Perturbation 41, 86, 228-229, 232, 247, 258, 302
Petrography 403
Phase 251, 253, 279, 313, 332, 357, 374, 400, 445
Phase space 416, 418, 420-424, 430, 432-433, 437, 444, 449, 451
Phylogenetic process 349
Physics, Physical
   hydrology *see* Hydrology
   law/principle 33, 57, 219, 262, 285
   mechanism 18, 449, 468
   model *see* Model
   process 124, 128, 208-209, 246, 248, 280
   space 122, 142, 172
   system 71-72, 463

Piece-wise linear 280
Pleistocene 441
Poincare section 430
Point process 19, 21, 129, 172
Poisson 32, 120, 173
    distribution *see* Distribution
    multifractal measure 170, 172
    point process *see* Poisson
    rectangular pulse model 20
    white noise 19
Polar caps 400
Polar coordinate 123
Polynomial
    cubic 294
    Hermite 443
    local 42, 45, 422, 438
    order 438
    trend 195, 218
Population 68-70, 73-75, 84, 89-90, 92, 98, 251, 257, 305-306, 349-351, 354, 356, 469
Porous media 107, 199, 228, 232
Power law 125, 148, 155-156, 161, 164-165, 167, 193, 196, 200, 209, 212, 217
Power spectrum 186, 200, 228, 385, 388, 415-416, 424, 434
Precipitation
    amount 23-25
    antecedent 289, 331
    burst 20
    convective 247, 248
    cumulative 21
    data 24, 255, 403
    depth 20
    disaggregation/downscaling 31, 37
    dynamics 437
    event 20, 22-23
    error 258-259
    field 9, 249, 403
    forecast/nowcast 7, 245, 248-250, 254, 262-263, 304
    index 251, 289
    measurement 252, 261
    mesoscale 177
    model/modeling 2, 19, 21, 24
    occurrence 21, 24-25, 42
    process 17-18, 20, 23
    radar 245
    runoff/streamflow 32, 35
    simulation 19
    variability 251
Predictability 248, 250, 424, 437-438, 440
Predictand 289, 308
Prediction 4, 10, 24, 54, 58-59, 87, 93, 169, 173, 176, 212, 246-247, 259, 261
    horizon 358
    interval 363
Predictions in ungaged basins (PUBs) 6, 466-467
Predictor 42, 170, 288, 308, 362
Preprocessing 295, 329, 404
Pressure fluctuation 441
Principal component analysis 2, 10, 98, 464
Probabilistic structure 424
Probability 22, 25-26, 69, 88
    acceptance 91
    crossover 91
    density function 21, 25-26, 30, 87, 94, 167, 392, 394

506   Index

distribution 22-23, 25, 87-88, 112, 181, 332, 391, 393
exceedance 156, 159, 167, 294
genetic operation 354
mass function 28
transition 22, 441
vector 22
weighted moment 224
Process
   autoregressive *see* Autoregressive
   chaotic *see* Chaotic
   continuous *see* Continuous process
   evolutionary *see* Evolutionary
   Gaussian *see* Gaussian
   hydrologic *see* Hydrologic
   linear *see* Linear
   model *see* Model
   Markov chain *see* Markov chain
   nonlinear *see* Nonlinear
   periodic *see* Periodic
   physical *see* Physical
   point *see* Point process
   precipitation *see* Precipitation
   rainfall *see* Rainfall
   rainfall-runoff *see* Rainfall-runoff
   random *see* Random
   sediment transport *see* Sediment
   stationary *see* Stationary
   stochastic *see* Stochastic
   streamflow *see* Streamflow
Processing unit 95, 331, 335
Profile
   convective 248
   discharge 365
   humidity 248
   hydrometeor 251
   river 209, 218-219
Propagation 93-94, 246, 249, 304, 357, 363, 365
   error 250, 262
Proxy
   paleo *see* Paleo
   data *see* Data
   parameter 245, 252
Pseudo period 448
Pseudo random number 416
Pulse
   fractal sums of pulses 129
   model 176
   multiscale 107
   rainfall 216
   random 129, 217
   rectangular 20
   rescaled 217
   shape 129
   size 129-130
Pumping 380-381
   station 304, 320-321

Quadratic 27-28
Quantization 181, 190, 192, 194
Quickflow 323, 331-332

Radar 176, 202-204, 300
   precipitation/rainfall 203, 245, 248, 250, 254
Radial basis neural network *see* Neural networks
Radiation-hydrometeor interaction *see* Interaction
Radiative transfer model 251, 255

Radiometer 245
Rain, Rainfall 2, 20, 59, 122, 197, 199, 205, 355, 411, 426
  advection 199
  aggregate 21
  amount 19, 21, 23-24, 29
  analysis 21
  antecedent 288
  bare intensity 145, 147, 178, 208
  continuous 31
  convective 21, 253-254
  cumulative 19
  data 20, 180, 204, 261-262, 281, 426
  depth 66, 281
  detection 258
  disaggregation/downscaling 5, 6, 31, 176-177, 428
  distribution 6, 175, 300
  dressed intensity 147
  error 259, 263
  estimation 245-247, 249, 252, 254, 428
  excess 80
  extreme 23, 163
  field 176, 204-205, 245, 249, 252
  forecast 7
  frequency 221
  gage 176, 203, 205
  generation 5, 19-20, 42, 205
  infiltration 331
  input 221-222, 442
  intensity 20, 147, 164, 175, 178-180, 200, 202-205, 207-208, 226-227, 252, 426, 439
  magnitude 31
  measurement 203, 259, 261
  model/modeling 21, 22-23, 107, 146, 176, 227, 427
  occurrence 29
  overland 245
  pattern 276, 322
  prediction 170, 353
  process 4, 21-22, 177, 253, 428
  pulse 216
  radar 203, 248, 254
  rainfall-runoff *see* Rainfall-runoff
  rate 175-179, 226, 247, 394
  retrieval 251-252
  satellite 250, 254, 261, 263
  scale/scaling 123, 200, 202, 204, 443
  signal 426
  simulation 29, 31, 40, 42, 177
  space-time 169, 175, 204, 221, 300
  spatial 202, 258
  stratiform 253
  temporal 164, 170, 174
  time series 148, 175, 200-201, 206, 281, 413, 427
  uncertainty 250
  variability 206, 254, 437
  volume 221
  weight 429
Rainfall-runoff 8, 10, 436
  correlation 385
  forecast 302, 355
  mapping 310

model/modeling 8, 78, 86, 261, 276-277, 282, 289, 296, 305, 307, 317, 324, 331, 353, 355, 359, 385, 404, 411
  prediction 355
  process 4, 356, 413, 437
  relation 9, 392
  simulation 323
  transformation 445
Random, Randomness 5, 9, 20, 29, 147, 172, 219, 307, 348, 350-351, 365, 412, 415-416, 418, 424, 441, 444-445, 453-454
  cascade *see* Cascade
  constant 132, 134
  error/noise 93, 170, 257
  factor 120, 142, 145, 181
  field 117, 122, 228, 230
  function 116, 125
  matrix 123, 230
  measure 117-118, 130, 139, 215
  model 286
  number 217
  population 74
  process 108, 115-116, 118-119, 124-125, 127, 130, 213, 219, 415
  pulse 129, 217
  rotation 123, 232
  sampling 80, 444
  scaling 230, 232
  search 67, 73
  set 112-113
  structure 352
  variable 24, 28, 118, 120, 129, 136, 180, 197, 211, 223, 230, 287, 429, 443

variation 348, 350
vector 230
walk Metropolis 55, 78, 88
weight 131
Rational method 226
Recession 32-33, 62, 403
  limb 310
Reconstruction 10, 328, 376, 378, 380, 411, 413, 416, 418, 420-421, 431, 434, 437, 444, 448, 464, 468
  data *see* Data
  phase space *see* Phase space
Rectangular pulse 20
Recursive 5, 28, 93-94, 113-114, 157, 178
  estimation 5, 98, 170
Recombination 305
Redundancy 377
Reference system equation 448
Reflection 108, 112
Regime 77, 146-148, 200, 222, 225, 248, 421, 430, 435, 442, 449, 473
  flow *see* Flow
  scaling *see* Scale, Scaling
Regional climate model (RCM) 336
Regional flood frequency analysis *see* Flood
Regionalization 4, 6, 57-58, 464
Region of attraction 67, 416, 444
Regression 57, 70, 96-97, 259, 276, 300, 326, 361
  function 25, 356
  linear 33, 182, 184, 289, 296, 312, 330
  logistic 442
  polynomial *see* Polynomial

splines 438
tree 352
Regular, Regularity 89, 208, 376, 415
Regularization 83, 302
   Bayesian 314
   constraint 83
   function 83
   parameter 54-55, 83
   relationship 83, 97-98
   technique 83-84
   Tichonov 83
Reliability 95, 314, 425, 445
Remote sensing 2, 4, 6-7, 175, 245-246, 250, 260, 464
Renewal model/process 5, 21, 23-24, 29, 32
Renormalization 109, 115, 123-124, 126, 128, 203, 212, 220, 231-232
Reproduction 75, 305, 350-351
Resampling 26, 31, 40-44, 78
Reservoir 21, 32, 225, 276, 290, 321, 450
   inflow 300
   management 353
   operation 304
   optimization 324
Resilient propagation 305
Resolution 9, 112, 129, 132, 135, 137, 144, 147, 152, 158, 182
   coarse/low 175, 191, 247, 262, 336, 429, 455
   fine/high 7, 146, 156, 159, 162-164, 175, 177, 188, 191, 195, 246, 249-250, 252, 254, 261, 362-363, 429, 450, 455

multi 179, 377-380, 390, 393-394, 404
Response function 221-222, 281
Response surface 64-67, 70, 85-86, 90, 323, 443
Retention 305
Return period 161-166
Revised technological adoption life cycle (RTALC) 268-270, 274, 298
Reynolds number 146
Riparian 259-261
Rising limb 32, 310
River
   basin 208-210, 218, 220, 259-261, 263, 321, 429-436, 444
   basin classification 300
   catchment 164, 226
   course 209, 213-214, 219
   discharge 259, 281, 288, 300, 307, 325, 331, 404, 439, 445, 447, 451
   floods 107, 121, 220
   flow 4, 6, 9-10, 107, 197, 219, 221, 259-260, 281, 301, 317, 320, 334, 371, 411, 413, 429-436, 438, 440, 442-443, 447, 453
   level 281, 307
   network 6, 7, 107, 114, 199, 208-209, 214-216, 220, 225, 260, 467
   profile 209, 218-219
   stage 7, 281, 288
   system 439
Root mean square error *see* Error
Rössler equation 448

Rotating direction method 65
Rotation 70, 108, 111-112,
    122-123, 210, 230-232
    matrix 41, 230
Roughness
    landscape *see* Landscape
    surface 355
Routing 78, 225
Rule 61, 71, 90, 94, 114-115, 215,
    285, 297, 303, 326, 348, 350
    extraction 284, 297
    fuzzy *see* Fuzzy
Runoff 57, 66, 80, 225, 276-277,
    281, 288, 300, 316, 323, 351,
    353, 355, 359, 390, 413, 432
    coefficient 226, 436-437, 447
    continental 403-404
    direct 32
    hydrograph 310, 320
    kinematic 57
    peak 261
    rainfall *see* Rainfall
    snowmelt 429-430
    subsurface 32
    surface 32, 260-261

Sacramento soil moisture
    accounting model 56, 66, 77,
    79-80, 84, 313
Salinity 257, 333
Salt intrusion model 324
Sample 22, 63, 68, 185
    interval/time 356, 418, 436
    population 68-69
    sampling 54-55, 73-74, 76,
        80, 86-89, 92, 98, 252,
        443-444
    size 23, 26, 38, 44-45, 69

Satellite 7, 176, 202, 245, 249
    earth observatory 202
    geostationary 6, 245-246,
        251
    imagery 261, 528
    retrieval error 247, 249,
        257-258
Saturation excess 261
Scalar 26, 37, 40, 118, 400
    multiplier 84
    product 372
    scalarization 76
Scale, Scaling 2, 4, 6, 9-10, 34, 89,
    92, 106, 294, 296, 442
    anisotropic 122, 200
    asymptotic 162-166
    basin 56
    catchment 5, 82, 449, 468
    chaotic *see* Chaotic
    characteristic 119
    coarse 5, 169, 175, 177
    downscale *see* Downscaling
    dynamic 115, 123, 203-204
    exponent *see* Exponent
    fine 169, 175, 177, 246
    factor 120, 122, 129, 202, 211,
        377
    function 116, 169, 379, 380
    grid 248
    index 130
    invariance/invariant 107-115,
        122, 124, 126, 128,
        138-232, 440, 442
    isotropic 110, 122, 171, 204,
        205
    large 37, 46, 57, 111, 173,
        175, 177, 203, 208, 255,
        348, 361, 466

Index   511

limit 132, 159
local 82, 254, 466
long 77-78
macro 228
meso 177, 205-206, 226, 247, 307, 445
model 107, 176-178
moment *see* Moment scaling
multi *see* Multiscale
multifractal *see* Multifractal
operation 112, 115, 122
orthotropic 123, 203
parameter 201
point 43, 57
recursive assimilation 178
regime 147, 148, 200-201, 225, 421, 449
regional 247, 249
rescale 27, 111, 207, 217, 293
scalar/scalarization *see* Scalar
self-similar 225, 227
short 78
small 57, 82, 111, 121, 132, 159, 173, 178, 195, 208, 222, 255
space/spatial 178, 202, 206, 251, 348, 362
stochastic 435, 442
theory 467, 470
time/temporal 7, 17, 20-21, 23-24, 29, 31-32, 62, 77, 200, 221, 251, 348, 362, 371, 373, 376-377, 424, 428, 440
topologic 215
transformation 76, 107-108, 113, 122
upscale 108
watershed 82

Scatter/scatter diagram 134, 139, 418, 422
Schwartz inequality 399
Search space 83, 88-89, 307, 354
Seasonality 22, 34, 453
Sea surface temperature *see* Temperature
Sediment, Suspended sediment 7-8, 292
    budget 273, 279
    data/series 439-440
    discharge 9, 404
    dynamics 439-440
    estimation 304, 323
    flux 405
    model/modeling 301, 333
    concentration 323, 439-440, 444, 447
    load 439-440, 442-444, 447, 450-451
    simulation 404
    transformation 440
    transport 4, 10, 411, 413, 434, 439-440, 447
    yield 299
Segmentation 309
Self-affine/affinity 115-116
Self-interaction *see* Interaction
Self-organization 350, 473
Self-organized criticality 4, 209
Self organizing linear output map (SOLO) 312
Self-organizing map (SOM) 287, 350, 473
Self-organizing network 310
Self-similar, Self-similarity 109, 112-117, 127, 139, 145-146, 208, 210, 224-225

Semi-arid 24, 300, 319
Semi-distributed model 316
Semiparametric 42, 44
Sensitive, Sensitivity 30, 60, 67, 69, 78-79, 181, 247, 308, 450
    analysis 98, 272, 297, 322, 331, 447
    dependence 9, 454, 467
    to initial conditions 4, 9, 262, 412, 423, 441, 453, 464
Sensor 243, 249, 251, 253, 261
Sequence, Sequential 18, 25, 27, 147, 204, 252, 277, 324, 349-350, 379, 415-416
    data assimilation 54-55, 93, 257
    framework/scheme 78
    parameter estimation 55
Set
    functional *see* Function
    fuzzy *see* Fuzzy
    parameter *see* Parameter
Shifting mean *see* Mean
Shuffled complex evolution (SCE) 5, 55, 67-68, 73, 76, 86, 89, 95-96
Shuffled complex evolution Metropolis (SCEM) 55, 74, 89-90, 93, 95, 97, 307
Sierpinski triangle 110, 114
Signal 9, 181, 186, 195, 251, 256-257, 292, 371-373, 378, 385, 388, 390, 393-394, 399, 415
    chaotic *see* Chaos
    climatic *see* Climate
    hydrologic *see* Hydrologic
    periodic 430

    processing 9, 398
    raingage *see* Rain
    seismic 371
    streamflow *see* Streamflow
Signature
    measure 77, 79-80, 97
    pattern 97
Similarity
    hydrologic 57, 449
    self *see* Self-similarity
Simple, Simplicity 9, 22, 34, 36, 56, 64-65, 412, 419, 444, 450
Simple water balance (SWB) model 313
Simplex 65, 67-69, 73-74
Simplification 10, 443
Simulated annealing 67
Simulation, Simulator
    model *see* Model
    precipitation *see* Precipitation
    rainfall *see* Rainfall
    stochastic *see* Stochastic
    streamflow *see* Streamflow
Simultaneous approach/method 55, 72, 77-78
Simultaneous optimization and data assimilation 55, 93
Single step ahead 300
Singular/singularity 118, 168-169, 233
Singularity exponent *see* Exponent
Singularity spectra 10, 276, 471
Singular value decomposition 83
Skew, skewness 25, 45, 123, 139, 142
Slope 60, 80, 156, 160, 182, 218, 421, 439
    local *see* Local

Index   513

slope-area relationship 218
spectral 202
Smoothing 22, 27, 29-30, 39, 178, 186, 222, 391, 399, 437, 445
parameter 30
Snow
accumulation 292, 316
cover 7
hydrology *see* Hydrology
melt 292, 404, 429
pack 246
Socio-psychologic principle 306
Soft computing 300, 314
Soil, Soil type 57, 60, 78, 82, 97, 115, 208, 216, 220, 260, 442, 445
moisture 7-8, 21, 122, 199, 246, 248-249, 256-259, 281, 289, 443, 445, 464
moisture accounting model 56, 66, 313, 315
moisture deficit 323, 331
Solar cycle 403
Solar radiation 21
Solid earth 403
Solute transport 441, 449
Solution space 74, 287, 306, 308-311
Southern Oscillation Index (SOI) 445
Space, Spatial
averaging 202
complexity 81
correlation 258
data 18
dimension *see* Dimension
disaggregation 37
distribution 20, 81-82, 97, 216

dynamics 356, 441
embedding *see* Embedding
error structure 258
filling 42
Fourier *see* Fourier
interpolation/interpolator 362-365
model *see* Model
parameter *see* Parameter
patterns 82, 84, 441
phase *see* Phase space
physical *see* Physical
resolution 246, 255, 362, 428
scale *see* Scale
search *see* Search
solution *see* Solution space
space-based observation 246, 249
space-borne 249, 250
spatially distributed 53-57, 71, 77, 79, 81-82, 84, 87, 95, 96, 347, 351, 355
spatially extended 437
spatio-temporal 57, 97, 249, 261, 299-300, 437, 449, 451
variability/variation 37, 57, 208, 437
Specialization 2, 3, 306, 466, 467
Spectrum, Spectral
amplitude 372, 471
analysis 10, 209, 221, 372
coherence 298
density *see* Density
exponent *see* Exponent
Fourier 374, 392, 399
multifractal *see* Multifractal
power *see* Power Spectrum

power law *see* Power law
representation 121
sensor 251
singular 10, 215, 471
slope 202, 222
wavelet 383, 388, 396, 398, 399, 404-405
Spell length 23, 31
SPIGOT 45
Spiking layer 310-311
Spiking modular neural network 310, 312
Square integrable function/signal 373, 378-379
Stacking 301-302, 312, 314
Standard deviation 26, 33, 86, 186, 187-188, 192, 424
Standardization 32-33, 294
Stationary, Stationarity 108, 116
    assumption 395, 399
    cascade 147
    chain 22
    distribution 89
    increment *see* Increment
    measure *see* Measure
    model *see* Model
    process 121, 124-126, 128-129, 134-135, 142, 149, 170, 182, 212
    quasi 153
    signal 186
    weak 144
Steady state
    probability 22
    self-similarity 210
    solution 445
Steepest descent 65

Step-wise disaggregation 38
Step-wise linear regression 296
Step-wise method/strategy 72, 76-79
Stochastic, Stochasticity 4, 421, 464, 471
    approach/method 2, 4-5, 17, 32, 64, 412-413, 427-428, 430, 452-454, 464, 466 468, 471
    cascade 443
    dynamics 423
    equation 93
    error structure 249
    forcing 93-94
    generation 404
    hydrology 17-18, 25, 37
    model 5, 18, 32-33, 35, 108, 249, 272, 290, 427, 432, 443, 453
    process 169, 177, 395, 425, 454
    scaling 435, 442
    similarity/self-similarity 115
    simulation 17-19, 32
    system 422-423, 425, 437, 454
    theory 10, 452
    time series 423, 470
    weather generator 31
Stopped-training 301-302
Storage 32, 54-55, 317, 348
    basin 321, 436
    water *see* Water
Storm 19
Stratiform rainfall *see* Rainfall

Stream
    data 300
    downstream 213, 259-260, 307, 321
    ephemeral 35
    length 214
    main 210, 217
    order 215
    upstream 46, 210, 213, 218, 259-260, 307, 321, 329
Streamflow
    disaggregation 37, 435, 449, 455
    dynamics 435, 437
    forecast 7, 9, 34, 42, 44, 62, 262, 281, 315, 320, 432, 434
    generation 5
    measurement 66, 78
    model/modeling 4, 34, 302
    prediction/predictability 59, 437
    process 17, 21, 32, 35, 430
    recession 33
    response 59, 61, 70
    seasonal 33
    signal 430
    simulation 9, 19, 30, 32, 35-36, 39, 404
    time series 18, 34, 77, 82, 429, 433, 447, 468
    transformation 435
    unregulated 430
Structure
    branching 113
    correlation *see* Correlation
    covariance 34
    error 249, 257-259
    flood control 320
    geometric 424
    hydraulic 43
    identification 280
    input-output 319
    linear 319
    Markov 177
    memory 288
    model *see* Model
    probabilistic 424
    threshold 66
    topological 214
Subsurface 54, 413, 441
    flow 122, 233
    hydrology *see* Hydrology
    model 78
    runoff *see* Runoff
Summability 41, 44
Sunspot number 448
Superposition 132, 233, 316, 372
Supervised learning 336
Support vector machines 2, 10, 325, 464, 471
Surface
    area 113
    Brownian 196, 229
    flow *see* Flow
    forcing 247, 248
    hydrology *see* Hydrology
    land surface model *see* Land surface model
    land surface parameter *see* Parameter
    observation 251, 253

ocean 197
Pareto *see* Pareto
response *see* Response
roughness *see* Roughness
runoff *see* Runoff
soil moisture 249, 258
storage 32
temperature 7, 246
topographic 196-197, 209, 219
Surrogate data *see* Data
Survival of the fittest 305, 351
Suspended sediment *see* Sediment
Swarm intelligence/method 8, 306, 350
Symbolic
    knowledge 284
    rule 285
Symbiotic adaptive neuroevolution (SANE) 306
Symmetric/symmetry 27, 88, 123, 141, 191, 210, 232
Synthetic 4, 18, 37, 114, 221, 294, 322, 396, 404, 427, 449, 464
System
    chaotic *see* Chaotic
    classification *see* Classification
    complex *see* Complex
    equation 10, 411, 413, 448, 464
    hydrologic *see* Hydrologic
    identification *see* Identification
    stochastic *see* Stochastic
    theoretic model 280,
    water resources 17-18, 43

Taylor series approximation 27, 162
Temperature 18, 21, 114, 197, 247, 255, 288, 403-404, 413, 432, 442, 445, 447
    brightness 251, 256
    cloud top 252
    sea surface 300
    surface 7, 246
Temporal, Time
    averaging 205, 221
    coherence 398
    convolution 374
    correlation 33, 258
    data *see* Data
    disaggregation 37
    domain 372
    dynamics 356, 435
    evolution 393-394
    interpolator 363, 365
    memory 356
    propagation 356
    rainfall *see* Rainfall
    resolution 377
    scale/scaling *see* Scale
    variability 37
Terminal set 353-354
Terminology 474
Testing 44, 269, 272-273, 281, 285, 294, 302, 304, 322-323, 325, 329, 473
Theory
    chaos theory *see* Chaos
    fuzzy set theory *see* Fuzzy
    hydrologic theory *see* Hydrology

Index   517

Threshold 4, 66, 79-80, 168, 178, 208, 273, 283, 290, 310, 331, 391, 464, 467, 472-474
Thunderstorm 248, 254
Time of peak 261, 310
Time Delay neural network 333 *see* Neural network
Time series 38, 59, 170, 173, 372, 391, 402, 418, 420, 423-424, 441, 450
   analysis, 60, 288, 415
   approach/method 3, 321, 412, 452
   artificial 416, 444, 450
   chaotic 433
   conductance 403
   discharge 220-222, 357, 372
   forecast 31, 277, 438
   hydrologic *see* Hydrologic
   linear 276
   lake volume 413, 437
   meteorologic 403
   model 30, 38-40, 43, 276, 288, 318
   precipitation 403
   prediction 321
   rainfall 148, 175, 200
   resampling 41
   runoff 281, 390
   soil moisture 289
   stochastic 423, 470
   streamflow 34, 77, 82, 468
   synthetic 404
Time varying 78, 347-348, 351, 355
Topography 57, 65, 82, 97, 122, 209-210, 212, 218, 220, 226
   elevation 206
   evolution 209, 216

fluvial erosion *see* Erosion, Fluvial
Topology 107, 113, 312
   dimension 111
   information 312
   scaling 215
   structure 214
   surface 196-197, 209, 219
Training 272, 284, 286-287, 290-294, 296, 301-302, 306-307, 310, 313-314, 326, 332-333, 352, 356-357, 362, 365-366, 468, 473
Trajectory 94, 140, 257, 321, 416, 418, 444
Transboundary waters *see* Water
Transfer function 32, 54, 150, 268, 293-294, 296, 298, 361-362
Transformation
   data *see* Data
   function 84
   logarithmic 35, 59, 64, 76
   nonlinear 60, 84-85
   normal 24
   parameter 66
   scale/scaling *see* Scale, Scaling
   similarity 114
Transient 166, 399, 405
   feature 371
   order 400
   phenomenon/process 402-403, 405
Transition probability *see* Probability
Transition probability matrix *see* Probability
Translation/translator 108, 111-112, 117, 158, 210, 285, 374, 377-378

Transport 199, 362
  channel 226
  contaminant/particle/solute 5-6, 10, 413, 441, 449, 453, 464
  eolian 208
  evapo/convective 351
  process 353, 441
  sediment *see* Sediment
Transverse 232
Tree ring 42, 46
Trend 46, 62, 193, 195, 218, 224, 271, 278, 327, 336, 403, 442, 470
Tropical 203, 206, 246, 260, 404, 426
Tropical Ocean and Global Atmosphere (TOGA) program 202
Tropical Rainfall Measuring Mission (TRMM) 202, 245, 250
Troubleshooting 348
Turbulence 145-146, 199, 204, 390, 400, 402
  atmospheric 123, 197, 199
Typhoon 323

Unbounded *see* Bounded
Uncertainty 4-6, 45-46, 79, 82, 249, 259, 262, 332-334, 438, 443, 454, 464, 471
  model *see* Model
  parameter *see* Parameter
Uniform distribution *see* Distribution
Unit hydrograph *see* Hydrograph
Univariate 36, 318, 402, 432
Unsaturated zone 303

Updating 92, 94, 170, 173-174, 281, 305, 317-319
Urban hydrology *see* Hydrology

Vadose zone hydrology *see* Hydrology
Validation 28, 253, 352-353, 355-358, 365-366, 438, 467, 473-474
Variability 9, 18, 30, 32, 37, 46, 57, 82, 132, 184, 186, 188, 206, 246, 248, 251, 262, 403-404, 437, 450
Variable
  hydrologic *see* Hydrologic
  input *see* Input
  output *see* Output
  random *see* Random
  stochastic *see* Stochastic
Variance 25, 27, 29, 37, 66, 125-126, 136, 145-146, 150, 164, 179-180, 183, 186, 202, 204, 228, 388, 390
Variogram 209
Vector 26, 36-38, 41-42, 59, 72, 92-94, 122, 174, 230, 310, 418, 420, 428
Vegetation 7, 53-54, 56, 77, 82, 97, 208, 246, 257, 260
Velocity 146, 199, 307, 442
Verification 3, 295, 467, 473

Water 1, 53
  balance 62, 77-78, 80, 313
  budget 261
  collection 352
  consumption 276
  content 255

cycle 246, 249, 255, 263
demand forecast 320
discharge 439
distribution 352-353
equivalent 7
level 281, 307, 450
loss 216, 226
management 245, 259, 353
particle 214, 251
pollution 279
quality 4, 8-9, 42, 45, 277, 279, 352-353, 355, 359, 361-364, 403
quantity 352, 359, 361-364
release 450
resources 5-6, 17-18, 42, 245, 469
stage 435
storage 21
system 25, 80, 320
table depth 300
treatment 319
transboundary 7, 466, 474
Watershed 6, 32-33, 35, 54-55, 57-58, 61, 261, 277, 280, 385
    model 56, 58, 60, 65, 67, 77, 81
Wavelength 251
Wavelet 2, 4, 9, 46, 119, 135, 177, 371, 373
    coefficient 135, 182, 195-196, 374, 376
    coherence 371, 399-400
    Daubechies 384-385
    decomposition 218, 281, 376
    entropy 371, 393-394

    function 373-374, 376, 379
    Haar 135, 194, 205, 383-384, 390
    Mexican Hat 377, 384, 404
    Morlet 375, 377, 381, 384
    orthogonal 383, 390
    orthonormal 378
    transform 190-191, 194, 371, 373, 376, 378
    Wiener-Kintchine 398
Weather 254
    forecasting/system 249, 255, 260
    generation/generator 19, 31, 42, 44-45
    pattern 262
    prediction 247, 262
    simulation 41
    variable 42
Weibull distribution *see* Distribution
Weight 26-27, 40
    average 42, 314
    connection 331
    distribution 307
    function 30, 40
    moments 20, 224
    sum 109, 128, 130-131
Wet spell 23, 31
White noise *see* Noise
Width function 6, 164, 216-217
Wind speed 336
Window size 371

Zeros 25, 33, 178, 180, 413, 425, 428-429, 445, 450

GB 656.2 .H4 A38 2010